Research Methodology in Bioscience and Biotechnology

Kian Mau Goh

Research Methodology in Bioscience and Biotechnology

Research Mindset • Best Practices • Integrity • Publications • Societal Impact

Springer

Kian Mau Goh
Faculty of Science
Universiti Teknologi Malaysia
Johor, Malaysia

ISBN 978-981-99-2811-8 ISBN 978-981-99-2812-5 (eBook)
https://doi.org/10.1007/978-981-99-2812-5

This Springer imprint is published by the registered company Springer Nature Singapore Pte Ltd.
The registered company address is: 152 Beach Road, #21-01/04 Gateway East, Singapore 189721, Singapore

Preface

Research Methodology in Bioscience and Biotechnology: Research Mindset • Best Practices • Integrity • Publications • Societal Impact book is an all-in-one resource for new postgraduate students. The book is also suitable for newly hired full-time staff who have recently completed a degree, master's, or doctoral program or are still developing their research interests.

A successful researcher must take into account a number of factors. It could include things such as (1) strategy—things that can be seen or measured, like lab facilities, funding, a winning proposal (idea), reading plans, writing a thesis, publication, training student, networking, mentoring, short-term and long-term plans, etc. In addition to these strategies, there are other factors that we need to excel at research. Throughout this book, you'll learn everything about effective research practices, from the beginning to the end of the project. In this book, you will learn how to make an informed decision and become a more successful and impactful researcher.

Besides, other important factors to success are (2) attitude, mindsets, behaviours, effort, personality, hard work, habit, etc. Attitudes and habits are influenced by mindsets. This book tries to instil a correct mindset in the readers so that they will actually change and affect their research habits and behaviour in a genuine, positive, and manageable way. By reading the book, readers will be able to plan their research and publication in order to meet predetermined goals.

As a researcher, you are embarking on a journey that is not easy at all. If you are passionate about it, it can be enjoyable. The book is a very useful tool for anyone who is feeling lost in their research, or who is feeling disappointed, or who is still figuring out how they want to proceed with their research, I'm certain that this book will provide encouragement when they need it most. It is with this intention that I wrote this book. My goal is to prepare this book in a language that is easy to understand. The book is neither a textbook, a research manual, an experimental protocol book nor a motivational book. It is my goal to mix a variety of genres. Like a novel, this book is easy to read as it is written simply and concisely, but it contains helpful tips that you can follow just like a manual, and stories that sound like self-help books. This book can be used as a textbook since the content corresponds to what I have been teaching in class.

Throughout the book, I attempt to persuade you. Why do I use the word "persuade"? As you read this book, you will find the answer. I found Christ Anderson's definition of persuasive, which he provided in his book entitled TED Talks, to be very interesting. In a nutshell, it goes something like this: "The idea of persuasion is to convince people that the way we see things is not quite right, and we need to dismantle what is not working and rebuild it". I am persuading you to unlearn and erase the less correct research habits and mindsets and to relearn something more appropriate. Let's dive in, shall we?

Johor, Malaysia Kian Mau Goh

Acknowledgements

The support and understanding of my family and extended family is greatly appreciated. My wife deserves special recognition. My deepest gratitude extends to the faculty members and students at University Teknologi Malaysia for their support and for allowing me such autonomy in teaching and conducting research. I also would like to acknowledge the reviewer for his or her helpful criticism. Thank you to the chief editors of the journals that I serve as editor and reviewer for giving me such valuable experience and allowing me to gain a great deal of knowledge about the publication process. Thank you Amir Feisal Merican who inspired me to write this book. In particular, I would like to acknowledge Kok-Gan Chan, a collaborator, a friend, an excellent mentor, and a brother that I trust and whom I have trusted. He taught me that knowledge is power and that with hard work and dedication I can achieve my goals. Thanks to the following excellent people for contributing ideas related to this book, especially the last chapter: Lulie Melling, Lindsay Stringer, Maria Isabel Gonzalez Siso, Anna-Louise Reysenbach, Annarita Poli, Noor Hazarina Hashim, Zhang Tong, Neil Bruce, Rajesh Sani, Stephen Brian Pointing, Enketeswara Subudhi, Mohd Nazalan, Aik Choon, Heng Keat, Edgardo Rubén Donati, and Adrian Ho. Special thanks to industrial players that I know, Ramasamy Alagaratnam, Chee Keng, and Justin Lee. I am grateful to my friends, mentors, and research collaborators, who have provided advice and opinions that ultimately formed this book. They also showed me how to believe in myself and be confident in my abilities. Their support and guidance were invaluable in helping me complete this project.

Contents

About the Author

Kian Mau Goh During my secondary school years, my friends told me "You are such a problematic student. You are causing lots of problems". Whenever I could not find an answer, I threw the questions at my genius friends and my biology and chemistry teachers, so they could give me some answers. Sometimes, when my teachers and classmates were unable to provide an answer to a question, I received feedback that sounds like: "I believe the question does not fall within the scope of the syllabus or examination". Back then, I was very curious about the purpose of life. Why life existed, what is the role of each living things, how things work, and how each of them survive were always in my mind. Perhaps I have been thinking too much and bothering them with too many unusual questions—the basic scientific questions.

The year 1996 was important for us as we were preparing for an SPM examination (something like a national exam) in November. I can still remember a day in January. The headmaster invited one person at a time to his office. It was so panicky that I thought I might be caned, I was among the first dozen of student queuing up to meet him. Not at all. Curved writing was his thing. My headmaster made a handwritten bookmark on a nice white cardboard that he'd cut out, laminated, punched a hole in, and put a ribbon on for us to use. Each of us received a bookmark from him. On my bookmark it was written "Genius is 1% inspiration, 99% perspiration". Over the next couple of years, I continued to use the bookmark. It became my motto. My IQ is on the average score, so I know I'm not a genius. I am not talented. My goal has never been to become a genius. For me, the bookmark redefines what hard work is all about. Although I am only 1% smart, I need to work 99% harder. In light of inspiration I received from the headmaster, I know that if I truly wish to comprehend the purpose of each living organism, I must work hard to seek knowledge for the remainder of my life. The world is created by God with a purpose for which we can find answers in the bible. Throughout my life, I have been passionate about understanding how God created living things and the only way to do that is performing biological research.

I was constantly seeking answers to many unanswered questions in my head, and I hope to be able to uncover the truth of life and the mechanisms behind it during my time at university. Later, I began to realize that the information in scientific textbooks in high schools and universities is only the tip of an iceberg. Under the water, there remains a large portion of the unknown. It is the purpose of research to dive into the blue ocean to uncover the mysteries hidden beneath.

The first degree I earned was a Bachelor of Engineering. We learned about unit operations (machines in chemical engineering), making good assumptions, engineering calculations, mass balance, energy balance, system control, engineering problems and ways to solve them. I did not pursue a career in engineering, but rather chose to become a research biologist as my first two-years job. For a trained engineer, working in a biological lab was extremely challenging. Due to this decision, I need to be a student lifelong and keep learning biology as my background in science is limited.

My career as a research scientist lasted 2 years, after which I returned to university for my postgraduate program. I completed a PhD in Protein Engineering in 2007. I am now a biotechnology lecturer at the Science Faculty, Universiti Teknologi Malaysia, a research university. I am among the few engineers by training who take on this position in the Science Faculty. Students often express surprise when I share my background with them in class, as to why an engineering-trained individual can perform bioscience research and publish in bio-related journals. No one is born with the skills and knowledge necessary to conduct research and publish. I believe that any individual with genuine curiosity and motivation will be able to succeed. My research team focuses on extremophiles, primarily thermophiles and halophiles. Among my research interests are protein engineering, enzymology, metataxonomics, and metagenomics, as well as microbial comparative genomics and transcriptomics. I've published more than 80 indexed articles. I am an active journal reviewer and serve as the editor of Scientific Reports, Frontiers in Microbiology, and a few other guest editors for other journals.

I've been teaching a compulsory course (Research Methodology) for several years in which all postgraduate students within my department are required to enrol. The contents receive positive feedback from students. In December 2021, I shared the whole idea and intentions of my lectures to few stakeholders (alumni, industry and academic advisory panels) and sought for suggestions and improvement. One of the advisors, Prof Dr Amir Feisal Merican bin Aljunid Merican, a lecturer in the oldest university in Malaysia, threw me a wild idea that I never thought of—"Wouldn't it be cool if you turned all the lessons into a book? There're more people who'll benefit from it". My immediate response was...no. It would be better for me to spend more time managing my research and writing papers rather than writing a book. Six months after consideration, I began writing the first sentence. I intend to show the published book to Dr Amir Feisal and express my appreciation for his encouragement. Unfortunately, we lost him just prior to Christmas in 2022.

List of Figures

List of Tables

Science and Philosophy

1

Abstract

Chapter 1 demonstrates the differences between science and philosophy, as well as why researchers should have a philosophical mindset. In addition, we will discuss what a PhD is all about and the reasons for enrolling in a doctoral program. Additionally, we discussed the differences between fixed-mindset and growth-mindset as well as the difference between intrinsic motivation and extrinsic motivation. Ultimately, philosophical mindset, growth-mindset, and intrinsic motivation are some of the few essential elements for high-achieving researchers.

1.1 Science and a Degree in Science

The biotechnological and biological sciences are two branches of biology that fall under the large umbrella known as science. We know that scientific knowledge is formed through the analysis of quantitative experimental data or observation. But, what is 'science'? Let us start by looking at how four dictionaries define the word 'science' (Table 1.1).

Do we consider science to be a static field of knowledge? Certainly not. We have known for decades that microorganisms can be spherical, rod-like, curved, etc. Facts about bacilli, vibrios, spirillums, spirochetes, and other shapes can be referred to as science. The boundary for this scientific fact was broken when Anthony Walsby discovered a square-looking halophile known as *Haloquadratum walsbyi* (Burns et al. 2007). No one knew that microbes could exist in a square shape. Since 1961, enzymes have been classified into six Enzyme Commission numbering systems (EC 1 to EC 6). Later, scientists discovered that, oops, we were incorrect, there are seven groups, the latest being translocases (Fig. 1.1).

We have heard of the giant *Bacillus megaterium* (now called *Priestia megaterium*; simply put, megaterium refers to a big beast). Our understanding of

K. M. Goh, *Research Methodology in Bioscience and Biotechnology*,
https://doi.org/10.1007/978-981-99-2812-5_1

Table 1.1 Definitions of science based on different dictionaries

Source	Definition of science
Oxford Learner's Dictionary	Knowledge about the structure and behaviour of the natural and physical world, based on facts that you can prove, for example, by experiments
Cambridge Dictionary	(Knowledge from) the careful study of the structure and behaviour of the physical world, especially by watching, measuring, and doing experiments, and the development of theories to describe the results of these activities
Merriam-Webster Dictionary	Knowledge or a system of knowledge covering general truths or the operation of general laws especially as obtained and tested through scientific method
Collins Dictionary	Science is the study of the nature and behaviour of natural things and the knowledge that we obtain about them. A science is a particular branch of science such as physics, chemistry, or biology.

Fig. 1.1 Science is not everything

bacteria size in microbiology textbooks is often confined to very few micrometres. This is a past scientific fact that has never been challenged until 2022, when a discovery breaks this rule. The largest bacterium identified to date, *Candidatus Thiomargarita magnifica*, has an average cell length exceeding 9000 micrometres (Volland et al. 2022). This organism is visible to the naked eye. The Guardian News featured the news with the following headline: "Scientists discover world's largest bacterium, the size of an eyelash". It is about 50 times larger than what we normally learn from scientific facts.

Another example. The so-called science tells us that plastics, such as PET, take approximately 450 years to decompose. Researchers discovered that certain super-enzymes are capable of degrading plastics (Joo et al. 2018; Skaf et al. 2018). These enzymes are produced by the microbiome of the mealworms (as well as some other types of larvae). A number of research groups are working on this idea, including Stanford University, Beijing Institute of Technology, Beihang University, Aarhus University, University of Thessaloniki, and others. In the future, I hope that we will be able to rewrite the time it takes for plastic to degrade, not 450 years, but in days.

I would like to provide you with another example. An undergraduate textbook taught me that viruses are very small. Are you aware that there are giant viruses that have larger genomes than what an archaea has? There was a surprise in 2003 when the first giant virus (>1 Mbp genome size) was detected, and subsequently, more and more gigantic viruses were discovered (Brandes and Linial 2019; Rolland et al. 2021). Note that some of the archaea that we know had genome of size less than 1 Mbp.

All of the examples provided above are intended to demonstrate that science is not a perfect domain of knowledge because it is not a static field of knowledge. Science is not everything. Until somebody proves it to be incorrect or incomplete, the scientific fact remains valid. As science is ever-advancing, there is always a gap between what we know and what we do not know. In biological science or biotechnology research, one of the primary purposes is to find answers to questions which still remain unknown.

The English word 'science' is related to the Latin word 'scire' or 'scientia'. 'Scire' simply means know, or I know. In other words, science is an area of knowledge that we already know about: the natural, physical, biological, and chemical worlds. If you agree with this explanation, Bachelor of Science (Biology) simply means a program in a university that teaches what we already know about the biological world—emphasis on the phrase 'what we already know'. The second emphasis for Bachelor of Science is the delivery of knowledge is via teaching and learning what we already know.

Malaysian Qualifications Agency (MQA) is the accreditation body that ensures the educational program offered in Malaysia is of the highest quality. There might be some similar bodies in your country. The following statements were taken from the Malaysian Qualifications Framework (MQF), second edition (MQA 2018):

- Bachelor's degree students

> Learners will demonstrate a thorough comprehension of broad based and coherent body of knowledge and skills for para and full professional work embedding research, innovation and creativity in specialized areas. Demonstrate professionalism, resilience commitment to an ethical work culture, sustainability issues and an awareness of global citizenship in alignment with national aspirations.

I would like to draw some important conclusions from the complex statements shown in the preceding bullet points. Bachelor students are expected to have a thorough comprehension of broad knowledge. Now, below are the MQF statements for master's degree and PhD students.

- Master's degree students

> Learners at this level will demonstrate a mastery of knowledge in specific field/fields of study/work and through further learning, research, and/or professional practice; and Buttressed by a comprehension of strong theoretical knowledge, critical thinking, creative and innovative skills. They will also be able to generate new solutions to problems.

- Doctorate degree students

> They operate with confidence, knowledge and skills both in Malaysia/ASEAN as well as the wider world. Learners will demonstrate critical understanding of the most advanced knowledge at the frontiers of a field of study or professional practice. Independently conduct, manage, and lead advanced research which contributes to substantial, new and original knowledge, and/or professional practice. Produce research outputs in the form of a thesis, patents, products, new advanced professional practice or advanced technologies, creative models or works of art, or music. Demonstrate capacities to present and defend points of view, thesis and ideas in area(s) of expertise with knowledge and confidence.

For ease of comparison, here is the summary.

- Bachelor students—thorough comprehension of broad knowledge
- Master students—mastery of knowledge
- Doctorate students—demonstrate a critical understanding of the most advanced knowledge at the frontiers of a field of study. . . .and lead advanced research which contributes to substantial, new, and original knowledge

Two major differences can be observed. First, the level of knowledge increases. Second, a doctoral student is required to contribute to new knowledge. It is the responsibility of a PhD student to contribute to the advancement of knowledge by filling in the gaps. The role of a PhD candidate is not to reinvent the wheel, to reproduce identical or similar data, or to provide proof of what has already been discovered. In Chap. 2 of this book, I will discuss the concept of new knowledge in more detail.

1.2 Philosophy and a Doctorate of Philosophy

Table 1.2 provides an overview of the general meaning of philosophy. The word 'philosophy' is derived from the word 'philosophia' which means a love of wisdom. In Greek, *philo* means love and *sophos* or *sophia* means wisdom or wise. As a broader definition, philosophy includes an activity and attitude that helps people comprehend values, reality, beliefs, principles, knowledge, and wisdom about the world and its content including all living organisms and their relationships.

Table 1.2 Definitions of philosophy based on different dictionaries

Source	Definition of philosophy
Oxford Learner's Dictionary	A particular set or system of beliefs resulting from the search for knowledge about life and the universe. The study of the nature and meaning of the universe and of human life.
Cambridge Dictionary	The use of reason in understanding such things as the nature of the real world and existence, the use and limits of knowledge, and the principles of moral judgement. A group of theories and ideas related to the understanding of a particular subject. A particular system of beliefs, values, and principles.
Merriam-Webster Dictionary	Pursuit of wisdom. A search for a general understanding of values and reality by chiefly speculative rather than observational means. An analysis of the grounds of and concepts expressing fundamental beliefs. A system of philosophical concepts. The most basic beliefs, concepts, and attitudes of an individual or group.
Collins Dictionary	The study or creation of theories about basic things such as the nature of existence, knowledge, and thought, or about how people should live.

Most of us are aware that a master's degree may be earned by taking a taught course or by conducting a full research project. There are some universities that use Master of Science (MSc) for programs that combine lectures and small research projects. There are some universities that use the term 'Master of Philosophy' (MPhil) for research programs. Do you know of any biological science or biotechnology PhD programs available as taught courses? A doctorate of philosophy (PhD) is the highest level of education qualification. People joke that PhD stands for Pizza Hut Delivery. During my first year of undergraduate studies, the senior students told us that lecturers with PhDs have permanent brain damage (p-h-d). (Just joking).

Please review Table 1.2 again. The Collins Dictionary states that philosophy is the study or creation of theories concerning matters of basic nature such as the nature of existence, knowledge, and thought. Please put emphasis on 'Study or creation of theories…'. It is consistent with the statement of the MQF on the PhD program—lead advanced research that results in substantial, new, and original knowledge. Collectively, a PhD student (or someone with a PhD certification) is a lover of knowledge and wisdom, deeply familiar with a subject matter they have explored, and capable of generating new knowledge and science.

As a domain of knowledge, science is not error-free. The scientific fact remains valid until it is demonstrated to be inaccurate or incomplete. In an opinion article published in PNAS some time ago, the authors addressed the topic of why science needs philosophy (Laplane et al. 2019). Below are some of the main points described by the authors. Philosophical thought is responsible for improving the precision and utilization of scientific terms. Philosophical thought allows the development of scientific hypotheses and proving them using novel and testable empirical research. Scientific approaches with a philosophical perspective greatly influence the development of theories, experiments, and instrumentals, thereby leading to paradigm shifts that replace outdated ideas or assumptions (Laplane et al. 2019). The purpose of philosophical-minded researchers is to make a difference, push boundaries, explore the unknown, and make positive changes. You are a scientist with that mindset, aren't you? There is one subtitle of this book that encapsulates the theme of the Research Mindset. The ability to think philosophically is an essential component of a Research Mindset.

A few paragraphs earlier, I asked: Are there any biological science or biotechnology PhD courses available as taught courses? PhD programs may have some compulsory courses, but the main purpose of the program remains the generation (or discovery) of new knowledge through research. PhD programs are not designed to produce a wealth of experimental data or to repeat what has already been discovered (science, or information that we know). The doctorate in philosophy must demonstrate the philosophical attitude described in the preceding paragraph—love knowledge and wisdom and generating new knowledge.

1.3 Why Get a PhD and Become a Researcher?

1.3.1 PhD for Everyone?

Every postgraduate student in our department is required to take one of my courses. There were multiple questions in the first assignment, and the first one read something like "Why I register and study postgraduate study?". This question aims to provoke students to think critically about the motivations that drive them to study rather than work in the real world. Below are some common answers I received:

- To gain additional knowledge.
- To make a positive impact on society and the world at large after completing my studies.
- I am passionate about research.
- Due to my present (future) career requirements, I am undertaking a postgraduate degree.
- I wish to improve my theoretical and practical skills.
- Having a higher level of qualification.
- I find joy in doing research and enjoy discovering new things through research.
- To qualify as an educator in the future.
- Possessing a strong interest in a particular field of science.
- I look forward to contributing something new.
- It is not competitive to hold a bachelor's degree.
- Because of the pandemic and the transition to online learning, my knowledge and skills are lacking. I wish to acquire more knowledge.
- I am encouraged by my parents to pursue a PhD.

- I am realizing my dream.
- My parents want me to do.
- I have a passion for biology since I was in secondary school.
- I am not prepared to work.
- Enhance carrier development.
- To learn the latest technologies.
- To earn more money in the future.
- I need to obtain another certificate since a degree is too common.
- As a result of the closures associated with the COVID-19 pandemic, I have not gained sufficient skills and knowledge, and I need to reclaim what I lost while studying for my postgraduate degree.
- My goal is to be recognized.
- To obtain a better job, I need a master's degree.
- To discover and glorify God by discovering the science behind His creations.

In my opinion, a Bachelor of Science degree is suitable for everyone. I believe that most individuals can benefit from a master's in science or a master's in philosophy, but acquiring a PhD is an entirely different matter. When students are considering a PhD, they may not know what it means. A PhD student I knew almost got fired by the supervisor a few years ago because of slow progress in collecting data. According to the student, enough effort has been made and there is no reason for being terminated. During our long discussion, we came to the conclusion that the student had the assumption that spending hours reading journals, even until mid-night, and attending conferences to fill up cognitive knowledge was enough to earn a PhD. The student assumed that having read enough knowledge can be equated with a doctoral degree. In the past, the student had no research experience, but always performed well on exams, and thought that obtaining a PhD would be similar to that experience—understanding and memorizing information. In one sense, a PhD holder should possess knowledge; however, it goes beyond just knowing what other people already know. The pursuit of a PhD or the conduct of research generates new information that may reveal previously unknown facts/knowledge/theories.

One should have a compelling reason for pursuing a PhD before beginning it. I apologize for saying this, but if a person does not love new knowledge, is not interested in contributing something new, and does not have a very strong desire to learn more, he or she may want to speak to someone and determine if they are suitable for a PhD. To be brutally honest, postgraduate work, especially doing a PhD, is not a job for unemployed people or those who are not prepared to work in the real world. Besides, it is not the grade point average (GPA) achieved during the undergraduate years that determines a person's ability to conduct research. The grades received by an individual reflect their ability to memorize and regurgitate information for examinations, not their ability to conduct research. Do you agree?

We may have observed students who excelled in high school and undergraduate study and who are doing well in their master's program. It is unfortunate that some of them burn out halfway through their PhD or abandon it before completing it. Several factors can be attributed to this, but one possible explanation is the ambiguous motivation to pursue post-graduate studies.

You may be approaching the end of your PhD journey and considering becoming a full-time researcher for the next phase of your career. However, you may be uncertain about what the life of a principal investigator entails. It would be a good idea for you to read this article by Tregoning and McDermott (2020).

1.3.2 Extrinsic Motivation or Intrinsic Motivation

Earlier, I listed the reasons why students enrol in graduate programs. We can categorize the reasons into extrinsic motivation or intrinsic motivation if we examine them carefully. Ruth Gotian, in her book *The Success Factor*, described intrinsic motivation as the kind that comes from within, coming from the core values, interests, passions, mission, vision, and morals that we hold dear (Gotian 2022). Some people do PhD because it is their calling. An example of intrinsic motivation might include individuals with a strong mission, a mission in research, a desire to impact themselves before impacting their peers, a passion for discovering new things, an extraordinary super strong interest in science, or even a desire to discover God's creation. Researchers with intrinsic motivation usually have crystal-clear goals and stay very focused on what they need to accomplish.

Extrinsic motivation is another kind. Among the extrinsic motivations are the desire for obtaining a certificate, accolade, award, or peer recognition, the belief that PhD is a better degree than a degree and we will receive a higher salary, and registering a PhD program is an escape to working in the real world. Extrinsic motivations come from outside, for instance, it is my parents that encourage me to pursue a PhD. Some extrinsically motivated individuals may always strive to prove their abilities or intelligence. Extrinsic motivations easily lead to burn out or fail out. Earlier, I used several jargon. You may find the book *The Success Factor* by Ruth Gotian helpful in understanding extrinsic motivation and intrinsic motivation. In Chap. 8, we will discuss success in more detail.

1.3.3 Fixed-Mindset or Growth-Mindset

Are you familiar with the terms fixed-mindset and growth-mindset, or have you never heard of them before? You can easily find examples that illustrate the difference by searching the Google image search engine. As we are not permanently attached to either a fixed-mindset or a growth-mindset, we often juggle between them. Some individuals, however, tend to favour either side most of the time. The fixed-mindset person avoids challenges and fears failures and setbacks. Comments are not so welcomed since they view them as destructive criticism. When problems accumulate and they come to the conclusion that they are not good enough, they will eventually quit or let go very easily.

Growth-mindset individuals are always striving to improve their knowledge, wisdom, talent, or skill. Despite obstacles, they persist in overcoming challenges. They are open to receiving comments and criticism and consider them as opportunities for improvement. Research setbacks we often face include failures in grant/scholarship/fellowship applications, failures in experiments, and rejections from publications. Setbacks in research are not well accepted by anyone. In spite of this, growth-mindset people do not focus on the end result but on the process because they recognize failure as an opportunity to grow.

1.3.4 Research Mindset

There is a single subtitle in the book that encapsulates one of the themes of the book, that is, Research Mindset. First and foremost, one of the most significant components of a Research Mindset is the ability to think philosophically. Second, intrinsic motivation plays a vital role. The third essential component is a growth-mindset. These three elements form the DNA of an incredible researcher, akin to the incredible abilities possessed by the Incredible Hulk. However, having a good DNA of the Research Mindset is not enough, just like it is impossible for the Hulk to stay green for the rest of his life. The Hulk becomes stronger and more powerful when he becomes angry, heightened his abilities. Similarly, a positive DNA of Research Mindset requires excellent research strategies in order to be successful. Later in the book, we will discuss these strategies in more detail.

I believe that the group of individual with intrinsic motivation may include those who believe in the concept of a growth-mindset. PhD students or researchers with a PhD who have intrinsic motivation enjoy the whole process or journey, which eventually leads to excellent results. What is the reason for this? In the event that you disagree with that, skip the entire book and jump right to Chap. 8.

Please do not get me wrong. I am not implying that extrinsically motivated individuals with a fixed-mindset cannot pass and graduate with a PhD scroll. Additionally, you may find that some of them later work in universities or institutes. Have you come across anyone who is extrinsically motivated. The question arises for readers who are old enough and capable of following the progress: Do extrinsically motivated researchers outperform researchers who possess an intrinsic motivation, growth mindset, and philosophical mindset?

Our curiosity is often one of the driving forces that motivates us to investigate a topic. While some might consider data collection to be the most challenging aspect of obtaining a PhD, in my opinion, the most important aspect is our own mindsets, commitment, self-discipline, passion, and reasons to pursuing the degree or doing a research. The greatest hurdle is ourselves. We need to resolve a lot of challenges and problems.

Are you a PhD holder working in an academic or research field? Have you ever wondered why certain individuals achieve higher levels of success? One of the reasons is, in my opinion, Research Mindset. Some individuals treat PhDs as merely certificates and may choose only to do the bare minimum in their research careers.

The performance of PhD holders with a philosophical mindset, specifically intrinsically motivated individuals with a growth-mindset, is better than that of individuals who perceive the PhD as merely an academic credential for graduation. A high-performance individual will possess the following characteristics:

1. Being open-minded, aware, and seeking out new opportunities. A researcher looks beyond the point of view of other peers when examining a research topic.
2. Their boldness and willingness to try new things make them stand out from the crowd. A typical PhD holder would say, "No, we cannot do that". Philosophical-minded PhD holders, however, tend to say, "Let's plan something and try it. We have nothing to lose" and they find it difficult to remain interested in research activities that are focused on standard operating procedures or in skills that they have already acquired. In enzymology, for instance, the technique is very straightforward, if we know the methods of cloning, expression, purification, and characterization, we have a good understanding of almost everything. A similar process can be cloned to make another protein, and so on and so forth. Using the one protein, one paper concept, they publish on every single protein. However, it is very boring! A philosophical-minded researcher goes beyond the standard operating procedures and challenges themselves regularly, for example, they keep improving from the basic enzymology work and studying a new area they are unfamiliar with, such as protein structure modelling and docking. Once they have completely mastered that new arena, they begin to feel bored and seek something more challenging, that is, they move on to the actual crystallography of proteins. An important characteristic of high-performance researchers is their ability to learn outside of their comfort zone. A new challenge is always on the horizon for them.

To be a successful researcher, you should possess the gene cluster of philosophical mindset, intrinsic motivation, and growth-mindset. People with these Research Mindsets are usually the minority, but they are in the top tier. They can be considered as 'mutants' in a sense, as they are differ from norm. In some departments, faculties, or institutes, these mutants are sometimes seen as outliers. It is for this reason that I use the term 'mutant', just kidding. Nevertheless, having such a mindset is not the end all, but it is just a good place to start. These individuals achieve outstanding performance through careful planning, taking action, forming habits, and employing effective strategies.

1.4 Basic Research Versus Applied Research

You may hear these terms: fundamental science and applied science research. In case readers are not certain of the difference between them, you may wish to go online to learn more. In this section, we will explain it briefly. Research in fundamental (basic) science has the following characteristics:

- Contributes to the expansion of existing knowledge and proposes new theories and concepts.
- Motivated by curiosity, driven by a desire to understand the whys and hows of things.
- It is common for these projects to not have direct commercial objectives or at least not have immediate commercial purposes. Some of the knowledge may be put to use in the future in the form of certain applications.

The characteristics of applied science research are as follows:

- Providing solutions to real-life problems, industrial problems, and societal problems in the real world
- An activity that is associated with the production of a product, with procedures associated with the production of the product, or with activities to improve the product or procedure in the production of the product
- Client-driven, business-driven, or financial statement (money)-driven

Young researchers have a hard time seeing the difference between the two categories. If you are still confused, please keep in mind that basic research focuses on knowledge or what is known as the knowledge-gap. Anyway, there is a distinct difference between data, information, knowledge, and wisdom. Data is not knowledge. Knowledge is defined as the answers to the questions that need to be answered by using words such as 'why' and 'how'.

The term 'data' refers to the numbers, pictures, etc., that are collected during an experiment. High performance liquid chromatography, for example, is used to quantify an analyte; spectrophotometers are used to measure protein concentration, enzyme activity, or optical density of cells; and microscopes (i.e. FESEM) are used to view flagella; therefore, these numbers and photo are data. A piece of data such as the one stated above is not considered knowledge in any way. In the analysis of a pool of data, we may see trends between the data, we may observe that some cells are more biologically active, or some enzymes function better under some conditions. Such analyses are like a collection of information. The combination of data and a basic analysis will give us some useful information. However, the information generated from the experiment, however, is still not knowledge to me. We can learn from the information, that is, which one is better or faster among the samples we have, which one is bigger, what are the optimum conditions, if a species is present or not in the water or soil, whether a gene is upregulated, downregulated, or if it is not expressed at all. We talk about knowledge when we use the word 'why' or 'how' to refer to the questions that need to be answered. The basic science research field is related to knowledge; it is related to the questions relating to the 'why' and 'how' of science. A gene, for example, may be upregulated under certain conditions, but if you can provide an explanation as to why this gene is upregulated in a scientific way, then that is some sort of knowledge because you are providing an explanation to a Why question in the scientific field.

Try your best to put more emphasis on knowledge generation than just data generation alone. There is no doubt that you need data to support the thesis-argument of the knowledge. For a better understanding of what I meant by the term 'thesis', please refer to the following chapter. There are a great number of researchers that are very data-driven or who like to gather heaps of data. But they were not able to see the connections between them and were unable to understand the scientific reasoning behind the data. However, real philosophical researchers are always motivated by a desire to find the answers (new knowledge) because of their curiosity to know 'why' or 'how'.

1.5 Summary

Although science is not a perfect domain of knowledge, scientific facts remain valid until they are proven to be incorrect or incomplete. In fundamental science research, one of the purposes is to correct previous mistakes, advance understanding, propose new theories, and close the research gap. Eventually, research outputs may be converted into theses, journal articles, patents, and products and contribute to the socio-economic development of the country. Philosophical minds have a profound influence on theories, experiments, and technological instruments, leading to paradigm shifts that supplant outdated assumptions or theories. It is important to keep in mind that PhD programs are not designed to produce a large amount of experimental data or to duplicate existing research. This point will be further discussed in the following chapter. There is a strong correlation between your achievements and your inner self, are you driven by extrinsic or intrinsic motivation? Mindsets can be divided into two types: growth-mindsets and fixed-mindsets. Excellent and high-achiever researchers have these genes in their genomes: philosophical mindset, intrinsic motivation, and growth-mindset, and I named that as Research Mindset.

In closing, I would like to ask you two questions: Would you like to become a biotechnologist or a bio scientist who makes an impact? Is your goal to be a high achiever rather than an average achiever? Having the right mindset will determine your success in research outputs (grants, publications, etc.) as well as uncountable outcomes such as wisdom and scientific breakthroughs.

References

Brandes N, Linial M (2019) Giant viruses-big surprises. Viruses 11(5):404

Burns DG et al (2007) *Haloquadratum walsbyi* gen. Nov., sp. nov., the square haloarchaeon of Walsby, isolated from saltern crystallizers in Australia and Spain. Int J Syst Evol Microbiol 57(2):387–392

Gotian R (2022) The success factor, 1st edn. Kogan Page, p 19

Joo S et al (2018) Structural insight into molecular mechanism of poly(ethylene terephthalate) degradation. Nat Commun 9(1):382

Laplane L, Mantovani P, Adolphs R, Chang H, Mantovani A, McFall-Ngai M, Rovelli C, Sober E, Pradeu T (2019) Why science needs philosophy. Proc Natl Acad Sci 116(10):3948–3952. https://doi.org/10.1073/pnas.1900357116

MQA (2018) Malaysian qualifications framework (MQF), 2nd edn. Agensi Kelayakan Malaysia

Rolland C, Andreani J, Sahmi-Bounsiar D, Krupovic M, La Scola B, Levasseur A (2021) *Clandestinovirus*: a giant virus with chromatin proteins and a potential to manipulate the cell cycle of its host *Vermamoeba vermiformis*. Front Microbiol 12:715608

Skaf MS et al (2018) Characterization and engineering of a plastic-degrading aromatic polyesterase. Proc Natl Acad Sci 115(19):E4350–E4357

Tregoning JS, McDermott JE (2020) Ten simple rules to becoming a principal investigator. PLoS Comput Biol 16(2):e1007448. https://doi.org/10.1371/journal.pcbi.1007448

Volland J-M et al (2022) A centimeter-long bacterium with DNA contained in metabolically active, membrane-bound organelles. Science (New York, N.Y.) 376(6670):1453–1458

Student's Proposal, Dissertation, or Thesis 2

Abstract

Chapter 2 discusses the guidelines for writing a proposal, dissertation, or thesis and how to construct a good problem statement. Additionally, the chapter provides more than ten practical considerations for conducting a comprehensive and in-depth literature review. Good practices are also discussed, as well as common errors that students make. While the whole idea focuses on thesis writing, the information presented in this chapter is also applicable to writing research proposals aimed at securing research funding.

2.1 What Is a Thesis?

In this chapter, we will discuss good practices in preparing a proposal or thesis. In this instance, the proposal refers to the student's proposal rather than a research proposal that a principal investigator submits for funding. In any case, the concepts described here may also influence how you write a research proposal for funding.

The terms 'theses' and 'dissertations' are sometimes interchangeable in certain universities; therefore, please refer to your own institution if there are any significant differences. In the context of a taught course master program or a mixed mode program that consists of approximately 50:50 course work and research, some universities use the term 'dissertation' in place of a thesis.

As defined by dictionaries such as the Cambridge Dictionary, a thesis is a "piece of writing involving study of a subject for a college or university degree". Does a thesis consist of nothing more than a report to fulfil graduation requirements which eventually becomes a book and is placed on the shelf, or an eBook kept in a repository?

We might have overlooked another meaning of the term 'thesis'. The Cambridge Dictionary also defines a thesis as "the main idea, opinion, or theory of a speaker or

K. M. Goh, *Research Methodology in Bioscience and Biotechnology*,
https://doi.org/10.1007/978-981-99-2812-5_2

writer, who then attempts to prove it". According to the Collins Online Dictionary, "a thesis is an idea or theory that is expressed as statements and are explained logically". Collectively, a thesis for a postgraduate degree, especially a PhD thesis, represents much more than a review of the literature, a description of the methodologies, and a presentation of the experimental findings. An essential component of the thesis is the writer's (student's, or the whole research team's) opinion, theory, and idea. Further elaboration is provided in the following sections.

In this book, I will not discuss the format, typesetting, margin, number of pages, layout, fonts, or referencing style. Different universities have different styles. If we Google 'your university name' + 'thesis manual, or thesis format, or thesis guidelines', we will likely find a document describing thesis formatting. To obtain more information about preparing a thesis, you may wish to browse The Thesis Whisperer blog, or read a few books such as *How to Tame Your PhD* and *How to Fix Your Academic Writing Trouble*.

2.2 The Chapter One of a Thesis: Introduction

In a proposal or thesis, the first chapter is often referred to as the Introduction. Your institution may have different guidelines as to how subsections should be arranged. Typically, Chap. 1 may include the following subsections:

- Subsection-1 (i.e. 1.1). This section contains the background of the study or some thesis formatted as an introduction.
- Subsection-2 (i.e. 1.2). This section should include the problem statement, research gaps, aims, and objectives.
- Subsection-3 (i.e. 1.3). Scopes of the study.
- Subsection-4 (i.e. 1.4). The significance of the study, the outlines of the research, etc.

2.2.1 Background of Study

Let us assume that the first subsection of the Chap. 1 is entitled 'Background of Study or Introduction'. Below are my recommendations:

- The length of this section can range from 2–5 pages for a master thesis to around 3–7 pages for a PhD thesis. Please refer to examples of theses of your own institute to see what the average length is. If a student wants to write longer, no one can say no. However, if this section is too long, the reader may become lost and bored. Usually, 2–4 pages are good enough.
- It is recommended that students treat this section as a mini-chapter 2 that summarizes all the major points that will be discussed in the literature review part.
- This section should explain all technical terms with adequate explanations so that readers can understand them from the beginning. Ensure that the background

knowledge provides sufficient information to the readers or serves as a bridge to the subsequent section (problem statement, scientific missing gaps, objectives, etc.). In the absence of this, readers will be unable to understand any new technical terms in the section of the problem statement. It is better to introduce all the jargon in the Introduction and use them without explaining their meanings in the subsequent parts (problem statement, scientific missing gaps, objectives, etc.).

- Similarly, make sure to include all technical jargon in the title of the proposal/thesis in the Introduction section. If, for example, the title of the thesis includes the terms lipidomics, virome, metataxonomy, osteomicrobiology, ecological oncology, semi-supervised learning, random forest, cold plasma, quantum microscope, 4D printing, electrigens, geobatteries, biomimetic patches, leaf senescence, etc., it is best to describe these terms in the first section (Introduction), since not everyone is familiar with these terms. It is not sufficient to provide a definition; the explanation of the jargon should be combined with the findings of some recent publications. Do not let the jargon pile up, readers will get lost and eventually switch off.
- Do not include too much basic textbook information and, if we wish to be more strict, do not include any textbook information at all. It is not appropriate to state and explain topics taught in basic undergraduate courses. By detailing what PCR reaction is, photosynthesis, types of extremophiles, and so forth in lengthy paragraphs, the proposal/dissertation/thesis is judged to be poor in quality. If the candidate feels that it is appropriate to point out a few of those phrases, it is fine, but we should be cautious not to elaborate too much in several paragraphs.
- Do not describe methods that are too common. All laboratory equipment is quite standard when it comes to amplification, cloning, expression, electrophoresis, and purification of a recombinant project. It is unnecessary to elaborate in the Introduction. If, however, a very new technique is used, such as metagenome-assembled genome for prokaryotes, eukaryotes, or viruses, it might be better to explain in a few sentences since some readers may not be familiar.
- Consider who are the readers and examiners. If the student is affiliated with a department of microbiology, and the topic of the project is artificial intelligence or big data science in biology, it is likely that readers or examiners from the same department will not be familiar with the topic. If this is the case, it may be necessary to provide an explanation with sufficient background information.
- Please ensure that the information provided is current. It is better, for example, to include newer data or information in the first few paragraphs, rather than making generalizations about things we already knew 10 or 20 years ago (or even earlier).
- Please provide supporting evidence. Often, students use phrases such as 'there is a growing trend in ___' or 'more and more people are ___' or 'the price of X is increasing ___', etc. This is an ambiguous statement without sufficient recent evidence. The evidence here refers to some sort of statistical data. The use of the phrase 'in recent year ___' and ends with a reference dated 2010 would not make sense. The year 2010 isn't that recent, is it? I will further elaborate on this point in the following paragraphs.

2.2.2 Aims, Problem Statement, and Objectives

My institute usually arranges the problem statement and objectives in separate subsections of Chap. 1 of our thesis. Nevertheless, I would like to discuss the aims, problem statement, and objectives together since they are interconnected. Unless aims, problem statement, missing gaps, and objectives are clearly described, readers are not able to comprehend the whole writeup.

Diverse stakeholders have different perspectives. If we ask a student, why are you doing this project? In most cases, students will provide a very diplomatic response. The unexpected and straightforward response is that some just want to obtain certification or learn a new laboratory skill. An engineer may look for answers from the perspective of how the research can solve a specific engineering problem. Ideally, a social scientist or layperson would examine a biotechnology project from an impact to a sustainable or societal perspective. Entrepreneurs expect the results of a study to be commercially useful. However, a professor in a university is more likely to focus exclusively on research novelty, knowledge depth, and publications. Although the students have done their best and the supervisors agree completely with the writing style, it is still possible that the *viva voce* examiners are unsatisfied with the essay simply because each examiner approaches it from a different perspective. There is no one size that fits all. Anyway, the following subsections provide an overview of general good practices.

2.2.2.1 Research Gap

All engaged researchers closely follow the progress of their respective fields of study. After reading a sufficient amount of material, one can easily identify knowledge gaps, also known as research gaps or literature gaps. In other words, a research gap represents the absence of sufficient information or knowledge in a specific area. I have heard someone define a research gap as an area of scientific research where no one has explored enough a particular aspect of the topic. A knowledge gap refers to a state in which we know little or nothing. The word 'we' is used in the previous sentence to refer to more than just you and I, but to the entire scientific community within that particular research area. Of course, before we make a statement and declare it as a knowledge gap, it is essential to have thoroughly read and familiarized ourselves with the relevant literature. Having a solid understanding of fundamentals is crucial.

In Chap. 1 of this book, I described the meaning of philosophy. One of the learning outcomes of a PhD program is the ability to produce new knowledge. In Chap. 2, we began with the definition of thesis as the main idea, opinion, or theory of a person, and the evidence to support it. PhD and master's candidates should possess the ability to recognize the knowledge gap across the entire scientific community (and not just their own). Furthermore, they should have the ability to design experiments to collect data for hypotheses that have been made and utilize that data to confirm their idea, opinion, or theory. All aspects must be rigorously connected. The verb 'produce knowledge' or 'generate knowledge' may have different meanings for each of us. A more precise terminology would be to discover

Fig. 2.1 Research outputs are like pieces of a puzzle

new knowledge, since scientific knowledge already exists in the universe, and humans do not create new ones.

Imagine, for example, that we have a box of puzzle. Earlier researchers have managed to assemble 950 pieces; however, there is a gap of 50 pieces. In generating a new idea, opinion, or theory, we are generating a block that fits into the missing picture. When we contribute 5 pieces of block, the picture is now 955 and the next person will continue adding more pieces (Fig. 2.1). There is a possibility that when someone fit the last missing piece into the puzzle, we realized that it was actually a box containing 2000 puzzle pieces rather than 1000. Such is the nature of knowledge and research; as we explore further, we uncover new complexities and uncharted territories, highlighting the importance of continuous inquiry and the ever-growing puzzle of understanding. Please remember, the more we learn, the more we realize we do not possess enough complete knowledge.

To keep this section simple, I will restrict the discussion to basic research. Readers who are engaged in applied research may perceive the concept of missing knowledge or research gaps differently. Nevertheless, if the PhD project is oriented towards applied science, I believe that a knowledge gap still exists.

I found the following statement in an online article entitled 'The basics of writing a statement of problem for your research proposal' on the Editage insights website. A good research question should meet the following criteria:

- The study should address a knowledge gap.
- Research should be significant enough to contribute to the existing body of knowledge.
- The problem should be investigated rigorously through the collection of data.
- The study should lead to further investigation.

2.2.2.2 Common Mistakes

Following are some common mistakes I often encounter when reading a student's problem statement in their proposal/thesis or a researcher's proposals.

- Long, but without a clear problem statement. I have seen people write full 3–5 page descriptions (sometimes longer) of only literature. However, they do not describe what the research (knowledge, literature) gap is.
- Second, students often state a global problem or a regional problem that cannot be addressed by the current project. For instance, fossil fuel reserves are diminishing. How can the current biotechnology research project overcome the problem of fossil fuel depletion? Can we perform some biodegradation, fermentation, or enzymatic reaction and have the products transferred back to the earth and become fossil fuels? I doubt it. This is an example of a problem statement that cannot be resolved. Thus, as a reader, if I see a statement like fossil fuels are depleting, rivers are polluted, the air is polluted, etc., in Chap. 1's Problem Statement section in a proposal/thesis, it will be interpreted as general knowledge rather than as a real problem that can be solved by collecting experimental data. It is important to keep the scope of the problem as narrow as possible.
- Misplaced focus. Writers often like to give a detailed description of a topic, we'll call it A, for the sake of this example. Let's say that one whole paragraph is dedicated to the gaps in research number 1, number 2, number 3, and number n related to subject A. We expect that this project will address the problems of 1, 2, 3, ...*n*, to solve them in a meaningful manner. The writer then proposed using B, another topic of interest. B here has nothing to do with A. Avoiding the issue does not solve the problem.

I will illustrate 'misplaced focus' with an example. Drunkenness is caused by alcohol consumption. I see on YouTube that there are some kinds of medicines available at 7-Eleven in a country that can reduce drunkenness or force you to become more sober. Consider a scenario where one of these medicines has the problem of numbers one, two, and three. (Disclaimer. I have no idea of this type of medicine nor am I familiar with the field of medicine. I never got drunk anyway because I do not drink. Please take this as an example only. All the information is fabricated). Having said that, the proposal can, by right, study how one can modify the medicine (drug) and what are the mechanisms involved in modifying the medicine (drug structure for instance). Occasionally, however, people put the wrong focus on their research proposal. They are as if proposing why not drink Coke instead of beer, then we solve the problem of drunkenness. Yes, it sounds great, and it makes a lot of sense, doesn't it? However, wait a moment. What is the scientific problem, drunkenness or side effects (or efficiency) of the medicine? For this example, I believe it is the latter. The purpose of a scientific problem is to solve a particular problem, not to avoid it by substituting some other alternative, that is, drinking soft drinks. Take a look at your past writing and see if you have misplaced a focus. If you've done that before, you weren't sober!

Here is another example. In general, baldness refers to an excessive loss of hair from the scalp. A person who is bald in my country is called a *botak*. There are a number of reasons that may lead to *botak*, such as heredity, oily scalp, stress, food factors, medication treatment, etc. When we assume someone mentions the issues of *botak* in the problem statement of a proposal, we will automatically assume they are attempting to understand the real mechanism that causes *botak* (basic science) or maybe they will formulate some kind of shampoo or hairspray that will prevent hair loss (applied science). In what way are we putting our focus in the wrong direction? The project involves redesigning a new wig hair using some cheap and recycled materials. The research problem is *botak*, not wig.

2.2.2.3 Ways to Write a Research Problem

There is no standard template (SOP) for writing a problem statement. Be creative from time to time and never stick to the so-called best way to write. It is not true that there is a general practice or the best way to draft a problem statement for a thesis or a grant; there will always be better ways to articulate our points when preparing a research proposal or a thesis. I recommend that you always try to be a little different every now and then and experiment with which way of writing will grab the attention of the readers most effectively. Below are few examples that we may consider.

Example 1. A Paragraph with Four Sections

Consider that we decided to write only a paragraph for the problem statement. Often, we write more than one paragraph. The paragraph can be divided into four parts:

Part 1: An overview of what we already know or have done by other researchers or our team.

Part 2: Write about the knowledge/research gap. Explain in detail the unknown knowledge, theory, concept, method, model, process, or anything else relevant.

Part 3: Explain why it is necessary to address the issue mentioned above. Try to write answers to a question like 'so what?' or 'why'.

Part 4: Describe what is the general approach to achieving the goal.

In the entire paragraph, all the parts mentioned above were arranged continuously in sentences. The length of each part can be adjusted according to a certain ratio, for instance, two sentences for each part resulting in a final ratio of 2:2:2:2. There is no fixed ratio but if we have elaborated enough literature in some previous sections, we may decide on a 2:4:1:1 ratio, or 2:6:2:1, or any ratio. Do not be too rigid. In my opinion, it would be better to emphasize part 2 (research gap) rather than explain other parts, that is, the experimental approach.

Example 2. One Objective Approach per Paragraph Let us assume that our project has three objectives. We can then write three paragraphs, one for each objective. There may be two major parts in each paragraph (see below for the first paragraph dedicated to objective 1).

Part 1: Description of what is already known or has already been done within the scope of objective 1 from different research groups or published articles.

Part 2: Write about the knowledge/research gaps. What is the unknown knowledge, theory, concept, method, model, process, or anything related to objective 1 that is unknown?

For the second example stated above, assuming that the paragraph has ten sentences, we can probably write five sentences for part 1, and five sentences for part 2 (5:5). If possible, we may use 3:7 or 4:6. Would you recommend that we use a ratio of 9:1 where we only use one sentence to describe what the problem statement is instead of using multiple sentences? By using nine sentences to explain what is already known and by concluding the paragraph with just one sentence on the missing gap (9:1), it is possible the readers will not interpret the message correctly. It is important to note that the ratio of each part here makes a noticeable difference in the way the reader understands what we have written. The ratio of known and unknown makes a big difference. Following completion of the first paragraph, we may move on to the second and third paragraphs, which each address objectives 2 and 3.

Example 3. Research Problem for Verifying Earlier Theory/Assumption/ Finding Assume that there are two main schools of thought that are contrary to each other and that they provide inconsistent evidence concerning the issue. For instance, the role of an amino acid at the center of the catalytic pocket of an enzyme in product specificity remains a subject of conflicting evidence and contradictory findings. While a research paper has demonstrated its significance in influencing product specificity, subsequent studies have reported inconsistent results, casting doubt on its actual contribution. In order to establish who is correct and why the discrepancy exists, a new project is developed. An example of the problem statement would be as follows:

Part 1: An overview of what is already known and what has been accomplished for both schools of thought.

Part 2: Discuss in detail the inconsistencies and contradictions between both theories or findings.

Part 3: Write out the speculations for why there are inconsistencies and contradictions, and what is the missing gap? Explain why more research should be done.

Part 4: Describe the general approach (a new angle, strategy, approach, technique, etc.) towards achieving the goal.

Let us see one fabricated problem statement below; please focus on the writing and not the science since I am not an expert in this field. "The mammalian brain uses Gamma-aminobutyric acid (GABA) as an inhibitory neurotransmitter (reference 1). Research has established that GABA has stress-reducing and sleep-enhancing properties (reference 2). There has been some research on GABA effect on sleep and stress, but results have been mixed. Some studies have demonstrated that GABA can induce sleep and ____ (reference 3–5), while other studies have shown that such a hypothesis is misleading (references 6–8). It has been speculated that inconsistencies and contractions are due to ____ and _____, but it has not yet been verified. In order to address this issue, we have developed a new approach to _____. We aim to ______".

2.2.2.4 Create a Clear Transition

I recommend that a clear transition be demonstrated between the two major parts (what is known) and the second part (what is unknown) when writing problem statement. If necessary, we may use words such as however, nevertheless, yet, despite, yet, even so, therefore, as a consequence, even though, etc., between the two parts or wherever is appropriate. Let us examine an example.

> Earlier microarray research suggested that the genes *x*, *y*, and *z* are responsible for regulating the adaptation response of cell type A toward ____ stress (reference 1). This finding concurs with a recent transcriptome study that used cells of types B and C, which are taxonomically related to cells of type A (references 2 and 3). Even so, it is not known whether hypothetical proteins positioned in between the gene cluster would have any effect on cell adaptation. Research has not been conducted to date. Because these genes have not been cloned and analyses of their biochemical functions have not been conducted; therefore, it is difficult to determine ____. The closest protein homologue in the Protein Database (PDB) is less than 20%, and as a consequence, we are unable to predict the structure and thus unable to ____. We know about ____ (reference 4); however, researchers are uncertain ____. There is no information available on how or to what extent these hypothetical proteins would function as ____or ____. Despite some publications indicating ____, it is not known, in this case, whether ____. Without these information, scientific community is unable to ____. The purpose of this study is to ____.

For demonstration purposes, I fabricated the problem statement provided and please acknowledges the need for improvement. As this is a fabricated problem statement, there may be some exaggerations in the earlier paragraph. I would like you to pay some attention to a couple of points. Firstly, the use of transitional words (such as 'even so', 'therefore', 'however') in the example above serve to emphasize certain information. Secondly, could you be able to calculate a ratio of 'what is known' to the total number of sentences representing 'what is unknown'? Please re-read the paragraph above and note that the empathic focus is the missing knowledge (knowledge gap or the unknown). It is not easy to write a good problem statement. The first draft of the problem statement should be a simple paragraph that has a clear skeleton. Once the skeleton has been established, we can add more flesh (description and proof) to it. Here is an improved version of the problem statement:

> Earlier microarray research suggested that the genes *x*, *y*, and *z* are responsible for regulating the adaptation response of bacteria-A to osmotic stress (reference 1). In response to saline stress, proteins X, Y, and Z synthesize betaines and choline that act as osmolytes that enable prokaryotes to survive (reference 1). This finding concurs with a recent RNA-Seq transcriptome study that used bacteria-B and bacteria-C, which are taxonomically related to bacteria-A (references 2 and 3). There are two hypothetical proteins (namely, HPa and

HPb) within this gene cluster, arranged as *x-hpa-y-hpb-z* (please refer to literature review section for more information and illustration). Proteins X, Y, and Z have been extensively studied over the past 10 years (reference 1–6). Even so, it is not known whether hypothetical proteins positioned in the gene cluster would have any impact on cellular adaptation. The lack of research on HPa and HPb can be attributed to the absence of gene cloning and limited analysis of their biochemical functions; therefore, it is difficult to determine the biological role of hypothetical proteins HPa and HPb. The closest protein homologue in the Protein Database (PDB) is less than 20%, and as a consequence, unable us to predict the protein function of HPa and HPb. We know that the closest protein homologue is an enzyme that forms osmolyte ectoines (reference 4). Ectoines, another kind of osmolyte, could protect prokaryotes from saline stress more effectively than betaines and choline, according to one article (reference 4). Researchers have known for a decade that bacteria A, B, and C exhibit genes for betaines and choline, but never realized they may also encode for ectoines. It is unclear, however, whether hypothetical proteins HPa and HPb are related to ectoines. Even though the authors of an earlier publication (reference 4) made some assumptions, it is not known whether HPa and HPb play a role in saline tolerance because the author did not prove it by experiment. There is no information available on how or to what extent these hypothetical proteins would function as enzymes that form ectoines. Without this information, the scientific community is unable to understand the real biological role of these hypothetical proteins. The purpose of this study is to ____.

I want to clarify that the examples provided above are solely for illustrative purposes. They serve as sample statements to demonstrate certain concepts and techniques. It is important to recognize that these examples may not reflect specific real-world scenarios or research topics. Therefore, when crafting your own statements, it is essential to consider the unique context and requirements of your study. The following resources may be helpful if readers need more examples, templates, or ideas: (1) Usually, the authors of good research journal articles will describe the research gap or problem statement in the last paragraph of the Introduction. You may be able to get some inspiration for your own project if you observe and learn how prominent authors write that. (2) Another approach is to examine some digital thesis and review carefully. Pay particular attention to the writing style employed. This can offer insights into effective ways of articulating problem statements. (3) Besides, you can find more examples by asking Prof Google or chatbot like ChatGTP. (4) Consider exploring the examples provided on the Ref-N-Write website within their Academic Phrasebank section. It offers a collection of useful academic phrases and templates that can assist you in constructing problem statements.

2.2.2.5 Problem Statement Is Not Purpose Statement

There is a common misconception among young researchers and students that the purpose statement is equivalent to the problem statement; however, this is not the case.

Characteristics of a problem statement:

- Identifies a research gap that pertains to the absence of sufficient information or knowledge in a particular field for the entire scientific community.

- The scope is quite narrow.
- Focus on what we want to know (knowledge).

The following are the characteristics or contents of a purpose statement:

- Stress on an overview of the experiment's design
- Focusing on experimentation methods or approaches
- Emphasize the purpose of the experiment or idea or a global aim
- Concentrate on the action that will be taken in order to collect the data
- Focus on what we will do (action or experiment procedures)

To illustrate, we will rewrite the example of the problem statement provided in the earlier subsection and convert it into a purpose statement. In a proposal or thesis, the purpose statement cannot replace the problem statement. The paragraph below is a purpose statement, not a problem statement.

"Microarray studies have suggested that the genes *x*, *y*, and *z* are responsible for regulating the adaptation response of different cell types to osmotic stress (references 1, 2, 3). Several hypothetical proteins are positioned adjacent to the gene cluster that consists of genes *x*, *y*, and *z*. This study aims to examine the role of hypothetical proteins in bacterial adaptation to high saline media containing 5%, 10%, and 15% NaCl (w/v). By examining the RNAseq data, it is intended to compare transcript levels of genes encoding these hypothetical proteins. A medium containing 2% NaCl is used as a control for the experiment. Additionally, if one or a few genes are highly upregulated, the research will continue with cloning, expression in a shuttle vector system, affinity purification using Ni-NTA and gel filtration columns, and biochemical analysis of one selected protein. A crystallization of the protein will be followed by a determination of its X-ray structure".

We are still discussing the same project that involves hypothetical proteins. When a write-up focuses on experimental design, methods, approaches, or the steps undertaken to gather data, we consider this to be a purpose statement. Changing the wording and focusing on knowledge and scientific reasons will make the write-up appear more like a missing gap in the problem statement.

Here is a quick summary of the lengthy description explained earlier. Generally, a research gap, a literature gap, or a knowledge gap can be defined as areas that have not been explored or have been underexplored but that have the potential to be further studied. By shifting the emphasis towards the missing knowledge and scientific reasons behind the research, the write-up can better highlight the gaps in understanding. This lesson has taught you how to define and write one for your proposal or report. If you are still unsure where to begin, you may wish to read the following online article in Editage Insights: 'Don't know where to start? Six Tips on identifying research gaps'.

2.2.2.6 Drafting Objectives Using the BE-SMART Rule

Based on the same project that involves hypothetical proteins, a possible set of objectives could look something like the one provided below.

Examples of knowledge-based objectives are:

1. To assess the transcript responses and relationships of hypothetical proteins in cells exposed to different levels of osmotic stress
2. To discover the biochemical roles of hypothetical proteins in cellular adaptation
3. To analyse the protein structures of selected hypothetical proteins and inspect the importance of specific regions of such proteins

Examples of action-driven (experimental) objectives are:

1. To grow the cells in high saline media and extract the total rRNA for transcriptome sequencing
2. To clone, express, purify, and biochemically characterize hypothetical protein
3. To crystallize the purified hypothetical protein and perform site-directed mutagenesis using mega-primer PCR approach

Again, we are still discussing the same project that involves hypothetical proteins. As you carefully review those objectives, you will find that knowledge-based objectives will answer the how and why research questions. Action-driven objectives specify what will be done or what experiments will be used to obtain data, as their name implies.

We are probably aware of and familiar with Bloom's taxonomy. It was initially proposed in 1956 as a taxonomy model to describe the cognitive process of learning and the mastery of a subject. There are generally two levels of cognitive skill: low level (levels 1–2: knowledge and comprehensive) and high level (levels 3–6: application, analysis, synthesis, and evaluation). Should you not be familiar with Bloom's taxonomy levels, please conduct some online searching.

According to our earlier definition, a research gap is related to missing knowledge, and knowledge is classified under a cognitive domain. Would it be possible to incorporate Bloom's taxonomy into the objectives of theses/proposals that address research gaps? As indicated in the examples of knowledge-based objectives provided earlier, assess has a level of 6, discover has a level of 5, analyse has a level of 4, and inspect has a level of 4 in the Bloom's taxonomy scale.

Undergraduate students frequently use action-driven (experimental or research activity) verbs as part of their objectives. Some of them may continue writing in this style once they enrol in postgraduate studies. There are many action-driven (experimental) verbs in bioscience and biotechnology, such as to grow, to isolate, to cultivate, to separate, to extract, to amplify, to express, to purify, to clone, to crystallize, to run, to visualize, to stain, to identify, to sequence, to simulate, to dock, to do a questionnaire survey of the patients, etc.

The following are two paragraphs I hesitate to include, but I think they are necessary. Please accept my apologies for being slightly sarcastic by suggesting some interesting ideas for us to consider. Assume that the student has graduated and is being interviewed. The interviewer will often ask you what the objectives are of your research. "Ah well, my PhD is focused on the isolation of bacteria (objective 1) and extraction of genomes (objective 2)". Well, if I were the interviewer, my first impression would be that the entire PhD is aimed at isolating bacteria, but, what is the point of doing the project. In laboratories, autoclaves are a common thing. Are autoclaving and sterilizing also acceptable to be included in the objective of a project? "Oh yeah, my master is autoclaving the media"—do you think it makes sense?

I believe the readers will understand what I am saying. All these steps are part of our experiments: isolation, extraction, pH adjustment, autoclaving, running gels, running chromatography, building a phylogenetic tree, predicting protein structure with bioinformatics tool, etc. However, our primary focus should be on the knowledge, ideas, or concepts from these actions rather than solely on the action themselves. Research gaps can only be filled by objectives that deal with how and why questions. Let's not forget the essence of a thesis. The thesis serves as a significant platform for presenting the student's opinions, theories and ideas that is generated from the concrete data. It is crucial to establish rigorous connections among all aspects, encompassing the problem statement, hypothesis, research question, and objectives, and if possible literature review.

In terms of the master's and PhD thesis and a research proposal for getting a funding, I personally prefer to see scientific or knowledge-driven objectives rather than action-based objectives that describe what the candidate will do or has done (experimental or research activity). What I will explain here may not be the same as what you have understood in the past. As different universities, supervisors, and examiners may hold somewhat different schools of thought. Readers are advised to conduct some due diligence and consult with someone with more experience to see if it will work.

Objectives must be SMART, which stands for Specific, Measurable, Achievable, Relevant, and Time-Bound. In many instances, SMART is used to evaluate whether written objectives are appropriate. Measurable, in SMART, is sometimes misunderstood by people by being restricted to the action, variables, or the data of experiments (specifically action-driven objectives). If I could, I would modify it to be BE-SMART. The B stands for Bloom's taxonomy and the E stands for thEsis (opinion, theory, and idea). Measurables (M) in SMART should be linked to Bloom's taxonomy. Moreover, the associated problem statement—aims—objectives must be relevant to our thesis (opinion, theory, and idea, not referencing to the graduation report). Where does the new opinion, theory, or idea come from? Through valid and rigorous experimental data.

2.2.2.7 Drafting the Research Question and Hypothesis

When you do some online searching, you will probably discover that there are different types of hypotheses, such as simple hypothesis, complex hypothesis, directional hypothesis, non-directional hypothesis, null hypothesis, associative hypothesis, and casual hypothesis. Let me explain the term 'hypothesis' in a simple way. The word 'hypothesis' refers to an idea (a possible explanation or a guessing answer) that has not been proven for the research questions in your project. If a hypothesis has been proven, it becomes your thesis statement and is accepted as a scientific fact by others subsequently. Hypotheses and research questions can be written in a variety of ways. But in this book, we will limit our discussion to projects that are basic science with the aim to generate knowledge (idea, theory, or explanation) to fill a gap in the literature. There must be a clear link or connection between the following parts of a proposal or thesis: problem statement, research questions, hypothesis, and objectives. All of these components are interconnected (Table 2.2). An example of research questions and hypothesis based on the same project involving hypothetical proteins could appear as follows:

Research questions:

1. When cells are exposed to different salinities, how does osmotic stress affect the expression of hypothetical proteins (HPa and HPb) in the gene cluster *x-hpa-y-hpb-z*? (Or, how does salinity affect the whole gene cluster and hypothetical proteins transcript levels, and what is the underlying mechanism?) Which and how do the transcriptional factors affect the expression of these hypothetical proteins?
2. Hypothetical proteins are proteins with a very limited understanding of their functions. In the case that hypothetical proteins (HPa and HPb) are highly upregulated in saline-stressed cells, what is their biological function and how do they assist the cells in adapting to salinity? What is the purpose of these proteins? Why are they essential to bacterial survival? Are HPa and HPb enzymes involved in the formation of ectoines? How could researchers who have studied bacteria A, B, and C for a decade have missed these findings?
3. If HPa and HPb are enzymes that form ectoines, why are their sequences different from most other protein homologues (noted: $<20\%$ protein sequence similarity)? Despite the low sequence similarity, could the overall conformation be similar to other homologous structures? Why would the cells adopt a different amino acid stretch in comparison to the other homologues? How do such dissimilarities influence HPb and HPa's ability to produce ectoines?

Hypothesis:

1. Genes *x*, *y*, and *z* will upregulate when cells are under stress (reference 1–4). Since hypothetical proteins are sandwiched between gene clusters *x-hpa-y-hpb-z*, it is hypothesized that *hpa* and *hpb* genes will be upregulated under saline stress. Gene clusters *x-hpa-y-hpb-z* are probably expressed at similar amplitudes by the same transcription factor upstream of the gene clusters. We discovered that *hpa* and *hpb* genes are highly conserved in the genomes of almost all genera of prokaryotes associate to family-TT while preparing this proposal (unpublished data). We hypothesize the hypothetical proteins may play a role in saline stress adaptation of cells.
2. There are several ways in which prokaryotic cells adapt to saline stress. According to one review article (reference 6), the main osmolytes that protect cells are betaines, choline, ectoines, and the like. Our hypothesis is that hypothetical proteins (HPa and HPb) are part of the whole system that forms ectoines and

Table 2.1 Problem statement, research question, hypothesis, and objective

Problem statement (knowledge-driven)	Research question (knowledge-driven)	Hypothesis (knowledge-driven)	Objective (knowledge-driven)
Literature gap, knowledge gap, unknown, missing theory, missing explanation.	Asking questions that seek answers to a reasoning, relationship, explanation, etc.	A guessing answer to the research question. A guessing explanation to the problem statement. Idea or your explanation that needs to be verified by experiment.	Emphasize the purpose to get the answer to the knowledge gap/missing theory/literature gap. Do not tell how you do it experimentally.
Written in sentence form using numbering or bullet. Can be written in one or more paragraphs. Purpose statement is not a problem statement.	Written in question form. Can write in a paragraph. Using 'why and how' aims to get a scientific reasoning. The indirect questions (without why/how) should induce a why/how answer.	Possibilities of writing: It is hypothesized... We hypothesize... We assume... We think... We guess... It is likely that... It is possible that...	BE-SMART objectives

probably play a significant role in cell adaptation. In spite of low sequence similarity, HPa and HPb may be novel proteins involved in osmolyte ectoines formation.

3. Protein structure is more conserved than sequence, according to our hypothesis. We believe that the important domains, motifs, or amino acids should be very similar to well-studied homologous structures. We expect HPa and HPb enzymes to have very different enzymatic properties due to high variation in non-conserved regions.

Please remember that the problem statement, the research questions, the hypothesis, and the objectives must be interconnected. For example, objective 1 and research question 1 are connected to hypothesis 1. (Please read the text side by side and you will notice the connection). Table 2.1 and Fig. 2.2 summarize what we have covered so far. The cycle of doing research always begins with reading and reviewing literature (journals, book chapters, conferences, patents, etc.). We can only find the real research gap (knowledge gap, literature gap) by critically reviewing literature. We then make a wild guess (our hypothesis), which can be seen as an initial speculation. Our hypothesis should be validated based on

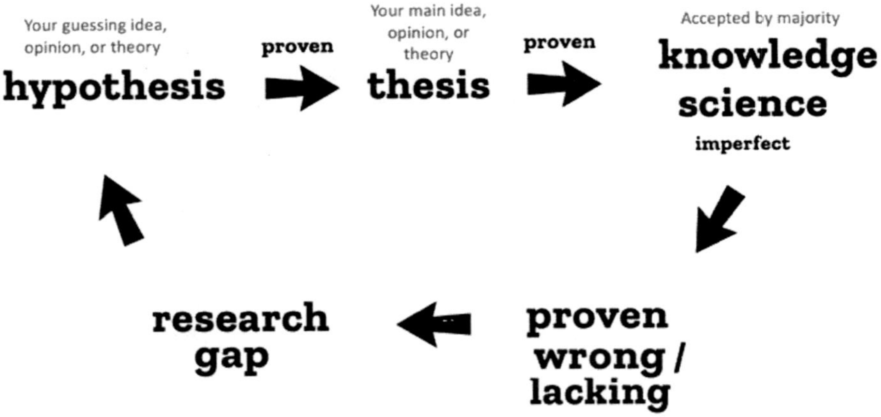

Fig. 2.2 The cycle of research

experimental data. If the hypothesis is supported by data, it becomes our personal thesis statement (not referring to the dissertation). As soon as we publish this knowledge, our idea, and our explanation in a journal, and when most other scientists agree with us, our idea becomes scientifically accepted. It will become a knowledge, theory, or explanation for the whole scientific community. Science is not a flawless field, as explained in Chap. 1. People can challenge our thesis argument if they prove it is wrong or lacking or ineffective, then a new set of research gaps open up. It is a cyclical process.

In my observation, many students and young researchers find it extremely challenging to establish systematic and consistent connections between the problem statement, objectives, research questions, and hypotheses. Some of us have tried using AI tools like ChatGPT for proofreading. But, do not ask ChatGTP to write problem statement for you. It's important to understand that it doesn't work that way. While ChatGPT can help align the problem statement, objectives, research questions, and hypotheses when provided with sufficient prompts, it cannot generate everything from scratch. You can utilize AI to your advantage by asking it to read your texts, explain your thoughts, and identify the specific stage or section where you are facing difficulties. Providing detailed instructions will enable the AI to assist you in refining your write-up and making it crystal clear. For example, if you have already written your proposal but feel that something is missing or lack proper connection, you can ask for advice such as:

- Is the message clear?
- How can I improve the clarity and connection between the problem statement and objectives?
- What else should I include?
- How can I make my problem statement more aligned with fundamental science (or applied science)?

- If you were a funder or reviewer, would you consider the write-up to be good enough?
- Are my research questions making sense? How can I rephrase them to begin with 'Why and How' instead of 'What'?
- Do you think my hypothesis aligns well with my objectives?
- What other steps can I take to enhance clarity the objectives?

By seeking guidance from AI in this manner, you can enhance your write-up and ensure that it meets the desired standards.

2.2.3 Significance of Study or Thesis Impact Statement

There are times when young researchers (students) are uncertain about what to write in their significance of study statement, which is also referred to as the thesis impact statement. To adress this, I would like you to ponder one simple question—what kind of use could the knowledge, analysis, discovery, or insight in this project be put to?

As another way of expressing this, thesis impact statement consists of a few sentences or paragraphs that briefly explain why this research is needed or important and how the study can contribute to society after completion. Depending on your preferences, you may consider writing the significance of study statements both within and outside of academia.

- Within academia. How does the thesis contribute to the current research area and other closely related research areas? To what extent does the study lead to further investigation? Does the finding clarify any misunderstandings that may have existed earlier?
- Outside of academia. In what ways will the thesis benefit social, public policy, commercial, public health, the environment, professional practice, standard, culture, quality of life, communities, and organizations? What relevance does the research have to the Sustainable Development Goals (SDGs)? Is the project relevant to the aims of the United Nations or World Health Organization? Does the research address fulfil the vision of the UN Climate Conference of the Parties (i.e. agenda in COP conference)? How do the findings fit into the main thrusts of the home nation's policies, such as the national biotechnology policy, the national biodiversity policy, the national medicine policy, and the national biofuel policy? If the project is highly applied science or specifically designed to reach commercialization, you can describe the scope of applications and who (the public) will benefit from it.

Let me elaborate a little bit further on the point outside of academia. Many nations have developed their own research and classification systems. In Malaysia, we have 10–10 MySTIE and the MRDCS (Malaysian Research and Development Classification System). You may have something similar in your own country. In the Social-

Economic Objective of MRDCS, research can be classified into five major groups: Environmental Sustainability, Economic Empowerment, Governance and National, Human and Societal Resiliency, and Advancement of Knowledge and Learning.

In the case of a research project that deals with food or animal feeds (e.g. new meat, proteins produced from black soldier fly larvae, etc.) utilizing biotechnology approaches, one can write an impact statement which indicates that this project is related to Economic Empowerment that supports halal industry, green economy, and agricultural sector reform. Research on animal feeds or food safety can also contribute to Human and Societal Resiliency, in particular, with respect to basic human and social needs. Another example. If a research project concerns machine learning, big data science, or advanced bioinformatics in biotechnology, appropriate social-economic categories would be the fourth industrial revolution or expertise building.

Each of the technical terms we read in the previous paragraph is a keyword in the MRDCS. We can open the webpage of MRDCS and compare it to the examples that I have provided above, and you will probably be able to make more sense from there. I am unfamiliar with the social-economic objectives of other countries. Upon searching the internet, some similar information is available for some countries. If you are unable to obtain such policies in your country, SGD may be an acceptable starting point. In Chap. 8, I interviewed an expert in the field of social science who has extensive experience integrating scientific research with social science perspectives. If you wish to learn more, please read the text.

2.2.4 The Chapter Two of a Thesis: Literature Review

In student dissertations or theses, chapter two is usually the literature review. Although it is not too difficult to write a literature review, writing a high-quality and reliable review is challenging.

2.2.4.1 Define 'Literature' and 'Review'

The Oxford and Merriam-Webster dictionaries provide multiple definitions for the terms 'literature' and 'review'. Here are some of the definitions for the word 'literature': all information related to a subject especially information written by experts, any writing that has lasting value, and written materials that promote an opinion. Examples of literature resources include journal articles, review articles, books, book chapters, conference proceedings, etc.

This definition of the word 'review' identifies its function: to view or see again; to examine and study again; to give a critical evaluation of; to summarize the most important facts of a piece of academic writing that you have read; to think or talk about something again in order to change it or to make a decision about it, etc. There are a number of synonyms for the word 'review' listed at thesaurus.com, including analysis, audit, check, inspection, revision, scrutiny, survey, reassessment, reflection, etc. Key home message: an extensive literature review includes a critical examination of the content; this is why the term 'literature review' is used rather than literature report.

2.2.4.2 Purposes of Literature Review

It is common for students and researchers to read prior publications, digest the information, organize it into genres or topics, and then write those into research proposals, theses, or the introduction part in journal articles. Such writing is known as a literature review. It is essential that we keep in mind the meaning of review. If we do not keep the following three points in mind when preparing literature reviews, we will end up with a document that looks like a summary without a purpose.

- Literature reviews enable us to visualize the scientific jigsaw puzzle, notice the missing piece, and consider how we may look for that missing piece (gaps in existing research, or the literature gaps).
- Through this process, we can gain a better understanding of what is already known, the principles behind the study, achievements made by other groups, and methods and techniques that are relevant to the research topic.
- An effective literature review will allow the writer and reader to determine where future research should be focused.

Students who spend sufficient time reading literature, reviewing information, analysing it, and reorganizing it in their proposal/dissertation will benefit from the following:

- Allows students to become familiar with the project before engaging in more serious activities
- Find research gaps
- Enhances critical thinking skills by filtering, reorganizing, finding connections, detecting contrasts, comparing findings, and making proper arguments
- Improves communication and writing skills
- Develops confidence

In the examination process, viva voce examiners use the literature review to evaluate the candidate's ability to research the topic, develop the position in the field, and convince the reader that the project is necessary. Examiners may also evaluate the students' ability to demonstrate that they have read and understood the academic literature on a particular topic. Remember that some reviewers will ask students to rewrite part, if not most, of the literature review section if the written work is of low quality. Besides, in the current era, your dissertation can be accessed online or assessed in a library. Your dissertation will be picked up and read by the next generation of students. After you have graduated, you may still receive praise or criticism depending on how well the document was written.

2.2.4.3 Constructing Literature Review Using Various Sources

The easiest way to assess literature review sources is through digital articles and review-articles. The major publishers in the area of biotechnology and biosciences include Elsevier, Springer Nature (including BMC and others), Wiley, MDPI, Taylor & Francis, American Chemical Society, Royal Society of Chemistry,

Table 2.2 Quick correction tips for Mendeley

Mistakes	Correction step	Corrected intended display
Non-italics in genus and species That is, Bacillus megaterium	<i>Bacillus megaterium<i/>	*Bacillus megaterium*
Wrong display of Greek letters or symbols That is, beta-cyclodextrin	&beta	β-cyclodextrin
Wrong display of superscript That is, *Nocardioides iriomotensis* IR27-S3T	IR27-S3<sup>T</ sup>	*Nocardioides iriomotensis* IR27-S3^{T}
Wrong display of subscript That is, H2O	H₂O	H_2O

Hindawi, Frontiers, Public Library of Science (PloS), De Gruyter, etc. It is important to note that each publisher has its own search engine. ScienceDirect, for instance, provides access to the book, article, review, and other documents published by Elsevier. However, ScienceDirect does not provide access to literature published by other publishers.

I recall many years ago when I was chairing a viva voce and the student exaggerated the novelty of the project. As the examiners dug deeper, we came to realize that the research gap and novelty claim were based on very few documents from a narrow set of keywords, and the candidate used only one publisher search engine, unaware that each publisher has its own search engine platform. Message no 1: Find documents using a different search engine and keywords.

In the previous paragraph, I listed at least ten major publishers and you are probably aware of others. The process of browsing all possible publishers and sitting in front of the computer to check each one by one will take a great deal of time. From our experience, we know that some publisher search engines have excellent user interfaces, some are precise, but we have probably had some poor experiences with oversensitivity to obscure or irrelevant documents. It may be that the algorithm of a search engine makes a different decision. Sometimes, the article cannot be found by the machine because the corresponding author has provided insufficient information, such as the title, abstract, or keywords. When I was a student, I liked to use publisher search engines. Today, I prefer to use an integrated search engine. Message no. 2: Use integrated search robot tools such as Google Scholar, Scopus, Web of Science (WOS), or others such as Lens.org.

According to the information provided in Wikipedia, Google Scholar started in 2004 and it indexes journals, reviews, books, conference papers, theses, preprints, abstracts, technical reports, etc. While Google Scholar may cover a great majority of online documents, it is possible that predatory journals are not excluded. Please refer to Chap. 5 on some issues of predatory journals. Besides Google Scholar, do try Semantic Scholar. We may try Lens.org. that uses PubMed, PMC, DOAJ (Directory of Open Access Journals), WorldCat, and multiple patent databases such as USPTO, WIPO, and EPO. I will write a little bit more on Lens.org. in some subsequent section. Google Scholar and Lens.org. are both free services, but access to the full paper is a different story.

As far as indexation and search engines are concerned, Scopus and WOS are probably the gold standard. In case you are unaware, Scopus and ScienceDirect are owned by Elsevier. ScienceDirect may be exclusive to its own publishers, but Scopus resources (documents, patents, etc.) are comprehensive and inclusive, so as long as a paper is indexed in Scopus, we will be able to find it regardless of the publisher. The WOS is also inclusive and comprehensive. WOS is owned by Clarivate, along with other products such as Publon, EndNote, and ScholarOne.

Message no 3: Learning and using Boolean Search. The search engine has a Boolean function that uses the following syntax: AND, OR, NOT, (), and "". Although I intend not to elaborate further, it is useful to know that if we get lost in search engines as a result of a large list of information or the output is not precisely what we are looking for, Boolean Search will come in handy.

2.2.4.4 Reference Manager

Correctly citing the original source of an idea or piece of data is not a new concept to us. If we do not provide a proper citation, we may be accused of plagiarizing. A manual typing approach resulted in a number of issues, such as being time-consuming, mistakes in typing, incomplete information, inconsistent style, unmatched or missing citations in the text and reference list, and leaving out the author names with special characters. Using a bibliography reference manager is the only way to make writing easier. RefWorks, EndNote, BibTeX, Mendeley, Zotero, Flowcite, Citationsy, and others are examples of reference managers. Some tools may have a fee, so check with your university as they may have a license for them. Some plugins are disabled due to M365 education license restrictions. Ensure that the software you intend to purchase is compatible with your computer and Word processing program before purchasing.

Almost all devices are compatible with Mendeley, except for Google Docs. Many of these programs are easy to use and have built-in reference templates for journals or theses. Mendeley, for example, provides 7000 different citation styles. Mendeley's citation style language (CSL) editor allows us to customize a citation style. Investing 30–60 mins into learning the basic operating procedures on YouTube will make our writing and referencing more effective.

Is the information retrieved by reference managers error-free? Certainly not. It is quite common to see inconsistent capitalization of the title, left out italicization, left out superscripts and subscripts, and missing volumes and pages. Manual correction in the Mendeley library is required for all of these issues. The correction is a bit time-consuming and tedious, but it is still preferable to typing from scratch in Word. Table 2.2 summarizes a few quick tips for correcting the most common mistakes found in Mendeley.

2.2.4.5 Advice in Preparing Literature Review

For chapter 2 (literature review) of a proposal or thesis, I am not in a position to provide a general recommendation based on the average number of pages or words, since the length is dependent upon the normal practice of our institute, peers' influence, and whether it is a dissertation or a PhD. The following are ten basic recommendations regarding the basic preparation of a literature review for a thesis. Subsequently, I have pointed out few additional tips that enhance the 'taste or favour' so that the readers will find the write-up interesting.

Primary, Secondary, or Tertiary Source?

A primary source for literature review refers to the original thinking or findings of new discoveries. Generally, the results section of a research article, thesis data, conference presentations, interviews, patents, government papers, and datasets in databases or repositories (NCBI, EMBL, Uniprot, PDB, etc.) are examples of primary data. It is important that we read, digest, and use primary sources when writing a literature review, and in fact, primary sources should be the main reference for writing chapter 2 of a thesis.

It is important to note that in addition to conventional resources, a large amount of government data and government-funded projects is now available on the Internet. It has been reported in the news that by 2025, all federally funded research will have its data made accessible to the public, according to the White House Office of Science and Technology Policy. Besides, several multidisciplinary data repository options are available, including Figshare, Zenodo, OSF.io, national Open Data, etc. These data are considered primary sources.

Generally speaking, secondary sources are documents or materials that analyse, summarize, interpret, reorganize, or restate primary sources. An example of a secondary source is a review, an editorial note for a special volume, a textbook, a book chapter, a discussion section of an article that refers to the work of another researcher, etc.

Secondary and tertiary sources are often difficult to differentiate as sometimes they are quite similar. Examples include fact books, encyclopaedia, Wikipedia, dictionaries, manuals, handbooks, textbooks (it depends), lectures, and reference works such as indexes and abstracts. One characteristic of tertiary sources is that the information cannot be attributed to one individual or group of individuals, and one example is Internet information. Internet sources can, however, be primary or secondary sources as well.

It has been my observation that young researchers entering a new area of research may find secondary and tertiary sources to be more easily understood, thus some of them will prepare their literature reviews according to the tertiary sources writing style they have been exposed to. There is no harm, in my opinion, in utilizing secondary and tertiary sources in the writing of a thesis. Nevertheless, I believe it is better to have the majority of your literature review using primary sources while a small percentage of it using secondary and tertiary sources. Often, examiners and readers are sceptical when student candidates cite Wikipedia (a typical tertiary source) or a web page in their theses. A good rule of thumb is to use articles from scholarly journals for our study and to fill out the majority of our references at the end of the thesis, proposal, or manuscript.

Can I Use Textbooks and Chapters in Books and Review?

Let me rephrase the question above a bit—Should we read textbooks and chapters in books when we are doing a PhD or even after obtaining a PhD? The answer is definitely yes because having a good theoretical foundation is essential. Let's get to the original question: Can I cite textbooks and book chapters in the literature review in a journal article and a thesis? My answer is it depends.

The first thing I would like to suggest is to avoid writing textbook information in a thesis or a journal article. Generally, I do not consider a chapter 2 that is filled with textbook information as a good literature review. Let me provide an example. It is likely that a student's literature review will include a chapter on microbial transporters and membrane proteins if the topic of the project is bacteria adaptation, heavy metal response, or chemical response. We do not advise writing thick and dense chapters describing basic lipid bilayer and membrane proteins (say two pages); supramolecular architecture of membranes that include fluid mosaic model, micelles, bilayer, liposome, bacteriorhodopsin, topology, etc. (three pages); active and passive transporter types (four pages); and other general theories undergraduate students are taking in Microbiology and Biochemistry classes. In this context, length and depth are of paramount importance. The inclusion of some basic terms and knowledge is beneficial, however, keep them to a minimum. Perhaps, instead of writing two pages on the basic lipid bilayer, why not just encapsulate the basic textbook information in a single short paragraph?

Visualize a cheesecake with many layers. The bottom or the base of the cake is usually the hard crust that is thin enough and let's compare that cake crust to textbook information. A delicious cream cheese spread, quark cheese, twarog cheese, or any other cheese type is placed on top of the cracker or cookie crust. A tasty cheesy feature on the cheesecake is the real deep literature reviews based upon recent primary experimental data and scientific findings published by academicians. If we were to read a section in Chap. 2 full of information from a textbook, that would be like eating a slice of cake that is 90% crust and 10% cheese. A majority of people prefer thin crusts and more cheese, which means scientific readers prefer in-depth and current scientific evidence that relates directly to the project scope.

Can we cite book chapters? The majority of chapters published by major publishers and edited by responsible editors are considered good references. In some cases, experimental data is published in book chapters rather than journal articles, and for such examples, it can be used as a reference. Avoid citing the book chapters if they are similar to a textbook.

Do we have the option of citing a review article or theses written by someone? Absolutely. However, beware of two potential pitfalls. The first is plagiarism. Please do not steal the texts, tables, figures, ideas, or even the sequence (arrangement) of information. Most reviews can be considered secondary sources. It is important to understand that not all writers for a review are experts in all areas and some are inexperienced. On occasion, we may notice a mistake in the secondary (or tertiary) data or description in a review article. Unless we exercise due diligence, we may receive incorrect information.

Can I Use Internet Sources or Magazines?

Answering this question definitively is difficult. Refer to the thesis writing guidelines or manual for your institute when attempting to answer this question. If it is allowed by your university, there is no reason to object.

There are a couple of excellent Internet sources for our research, such as some scientific blogs. A few publishers manage blogs, for example, Frontiers Science News or PLOS Pathogens. You may enjoy reading blogs on ScienceBlogs, RealClimate, etc. There are some prominent professors who manage blogs, but it is not common. As an example, the Thesis Whisperer blog is maintained by Professor Inger Mewburn from Australian National University. In addition to very interesting articles, you will find a lot of goodies there. In order to find a specific blog related to your field, you probably can search the Internet using the key phrase "X blog", where X is Microbiology, Virus, Biomanufacturing, Plant science, etc.

A number of publishers and societies publish magazines, for example, *Microcosm* by the American Society for Microbiology or *The Biologist* by the Royal Society of Biology. If the information is first-hand, a citation from a magazine may be appropriate. On the other hand, if we know that the statement contained in the magazine is a secondary source, we probably need to find the primary source.

My small advice is that if we cannot find a better research article, or if the online essay or magazine is too close to our heart, go ahead and add it to our literature review. As an analogy, these sources are comparable to the topping on a cheesecake. In case you missed the cheesecake analogy, please refer to the earlier subsection.

The Same Idea or Knowledge Was Discussed in Two Different Articles. Which Article Should I Cite?

If both articles are equally valuable, I would suggest citing both. If there are 20 articles that contain the same idea or information, it is not possible to list them all. While some may choose the oldest article that made the claim, others may choose a more recent one, while still others may choose a journal with higher metrics (which is sometime measured by its impact factor or CiteScore score).

Let us consider another scenario. There are two major schools of thought that contradict each other. Which one should you cite? Unless there is sufficient evidence to demonstrate one school of thought is incorrect, we probably cannot be biased and cherry-pick without strong scientific backing. I would suggest including both schools of thought in the literature review and critically analysing them. It may be necessary to discuss these concepts in detail in the literature review of your thesis if your project seeks to verify the concepts in each school.

Old Versus New?

I have given four recommendations in earlier subsections, and this is my fifth recommendation. For your information, all five questions are frequently asked questions that I receive when teaching literature review writing to undergraduate and postgraduate students in my department.

In my early career, I was taught to use older references or the earliest researchers who proposed the findings, concepts, or observations. There is no doubt that we should give them enough credit since the technology was less developed during those days, and they have provided us with a firm foundation. With time, I have been taught that we should use more recent publications instead, and that it is much better to cite publications published within the past 5 years. Would you vote for older or newer references, if you had 5 s to think about it?

Please do not overreact to this matter. There are numerous advantages to using the older or earliest references, and sometimes we might have no choice but to use them. If we intend to discuss the history and development of the topic, then we should include old references. Consider the following example. "Microarrays were developed by X in year ____ and short-read NGS-based RNA-Seq in year ____ (old reference). Recently, ISO-Seq, the full-length cDNA sequencing based on PacBio SMRT, has been used for ___ (a more recent reference)". As we may know, microarrays have existed for a few years prior to 2000 and RNA-Seq has been around for more than a decade. A latest reference is not appropriate in this context. Additionally, using new references indicates that we are using secondary sources rather than primary sources.

The use of outdated references in a literature review has multiple disadvantages. Those who examine theses or review journals may become bored reading the same information over and over again. It is difficult to identify research gaps if we read and cite only old references in literature. The data in some older references may have become obsolete or may no longer be convincing to readers regarding some important facts, trends, or arguments.

I have an allergic reaction when older references (say, references that were published 5–10 years ago) are matched with the following sentences.

- Example one: "There is a growing trend towards a methodology or a research field (old reference)". Using this example, trend implies the past plus the present and if the reference is too old, how can the authors claim that the current trend is identical to what was reported a few years ago?
- Example two: "The problem of ___ remains unsolved until today (citing an old reference that is 5 years ago)". There is a possibility that a scientific problem that was posted 5 years ago has been solved or at least partially solved.
- Third example: "Recently, the researchers have been ___ (old reference that is 5 years back)". The publication should not be considered recent if it was published more than 5 years ago, and I do not consider a report released 1–2 years ago as recent. Cambridge Dictionary defines recent as "happening or starting from a short time ago".

Bringing home the message is simple. As authors, we should determine the actual reasons for using a reference rather than just picking it up and adding it to the cart. The number one rule of thumb is that newer references always have an advantage, especially when submitting a grant proposal. In my country, a requirement for the evaluation of national research funding is that references must not be older than 5 years. The rationale is simple; the study should be new, recent, relevant to the current era, novel, fill a gap in existing research, etc. The second rule of thumb is that when discussing the development or progress of a topic, technique, or theory, this is where the older references should be aligned in accordance with the timeline.

Avoiding Reader Confusion

We may become lost without the assistance of a GPS (such as Google Maps or Waze) if we enter a new city with numerous junctions, divisions, or buildings that look identical. Some time ago in an oral examination, a thesis reviewer of considerable experience expressed confusion when reading the literature review of a particular thesis. A number of reasons were given, including the fact that the chapter is divided into parts (sections and subsections), which are similar to one another, and the terminology used is inconsistent. Writers tend to use synonyms when describing a subject, but readers may be confused due to our inconsistent.

A student may believe that a good thesis is one that is lengthy. Some students compete with each other in how thick their thesis is. However, this is not what the reader is thinking. The average number of television episodes per season is between 20 and 40. In South East Asia, the longest-running soap opera was broadcast for almost 2 years. The series consisted of 410 episodes! To be honest, I did not watch it all, but my grandma did. Generally, most people do not enjoy reading lengthy literature reviews, especially when it is boring, filled with textbook information, and messy.

It is possible to organize the literature chapter according to the funnel approach, where general information is presented first and then more specific information follows. For example, we discuss global issues before zooming in on country or regional issues. For another example, we examine broad antibiotic resistance and then focus on a specific strain. Another example would be the description of general hydrolase enzymes in the E.C. 3 group, then a description of carbohydrate hydrolases, followed by cellulases, and then endo-glucanase.

The funnel approach is the most straightforward approach. We find that students typically use a substantial percentage of the total pages (say $>70\%$ of chapter 2) to address general topics, simply because more general knowledge is easier to write about. However, the real core literature relevant to the project title, problem statements, and objectives is not critically analysed and documented. If this occurs,

Fig. 2.3 A paragraph with jumpy messages

Fig. 2.4 A paragraph with clearer messages

the thesis examiner will make a remark regarding the lack of depth of your literature review.

The literature information that is directly related to the project should be given more space. Defining a golden ratio between general information and project information is not possible. Nevertheless, 70–80% of general information may not be appropriate. Perhaps 60:40, 50:50, or 40:60 would be better. How would we measure that? If a literature review chapter contains 30 pages and about 24 pages of general information and textbook information, the percentage will be 24/30*100 = 80% (80:20).

The topical order is another organization style. As an example, the first problem statement is paired with the first objective, and it aligns with the first section of chapter 2 (literature review). If there are three problems and objectives, then the literature review will probably consist of three main sections. I have used such a strategy in writing a grant proposal. Personally, I seldom read a thesis that employed such a strategy.

It is probably possible to arrange a certain section or subsection of Chap. 2 using chronological order (though not the entire chapter). For example, we can create a section titled "History or Development of Gene Expression Studies in ____". We should probably begin with a discussion of the microarray concept and its key scientific findings. Afterwards, we can discuss Illumina-based high-throughput RNASeq, followed by a description of single-cell RNASeq, a description of full-length RNASeq, and finally, single-cell full-length transcriptome studies. The above arrangement resembles quite closely the current transcriptomic progress (at the time of writing).

There are certainly other writing strategies in addition to those described above. It is not necessary to choose only one approach and use it throughout the chapter; we may mix and match approaches provided the flow of the narrative (information) is logical and easy to follow.

Easy-to-Follow Storyline

Perhaps in the olden days when television was still black and white, most movies or dramas featured only one female and one male protagonist (leading characters), who would occupy almost the entire scene. Nowadays, movies have multiple leading and supporting actors and multiple storylines, and some of them are independent and some are intersecting. Unfortunately, writing a thesis or journal article is not the same as creating a film or television program. In scientific writing, the storylines tend to be limited and all of them must be connected.

Let us consider an example by assuming A is one theme and A1-A3 are sentences describing that theme. Let B be another theme and B1 and B2 are associated lines. If a paragraph has a lot of jumps and is hard to follow, it will look something like Fig. 2.3. After rearranging and arranging sentences with similar topics closer together, the paragraph becomes easier to read (Fig. 2.4). Similar information found in multiple subsections should be rearranged and grouped together. Once we have completed the first draft of the document, relocation is an important step in the editing process.

If there are ten subsections in a literature review, we need to find a way to interconnect them so that the information flows logically and makes sense. There should also be some continuity between paragraphs within a subsection, else it will appear disjointed. Our brain will be confused if the information is disjointed. In general, jumping between stories is not a problem in movies, but it can lead to serious misunderstandings when it comes to thesis writing. Keep it focused.

Cite as We Write

I have introduced you to tools such as Mendeley and EndNote in the earlier section. I recommend that you add references as soon as you write. Adding a reference can sometimes cause you to lose track of what you were writing. To overcome this problem, it is probably at the end of the page or subsection that the references should be added.

Use Visual Aids

The visual aids include motion pictures, charts, maps, 3D models, etc. The most effective visual aids in written documents are figures and tables. The value of an image cannot be underestimated. Most publishers usually allow students to use their copyrighted figures published in their thesis, provided that fair-dealing principles are adhered to and proper citation of the author's work is included. Students may also draw their own figures when necessary.

Consider the following scenario: the project is concerned with microalgae, and part of the literature review covers different types of microalgae and how to grow them. It would be beneficial to include microscopic images of various species and photographs or illustrations of open-air ponds and various types of photobioreactors. The reader may find it easier to comprehend the text by viewing visual aids rather than simply reading it.

Students are always encouraged to prepare a long table that summarizes the data, information, and references related to a topic. We often see such tables in review articles. It is possible to develop our own table and update it periodically. Taking the microalgae example, we can compare the type; size; growing conditions; medium;

yield of products (e.g. whole-cell biomass, lipids, carotenoid); flocculation strategy; references; and other details related to the project, all represented in a table.

All tables and figures should be cited in the text of the dissertation or journal article. Text should be used to explain tables and figures. It is a bad habit to leave the figure or table alone and imply to the reader, this is it, read and comprehend it on your own.

Using Own Words in a Clear and Readable Manner

Scientific writing and thesis writing require original wording. A good practice is to digest the information that you have read, blend it, and then rewrite it. At first, it may seem difficult. It is common for inexperienced students to use the contents of a single journal article for one paragraph, and then summarize another journal article in the following paragraph. There is a possibility that the student may have plagiarized, not necessarily the sentences, but there is a possibility of plagiarizing ideas. It is a bad practice because readers will notice the problem and will find it difficult to follow the writing as it seems discontinuous. In case you are not sure what this means, you may want to read more excellent review articles and observe how they write. The information in most good review articles is blended in some nice way and presented in a theme rather than as a collection of disconnected paragraphs. Often, due to a limited number of words, the explanation in review articles is too brief, but in a thesis you can elaborate to ensure the information is clear. If you want to make your writing more effective, combine the information you have read, repackage it, and use your own words and phrases.

Concise language is preferred in academic writing. Simple language is not always informal. Complexity refers to ideas, concepts, hypothesis, and analyses in a report. Our writing must be straightforward in order to effectively convey our ideas. The best style is one that conveys the most meaning in the fewest words possible. There may be some confusion when a longer word or series of words is used instead of a simple expression. In a separate chapter (Chap. 6), we will discuss how to improve writing clarity and tools that we can make use of.

Jargon refers to a language or term used by a specific group of people that is not understood by most other individuals. There is not much we can do about biological standard technical jargon. A chapter entitled 'Avoiding Jargon' in the book entitled *How to Write and Publish a Scientific Paper* suggests that we avoid using confusing jargon (Day and Gastel 2006). Further, we should avoid jargon that is characterized by circumlocutions and long words. As defined by the Merriam-Webster Dictionary, circumlocutions are "the use of an unnecessarily large number of words to express an idea". A six-page example of circumlocution jargon was provided by Day and Gastel. As an example, we can replace the circumlocution jargon 'has the potential to' with a single word 'to' or 'can'. For instance, 'The newly discovered protease has the potential to' can be shortened to 'The newly discovered protease can...'.

In journal writing, it is not necessary to provide definitions because we always assume that readers have some kind of background. Nevertheless, we cannot pre-determine who are the readers of a dissertation or a thesis; therefore, I suggest that we can give some proper definitions to rarely used technical jargon. But if we keep giving a definition, readability may be disturbed. We can try to give some hints in the sentence we construct.

Let us examine one example. "During heterotrophic cultivation of X, organic carbon sources such as ___, ___, ___, or ___ are utilized by X as both energy and carbon sources. In mixotrophic cultivation, some X species can perform ___". The first sentence is easier to understand, since many of us understand precisely what heterotrophic means, and organic carbon in the second part of the sentence hints us something. However, the second sentence may not be clear to all readers since mixotrophic refers to organisms that utilize multiple sources such as photo-, chemo-, litho-, organo-, auto-, and heterotrophic. In order to clarify, at least give some hints as to what is meant by mixotrophic in this context.

Moreover, when abbreviations (PBR, ARG, ARB, etc.) are used, we need to clarify them when first using them. The definition of the abbreviation can be made in each chapter of the dissertation if the writer is concerned that the readers may miss it. It is not necessary to define common abbreviations such as DNA, RNA, and GPS. By the way, PBR is short for photobioreactor, ARG is short for antibiotic resistance gene, and ARB is short for antibiotic resistance bacteria.

When I began writing my first manuscript and thesis, I was frequently reminded of two common blind spots. The first problem is an ambiguous sentence, and the second is an exaggeration. There are times when these twins come together. As an example, let's take a look at:

- Nanotechnology is a new recent field of science.
- In light of the increasing price of oil, there is an urgent need to develop alternative fuels, such as biofuel generated from waste.

As far as language is concerned, neither of the sentences above contain any grammatical errors. The sentences are clear, self-written, and do not contain plagiarizing language. Let us examine it in more detail. Firstly, there is an adjective and the word is new. How would new science be defined here if we were to think

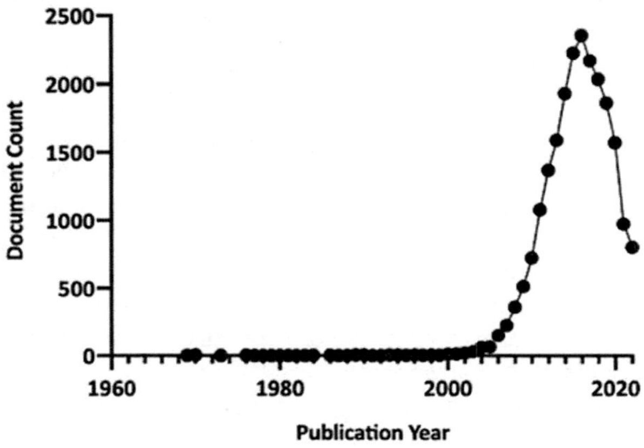

Fig. 2.5 An example of total publication trend related to 'ABC'

carefully? What is considered new? And how new is new? While we are aware that recent technological developments have pushed nanotechnology to new heights, we tend to forget that nanotechnology was first proposed in 1974 (according to Wikipedia). A student who has not been exposed to nanotechnology during their first degree will believe that it is an entirely new field. In the second sentence, the exaggerated word is 'urgent need'. I believe the sentence would sound better if we used the word 'important' instead of urgent. You should avoid ambiguous sentences and exaggerated statements.

In some situations, we may be stating a fact, but the reader may perceive it as an exaggeration. As an example, if you discover something for the first time in your country or in the entire world, you must be very proud. And I also feel very happy for you. In scientific writing (thesis or journal), try to avoid expressing things like: This is the first study that has found ____. There are some readers who get a big allergy when they read the phrase 'first study' or something of the kind. In spite of the fact that this is a true statement, there is no exaggeration, readers do not feel quite comfortable reading it. It is possible to present that in a more polite manner in dozens of ways.

Bibliometric Analysis

According to one article, bibliometrics is closely related to informetrics and scientometrics, though the terms have slightly different meanings (Ellegaard and Wallin 2015). The term 'bibliometrics' refers to the application of mathematical and statistical methods to books, articles, or any available documents that were published (Mejia et al. 2021). It is not difficult to analyse the literature documents available in databases using bibliometric methods. There are bibliometric functions built into Scopus, WOS, Dimensions, and Lens.org.

Figure 2.5 is a publication trend taken from Lens.org using the keyword 'ABC' (The abbreviation ABC refers to a bioscience research field, which I will not reveal). The results of this simple search indicate that 2013–2016 was the golden age for this research field. Apart from showing the total number of documents for each year, Lens.org statistically classified these documents and gave us information such as what are the top fields related to the keyword, the most active institutions, the most active authors and the countries where they are from, the most active publishers, the top journals that publish such topics, and so on (not shown in Fig. 2.5). Lens.org webpage automatically displayed charts to display the summaries. It should be noted that the following figure was not generated by Lens.org. Third-party software was used to generate the graph from the CSV data downloaded from Lens.org.

Readers who want only Scopus- or WOS-indexed documents can also access both bibliometric functions through these databases. Once a search has been conducted using appropriate keywords and Boolean functions, we can export the metadata information from the database and display the results using third-party software, such as VOSviewer. This software is used by researchers for displaying network data, such as co-authorship networks, text mining, etc.

I must emphasize that bibliometric analysis is not the same as a literature review. Bibliometric analysis can be viewed as a bird's eye view of an area of research. If that is the case, what are the benefits of bibliometric methods? The following are some reasons for doing a bibliometric analysis.

- Observing the trend from the past to the present. Often, our eyes and minds are deceived. As an example, if we intend to purchase a specific car model or colour, it appears like the road is full of such vehicles. We are surrounded by what we read (or think) when we read about a particular research topic. This is something I have noticed students and writers tend to use: "There is a growing trend in ABC research and period". The period means full stop and no proof was used to support the statement. As we can see from the bibliometric figure provided above, the total number of publications for ABC has decreased since 2017. Accordingly, it is inaccurate to state that this type of research is on the rise. You should provide evidence next time when you write something like "there is an increasing trend in so and so", or "more and more researchers are paying attention to so and so".
- We may be able to identify real experts through bibliometric analysis. Using a few clicks of the mouse on the Scopus or WOS website (on the left side, after performing a search), we can see who has published the highest number of documents relevant to a given keyword. Identifying the real experts allows us to formulate the following course of action: (1) The first step would be finding their papers, reading them, and analysing what are the major contributions and where the gaps in research exist. (2) Second, make acquaintances with a few of them, fostering a friendship and eventually leading to collaboration or mentoring.
- Make the thesis more interesting. It is incorrect to cover the entire Chap. 2 (literature review) with bibliometric analysis. Whenever possible, I think that blending a few figures obtained from Scopus, WOS, Dimensions, Vos viewer, and Lens.org would be beneficial. This could make the content more engaging. I suggest you do not overdo it, as not all researchers are familiar with bibliometric analysis. Some journals even specifically state in their Instruction to Authors that they do not favour bibliometric analyses. What matters most is the scientific content, not statistical or library information.

Social-Economic Factors

In the previous section, where we discussed thesis impact (significant impact), I suggested that we can discuss how the current project engages topics in SDGs, the United Nations, the World Health Organization, home nation policies, or any other topics relevant to social economy. In my opinion, social-economic factors are not part of the real literature review for a biotechnology or bioscience project. Yet, adding a few sentences or some short paragraphs to the literature review may make reading more interesting and help readers understand the significance of this project to society, the environment, and the economy. In many funding applications for research grants, the writer is required to incorporate social and economic factors into the proposal. Regardless of whether it is a mandatory requirement or not, it is a good idea to consider doing it. Social economy is not well known to most biologists. It is becoming more common for multidisciplinary projects to include social-economic aspects. You can learn some basic concepts by reading some description in Chap. 8.

Patent Files

For those unfamiliar with patents, please read Chapter 4 in which I provide a brief explanation of the basic concept. Google Patents is a great source of patent information. It is recommended that students who are working on applied science and commercial projects perform patent searches and incorporate such information into the literature review.

Open Data

Often, writers refer to specific numbers such as the numbers of patient of a certain sickness, market prices of compounds, sales of biotech products, amount of grant money invested in a specific area, etc., that are reported in some journals. Unfortunately, this information is often scarce, more recent data is difficult to locate, and the data may be restricted to particular countries. In addition, our own country's data is often missing.

In a literature review, we may write something like: The number of patients with ___ disease was recorded to be ___ in the year 2018 (reference). Readers or viva voce- or journals-examiners may inquire: Are there any updated data? Can you provide information about the number of cases in our country or neighbouring countries? These questions are difficult to answer.

We are living in the age of big data. Published data in journals is just one of many sources that are available to us today. The metadata from many countries is now publicly available. If you are interested in finding this type of information, you can search the internet using the keyword 'X open data', where X is the name of your country. Examples are provided below. In searching for Malaysia open data on Google, I will find the website data.gov.my where I can view national statistics, international trade, education, agriculture, the environment, health, and many other types of data. In this way, we are able to assess government data.

My quick search for 'X open data' works well for Malaysia, Australia, Singapore, New Zealand, India, Canada, China, Africa, Uganda, the US, the UK, the EU, and many others. To find the information we need, we must sit down and gather information manually, which takes a lot of time. If the data are prepared in an Excel file or CSV format, then we may need to use the raw data, convert it to an appropriate figure, and then include it in the thesis.

2.2.5 The Chapter Three of a Thesis: Materials and Methods

Due to the varying nature of bioscience and biotechnology research studies, I chose to state quickly some good practices in writing the methodology (or materials and methods) instead of explaining the experimental design.

- A proposal is written in the future tense. A thesis or a journal manuscript, however, uses the past tense.

- Make sure that the language is used in a manner that does not resemble an instruction manual. This sentence, for example, is not appropriate: "Later, centrifuged for 5 mins and the supernatant discarded".
- Include a flow diagram of the experimental process in the proposal and thesis. If possible, indicate where each objective fits within the diagram.
- Within one page, the flow chart should be as detailed as possible. Add some figures or use an infographic style to the flow chart. The chart shape is not mandatory, but some readers prefer to see the correct use of the chart shape, such as rectangular (process), rounded rectangular (terminal—start and end), diamond (decision), etc.
- Whenever a project has various phases or stages, it is important to communicate it clearly through text, flow diagrams, and Gantt charts.
- Provide field work sampling detail: GPS coordinate, depth, how many grams or volume, sample type, ways to preserve them, etc.
- Consistently use units, dimensions, and symbols. For example, mg/ml or $mgml^{-1}$; cm or ft.; or others.
- Name the brand/manufacturer, town, and country consistently. Example: TissueLyser II (Qiagen) or TissueLyser II (Qiagen, Hilden, Germany).
- It is important to disclose the version of the software and its settings used for the work.
- The phrase '...using the default parameters or settings' may not be adequate. An elaboration of a bioinformatic protocol is often necessary. These include every parameter, threshold, iteration, calculation strategy, etc.
- The methods used should be in alignment with the research questions or objectives.
- Indicate the number of replications and statistical approaches.
- Explain how calculations were conducted, that is, enzyme activity, enzyme kinetics, etc. An example of the calculation can be included in the appendix.
- Please provide references.
- Write the variables clearly.
- Clearly state what the controls are.
- Abbreviations should be spelled in full when first introduced. Avoid creating new abbreviations that will not be used frequently throughout the document.
- For proposals, the methods can be less detailed or more or less comparable or slightly longer to those used in journal articles. Nevertheless, some evaluators prefer a detailed write-up. During the writing of the thesis, the methodology must be as specific as possible so that the next students who read the thesis will be able to follow it and reproduce the same results.
- The advantages and disadvantages of different methods should not be discussed in detail. However, state clearly and briefly the rationale for using a selected approach.
- If possible, organize it according to the order of the results.
- Clearly state the ethical approval or permit, if any.
- Avoid unnecessary detail, that is, the brand or colour of box or container.

- Include a Gantt chart in the proposal. In many cases, the Gantt chart will not be included in the final thesis.
- The results should not be included in this chapter.

Depositing data or biological samples in community-recognized repositories or databases is common practice. The thesis report should include this information in a clear and concise manner. It is a requirement of most journals that authors deposit data or samples prior to submitting their manuscripts. Microbiology researchers are required to deposit new type strains in two separate culture collections within separate countries by the International Committee on Systematics of Prokaryotes (ICSP) and the International Journal of Systematic and Evolutionary Microbiology (IJSEM). The ATCC, JCM, DSMZ, KCTC, etc., are examples of culture banks. Plant specimens are deposited in a herbarium. The sequences of DNA and RNA are deposited in the DDBJ (DNA Data Bank of Japan), NCBI (National Center for Biotechnology Information), or ENA (European Nucleotide Achieve, EMBL-EBI) databases, which are linked by the International Nucleotide Sequence Database Collaboration (INSDC). The Genome Sequence Archive (GSA) is another data repository maintained by the China National Genomics Data Center. A number of publishers, such as Springer Nature, Elsevier, and others, seem to be accepting depositions to the GSA.

Other possible commonly used repositories include ArrayExpress, GenomeRNAi, UniProtKB, Protein Circular Dichroism Data Bank (PCDDB), Crystallography Open Database (COD), Electron Microscopic Data Bank (EMDB), Protein Data Bank (PDB), Database of Interacting Proteins (DIP), Genome Expression Archive (GEA), PeptideAtlas, Global Biodiversity Information Facility (GBIF), and many more. In case readers wish to learn more about other repositories relevant to your projects, such information can be found on the Data Repository Guidance webpage of Scientific Data journal.

2.2.6 Chapters Four and Five of a Thesis: Results and Discussion

Some universities prefer to separate the results and discussion chapters, while others allow the two to coexist. It is common for students to feel comfortable describing the observations they made and the data they collected during their experiments. Despite this, some students have difficulty writing the Discussion section (or blending it into the Result section). The purpose of this section is to provide advice on how to write a discussion.

The following information was found online in Scribbr under the heading 'How to Write a Discussion Section': "An effective discussion emphasizes the significance, relevance, and meaning of the results. Discussion should explain what the significance of the data is and how it relates to the research questions".

Additionally, in a PLOS webpage entitled 'How to Write a Discussion and Conclusion', the essay suggested that the results should be accompanied by an explanation of the significance of the results. The limitations and strengths should

Table 2.3 Examples of possible ways to include a discussion

Experimental data	Possible ways to include a discussion
Phenotypic characteristics of bacteria (e.g. optimum temperature, doubling time, etc.).	• Compare your results with those of other studies including similar strains, or other genera within the same family. • Explain the possible reason why a certain bacterium is an outlier with a distinct phenotype. Is it possible that this may be the result of the location where it was isolated or is it the result of a horizontal gene transfer from another species?
Genome-to-genome comparison (e.g. percentage of conserved protein, POCP; average nucleotide identity, ANI; digital DNA-DNA hybridization, dDDH; etc.)	• Answer questions such as whether the niche where the bacteria was isolated may have influenced its genome design. • Examine the pan-genome, genome shuffling, gene gain-loss, etc., and determine whether homologous recombination, gene conversion, transposition, illegitimate recombination, or prophages invasion are responsible for DNA rearrangement and shuffling.
Global transcriptomic response or gene expression study (e.g. comparing transcripts of a case and a control, and discovering which genes are upregulated or downregulated).	• Explain whether or not the results were expected and provide explanations. • Have similar results been reported in other publications using a different cell type or a slightly different experimental setup? • Explain the biological pathways in greater detail. Are there any new possible correlations that were previously unknown? • Could hypothetical proteins have a role? • Create a cell schematic diagram to illustrate the data, and how do you feel that compares to what we currently know? • Can the transcriptomic provide us with information on how the bacteria survive in the natural setting, for example, halophiles in the sea, or hyperthermophiles in hot springs, or why the cells prefer xylose to glucose, etc. • Give possible explanations for an unknown circumstance and support them with strong arguments.
Protein characterization (e.g. activity, kinetics, sequence-structure analyses, X-ray structure, molecular dynamic, docking, etc.)	• Comparing with closely related or far away sequences published in the literature, UniProt, or PDB. • From a sequence or structure perspective, why does your protein have less inhibition effect, higher thermostability, certain substrate specificity, product specificity, etc.? • Based on molecular dynamics, deduce how does the structural change affect the function of the protein in the cell? • What makes the protein cell-bound and does it have any relevance to the cell's survival?

(continued)

Table 2.3 (continued)

Experimental data	Possible ways to include a discussion
	• Is there a relationship between the length of loops, flexibility, size of the binding pocket, specific amino acids, motif, signature, or domain of a protein and its biochemical properties? • Is your observation controversial to earlier work and why?
Study of single nucleotide polymorphisms (SNPs) in genes relevant to specific ethnics on a national level	• Interpret the relationship between SNPs and ethnics groups. • How the findings address the scientific gaps within the local ethnics committee. • Compare the findings of the local ethnics group with those of a similar group located in another country. Does the general conclusion remain the same?
Experiment comparisons (e.g., different bioinformatic settings for the same data input; comparison of MAGs generated by long-read alone, short-read alone, and hybrid assembly; comparison of different biomarkers, comparison of extraction approach, etc.)	• Finding patterns between two datasets and comparing them to published reports. • Describe why one methodology is superior to the other. • Does the conclusion apply only to the case studied, or can it be applied to most conditions in general? Why?
The diversity of life (e.g. metataxonomy, population shifts, etc.)	• Based on statistical analyses such as PCO, NMDS, heatmap, etc., are there specific types of groups in the taxa in the whole population? • Are there any similarities between this dataset and other datasets? • What effect do physicochemical factors (temperature, pH, chemical content, water table, and gas content) have on the relative abundance or shift? • What is the role of the predominant population or the minority taxa?

also be discussed. In the report or thesis, we should discuss the progress we have made towards filling the research gap and identify what the next steps will be for any outstanding questions. It is also important to clarify what the primary takeaways are. Our goal in discussion will be to identify correlations among the data and determine how they relate to prior knowledge. In addition, we can further discuss whether the results align with our expectations/hypotheses, and what could be the reasons for this. If we search for advice on the Internet using a phrase such as how to write a thesis discussion, we will find many suggestions.

Collectively, these are the main purposes of discussion based on the description provided above: (1) Considering the collected data, what is the answer to the research questions, hypotheses, and objectives? (2) In what ways do the collected results relate to the findings of other published studies? (3) Why is that so? We can name more, but we will focus on these points by giving you some examples (Table 2.3).

Avoid hiding or covering up data that does not seem to fit. The evaluators of theses and manuscripts are often able to identify weaker data and are interested in seeing how we respond to these weaknesses. In the discussion section, we describe the strengths and limitations of the data or experimental results. As a good practice, it is important to mention how the limitations of an experiment might affect the conclusion of the study. In my opinion, there are no shortcuts or tricks to writing an effective discussion. Reading enough good publications and having a good understanding of existing knowledge in the field is the only prerequisite for developing good writing skills in the discussion section. These are the key takeaways:

- Interpretation (finding relationships, correlations, or patterns; and what the data means).
- Comparison (compare with earlier work).
- Explanation (explain why this is so and why that is so. Make justification).
- Closing the knowledge gap (state clearly how the findings contribute to closing the knowledge gap).
- Do not interpret the results in an exaggerated manner. It is imperative that claims are supported by sufficient evidence. Ensure that the claims are not excessive.

2.2.7 Drafting Abstract

Abstracts are often found on the first few pages of a thesis or dissertation report. It is recommended, however, that this part be written or polished after the majority of the other chapters have been completed. There may be some detailed guidelines provided by some universities; therefore, readers are advised to take a close look at them. The abstract is usually written in a single paragraph on a single page or within a predetermined word limit. The inclusion of references is generally prohibited. As described below, an abstract consists of five main components.

- An introduction. Describe the general background and/or the specific background in one to three sentences.
- The purpose. Provide the problem statement, the purpose statement, the objective, and/or the significance of the research. Please write one to four sentences that summarize the major reasons or purposes for conducting this study.
- The methodology. Provide a brief overview of the approach, strategy, or method in two or three sentences.
- Findings. Write the major results and the meaning of the results. This part should occupy approximately 50% of the abstract.
- The conclusion. Please limit the writing to two sentences maximum. It is possible to place the significance of the research at the end of the paragraph.

2.2.8 Setting and Using a Word Template

Some universities have created Word templates in the .dot, .dotx, or .dotm format. Some journal publishers may also provide templates. Having such a template will allow us to make use of the 'style function' within Word. We may highlight the sentences or paragraphs in Microsoft Word, pick and select the appropriate predetermined style, and it will be edited according to the spacing, font, alignment, paragraph spacing, indentation, etc. A template function can be used to generate the table of contents, lists of tables, lists of figures, etc.

If you cannot find a template for your university's website, you can probably create one. Watching a few YouTube videos will allow us to create and learn the correct way to format our writing using a Word template. Rather than using the default Word document, we should use a template, which will make writing and editing more straightforward.

2.2.9 How to Get Started and Keep the Ball Rolling?

There are some people who enjoy lab work, but hate writing. There is a name for it called thesis blue. Beginnings are always challenging. Nike has a trademark or slogan entitled 'Just Do It'. During the writing of my thesis, I kept reminding myself 'Just Do It'. Research beginners do not enjoy writing for a variety of reasons. As time passes with more and more experience, writing becomes relatively easier eventually. You need not worry if you are experiencing difficulties with English, grammar, tenses, or vocabulary, since I will explain how we can deal with those challenges in Chap. 6.

It is a very bad habit for some students to only start reading journals when they are writing their thesis. Some might argue that they cannot memorize everything if they read the articles months or years ago. Of course, no one is going to remember that as well. It is important, however, to read journal articles regularly. Let me tell you a story that I heard from a radio station. A grandfather and his grandson were talking about religions. He asks his grandpa to tell him how many percentages he can recall by reading the religion's doctrine each day. Before getting the answer, the child was instructed to use a bamboo basket to collect water from a pond. After a number of trials, the child realized that the basket had holes and could not retain water. By repeatedly dipping the basket into the pond, the initially dirty basket became cleaner. The act of reading journals regularly is like collecting water in a bamboo basket. If we do it frequently, our minds become clearer so that when we begin to type a manuscript or a thesis, it becomes easier.

Make use of technology to assist us in taking notes. A journal reading log with a summary may be organized on a Google Doc or Google Sheets. Avidnote is an alternative tool for taking notes for researchers. Using Avidnote, we can open a PDF file, make a note as we read, and keep track of all our notes. Another option is to use Evernote. Whatever platform we choose, we must be consistent in using it from the very beginning.

Reading is one of the prerequisites for writing a thesis. Here are some recommendations for writing a thesis and keeping it on track.

- What is the best way to begin? Develop the backbone first. List the titles of the sections, subsections, and sub-subsections. Fill in the main keywords or phrases for each section. Subsequently, include to-do lists for sub-sections, such as adding a table, a figure, SDGs, patent information, and bibliography information alongside the findings in papers.
- If possible, prepare the tables and figures before writing each section.
- Set the number of pages for each section; subsection; and sub-subsection based on keywords, phrases, or major points.
- We should begin writing now. Let's say we have decided to write two pages for a subsection. Once we reach the quota, we will stop writing and decide. Shall I add a new page, or should another section be moved first?
- Before we move on to the next subsection, we should read it three to five times for correcting expressions, tenses, spelling, italics, format, and references and adding examples.
- The cycle will continue with the next subsection.
- If we are stuck somewhere, we may wish to skip to another subsection or take a break.
- If we are experiencing difficulty moving forward, take a break, talk to someone such as a mentor or close friend, go to the gym and sweat, and then return to square one and do some reading, watch some relevant conference recordings, etc. And 'Just Do It' again.
- Once approximately 50–70% of the basic information is located in Word, it is necessary to review all major references and a few recently published articles again. Perhaps we have missed something important, and it is time to include it in the draft chapter.
- We may want to consider incubating the section by leaving it for a week or more and then re-reading what we wrote. Make it more readable.
- Consider moving sentences or paragraphs, creating new subsections, or consolidating redundant text. Make it more readable. Avoid using A-B-A-B in a paragraph, and group all sentences relating to theme-A together, while theme-B is addressed in another paragraph of a subsection.
- It is important to check the position of references. Referencing may be best placed at the end of the sentence if we are using the author's name and year style. It is possible to interrupt the reading process if there are references in the middle of a sentence, especially multiple references. In the case of numbered references, they may be placed in the middle or at the end of the sentence.
- Make sure you read it again and again. It is common for students to only read back once or twice, if not three times, what they wrote. This is not enough. As there is no absolute golden number, the idea is to keep checking back on what you have written to increase clarity and correct any errors.
- Take action to manage stress.
- The final step is to read it again and again. And remember 'Just Do It'.

2.2.10 Summary

The purpose of this chapter is to discuss good practices in the preparation of a proposal or thesis. A thesis goes beyond meeting graduation requirements since it explains the writer's main idea, opinion, or theory based on valid experiments and rigorous analysis. Students who are reading this, please keep in mind that your thesis (dissertation) should consist of your thesis (idea, opinion, or theory). There is more to your thesis (dissertation) than a pile of text, information, tables, and figures.

An area of research that is lacking sufficient information or knowledge is known as the research gap. Before declaring a gap as a knowledge gap, we should have read enough so that we can recognize the knowledge gap across the entire scientific community. The problem statement is related to the knowledge gap. The project objectives should be connected to the problem statement and well aligned with the research gap. It is recommended that the written objective uses the BE-SMART concept (Bloom's taxonomy, thEsis, Specific, Measurable, Achievable, Relevant, and Time-Bounded). A thesis impact statement can include contributions to academia as well as contributions to society (such as the SDGs).

The purpose of writing literature reviews is to (1) find missing research gaps, (2) obtain sufficient background information, and (3) deliver the information to readers (examiners, other students, etc.). Avoid lengthy general knowledge. Use reference managers and cite as you go. Analyse the citing references critically and select the strongest ones for rewriting as literature reviews. When the literature review is poorly prepared, it will be nothing more than a discontinuous summary with a plagiarism issue. Real core literature should be relevant to the project title, problem statements, and objectives and emphasize on science. When writing a literature review, avoid using information from a textbook or anything that is too general. The Results and Discussion chapters should include a substantial amount of interpretation of the data, a comparison with earlier work, rationale for why things are the way they are, and explain how the findings contribute to closing the knowledge gap. Finally, just do it and graduate.

References

Day R, Gastel B (2006) How to write and publish a scientific paper, 9th edn. GreenWood Publishing Group

Ellegaard O, Wallin JA (2015) The bibliometric analysis of scholarly production: how great is the impact? Scientometrics 105(3):1809–1831

Mejia C, Wu M, Zhang Y, Kajikawa Y (2021) Exploring topics in bibliometric research through citation networks and semantic analysis. Front Res Metr Analyt 6:742311

3 Research Integrity and Publication Ethics

Abstract

Chapter 3 addresses integrity and ethics issues in an attempt to create awareness before the readers fall into the trap and damage their reputations. Among the topics covered are plagiarism, self-citation, falsifying data, exaggerating findings, authorship in publications, the Contributor Roles Taxonomy (CRediT), Committee on Publication Ethics (COPE), conflict of interest, and the requirement for obtaining permits and approvals for specific types of research. Key performance indicators (KPIs), particularly publications, shall not be used as a reason to erode research integrity and ethnicity; therefore, a few common mistakes are discussed in detail too.

3.1 The Need for Integrity and Ethics

The meaning or synonyms of integrity include honesty, morality, accountability, virtue, or having strong moral principles and a sense of fairness. Ethics and integrity should be incorporated throughout the entire process of conducting research and publishing, including drafting a proposal, procuring materials, collecting and analysing data, as well as writing, publishing, disseminating information, IP protection, and commercialization.

The standard of integrity varies from one individual to another. A breach of integrity damages the research process, corrodes relationships between researchers, undermines trust in and the credibility of research, or poses a threat to society or the environment. Ethics and integrity are the responsibilities of many stakeholders, including the supervisor, student, principal investigator, funder, author, reviewer, editor, publisher, institution, university, and biotech company.

In case readers are not aware of Theranos, the recent biggest biotech scandal, you can gain insight by watching an HBO documentary or viewing some videos on

K. M. Goh, *Research Methodology in Bioscience and Biotechnology*,
https://doi.org/10.1007/978-981-99-2812-5_3

platforms such as YouTube, Yahoo Finance, Business Insider, CNBC, or CNN channels. A jury convicted Theranos' founder in January 2022 of defrauding investors and sentenced her to 135 months in prison and USD eight million in November 2022. The case was reported by the news media and appeared in the journal *Nature* as a news article (Waltz 2022). Besides that, we may have heard from the news that some researchers or academics are being dismissed because their research integrity has been compromised. Some were even jailed.

In order for universities to make a significant contribution to research integrity and ethics, a unit or office must be established to develop policies, procedures, and regulations related to the detection, investigation, and prevention of research misconduct. Furthermore, universities and institutes should protect whistle-blowers as well. As an example, the Singapore Nanyang Technological University (NTU) has established the Research Integrity and Ethics Office to maintain NTU's standards in research integrity and ethics. All new researchers (postdoctoral fellows, research office staff, and research assistants) are required to participate in the research integrity training program within a few weeks of joining the university. Similarly, all NTU new research students are required to complete some similar course and pass all assessments.

The majority of reputable publishers take scientific misconduct seriously. For example, the Springer Nature Editorial Advisory Group (SNEAG) oversees the journals' editorial standards and policies, and resolves publication ethics concerns. The SNEAG team also provides training to its office staff, external journal editors, reviewers, and authors. Elsevier provides editors with an online resource called the Publishing Ethics Resource Kit (PERK) to assist them in handling cases (https://www.elsevier.com/editors/perk).

3.2 Scientific Misconduct

3.2.1 General Meaning

The US Office of Research Integrity was established in 1993. The following text is taken the from Code of Federal Regulations, Title 42/Chapter 1/ Subchapter H/Part 93/Subpart A/§ 93.103 (available online: https://www.ecfr.gov/).

"Research misconduct is defined as fabrication, falsification or plagiarism in proposing, performing or reviewing research, or in reporting research results.

(a) Fabrication is making up data or results and recording or reporting them.
(b) Falsification is manipulating research materials, equipment, or processes, or changing or omitting data or results such that the research is not accurately represented in the research record.
(c) Plagiarism is the appropriation of another person's ideas, processes, results, or words without giving appropriate credit.
(d) Research misconduct does not include honest error or differences of opinion".

In a broader sense, the European Code for Conduct for Research Integrity states other unacceptable practices. Among these are manipulating authorship (to be discussed in detail below) and self-plagiarism, including publishing in different languages. On their website, the National Institutes of Health define research integrity as the use of honest and verifiable methods when proposing, conducting, and evaluating research as well as following generally accepted professional codes or norms when reporting research results.

3.2.2 Plagiarism

Many universities use Turnitin as the plagiarism detection software, but the majority of publishers rely on iThenticate. However, it is important to note that plagiarism is not simply limited to using someone else's sentences without proper citation. I liked the way plagiarism was defined in this slide shared by Catriona Fennell at online Elsevier Academy (https://researcheracademy.elsevier.com/): "Plagiarism is the appropriation of another person's ideas, processes, results, or word without giving appropriate credit, including those obtained through confidential review of other's research proposals and manuscripts", where the original text was taken from Federal Office of Science and Technology Policy, 1999. According to Merriam Webster, plagiarism means "to steal and pass off (the ideas or words of another) as one's own". Plagiarism is indeed the act of stealing someone's work or idea and claiming it to be your own. The practice of using another person's data without their consent is very serious offense.

Software that detects plagiarism is not omniscient enough to detect all forms of plagiarism. Various common cheating techniques can be employed to surpass the plagiarism threshold that has been set by the stakeholders (publishers or universities); however, this should not be the knowledge we are looking to acquire. Regardless of whether theft goes unnoticed or unpunished, it remains an act of stealing. Similarly, plagiarizing someone's work, ideas, and data without being detected or caught by plagiarism detection software is still plagiarism. It is essential to remember the commandment "Thou shalt not steal" from the Ten Commandments and uphold integrity by appropriately crediting the intellectual contributions of others.

3.2.3 Publication and Citation Interfere

The manipulation of citations, such as the practices like self-citation of irrelevant references, requests for the inclusion of publications by reviewers and collaborators, and the demand by the editor of a journal to include a certain number of their own articles are examples of unacceptable practices. But, why do people make such requests? One of the reasons for this trick is to boost citations. Besides, it is important that funders of research should not interfere with research processes or publications.

3.2.4 Approval and Permits

Research integrity encompasses not only the proper conduct of research but also includes respect for the subject (such as patients or animals) and human. Certain approvals are required before a research project may begin. If the work involves the use of human subjects, an informed consent form and statement of compliance is required. Stakeholder groups involved in the research process may include entities such as ethnic committee in the ministry of health, the ministry of science, hospitals, patients, universities, etc. If you are conducting experiments involving live animals, including mammals, fish, and invertebrates, please read and follow the ARRIVE guidelines v2.0 (Animal Research: Reporting of In Vivo Experiments) (du Sert et al. 2020).

Forming research partnerships between high-income and low-income settings can be highly beneficial for both parties. In order to avoid discrimination, double standards, or exploiting parties from lower income levels, a general suggestion with 23 articles has been drafted by the members of TRUST consortium. The full text of the document is available online under the title 'Global Ethics Code of Conduct for Research in Resource-Poor Settings'.

Many bioscience and biotechnology research projects involve field sampling of soil, water bodies, wastes, air, plant parts, insects, etc. Permits may be required for certain types of samples. To ensure compliance with regulations, it is important to refer to the Convention on Biological Diversity, which has been signed by 150 countries (https://www.cbd.int/). Additionally, the Cartagena Protocol and the Nagoya Protocol are also related documents or protocols to consider. The Nagoya Protocol deals with the issue of access to genetic resources and benefit sharing. Besides, it may be necessary to obtain a license or permit for the import and export of biological materials and samples. Each countries may have their own legislation, regulations, and practices in place, and it is crucial for readers to consult with the relevant stakeholders in their respective regions. Readers should be aware that breaching the law comes at a price.

3.2.5 Falsifying Data and Exaggerating Findings

The process of processing or analysing raw data in order to form final presentation data must be reported honestly. Besides falsifying data, exaggerating findings are poor practices. In order to impress funders, peers, or even journal reviewers (i.e. in the event of stating the novelty of a project) and viva voce examiners, some authors give statements based on misinformation or exaggerate part of the data to make something seem larger, more important, better, or worse than it really is. Let data speak for itself instead of selectively picking out data that fits the initial hypothesis or assumption while ignoring the other massive dataset that goes against it. As a rule, researchers should do the right thing and do things correctly even when no one is watching.

3.2.6 Misconduct Stain Our Record Versus Retraction with Honour

I wish to reiterate that breaching the law has a cost. Misconduct in the scientific field is not a shortcut to generate data and publications and should not be groomed as a habit. Engaging in research misconduct can have long-lasting effects on a person's reputation and career. Once a whistle-blower report is made and proven to be true, it will permanently stain a person's record. A university may, for example, withdraw the award of certification to a graduate who is found to have committed research misconduct even after he has graduated. The head of the department and university's top management will be notified if a staff had misconduct. A job may be terminated. Cases of research misconduct may be covered by national or international media. Published papers can be retracted with explanations given.

In some cases, published papers that contain fraudulent or erroneous data may be retracted with a formal notice from the publisher. You can find an example of this in the retracted articles, where the original article title and names of authors can still be found on the Retracted Note provided by the publisher. These retractions often include a statement such as: "The authors have retracted this article because they do not have ownership of the data they report. All authors agree with this retraction". There is no doubt that the authors made a grave mistake. A retraction such as the one described above is a type of misconduct that tarnishes reputation. It is important to note again that plagiarism is the act of stealing someone's work, idea, or sentences and claiming it to be your own.

Retraction is not necessarily a bad thing. The readers are encouraged to visit an online post entitled "Retraction with honour". The journal *Evolutionary* published an article about amoeba-bacteria symbiotic experiments that proliferate throughout many generations. After the publication was available online, the authors decided to sequence the microorganisms and discovered that contamination had occurred unexpectedly. No one would have noticed the mistakes if the authors decided to remain silent. Definitely a thumbs up to the group because they reported in honestly and retracted the article, and I am sure that it was a difficult decision that required a great deal of consideration and effort on their part. Suppose you were one of the co-authors, what would you do? Would you like to accumulate citations based on an article with mistakes, or would you like to have your article retracted honourably?

3.3 Authorships

3.3.1 Honorary Authorship

In a news article published in *Nature*, the concept of hyper authorship was discussed (Singh Chawla 2019). In the 5 years between 2009 and 2013, there were 573 articles with 1000 or more authors. However, this number has increased to 1315 papers in the following 5 years. This trend is particularly noticeable in fields like global epidemiology and climate change, where research projects involve hundreds or even thousands of authors.

In the fields of microbiology, bioremediation, genomics, transcriptomics, protein science, plant science, and some other aspects of bioscience and biotechnology, it is common to read articles that have been written by three or more authors. Unless the content involves a serious subject matter, hyper authorship is not often seen in articles on biotechnology. In my capacity as a reviewer and editor, I sometimes encountered "mini hyper authorships" which contained dozens of authors in relatively simple works, which caused me to question the existence of honorary authors. The term 'honorary authorship' refers to authors who are listed on the authorship list despite making no significant contribution to the research. There is a lack of integrity and ethics in such a habit.

3.3.2 Contributor Roles Taxonomy

The Contributor Roles Taxonomy (CRediT) was developed with the aim of recognizing the contributions of individual authors, reducing authorship disputes, and facilitating collaborations (https://credit.niso.org/). The idea originated from a discussion between Harvard University and the Wellcome Trust in 2012, with input from researchers, editors, and publishers. Currently, the majority of journal publishers now request authors to clearly identify and describe the contributions of each author. Which individuals should be considered authors? They must have significantly contributed to at least one of the following 14 roles (Table 3.1). The honorary authors who have not contributed should be removed.

According to the CRediT, authors should have significantly contributed to at least one of the 14 roles listed above. For more information, it is recommended that readers read an article that elaborates on the topic of scientific publication integrity in more detail (McNutt et al. 2018). In November 2022, *Scientific Reports'* chief editor sent an email to all editorial board members, urging us to actively promote good research ethics, with a specific focus on publication ethics. Here is a statement taken from the Scientific Reports website:

"**Each author** is expected to have made substantial contributions to the conception **or** design of the work; **or** the acquisition, analysis, **or** interpretation of data; or the creation of new software used in the work; or have drafted the work or substantively revised it

AND to have approved the submitted version (and any substantially modified version that involves the author's contribution to the study);

AND to have agreed both to be personally accountable for the author's own contributions and to ensure that questions related to the accuracy or integrity of any part of the work, even ones in which the author was not personally involved, are appropriately investigated, resolved, and the resolution documented in the literature". (source: *Scientific Reports* journal's webpage).

In summary, adherence to rigorous research practises, maintaining the integrity of publications and upholding ethical principles are critical to scientific progress. By meeting the contribution criteria outlined by CRediT and promoting transparency, accountability and integrity, we enhance the credibility and reliability of scientific research, which benefits the global scientific community and society at large.

Table 3.1 Roles of authors

Role	Description
Conceptualization	Formulation or evolution of objectives, aims, and goals.
Data curation	The curation of data involves activities in which metadata are produced, data is filtered, and research data is maintained for initial use and re-use.
Formal analysis	Analyse or synthesize study data using statistical, mathematical, computational, or other formal methods.
Funding acquisition	Acquisition of funding for the project that led to the publication.
Investigation	Performing a research and investigation process, such as an experiment or collecting data or evidence.
Methodology	Developing a methodology or creating an underlying model.
Project administration	Responsible for the planning and execution of research activities within the project.
Resources	Providing resources such as study materials, reagents, kits, patients, lab samples, animals, insects, microbes, instrumentation, computing resources, and other analysis tools to researchers.
Software	This includes programming, software development; designing computer programs; implementing the computer code and supporting algorithms; and testing the functionality of the code.
Supervision	Overseeing and leading activity planning and execution, as well as providing mentorship to individuals outside the core team.
Validation	This involves the verification of the overall replication/reproducibility of the results/experiments and other research outputs.
Visualization	Prepare, create, and present a visual aid or a data presentation.
Writing—original draft	Preparation, creation, and/or presentation of the published work, including the creation of the first draft (including substantive translations).
Writing—review and editing	Responsible for the preparation, creation, and/or presentation of published work, including pre- and post-publication phases.

3.3.3 Committee on Publication Ethics

Besides CRediT, many publishers now follow the code of practice suggested by COPE (Committee on Publication Ethics). Are you aware of COPE (https://publicationethics.org/authorship) that has been active for the past 25 years? This was written on their website: "COPE is committed to educating and supporting editors, publishers and those involved in publication ethics with the aim of moving the culture of publishing towards one where ethical practices become a normal part of the publishing culture". Please review their website on matters for more information.

3.3.4 Arrangement of Authorship

It is an art to arrange names in a sequential manner in the authorship list. The first author should be the author who made the most substantial contributions to the work described in the article and was responsible for conducting the majority of the underlying experiments. Editage Insights suggests that the first author of the work is usually the one who has made the most significant intellectual contribution to the work, in terms of designing the study, obtaining data from experiments, analysing the data, and writing the manuscript. Several publishers allow dual or shared first authors. PhD candidates (or master's candidates) at some universities are required to publish a first-authored article in order to graduate. I have heard a dispute between a supervisor and a student in which both parties claim the first authorship. Before drafting a manuscript, it is a good idea to discuss authorship in order to avoid authorship disputes.

3.3.5 Corresponding Authors

In another aspect, have you encountered any students who are fighting to be corresponding authors for whatever reason? All notifications from the journal are sent to the corresponding author, including the status of the manuscript, reviewers' comments, and the final decision. As a rule, the corresponding author is either the group leader (i.e. principal investigator of a grant) or a senior researcher whose contact information will not change in the near future.

Collectively, it is recommended that the student can be the first author and the corresponding author should be the supervisor or the principal investigator. When the primary contributor of the manuscript is also the leader of the research group, he or she can serve as both the first and corresponding author of the study. As an example, when I am writing a review article, I will be the first and corresponding author if I do most of the work, such as collecting information and writing.

3.4 Peer Review Misconduct

There should be no conflicts of interest in the peer review process. Some advice are listed below:

- Do not submit identical manuscripts to two journals at the same time. Publishing houses may utilize simultaneous submission programs in order to identify duplicate submissions.
- Duplicate findings published in different journals in different languages are not accepted.
- It is incorrect to create several publications from the same research using the 'slice by slice' approach, also known as salami slicing (salami science). In case you do not know, salami is a fermented air-dried sausage, and salami slicing means cutting the whole salami (project) into very thin slices. Some researchers have taken this approach in order to maximize the number of publications, but the oneness of the whole science has been compromised as a result.
- Avoid reviewers and editors from the same institution or those we know.

- Reviewers and editors who are our recent collaborators or co-authors or students should not be nominated when we make a manuscript submission.
- Do not fabricate reviewers and their emails that do not exist.
- We do not review submissions from authors who are direct competitors.
- It is best not to ask your recently graduated PhD student to serve as the reviewer for your own manuscript.

3.5 Viva Voce Misconduct

The majority of universities conduct viva voce examinations with both internal and external examiners. A chairperson leads the examination. During the oral examination, students must defend their research orally. Neither the chairperson nor the examiners should have any conflict of interest. Family relationships, or any private, personal, or commercial interests in a current or previous personal relationship with a student are examples of conflicts of interest. Both the chairperson and the examiners should not be involved in the grant or project, and they should not be co-authors of the student's publications. If a situation arises that compromises the professionalism of the examiners or affects their actions, judgements, or decisions, an examination board may have a conflict of interest.

In some systems, the supervisor is responsible for nominating a few names that may be suitable for recruitment as examiners. In order to ensure that only the most qualified individuals are appointed as examiners, there should be some type of mechanism or committee to look after this matter. The appointment of easy-going examiners who are not subject matter experts is less appropriate.

In most cases, when we discuss potential conflicts of interest, we refer to past or current relationships or activities. For examples, the reviewer was one of my students, or the reviewer was a former collaborator—these are past or current relationships. Let us examine one case and determine whether it constitutes a conflict of interest. There will be a viva voce examination for Student-A, a master's student, within a few weeks. Two examiners were formally appointed a few months prior to the submission of the dissertation for review. During the course of time, one of the examiners decided to accept Student-A as the PhD candidate once he/she had passed the current viva voce. It was decided a month before the viva voce that Student-A would register as a PhD candidate with one of the examiners. Initially, it was not anticipated and not planned at all. Is there a conflict of interest here since the current examiner will supervise the evaluated student the following semester? This is one example of a potential conflict of interest that is not related to the past relationship or activities but rather relates to a problem that may arise in the near future. My opinion is that there is a conflict of interest. Considering myself the examiner, I might make a biased decision. As an example, I might feel an urgent need for the student to complete the master's degree and enrol him/her in the PhD program as soon as possible in order to collect data for the grant. The student may not be happy if I make too many comments and decide not to join my team. If I comment too much, there might be too many things for the student to correct the dissertation, causing the PhD enrolment to be delayed. This seems to be a conflict of interest, doesn't it?

3.6 New Threat in Research Ethics and Misconduct

3.6.1 Emerging Artificial Intelligence

Artificial intelligence (AI) is a double-edged sword. There is great potential for the use of AI in the research fields of medical biotechnology, agriculture, bioinformatics, and other areas. Misuse of AI is a new area of concern from the standpoint of misconduct. An example of this is the fabrication of images for the purpose of manipulating research results (Wang et al. 2022). The use of AI tools has also been discussed in another chapter, and one of the applications of these tools includes summarizing long documents, preparing a journal manuscript or dissertation, and generating an abstract from a full text. Yet, there is a concern that some individuals may become lazy and irresponsible, relying solely on AI without actively reading or verifying the information. This could lead to the misuse of AI tools, where proposals, theses, articles, or journal reviews are generated without proper understanding or critical assessment of the content. Moreover, it is essential to acknowledge that AI may be misused to create drawings or figures that lack proper experimental basis.

In 2018, the GPT-1 model was released, followed by the GPT-2 in 2019, and the GPT-3 in 2020, which became widely known by the end of 2022 or the beginning of 2023. GPT is an acronym for Generative Pre-trained Transformer version 3 by the OpenAI. As compared to previous versions, GPT-3 makes use of the most current state-of-the-art language model as well as a much larger training library. A wide range of applications has been developed utilizing this system, including language translation, summarization, and question answering. We can expect more advanced GPTs in the coming years. Please refer to the Appendix of this book, where I have listed some AI tools that you may find useful for various research activities such as writing, literature review, oral communication, slide design, coding, and more.

In an article published in *INSIDER* on July 20, 2022, the news stated that AI GPT-3 could write an academic thesis within 2 h by itself. Is it necessary for universities to set rules that prohibit students from using AI in literature reviews, finding trends, solving their problems, writing a thesis, writing codes or a program, generating a figure, or any other activity that we can think of? Is it a good idea to ask students to use GPT to learn the latest information rather than just reading lecture notes or Wikipedia? Should academicians simply keep quiet and not bother with this issue? It is a difficult question, and I believe the university senate members will need a lot of wisdom to decide. GPT and AI will likely create more storms in universities and research in the beginning, but eventually everyone will get use to all these AI platforms.

In fact, 99.9% of this book was completed before I began playing with ChatGPT. Anyway, I wanted to see how powerful GPT is, which is why I spent close to an hour for the first time asking some general and very specific questions to GPT. The first question I asked was, what are the genes that cause cyanobacteria to produce cyanotoxins? The ChatGPT answer was quite satisfactory. By twisting the question, I eventually asked three different related questions, but the answers were more or less similar and did not quite address my main point. Due to the lack of literature pertaining to the question I posted, the machine has no information from which to learn and answer the question. However, after several months, I revisited the same questions and noticed that the AI had become significantly smarter. Its ability to answer the questions accurately had improved, and it continues to enhance its capabilities.

Many chatbots, like ChatGPT, perform exceptionally well in non-research knowledge. I asked a question: How to make a good presentation at a conference? During the course of a few seconds, the chatbot wrote information that I had outlined in the entire Sect. 6.3.2. I spent many days writing that section and reading the book *TED Talks* to ensure that what I wrote made sense. However, ChatGPT took a few seconds to complete.

Another question I had for ChatGPT was: Some reports claim that GPT can write a thesis in 2 h. Can this be true? The following is the answer provided by GPT, word for word, without any modifications:

"It is unlikely that a high-quality thesis could be written in just 2 h using a language model such as GPT. While language models like GPT have been trained on a large amount of text and can generate coherent and contextually appropriate responses, they are still limited by their training data and the algorithms that generate their responses.

Creating a thesis requires a significant amount of research, analysis, and critical thinking, and it is unlikely that a language model could generate a thesis in a short amount of time that would meet the standards for a graduate-level or research-level degree.

Additionally, a thesis is not just about generating text but also about understanding the subject matter, presenting original ideas and arguments, and demonstrating a deep understanding of the research literature. These skills cannot be replicated by a language model in a short amount of time.

In short, while a language model like GPT can certainly generate text that could be used as part of a thesis, it is unlikely that it could generate a complete, high-quality thesis in just 2 h".

Despite ChatGPT being capable of producing convincing text, it tends to be inaccurate, so it can distort scientific facts and spread misinformation at this point. However, it will certainly become very powerful very soon. With continued development, its potential will be unleashed.

According to a news article published on 18 January 2023, 'ChatGPT listed as author on research papers: many scientists disapprove', there are at least four publications where ChatGPT was listed as an author. It has been agreed by *Nature*'s news team that AIs, such as ChatGPT, do not fulfil the criteria for being a study author since they are not responsible for the content and integrity of the study. What is your personal opinion on that?

I would appreciate it if you would read the Instructions to Authors for the journal you intend to submit a manuscript to and see if you can find any description of ChatGPT/GPT-3 or any newer AI machine learning system as an author. As gatekeepers and goalkeepers, editors and reviewers of each journal should play an equally vital role. As a matter of personal opinion, I voted no to include AI as an author. Even if you voted against AI, I do not believe that would significantly affect writers who intended to utilize it in an unethical manner. It is likely that as long as the manuscript is published, they do not really care about the 'feeling' of AI, given that the machine learning system will not feel sad, cheated, or feel that its idea or writing is being pirated by anyone else. Bad news: If you cheat a real human collaborator, you will never dare to interact with them again, but good news: once you betray a machine learning system (i.e. ChatGPT) by stealing its work, you will gladly retry it. The name of ChatGPT can certainly be controlled in the list of authors, but there will never be a way to catch or prevent people from using a machine learning system quietly to write and publish. To conclude, I am not against the use of AI to assist in writing, but I strongly oppose its unethical usage.

One of the sections of Chap. 5 discusses the dos and don'ts of writing a review article. There may be some authors who are not experts in the field or topic publishing a review. In general, they do this for multiple reasons, and it is not acceptable if the intention is to achieve KPIs. It is fine if the writers are experts; however, if they are not, it may result in a poor-quality review. It was once necessary to read a large amount of material, analyse it, arrange themes, and write a review article in the past. Now, does AI pose a threat in this scope? Many non-experts are already producing reviews of little value, and if these authors start using AI to make their reviews appear more sophisticated, it raises ethical concerns. Although I am not opposed to the use of AI, I am concerned about what will happen to the next generation of students and researchers who read review articles that have not been prepared by true experts, but rather by non-experts chasing publication KPIs while blindly utilizing AI tools. The question arises: Who wrote the article or review—a human or an AI robot? Does a robot-writer qualify as an author or a ghost writer? As a side note, a ghost author is someone who contributed significantly to a manuscript without acknowledgement.

Most of us may not be aware that AI has been used in some editorial systems to assist with the submission of manuscripts and initial quality checks. The question arises: Will a machine learning system eventually replace humans in making decisions (rejection or acceptance)? It is well-known that reviewers and editors can exhibit bias, which is challenging to completely eliminate. This prompts us to consider whether AI has the potential to perform better in this regard.

Biological and biotechnological research cannot be separated from the emerging technology that is AI. It is likely that many researchers will be reluctant or have a difficult time accepting this new trend, especially those who are more conventional. In general, humans tend to be reluctant to accept brand-new ideas. In retrospect, some supervisors may have been hesitant to purchase molecular kits when they first appeared on the market with the excuse that students would not learn anything if they owned the kits. However, those supervisors may have been misguided. A decade ago, some people were unhappy when one publisher created a journal with a 500-word limit, but now more and more publishers are doing the same. Human tends to be reluctant. Time will tell what happens to GPT and other upcoming AI machines. What role will they play? Will they be a master, a servant, or just a tool? There is no possibility that one would be able to completely stay away from AI, perhaps not right now, but at some point in the close future.

3.6.2 Grey Business

The production of journal publications by 'paper mills' may also be an emerging problem in the close future. Althought I used the word 'emerging' here, but it is possible that such an industry has existed for some time in some countries. These companies are profit-oriented enterprises that manipulate manuscripts, fabricate data, duplicate published images, and otherwise fabricate papers without doing sufficient research. There are some paper mill companies that write theses for students. Based on an analysis conducted in 2020, more than 1300 suspicious articles could have originated from paper mills (Else and Van Noorden 2021). For more information, it is recommended that readers read a systematically written comment from COPE; the online document is titled 'Paper Mills. Research report from COPE & STM'.

A paper mill, as described in the previous paragraph, is a company that acts as a ghost writer for journals. In layman's terms, 'paper' refers to articles, reviews, etc. I would like to create a new term, Book Mills. Please do not search it on Google since you will not find it since I invented it after receiving an email from someone I do not know. I have never seen an offer of this nature before. The following is an exact copy of the original message that was emailed to me:

"I'm XX, an editorial assistant from a publishing business in the XX, and I'm contacting you with a recommendation from our editorial department. We'd like to write a book for you based on your excellent contributions to the scientific community. Researchers like you contribute so much to the scientific community, but you don't get nearly enough recognition. You will remain unknown to the general public regardless of how many papers you publish in prestigious journals. We devised this one-of-a-kind method to address this issue. For instance, if you did some exceptional study in the field of 'Diabetes,' why not publish a book about it?

Writing a book will be difficult for you due to your hectic schedule. Our staff of skilled writers, on the other hand, may write a book for you and send it to you for review. Also, no one will buy a diabetes book written by an unknown author, but if it is written by an expert like you, it will attract a lot of attention and you will have complete control over what goes into the book. Following that, your book will then be listed on well-known book websites such as Amazon, eBay, Goodreads, and a slew of others. As a result, you will be well-received, and your hard work and scientific achievements will be recognized. Last but not least, your book will be published in Google News, Yahoo, and other big news outlets after it is released. What more could you want? We only require your book writing contract in order to develop a proposal and theme. I'll be waiting for your response".

Yes, I responded. My only response to this email is to mark it as spam and block it. We would like to remind our readers not to attempt any such offer. If you step into the mud, you will soon realize that it is a pile of manure. Getting out of it quickly is the best course of action! Free lunches do not exist, and there will never be a free lunch.

Human writers have traditionally been employed by paper mills. Before AI such as GPT became so powerful, can we assume that ghost writers are real humans with some knowledge (or they are real experts in the field)? What about now? Would it be reasonable to assume that these paper mills are experimenting with AI writers? AI is more powerful, faster, and cheaper than a human, isn't it? For most businessmen, faster means saving time and making more money. It seems likely that conventional ghost writers will soon lose their jobs unless they start learning artificial intelligence, don't you think so?

Another emerging threat is the selling and buying of authorship. Did you ask, 'What? Are you kidding?' Yes, it is true. The business idea involves paying someone to have their name included in a publication that they did not participate in. Of course, there may be other business models that we are unaware of. You can read more about the paper authorship trade by searching the full article online under the headline 'Multimillion-dollar trade in paper authorships alarms publishers'.

3.7 The Dilemma App of Research Integrity

Ethics and integrity can be talked about relatively easily on paper; however, when a real challenge to research integrity appears right in front of us, we may have difficulty making a timely decision. The Erasmus University Rotterdam (the Netherlands) has introduced a card game called Dilemma Game on its campus. Upon entering 2020, it went digital and we can download the app (from Play Store and App Store) to our mobile devices.

According to the website, the app is intended to use real cases to "stimulate awareness of, and an open and critical discussion about, integrity and professionalism in research". Having read a scenario, the reader can determine four possible actions and post an answer. We can also see the percentage of votes that have been cast by all participants.

One of the questions in Dilemma Game was something like this in a simplified manner: "A PhD candidate had his/her manuscript rejected multiple times. One of the authors suggested his/her name be withdrawn, but will act as a reviewer for the submission and submit the manuscript to a journal for which the senior collaborator is serving as editor. In such a situation, what should the student do?

1. The student may discuss it with the supervisor.
2. Decline it and choose a lower-ranked journal.
3. Take the offer.
4. Complain to the research integrity committee".

How would you respond if you were in the student's position? We should keep in mind that it is the policy of many universities that publication is a requirement before a student may submit his or her thesis for examination.

Rather than suggesting a definitive right or wrong course of action, the intent is to get us to think about why we reached a particular conclusion. We can also read a comment by the Erasmus University Rotterdam on the issue. Researchers who teach ethics and research integrity, principal investigators who are responsible for major decisions, especially hard-to-make ones, and students who are one of the most important stakeholders in research activities can benefit from the Dilemma Game app. Download and try it yourself.

3.8 Summary

Students and researchers are under pressure to meet KPIs. We can talk about ethics and integrity reasonably easily; however, when a real challenge to research integrity arises right before us, we may find it difficult to make a decision. Ethics and integrity should be ingrained deep in our hearts, and we achieve this by practising them. Be careful not to commit plagiarism, exaggeration, fabrication, or falsification. Please pay close attention to the guidelines of CRediT, COPE, US Office of Research Integrity, European Code for Conduct for Research Integrity, Convention on Biological Diversity, Cartagena Protocol, Nagoya Protocol, and local authority bodies. There is a wealth of real-life examples present in the Dilemma Game app that can assist in provoking our thinking and provide excellent information for teaching and sharing purposes. Make sure that KPIs do not slowly erode our ethics and integrity. Remember, misconduct stain our reputation.

References

du Sert NP et al (2020) The ARRIVE guidelines 2.0: Updated guidelines for reporting animal research. PLoS Biol 18(7):e3000410

Else H, Van Noorden R (2021) The fight against fake-paper factories that churn out sham science. Nature 591(7851):516–519

McNutt MK et al (2018) Transparency in authors' contributions and responsibilities to promote integrity in scientific publication. Proc Natl Acad Sci USA 115(11):2557–2560

Singh Chawla D (2019) Hyperauthorship: global projects spark surge in thousand-author papers. Nature. https://doi.org/10.1038/d41586-019-03862-0

Waltz E (2022) Elizabeth Holmes verdict: researchers share lessons for science. Nature 601(7892)

Wang L, Zhou L, Yang W, Yu R (2022) Deepfakes: A new threat to image fabrication in scientific publications? Patterns 3(5):100509

Good Research Practices

4

Abstract

As part of the postgraduate program, defending a proposal is one of the most significant milestones, and defending a thesis is another. For academicians, postdoctoral fellows, and scientists, writing a grant proposal and publishing an article represent the beginning and closing of a typical research endeavour. Is there anything else we need to consider between the beginning and end of the project to guarantee a successful outcome? Besides collecting good data, is there anything else we should be watchful of? In this chapter, we will discuss some good practices that may assist you in becoming a better biologist or biotechnologist researcher. This chapter covers topics like basic sampling approaches, maintaining biological samples, reproducibility, validation of data, using multiple approaches for experimental data collection, managing digital resources, generating figures for reports, IP protection, developing reading habits, etc. Besides, this chapter emphasizes the notion that excellent researchers are also excellent planners and are able to translate plans into action. Additionally, the chapter includes more recent upskilling tools for readers to be aware of, such as the Nature Masterclass (and the likes), Massive Open Online Courses (MOOCs), etc.

4.1 Managing and Keeping Biological Samples

For the management and preservation of samples relevant to biological studies, there is no one size fits all protocol. In general, there are a few main sampling strategies. Two major categories are probability sampling methods and non-probability sampling methods. Examples of the former category are provided below.

K. M. Goh, *Research Methodology in Bioscience and Biotechnology*,
https://doi.org/10.1007/978-981-99-2812-5_4

4.1.1 Simple Random Sampling

Using this method is simple, faster, and does not require a person to be familiar with the data being collected. In a simple example, a student wants to isolate some bacteria and fungus from forest soil, garden soil, lake, effluent, or peat. A simple random sampling is sufficient if the project does not intend to investigate the relationship between microbes and abiotic factors, geographical influences, or other more complicated scientific questions. A single-point random sampling provides a snapshot of the sample that can be used for preliminary analysis. The only thing to do is to scope a few grams or ml of samples randomly from a number of sites and to record the location with a GPS. The actual nature of the samples should be understood, however. For instance, soil and microbial mats have a large degree of heterogeneity. The results of a single random sampling may not be representative of the whole picture.

In another example related to plant seedlings, say 100 seedlings are grown in polybags and each seedling is identified with a number between 1 through 100. If our sampling size is 10% (N = 10) of the total population, then we can simply use any online random number generator tool to generate ten random numbers and then we pick the polybags accordingly.

4.1.2 Cluster Sampling Strategy

When necessary, a researcher can divide the entire population into clusters that represent a population. Let us consider an example. The intestinal tract of chickens contains microbiomes that are essential for digestion, nutrient absorption, immune response, and pathogen exclusion (Shang et al. 2018). Suppose we have a small poultry farm consisting of a number of pens for the chickens, perhaps we could select a few pens to represent the entire farm in a microbiome sequencing project. In this case, one pen is equivalent to a cluster.

4.1.3 Systematic Sampling

We may consider a systematic sampling approach to be sampling every *x*th, where *x* may be length, time, counts, depth, etc. Say we have 10,000 samples collected from COVID-19 patients. In this approach, we can decide to sequence virus RNA every 100th sample. As another example, we may be monitoring changes in biomarkers at a particular site (biofilm, water, soil, etc.) every 3 months. In this case, systematic sampling would mean taking a sample at every time interval (e.g. every 3 months for 2 years). Another example would be sampling every *x*-meters (say $x = 50$ m) along a seashore and using a biosensor technique for pollutant detection. Usually, the oil palm trees are planted in triangular patterns with an interplanting distance of nine meters. In systematic sampling, we can collect some roots or soil for every oil palm tree planted at a distance of x-meters (say 27 m) for fungus *Ganoderma* detection.

4.1.4 Stratified Random Sampling

This technique is commonly employed in studies of public health or health research. It may be of interest to researchers to understand more about specific subgroups within populations, for example, different ethnic groups or age groups. If so, stratified random sampling would be a more appropriate approach. For instance, in one study, Alabama city was divided into smaller groups according to their area or location (Howell et al. 2020). By dividing the population into smaller groups based on areas or locations, researchers can ensure that each subgroup is represented in the sample. Employing a stratified sampling approach enables researchers to obtain a sample that is both accurate and representative of each subgroup.

4.1.5 Labelling Samples

Bioscience and biotechnology experiments usually involve a large number of specimens, samples, or tubes. Our samples in the experiment of protein purification included crude enzyme expressed by a host cell, concentrated crude enzyme (e.g. using a concentrator or salting out procedure), flow-through tubes collected from a chromatography, and separate elution tubes collected from the chromatography system. It is likely that we would end up with more than dozens of tubes of solutions during a single experiment. This is also true in cloning, where each step involves several tubes. It is ineffective to label tubes or samples according to dates alone since this will likely result in a loss of memory after a period of time. On a petri dish, it is possible to write enough information due to a large surface area; however, on a round cap PCR tube, it is extremely difficult.

There are a number of commercially available solutions, such as the portable or handheld label printer. Some good ones are resistant to solvents, waterproof, and suitable for samples stored in freezers. Expensive printers are equipped with barcode generators and scanning devices. While most laboratories continue to label tubes or bottles using a marker pen, it is likely that some laboratories with larger budgets will choose to purchase printers.

We could develop some simple with a permanent marker pen. For instance, I refer to each run of chromatography as R1, R2, and R3, which refers to different runs of the experiment. The R1 FT code represents the flow-through sample for run-1. Run-1 fraction 1 (elution tube) is indicated by R1 F1.

Let us look at another example. The ultra-low freezer at -80 °C may be filled with glycerol stocks of bacteria. It is my habit to name isolated bacteria by combining the abbreviation of the sampling site, a number, and some other special features. CL10P on the label can be interpreted as a bacteria isolated from Clemente soil (prefix CL) and it is the tenth bacteria picked from the spread plate, and P may be interpreted as either the purple pigmentation or the student's name, for instance, Paul.

In order to avoid mistaking irrelevant tubes as positive samples, we should discard them as soon as possible rather than keeping them in freezers. Few years ago, one student I know was struggled for months with gene amplification. He later realized that he had made a mistake as the *E. coli* stock contained only a plasmid without the gene of interest. The error occurred because he had forgotten to discard failed experiment tubes.

It is important to have some systematic labelling of runs and samples according to your preferences, and the labels should be informative enough that you are able to recognize their content as soon as you see them. Regardless of whether the barcodes are printable and scannable or handwritten, we still need a proper respiratory library for storing all the details, for instance, a logbook or simply Google Sheets.

4.1.6 Storing Samples

It is common practice to store materials at lower temperatures. For a short time, buffers, ampicillin, and anything else can be kept in a refrigerator (2–8 °C). A freezer at −20 °C is ideal for storing DNA polymerase, ligase, genomic DNA, plasmid DNA, RNA, PCR products, dNTPs, antibodies, antibiotics (not all), and purified protein. The ultra-low freezer at −80 °C is used to store bacteria stocks, competent cells of *E. coli*, RNA, and whole blood. Cryogenic storage (−150 to −190 °C) is intended for the preservation and storage of biological materials at extremely low temperatures. Samples of these types include human samples, stem cells for autologous transplants in cancer patients, adipose tissue, umbilical cord blood, and mesenchymal stem cells for regenerative medicine, among others. Typically, seeds are frozen at −18 °C under humidity control or in liquid nitrogen at −196 °C. The seeds tend to lose their ability to germinate even after they have been frozen. This is the reason why new seeds must be produced to replace aging samples.

Wrong storage causes materials to lose activity or deteriorate, resulting in substantial costs. Large, well-managed national laboratories utilize radio-frequency identification (RFID)-based or other management systems. Laboratory cold storage machines are a major source of energy consumption in a laboratory. Hence, the International Institute for Sustainable Laboratories and My Green Lab launched a program called the Freezer Challenge, which aims to ensure energy reduction. This program exposes us to best practices in energy management. We can, for instance, use up to 30% less energy if we set −70 °C instead of −80 °C. The Freezer Challenge webpage shows a list of things we can do. Awards are given to those laboratories and research organizations that have saved the most energy and improved their material storage. The winners for 2021 were the University of Alabama Birmingham, AstraZeneca, the National Institutes of Health, and the Children's Hospital at Westmead. Well done and congratulations. Is your institution the next winner?

Transporting samples requires our attention as well. If the sampling site is located at a distance from the lab, some individuals prefer to keep it in the same condition as the sampling point. To maintain the temperature of hot spring samples, some researchers place them in thermoses. Depending upon the purpose of the sampling, the reader may wish to consider cooling the sample using a block of dry ice and bringing it back to the laboratory instead of keeping it warm. Remember to ensure that the procedure we use is scientifically sound and complies with our downstream work.

It is common to send DNA from one lab to another. Plasmid DNA is often dried on filter paper that is marked with a pencil and placed inside a plastic bag before posting. Depositing microorganisms in culture banks is also a common practice. For more information regarding how to treat the microorganisms, please contact the culture bank, for example, DSMZ. For those of you who are working with DNA or RNA, ZymoResearch's DNA/RNA Shield and ThermoFisher Scientific's RNAlater are excellent products. You may wish to read more about it if it is relevant to you.

4.2 Reproducibility and Validating Experimental Data

As part of the Scopus and Web of Science databases, there is a type of document or record referred to as 'retraction'. Retraction is essentially a record of papers that have been withdrawn due to major unintended mistakes, ethical issues, or misconduct such as falsifying data. Retraction Watch is a nonprofit organization devoted to monitoring retracted papers. As of the date of this writing, Retraction Watch listed more than 200 retracted documents related to COVID-19.

Essentially, reproducibility means that two honest researchers will get the same result if they do the experiment separately. It is considered reproducible in bioinformatics when the same results are obtained using the same input data and software settings. The concept of reproducible research should include reproducible methods, reproducible results, and reproducible inferences (Plesser 2018). The *Nature* editorial previously reported a survey of 1576 participants, of which 52% agreed that there is a significant crisis in reproducibility (Baker 2016). No matter whether the issues surrounding reproducibility can be characterized as a crisis, Shiffrin and coworkers observed that the current emphasis on reproducibility overlooks a more pressing concern: the validity of reports. A reproducible result does not imply validity (Shiffrin et al. 2018).

Studies that use poorly characterized reagents, antibodies, cell lines, chemical agents, and animal models can lead to poor reproducibility and the ability to make accurate predictions when extrapolated to humans (Flier 2017). Collaborations today cross scientific boundaries, bringing together disparate approaches and techniques, ensuring reproducibility becomes even more challenging. Besides, obtaining data from service providers is also a concern since we may not know all the details. Ensuring reproducibility in research requires service providers to adopt transparent practices and provide detailed documentation of their data generation processes. This transparency enables researchers to confidently reproduce and validate results, thereby strengthening the reliability of research across various domains. There are other new challenges described in this publication (Flier 2017).

Researchers should ensure their findings are reproducible and valid. My opinion is that the principal investigator or supervisor should verify the raw data provided by the service provider, student, or postdoc rather than accepting anything presented to them in nice diagrams or charts. It is an important first step. Second, statistical analysis should be performed properly, and only reproducible and valid data should be published. Responsible reviewers and editors should be the third gateway. Readers constitute the fourth gateway. I recently attended a webinar in which the speaker suggested that publishers should consider implementing a feature that would allow readers to provide both positive and negative feedback openly and freely. It is difficult to predict if this will work, but the suggestion is a double-edged sword, with the potential for both constructive discussions and the dissemination of inaccurate or misleading information.

4.3 Polyphasic Experimental Approaches

A polyphasic process is defined by Merriam-Webster as one consisting of two or more phases. Prior to the advent of molecular genetics, microbial taxonomy was determined solely on the basis of phenotypic traits. Microbial taxonomy is now using a polyphasic approach, which refers to the systematic evaluation and integration of chemotaxonomic, phenotypic, and genotypic data to arrive at a taxonomic proposal or determination (Schleifer 2009). In the past, DNA-DNA hybridization was considered the golden standard for determining genetic similarity between genomes. Digital genome-genome comparison is commonly used in this decade (Chun and Rainey 2014). Technology continues to advance, and the technique we selected for our research will also evolve.

Bioscience and biotechnology projects have traditionally been mono-area in nature. Microbiologists, for instance, rarely collaborated with plant scientists in the past. Today, we are familiar with terms such as plant-microbe interaction, plant microbiome, etc.

Many laboratories have transitioned from mono-area approaches to polyphasic approaches or interdisciplinary (or multidisciplinary) approaches. For example, -omics approaches that include genomic, transcriptomic, meta-taxonomic (e.g. 16S rRNA, 18S rRNA, and other markers), metagenomic, metagenome-assembled genome, proteomic, metabolomic, lipidomic, and epigenetic analysis can be integrated with the majority of biological research. For instance, a paper by Hasin et al. reviewed multi-omics approaches to disease studies (Hasin et al. 2017). A multi-omics approach is also applied in many other aspects of research as well.

Bioinformatics is defined as the application of tools for computation and analysis to capture, analyse, and interpret biological data. Many of us are familiar with basic bioinformatics applications such as the NCBI database, protein structure prediction, molecular dynamic simulation, and DNA-RNA-protein sequence analysis. Bioinformatics includes much more than these applications. The field of bioinformatics can be used to study genomics, proteomics, transcriptomics, computational intelligence, pattern recognition, clustering and classification, data mining, biostatistics, machine learning, data science, model design, and many more disciplines. Embracing the field of bioinformatics enriches the scope and magnitude of your research, enabling you to delve deeper and wider, ultimately resulting in comprehensive findings of great significance. It is through the integration of bioinformatics that you can pave the way for transformative discoveries with far-reaching impact.

Furthermore, conventional bioscience and biotechnology research may have too little connection to social science, sustainable development, economics, policies, engineering, business, and management. However, it is important to take bold steps and explore new avenues beyond the comfort zone of traditional research practices. To the best of my knowledge, there are several countries or research grants that encourage biologists to integrate social science elements into their research proposals. To summarize, I would suggest that readers might consider incorporating -omics, bioinformatics, machine learning, social science, and other new scopes in their current or future projects. Embracing these diverse disciplines can foster innovation, deepen understanding, and contribute to transformative breakthroughs in biological research.

4.4 Managing Digital Resources

Biologists should recognize the critical importance of effectively managing digital resources. Digital resources could mean materials that are created digitally or by converting a hardcopy material to a digital format. Digital resources in teaching and learning may also include websites, software, or apps that support students and teachers. As a researcher, digital literacy is an essential skill. Being productive in this domain does not imply that one must be an expert in the field of IT. By investing time in internet searches and acquiring knowledge on efficient digital handling techniques, one can discover more efficient and intelligent ways to manage digital resources. As we proceed with this section, we will cover a few skills that will improve our digital competency and, thereby, save our time eventually. Improving our digital task management efficiency enables us to devote more time to critical research activities such as thoughtful analysis, strategic planning, and impactful writing. Optimizing our digital workflows empowers us to allocate greater energy and resources to the core intellectual endeavors that drive scientific progress.

Try enhancing Gmail settings. Start by adjusting the default Undo Send timeout from 5 s to 30 s; if we are prone to making mistakes, such as forgetting attachments. Take advantage of features like Smart Compose (as well as Smart Compose personalization), which can save time by formulating a sentence. You should be aware of the Schedule Send feature, which allows you to compose a message and have it automatically posted on a specific date and time. Try to make a signature that includes name, position, address, links to Google Scholar, etc. Emails will look more professional if we do so. I have seen some email signatures that are too lengthy, out of date, or contain other unnecessary information that may cause misinterpretation.

Lastly, do we spend too much time clearing junk mail and messages we will never read or open? If so, consider unsubscribing, blocking, or muting irrelevant senders. Additionally, students and researchers from non-native English-speaking countries often overlook the importance of proofreading. Tools like Firefox's spell-check feature, Safari's check spelling while typing feature, or add-ons like Grammarly can greatly assist in ensuring error-free communication. Remember, writing in full capital letters can create a negative impression, so it's best to avoid it. It is true that familiarizing yourself with Gmail has little to do with completing a biological research project; however, it does make us appear more professional and helps us manage our time more effectively. You should spend 10–20 mins watching a few YouTube videos on advanced tips for using Gmail. This investment will make your work in the future easier.

Try getting some useful add-ons. If you are a committee member for a conference or workshop, and plan to send participants some customized emails or documents (e.g. participation certificates), mail merge tools can be useful. It is common for guest editors or associate editors of special topics or volumes of journals to use mail merge to contact potential contributors. There are several add-ons available in the Google Workspace Marketplace, including Mailmeteor for Gmail, Mergo, Yet Another Mail Merge, GMass, etc.

I also find it helpful to have add-ons that allow me to draw simple diagrams, such as flowcharts, for our research proposal. Many of these tools are easy to use and allow us to add collaborators to edit the diagram on the respective computer. Some examples include Lucidchart Diagrams, diagrams.net, Draw Diagrams, and SmartDraw Diagrams. SketchUp, SketchUp for Schools, and AutoCAD web apps are some examples for those requiring 3D drawing. Lucidchart Diagrams and diagrams.net are alternatives to Visio. These add-ons are available in the Google WorkspaceGoogle Workspace Marketplace.

A Gantt chart is also frequently included in proposals. Many free Excel templates are available online for creating nice design Gantt charts. In case you are concerned about possible bugs hidden in the downloaded file, you may obtain some software in Google Workspace. These include GANTTophant and Gantter.

4.5 High-Resolution and Effective Figures

It is said that a picture is worth a thousand words. The art and science of producing high-quality figures are inextricably linked. Once the culture is established in the laboratory, the other students will develop the habit of producing excellent figures. The majority of us will state that we are scientists, not artists; therefore, we should not place a great deal of emphasis on the artistic illustration. Many laboratories possess excellent scientific talents but lack the artistic ability (e.g. poor colour perception, etc.). We have seen scientists collect valuable scientific raw data, but present it in an ineffective or strange manner. Artwork or visual aids are essential for scientific communication. The requirement for figure size, lettering on the artwork, and resolution are usually written clearly in Instruction to Authors. We shall cover these three elements in this section.

4.5.1 Size and Resolutions

It would be a good practice to determine the desired size of each figure in advance. Most journals often print their texts in two columns (left and right). If the artwork is small, we can format it so that it fits nicely within the column's width. Using this format, for example, we can make the width of the image less than 9 cm, which will make the image fit nicely to half the page horizontally (a single column width). In cases where a larger figure is necessary, it is important to note that there is typically limited space available, usually no more than 19 cm in width. Therefore, it is recommended to set the figure to a maximum width of 19 cm to ensure it can be accommodated within the journal's layout.

Once the size is determined, it becomes easier to standardize the lettering on all the artwork. Lettering here refers to adding numbers or words to the figures, including (1) subpanel numbering, that is, (a), (i), or (I); (2) axis of the graph; (3) sample name; (4) grouping name; etc. If we do not size the figures to our preference, it is possible that the typesetting staff in the publisher's department will resize (usually reduce) the figures at their discretion in order to reduce the number of pages. Sometimes, resizing can result in inconsistent letter sizes across figures. If shrinking was excessive, the lettering would be too small to read, and thus the entire figure would be meaningless. The final size of the lettering is usually between 7 and 10 pt. depending on the publisher.

The majority of journals and publishers have strict requirements regarding the resolution of bitmap images such as graphs, photos, and illustrations. Lettering resolution should be considered too. Figure or graph pixels at more than 300 dpi are usually sufficient. With a few clicks of the mouse, it is possible to determine the pixel density of an image. If you are uncertain, do a quick check by asking Prof. Google, Siri, Alexa, or Google Assistant. A small figure (300 dpi) fitting into a single column should have a pixel size of more than 1000. By considering 300 dpi resolution, full-width bitmap images will require more than 2000 pixels.

4.5.2 Basic Graph Using Excel with the Aid of Photoshop

In my experience, many students prepare their figures with Excel, print screen them, and crop and resize them by pulling the edge of the figure after inserting them into Word. The results are not always visually appealing. The lack of quality in the figures is not appreciated by reviewers and editors. A low-quality figure in the manuscript will cause me to be less confident with the manuscript, if I were the reviewer. In the following paragraph, you will find suggestions on how to prepare figures that are more professional.

It is recommended that you do not use the default settings in Excel. The gridlines, for example, are ugly. In another example, if we create more than two plots with different volumes of data or parameters, the spacing between bars will be inconsistent. Furthermore, the default colour schemes in Excel may not always appear professional. If our dataset is large, editing the colour of each graph in Excel will be challenging. When we have multiple figures in a microbiome project and the bar plot contains more than 30–50 taxa, it becomes a nightmare. Keeping the colours consistent across numerous graphs is important to make the whole report (thesis or manuscript) appear neat and easy to read. Additionally, adjusting the font type and size is an arduous process if we have five to eight figures in a manuscript or more than 20 in a chapter of a thesis. There are other problems, but I will refrain from describing them.

Even if Microsoft Excel is our only tool, we can still create some high-quality graphs with it, for instance, pie charts. It is important, however, to ensure any manual adjustments are applied consistently across all figures in the same write-up. To be clear, manual adjustment is not data manipulation and adjustments here refer to editing the size, font, colour, line thickness, etc.

Let's see an example. It is common to have a figure with multiple subpanels, such as a figure that includes (a) a site photograph, (b) a SEM micrograph, (c) a bar plot of bacteria relative abundance, and (d) a bar plot of archaea relative abundance. People often insert figures (for this example four images) into Word or PowerPoint and save them as images or PDFs. This approach does work if we know how to make them look neat, consistent, and high resolution. My observation, however, is that many of the time, the compiled figures are unsatisfactory. My recommendation is to export the graph in a format (i.e. tiff format) that is recognized by Photoshop, GIMP, etc. Ideally, the image files should be saved at the highest possible DPI resolution. It may work better with a pie or bar plot, but line charts are more challenging. Make sure to set the size of the canvas and the figure in Photoshop, and type in the lettering in Photoshop (or GIMP, or others) rather than Excel. Combining Excel and Photoshop is a magic trick for ensuring that the size, resolution, and lettering comply with a publisher's requirements. For those seeking a foundational tutorial of Photoshop, YouTube is the best video resource.

4.5.3 Excellent Graphing Tools

It is highly recommended that readers try GraphPad Prisms for their data analysis needs. It is more accurate to use Prisms to compute best-fitted lines and other types of line graphs. Of course, Prisms can handle other types of graphs. For biologists who may lack formal training in statistics, I encourage you to take advantage of the tutorial videos available in Prism Academy, which cover biostatistics topics. Prisms software allows users to easily use statistical analyses (such as t-tests, ANOVA, and many more).

Prisms displays graphs in a pleasing colour scheme. There are several graph types that are common for biologists, including subcolumn graphs, violin plots, box plots, etc., and all can be created with a few clicks. The Prism graph software has a very unique feature called Prism Magic—once certain attributes of a sample graph have been set (e.g. size, colour, fonts, angles, ranges, line thickness, etc.), all subsequent plots will be formatted correctly. I also like Prism because the images are of excellent resolution. My personal preference is to keep data in Excel and use the graphing function for simple viewing. For publication, I suggest you import data from Excel and generate graphs using Prism.

The Originlab tools are an excellent platform for scientific and engineering graphing. Within Originlab, you can visualize the physicochemistry of water using a Piper diagram and analyze soil parameters using a soil and rocks classification diagram. Heat maps with dendrogram, Venn diagram, volcano plot, and enzyme kinetics calculation are also available in Originlab. Furthermore, OriginLab contains statistical tools such as principal component analysis (PCA), Chi-Square test, and many more.

The JMP (pronounced jump) software is another excellent tool if we have huge datasets. Besides, we can also consider SigmaPlot and MATLAB. Prism, Originlab, SigmaPlot, MATLAB, and JMP are all paid programs. Signing up for an OriginLab learning edition allows students to use the software for free for 6 months, and an academic license at a reduced rate is also available for some of these software.

No budget? No worry. We can generate graphs freely by using the tool function on the website of the National Genomic Data Center (NGDC). The system provides a wide variety of plot types, including bar, box, comutation, dot, forest, heatmap, histogram, lollipop, MAF summary, Oncopoint, survival, Maftools Titv, Upset, Venn, Manhattan, and Q-Q plots.

Readers who possess a basic understanding of computer language or are willing to master basic command lines can use R. The R-package is a free software environment for statistical computing and graphics. For example, a complete R Graph (https://r-graph-gallery.com) can generate the following chart types: violins, histograms, boxplots, Ridgelines, scatterplots, heatmaps, bubble plots, lines plots, 2D density plots, bar charts, radar charts, parallel coordinate charts, lollipops, pie charts, dendrograms, circle charts, various types of maps, Chord diagrams, network plots, Sankey Diagrams, animated data plots, interactive graphics, and 3D charts. The names of these plots may not be familiar to us, but you must have seen some of these in good journals and may have wondered earlier how to create such graphs. R is the answer. To begin, readers can explore ggplot2, CRAN R, or others. The R language can be learned online through tutorials. Join R consortium that runs programs like R-hub, R-ladies, RUGS, and R-Idea if readers require more assistance and wish to be more active in the community.

4.5.4 Tools Related to Biological Figures

In recent years, there have been more and more tools to draw biological figures such as cells, cell structures, epithelia, membranes, and transporters, or to depict biological pathways in an illustration. You may wish to try the free education version of BioRender. Please note that the free version does not allow us to publish the figure in a journal. There are additional features available in the paid version of the software, including high-resolution export without a watermark, as well as the ability to publish in journals.

It is common to build a phylogenetic tree or a dendrogram. Before creating a tree, please ensure you understand the differences between UPGMA, neighbour joining, maximum parsimony, maximum likelihood, and others since each method is designed for different applications. PHYLIP, phyloT, EMBL-EBI tools, iTOL, and many others are some of the common approaches for constructing trees. It is sometimes necessary to improve the default display tree. As such, I recommend that readers use software that allows us to adjust the thickness of lines, fonts, resolution, or other parameters in such a way that the tree meets the basic figure requirements of journals. MEGA is an example of an easy-to-learn software. MEGA, however, has some limitations in terms of lettering. For example, some font types required by some journals are not available in MEGA. If this occurs, you will probably want to save the tree in nwk file format and open it with another application, such as Figtree. MEGA does not permit users to combine italicized and normal lettering; however, sometimes we need that feature when naming genus, species, and strains. We may consider creating the tree in MEGA and doing the lettering in Photoshop, GIMP, or other tools. If readers would like to follow this approach, they must do so in an ethical manner in order to ensure the accuracy of the information. The original tree must be kept if there is a question about it.

The concept of motif and domain is well known to protein scientists. InterPro is not new to them when it comes to determining the position of domains in a query protein sequence. These domains are typically displayed as rectangular boxes. The screen capture of InterPro result does not provide us with a good representation. Similarly, it is common for researchers to show the direction of genes or gene clusters with arrows. Online RAST is one of the common tools used for this task. Again, screenshots should be avoided. Redrawing such figures to the scale is therefore recommended. How can this be accomplished? We can do this with Word or PowerPoint, but Photoshop or the like would be better.

4.5.5 Publisher and Third-Party Artwork Service

A variety of publishers and companies offer artwork services, including Springer Nature, Elsevier, ACS Authoring Services, Editage, LetPub, bio-graphics scientific visualizer, SciStyle, FlatWorld, Arkitek Scientific, Fiverr Graphical Design, Medical Illustration Company, Mind the GRAPH, and Scigraphix. Another service provider, MolGraphics, collaborates with authors to prepare cover designs in *Nature Materials*, *Nature Chemistry*, *Nature Catalysis*, *Cell Systems*, etc. Please conduct your due diligence before performing the transaction.

4.5.6 Artificial Intelligence Artwork

We have probably heard about Alphabet's or Google's AlphaFold, an artificial intelligence program that predicts protein structure. It is likely that most readers were not aware of Google's artificial intelligence, Imagen, which draws a figure from the text. It is unclear whether Imagen can generate biological artwork at this point in time, but we can never tell when artificial intelligence might become our best friend in scientific graphing and imaging. If you are interested, you may find more information on Imagen, DALL-E, LAFITE, GLIDE, and few other tools listed in the appendix of this book.

4.6 Intellectual Property Protection

Intellectual property (IP) is not commonly placed in the planned tasks or achievements for most researchers in comparison to journal publications. The Cambridge Dictionary defines IP as "someone's idea, invention, creation, etc., that can be protected by law from being copied by someone else". World Intellectual Property Organization (WIPO) defines it as "Intellectual property (IP) refers to creations of the mind, such as inventions; literary and artistic works; designs; and symbols, names and images used in commerce". Examples of IP are copyright, patents, trademarks, industrial designs, geographical indications, and trade secrets.

4.6.1 Copyright

A copyright is granted to the creators of creative, artistic, or literary works which include academic articles. Are you aware that there is a transfer of copyright between the authors and the publisher for conventional models of publication (subscription models)? Most publishers permit the authors to retain the copyright in an open access model to a certain extent. An article that has a CC BY 4.0 note, for example, indicates that the authors are free to share the content in any medium or format. Moreover, the authors have the right to "remix, transform, or build upon the material for any purpose, even commercially". Visit this website (https://creativecommons.org/licenses/) for more information about Creative Commons licenses. Additionally, for most biotechnological research, copyright is not too relevant for most research scopes, unless we are creating legal protection for a logo which we are creating for a product. Viscozyme©L, Promozyme© D2, and Optimax© L-1000, which are all commercial enzymes, have the copyright logo. Similarly for trademarks, it is not too relevant for most research. In addition, it is crucial to be aware of fair use, a concept that allows limited use of copyrighted material without explicit permission. Fair use specifically pertains to activities like teaching, research, and thesis writing. It permits the incorporation of excerpts, quotations, or references from copyrighted works, striking a balance between the requirements of educators and researchers and the rights of copyright holders.

4.6.2 Industrial Design and Geographical Indications

A design that is protected by the law is known as an industrial design; however, such protection is relatively weak. A geographical indication is another IP type. Geographical indications cover agricultural products, foodstuffs, wines, handicrafts, and industrial products that are connected to their place of origin, for example, Darjeeling tea (where Darjeeling is located in the Eastern Himalayas) or Sabah tea (where Sabah is located in Malaysia). It is my opinion that none of the findings of laboratory-based bioscience are relevant to geographical indication protection. A trade secret is another type of intellectual property. Once again, I am unable to see any connection with common biological research.

4.6.3 Patent

The patent is probably the most relevant type of IP protection for most biotechnology researchers among the six types discussed briefly above. Researchers involved in gene cloning and enzyme research can protect the genes and protein sequences of native and protein-engineered sequences developed by the inventors. Patents of this type include US9695407B2 and US9737597B2. It is possible to patent a composition, for instance US9322012B2, which claims a composition which consists of enzymes and functionalized nanomaterials. It is also possible to patent a method or process. For example, the inventors of WO2017203440A1 are protecting the method of producing probiotic biofilms that contain *Bifidobacterium* and *Lactobacillus*. In the patent application US977037B2, a method is described for a fermentation product that, when applied to plants, results in higher growth. The patent application US9474776B2 purported to describe a method for improving bee survival by integrating fungi. Not every aspect of biotechnology, however, is eligible for patent protection. Discoveries of naturally occurring substances or simple scientific principles, for instance, are typically not patentable.

Patent laws and procedures vary from country to country. If the maintenance fees are paid on time, a patent can remain valid for up to 20 years from the filing date. Applicants should file a patent application before making any public disclosure, including publication in a journal, submitting data to the NCBI database, or presenting at a conference. A published journal article, for example, can turn out to be prior art that prevents a patent application. Yet, Singapore, Malaysia, Canada, Argentina, Japan, and the United States all have a grace period of 12 months, while some countries may operate a grace period of 6 months. Therefore, researchers usually submit patent applications before they submit very similar findings for publication in a journal. It is possible that we may face some difficulties if we reverse the process, particularly if we have exceeded the grace period stated earlier. If you are unsure about the message here, it is advisable to contact your patent officer at your university and they may be able to provide you with better answers and instructions.

Patents have traditionally contained novel elements, involve invention procedures, and have an industrial application or solve some existing problem. It is assumed that inventors are human; however, this may be debatable today. An article published in *Nature* indicates that AI systems have been included in patent applications as the inventor. Perhaps it is time for patent offices, governments, and all stakeholders to come together and enact patent legislation regarding artificial intelligence (Alexandra George and Toby Walsh 2022).

4.7 Develop Reading Habit

4.7.1 Chefs Eat Researchers Read

It is common to see photos of celebrities and celebrity chefs displayed on the walls of hawker stalls, restaurants, as well as on stalls selling street food in Singapore and Malaysia. If the owner notes that the customer is a famous individual, it is the ideal time to take a selfie, print it out, and use it probably for publicity. There is a land bridge between Johor Bahru (JB) town (Malaysia) and Singapore, and it usually takes 1–2 h from Singapore to JB during a normal working day; however, the time can increase substantially during a long weekend or public holiday. Celebrities and celebrity chefs in Singapore enjoy travelling and are willing to travel across the country in order to taste JB's food. I recall watching a cooking show where the chef said: "I love eating food prepared by someone else and I learn how to be creative and improve my own cooking from there". I believe, for a chef, a cook, or a housewife who does most of the cooking, eating outside food, especially a variety of inexpensive nice food, will significantly improve their cooking abilities. This willingness to explore different culinary experiences reflects their passion for gastronomy. Would you agree? How about students and researchers? The cultivation of a reading habit is essential for both students and researchers. Regular reading increases knowledge, vocabulary, and critical reasoning abilities. We should make reading a habit.

At a workshop I participated in, a young academic staff asked me for advice or tips regarding conducting research and publishing. Reading is my first recommendation. Postgraduate students who do not read enough have difficulty writing their theses. Some students take up to 2 years just to write a thesis or dissertation. My thesis was 200 pages long and I wrote it in about half a year. I was limited by the language, but not by the knowledge. A lot of students have good command of English writing, but lack knowledge, and therefore, they read as they write. An individual who writes poorly in English and is not accustomed to reading journals will be in the worst situation. A principal investigator, postdoctoral fellow, or any researcher who reads less will certainly feel challenged when writing the discussion for his or her manuscript.

My opinion is that reading is a portal to knowledge and any reading materials are mirrors and quiet places for reflection. Students and researchers ought to have a habit of reading and we should enjoy reading. It is analogous to a situation where a chef is hungry for nice food and is willing to travel great distances while stuck in bad traffic to obtain a bowl of fish ball noodle soup, *kway teow* soup, roasted duck, curry fish head, spicy nyonya meal, *roti canai*, *nasi lemak*, *nasi beriani*, *sup tulang*, or a box of banana cake. I have to admit that I love food, but I am a reader more than a food lover. Just like a hungry food lover seeking food, I will search through any available resources (journals, books, etc.) to answer a scientific question, to solve a technical problem, or to satisfy my curiosity. In the News section on Nature Index, published on 8 September 2020, the title of the news is "This scientist read a paper every day for 899 days" (available online). According to Olivia Rissland, who wrote the sharing, reading makes her a better scientist. Rissland used Google Sheets to keep records and felt gratified after accomplishing the goals of each day.

4.7.2 Reading Information Provided by Suppliers

Students are encouraged to read supplier manuals, eBooks, or online materials in addition to journals. In addition to journals, these resources provide valuable information for their studies. Inexperienced students may have difficulty understanding the content of journals due to technical jargon or the lack of prior knowledge in the field. To overcome this, students can supplement their reading with textbooks, which offer explanations of basic concepts and terminology. Websites of suppliers provide a wealth of information. Several resources are available on the Illumina, PacBio, and Nanopore official websites if your project involves next-generation sequencing. Students can find resources on Promega, New England Biolabs, and ThermoFisher Scientific for molecular and protein-related projects, to name a few. We will find everything we need on the GE Healthcare website if we are new to protein purification.

Furthermore, I read documents that explain troubleshooting because they help me enhance my critical thinking and problem-solving abilities. For example, we could conduct a Google search using the following keywords: troubleshoot PCR and RT-PCR, troubleshooting HPLC, troubleshooting GC, troubleshooting flow cytometer, or anything relevant to us. We will learn what the common problems are and how to solve them.

GEN Protocols is an online resource for academic researchers interested in learning about the latest developments in biotechnology. It provides discussions, opinions, protocols, software tools, examples of applications, and tutorials in the fields of drug discovery, analytical testing, cancer research, computational sciences, infectious disease research, -omics, translation medicine, and genome editing.

I also make it a habit to read Instructions to Authors before submitting a manuscript. It seems that many people overlook this important step.

4.7.3 Are We Hypocritical?

Developing Reading Habit is the title of this subsection. In the past, academic articles were printed, compiled, and placed in libraries as hard copies; today, articles are published online. Access to journals was difficult and time-consuming in the past. Now, the issue is that there are far too many published articles and people are unwilling to spend time reading them for whatever reason (or excuse) such as "I have got too much on my plate", that simply means too busy, too many things to deal, or have too many things to worry about. The primary issue seems to be a lack of reading habits.

Please accept my apology for writing this. We may scold our students with the following phrase: "You are lazy and you do not read journals". A finger is pointed towards the student while four fingers point towards our direction. Dear supervisor (including myself), do not be hypocritical. It is easy to tell whether a student has read enough or not. Similarly, as we talk, students will sense that we have not done our parts in reading. It is with regret that I write something that may come across as a bit harsh, but my intention is to create awareness. Rather than pointing fingers solely at students and labeling them as lazy, let us reflect on our own role in shaping their learning experience. We should encourage students to explore a wide range of resources, including journals, to deepen their understanding and enhance their research skills. Let us break the cycle of the blind leading the blind by prioritizing our own reading habits and nurturing a passion for knowledge within our students.

4.7.4 Making Ourselves Read

The following suggestions may assist in developing reading habits, especially for busy lecturers and researchers. It is a good idea to print out research and review articles and keep them in your car or bag. During a breakfast or afternoon tea break, we can probably read a few paragraphs while waiting for the food to be served. It is also possible to read while travelling in public transport to and from our home or lab. Some may say that it is too noisy. In the past, I was easily disturbed by noises, and I was unable to concentrate on reading, but now I have developed a habit of being able to pay attention, regardless of the noise around me. It is likely that we can complete an article within a few days if we spend 15–30 minutes a day reading and at least one or two journal articles each week if we do so. Many readers may ask, "Huh, print? Can I read online materials?" I belong to the older generation that prefers paper documents. Print engages the senses and makes readers more focused. No problem, you can go digital if you wish.

Be keen to accept an invitation to participate as a reviewer. In my younger years, when I had ample time, I would accept any invitation from relevant journals ranging from non-indexed journals, Scopus, to submissions, to the WOS Q1 - Q4 journals. My highest review rate was about 50 manuscripts annually, but now I am more selective and review approximately 20. Am I too free? No, definitely not. I am forcing myself to read the first-hand information directly from the authors' source. It can be painful to read poorly written manuscripts; however, it made me reflect on why I considered such a write-up as weak and how I can prevent those errors in the future when I sit down to write my own manuscript. Consequently, I am training myself to become a better author.

Start by serving as an associate or guest editor for a decent journal if we are qualified to become one. Besides the benefits of advancing our academic career and potentially expanding our professional network, working as an editor allows us to read relevant research articles in advance of other colleagues. The principle of 'it is more blessed to give than to receive' holds true here as well. However,recent news on Nature Briefing (2023) has highlighted a concerning under-representation of female scientists in editorial roles, with less than 15% serving as editors and less than 10% as chief editors (online source: 'Largest-ever study of journal editors highlights "self-publication" and gender gap'). It is not a good sign, but I am aware that most journals' editors are trying hard to ensure gender equality in the journal. Do you mind finding a way to become a member of the team, ladies? Your participation will benefit not only you but also the entire scientific community. By taking on these roles, you can help bridge the gender gap in scientific publishing and ensure a more diverse and inclusive representation of perspectives.

Writing a review is another effective method for forcing myself to read. For additional information on what a review is and who should write it, please refer to the chapter on Publication 101. In order to write a decent review, I usually read 100–150 documents. Reading will become a habit once the ball starts rolling.

As I browsed through the internet, I found some other suggestions for developing a reading habit for busy individuals. Among the ideas are to listen to audiobooks, to purchase an e-reader or tablet, to commit to reading 30 minutes a day, do it consistently for a minimum of 21 consecutive days to establish a new habit, to look for breaks in our schedule to read, to read before going to bed, to join a reading challenge, to always carry a book, to unsubscribe from Netflix, and to get away from social media. If I may rephrase it, do consider reducing the amount of time spent on social media. In recent years, there have been few publications that address social media addiction and the implications for future research, job performance, self-esteem, and life satisfaction. Since I am not a social science scientist, I am unable to suggest reading materials. Nonetheless, I am aware that social media is taking away our time for reading. Taking a look at our children, we are probably no different. A PhD candidate shared his experience in a Career Column published in *Nature* on 15 February 2022 concerning TV and telephone addiction which caused him great difficulty. I am glad that he walked out of the trap after taking time to reflect and seek counselling. In case you are interested in reading more, the title of the article is 'Turning off my phone improved my science'. This article has been recommended to many postgraduate students in my department. Several students emailed me to say they are experiencing the same problem of phone addiction, which is causing them to pay less attention to scientific matters. Feed our minds with knowledge, not garbage. It is recommended that you read Cal Newport's book *Deep Work: Rules for Focused Success in a Distracted World* if you have serious problems with that.

Besides journals related to our research field, we should broaden our reading scope to include self-help books and books that expand our horizons. Some of the books include *The 7 Habits of Highly Effective People*, *Atomic Habits*, *GRIT*, *The Art of Thinking Clearly*, *The Power of Positive Thinking*, *Blue Ocean Strategy*, *The Success Factor*, *Unreasonable Success and How to Achieve It*, *Surrounded by Setbacks*, *The Bullet Journal Method*, *High Performance Habits: How Extraordinary People Become That Way*, *Organize Tomorrow Today*, *TED Talks*, etc. Another interesting book is *So Good They Can't Ignore You: Why Skills Trump Passion in the Quest for Work You Love*. You might also be interested in *Deep Work* by Cal Newport or *The Power of Full Engagement* by Jim Loehr. The book *How to Stop Worrying and Start Living* by Dale Carnegie is old (1948), but the content is still relevant today. Apart from these books, *GRIP: The Art of Working Smart* by Rick Pastoor is one of the books I highly recommend. In the following sections, I will elaborate on a few of the most important advice from these books.

Even if you have no budget to buy books to read, no worries, if you search the names of these books on YouTube, Audiobook, or on certain podcast platforms, you will get some information. Without reading the entire book, you will at least be able to grasp some basic concepts. After watching YouTube, if you decide to buy the book, go ahead. As a YouTube user, I like the fact that if I watch a video clip, say about *The Bullet Journal Method* or *Atomic Habits*, etc., the machine records it, and if I watch another similar video on the same motivation or time-management theme, YouTube will suggest some other videos. If you do not know what book to read or watch on YouTube, try browsing the channel 'Productivity Game' on YouTube. We should feed our minds with knowledge rather than garbage. As soon as we search for garbage in YouTube or any social media, the AI behind it will remember our habit and suggest us more garbage based on our previous searches. It is difficult to define what garbage is here, but I hope it refers to anything that is not helpful to us, taking too much of our time, and making us lazy or addicted to it.

In the age of digital media, where video content has become increasingly popular, it's essential to find alternative avenues for expanding our knowledge and embracing the joy of learning. While some may prefer videos over traditional books, I encourage those individuals to explore the captivating realm of educational YouTube channels. Among the vast array of options, two channels that stand out are 'Life in Academia' led by Matthias Rillig and 'Andy Stapleton' channel, named after the person in charge of the channel. each offering a unique blend of knowledge and inspiration in their respective fields. Matthias Rillig is a distinguished professor of ecology at Freie Universität Berlin. This channel serves as an invaluable guide for aspiring researchers at all stages of their academic pursuit. Matthias provides a wealth of knowledge and practical advice, covering topics ranging from publishing and writing to research strategy and overcoming challenges. With tailored playlists for doctoral students, postdocs, and new PIs, there's something for everyone seeking to thrive in the academic realm.

4.7.5 Using AI to Make Summaries

The act of reading makes us a better scientist. Students and young researchers are advised to read conventionally and not to take shortcuts. However, there are situations where shortcuts or assistance may be warranted. For instance, if a researcher has previously read a full journal article but cannot recall the main points and lacks the time to reread it, artificial intelligence (AI) can provide support. AI assistants can help retrieve relevant information, but it is important to exercise caution. AI is still in the process of learning, and its suggestions may contain scientific inaccuracies or biases. While the text suggested by AI may appear faultless at first glance, its scientific validity must be evaluated critically. Dear readers, I was hesitant to write this subsection, but I ultimately decided to include it. The earlier sentence suggested a boundary of how AI can be used; however, if you think carefully, an individual can manipulate AI assistants to take many shortcuts that are not ethical. By establishing a balance between embracing technological advances and cultivating our own scientific acumen, we can continue to maintain the integrity and excellence of our research.

Several AI tools are available for summarizing long documents. For example, the system allows us to upload a journal article to the site, and the tool highlights the possible points of importance. Examples are ReadTune, Iris.ai platform, ChatPDF, and many others, some of which are included in the book's appendix. Artificial intelligence is still a machine that continues to learn, and it may not be able to detect our needs or what the scientific community might require in terms of knowledge. It is highly likely that AI will misunderstand the original text or miss the important points. It is important to stress that artificial intelligence is only a tool, and it cannot replace the process of reading and learning.

4.8 Translating Planning to Action

4.8.1 Power of the Calendar

GRIP, written by Rick Pastoor, was originally published in Dutch in 2019 before being translated and published in English (Pastoor 2022). The topic of this book has absolutely nothing to do with bioscience or biotechnology research. Instead, the author discusses ways to work strategically and transform planning into action. The book is well worth reading. I realized that I have been using some of the techniques described in the book since my PhD and continue to use them today. In addition to *GRIP*, there are other books that teach a similar strategy, such as *The Bullet Journal Method* and *Organize Tomorrow Today*.

Rick Pastoor's first important suggestion is to place our plans on a calendar and to follow them. In the early stages of my PhD, the laboratory supply companies provided us with table calendars or diary calendars, and digital calendars were not available at that time. I have made it my habit, since then, to write my short-term (3–6 months), midterm (6–9 months), and long-term goals (about 12 months) on the table calendar and place it on the lab bench. My undergraduate background was in bioprocess engineering (B. Eng.), and my postgraduate project focused on protein engineering, which involved basic bioinformatics such as sequence analysis and primer design, gene cloning, protein expression, and protein purification, as well as biochemical characterization of the native and mutant enzymes. As I have only previously trained in protein purification, I have much to learn. As a midterm goal (6 months), I have decided to learn molecular cloning techniques and theories. I have placed these goals on the table calendar. I have set priorities in regard to action goals such as when to make a purchase, when to begin an experiment, and how much time I need to devote to self-reading before learning a new laboratory skill.

Here is an example of the goals and please try to visualize how it would look on a table calendar.

- 1 January–28 February. Finish reading Sambrook's *Molecular Cloning* book (theory and protocol) and read Promega's protocols (at that time, our laboratory was using only Promega and Fermentas products).
- 15 March. Selecting restriction sites and designing primers. Place orders.
- 30 April. Learn the proper use of the PCR machine, electrophoresis, PCR purification, restriction digestion, ligation, and transformation.
- 1 Jan–30 May. Read about the concepts and approaches in protein engineering, including site-directed mutations, random mutations, chemical approaches, etc.
- 30 June. Completion of basic training on IPTG induction.
- 30 Sept. Conduct mega-primer PCR and overlap extension PCR for mutagenesis.
- 31 Dec. Try to get five mutant proteins.

Initially, the plan was (1) I would need about 3 months to learn the basic molecular skills, (2) another 3 months to decide which amino acids I should modify, and (3) at the end of the first year, I should have some mutant proteins in my hands. Even my wedding date and my honeymoon date were noted on my calendar, along with the dates on which I would be able to take a short break. I became accustomed to using my calendar as a navigational tool.

Surprisingly, I completed the construction of 12 protein mutants in 9 months and that was 3 months sooner than I had anticipated. Another group in another neighbouring university was working on a similar project, but with a different protein. It was unclear to them how 12 mutants could be generated in such a short period of time. Such a feat is not due to luck, as I do not believe in luck. As our laboratory facilities were not that advanced at the time, many protocols were conventional in nature and prone to failure. Since I was lacking approximately 90% of the skills I need for the project, I have to admit that I was not smart.

Two reasons contributed to the shortening of my milestones: (1) hard work, since I worked on weekends and (2) putting plans on the calendar and turning them into action. By using a calendar, I was able to set priorities and maintain focus. My calendar was filled with activities; however, there are a few gaps in between for buffering purposes. If I fall behind schedule, my only alternative was to make additional efforts to catch up. Using a physical calendar is no longer an option for most readers. You can use any digital calendar such as Google Calendar if you wish.

The importance of writing down schedules, to-do lists, meetings, etc., in a calendar cannot be overstated, regardless of whether you are a student, postdoctoral fellow, academic, or full-time researcher. Do not let things float in our minds and burn our memories. Put them in writing.

GRIP, written by Rick Pastoor, suggests that we leave 20% of each month, week, or day open. In this way, we can accommodate unexpected meetings or events. In addition, we need to learn what the Eisenhower Matrix is and how to prioritize our tasks. (Note: In the Eisenhower Matrix, there are four categories: urgent, not urgent, important, and not important.)

Rick Pastoor reminded us in his book that goals should be SMART—Specific, Measurable, Achievable, Realistic, and Timely. This quote is taken from Rick Pastoor's book on page 132: "Lots of people have big plans; not everyone manages to bring them to fruition. The trick is to practice breaking down big ambitious endeavors into little doable tasks. You'll be amazed at how far this simple skill will take you and how much you'll accomplish". It does make sense, doesn't it?

4.8.2 Think Win-Win

The 7 Habits of Highly Effective People by Stephen R. Covey is an easy to read and understand book. The original version was published more than 25 years ago. Whether we are students, researchers, or principal investigators, we may find it useful to adopt some of the habits described by the author. Here are some examples. Habit 2 is to "begin with the end in mind". This concept is in accordance with GRIP. In addition, in a subsequent chapter, titled 'Publication 101', I share with you how to plan a publication from the very beginning.

"Think Win-Win" is Habit 4. The author Stephen R. Covey believes that high-trust relationships facilitate collaboration more effectively. The work of a good researcher goes beyond the lab apparatus, equipment, kits, data, and publications. Interactions and collaborations with other people play a significant role as well. Do not limit collaboration to co-authorship, co-funding, or co-applying for grants. Consider Win-Win as a bigger picture. We may wish to consider the following human relationships in the research journey, for example, student—supervisor, student—PI—support staff, PI—supplier, one PI to another, and the relationship between a student and another student. If we have a genuine sense of Win-Win, things will run more smoothly. Put into practice what we have learned about the Seven Habits. It is not surprising that people who are highly productive and those who are less productive have distinct habits. Try James Clear's book *Atomic Habits* if you are trying to create a positive habit but do not know where to start.

4.8.3 Keeping *Blue Ocean Strategy* in Mind

In late 2007, my employment as an academician was just starting, and I was deciding what path to pursue for my research until my retirement. The former vice chancellor of my university is a big fan of the book *Blue Ocean Strategy* written by W Chan Kim and Renée Mauborgne. He shared his thoughts on the Blue Ocean Strategy for months and I was deeply influenced. The term 'Blue Ocean Strategy' refers to a market for a product in which there is little or no competition. Drawing inspiration from this concept, I began contemplating the idea of conducting Blue Ocean research, avoiding areas where my department's staff and other researchers in the country were already heavily involved. I avoided the red ocean research areas in which many other people have embarked. If I step into those areas, I may lose out. We can attempt to build out our own distinct niche and avoid needless rivalry that can impede our success by avoiding oversaturated research areas. With this strategy, we may explore uncharted territory and carve out our own niche in the field of research, guaranteeing that we have the freedom to contribute significantly without being hampered by rivals already in the market.

The Scopus database contains more than 200,000 documents related to the bacteria *Bacillus*. There are hundreds of *Bacillus* species listed on the LPSN.dsmz.de website. When beginning a new research project, if we had just obtained an isolate of *Bacillus*, the first question we would ask is whether this isolate is worthy of further research. There are several other factors to consider, such as how likely it is that bacterium is a new species, have others carried out similar research to what I am planning to do, and how easy it will be to publish? In my opinion, we can classify *Bacillus* in the red ocean according to the microbiology aspect. There were several *Bacillus* isolated by my students, but they all ended up in the sauna in my autoclave machine. Anyway, it is still possible to find areas of blue ocean in *Bacillus* if we carefully examine the research gaps. If you have been working with *Bacillus* stock cultures for decades, then do not throw them all away in the trash cans prior to autoclave. If you have been working with them for decades, then they are diamonds. However, if you are leading a new lab for the first time, you may want to isolate some other genera to begin with.

Whenever my students isolate a bacterium, sequence the 16S rRNA, and cannot pronounce the genus name, they become concerned because there have been few publications. Nevertheless, I would be delighted. It has been many years since one of my students isolated a halophile, but the number of publications found in Scopus and Web of Science was less than ten. She was so apprehensive about working with something that was relatively unknown. This is a blue ocean area. The majority of students will have this kind of fear; even young PhD researchers will have this kind of concern as well. It took me a while to persuade her to pick up the project and start working on it.

Many research gaps, applications, or potential commercially motivated research opportunities exist in the red ocean fields. But we must understand that the competition is relatively high, from securing a grant to publication. The novelty of our idea may soon lose its novelty, as someone in another part of the world has the same idea and may publish it faster and do it more efficiently than we do simply because they have more resources. Doing a similar project within a faculty or university is a double-edged sword. We may be able to work together as a team and that is great. If two principal investigators (PIs) are not cooperating, it is more likely that students are experiencing stress due to competition, protecting ideas from being stolen, one of the PIs may be the viva voce examiner as they have similar areas of expertise, etc. A healthy dose of competition among PIs within an institution can keep things on track if the PIs are of high integrity and ethical behaviour. There are, however, some disadvantages as well.

In Rice Stadium on 12 September 1962, President Kennedy delivered his Moon speech. It was meant to convince Americans to support Apollo—the Nation's Space Effort. The text below is part of his speech, and you can search the whole thing online. "We choose to go to the moon in this decade and do the other things not because they are easy, but because they are hard. Because that goal will serve to organize and measure the best of our energies and skills, because that challenge is one that we're willing to accept. One we are unwilling to postpone. And therefore, as we set sail, we ask God's blessing on the most hazardous and dangerous and greatest adventure that man has ever gone. . . .". I personally like this phrase ". . .not because they are easy, but because they are hard".

There is so much fun, challenge, and attention required when working on blue ocean projects. Do I mean to imply that the best way to conduct research is to concentrate our entire life on exploring blue oceans? No, definitely not. Despite the benefits, there are several drawbacks. Funders and peers need to be convinced why this project is so special because most of your peers may not be familiar with the term, scope, or subject matter you are referring to. It is likely that the journey will be a bit lonely since there may be fewer people with such an interest; therefore, finding a collaborator may sometimes be challenging at the start. As a reminder, being different can be very isolating, and working towards the goal can be a difficult task.

In view of the novelty of the work, publication is relatively straightforward if the experiment is well planned and the manuscript is well prepared. Please be aware that we will not receive immediate citations as very few people work in that area. Does the list of cons seem to outweigh the list of advantages? That depends largely on your interests, short-term plans, and long-term goals. In order to be on the safe side, readers (new researchers or academicians) might consider balancing the blue- and red-seas. At the beginning of my career as a PI, my resources were split almost equally between red- and blue-ocean projects. As I moved on, I spent close to 70% of my time on blue-ocean projects.

If you are newly appointed as an academician or researcher, you are going to have to choose between two options: either (a) do something easy to generate data, but hard to publish or (b) take on a difficult to complete blue-ocean project, but relatively novel to publish. There is no one who can make a decision for you. Make sure that you read widely, seek guidance from your mentor, and perform all the necessary due diligence.

To summarize, be daring to dream, and to translate those dreams and plans into actions. While dreaming is important, it is meaningless if there is no action taken. We shall transform our plans into realistic goals.

4.8.4 Determine Your Own Fate

You may have heard the phrase SMART goals numerous times. Most of the time, it remains a slogan, doesn't it? Do you have unmet research goals for your yearly resolutions? Or do you not even have a resolution set at the beginning of the year? The goals I refer to in this paragraph are not specific to a particular project or grant but refer to your career research goals as a whole. For those who do not feel that they have control over their research goals, I recommend that you read the book *Your Best Year Ever* by Michael Hyatt. The author recommended SMARTER rules because the normal SMART goals is not good enough. Goals must be Specific, Measurable by setting a habit, Actionable, Risky (to stretch the ability to achieve our goals), Time-keyed and Triggered by setting deadlines, Exciting (clear motivations), and Relevant (Reality check). In some instances, we have too many research goals, and we must ask ourselves, is this the right time to pursue this goal? If the answer is no, wait until a better time is available.

Do you, dear readers, have control over your own destiny in research, especially if you are a green PhD who has just graduated and is working in a particular organization? Put an end to wasting your life and take control instead. The fate of a researcher could be classified into three categories (Fig. 4.1).

The first category (a) consists of planning a goal, developing a strategy for it, and achieving the goals you set. The life you lead is completely in your hands. In the second case (b), you may have a dream but do not dedicate enough time and effort to accomplishing it, or you are lacking skills, and as a result the goals will remain as a daydream. There is a third category (c) of people that do not have a goal and a plan, and it is those people who just do nothing and someone else will take up their time and force them to work on a project that they are not interested in. The life you lead as a researcher is not controlled by you, and you do not live a life for yourself. Your time will be spent fulfilling other people's goals rather than working towards your own. How would you classify yourself, a, b, or c?

If you feel that you fall within category (a) as depicted in Fig. 4.1, congratulations. Probably, if you fall into case (b), you have always felt lost in the wilderness and never reached the promised land, then something is wrong with one of the possible reasons: self-belief, planning, execution, preservation, skills, time management, or something else. Those who hit at least two jackpots should read the following section and Chap. 8. In the third category (c), if you do not mind, your colleague or boss does not mind, and the KPI system does not mind, there is no problem. In my conversations with young researchers, many believe that they have no control over their destiny. It is common for them to claim they are in category (b) because of the amount of work they are given. The reason they end up in category (c) is that there is a powerful big boss and it is nearly impossible to escape. It is also

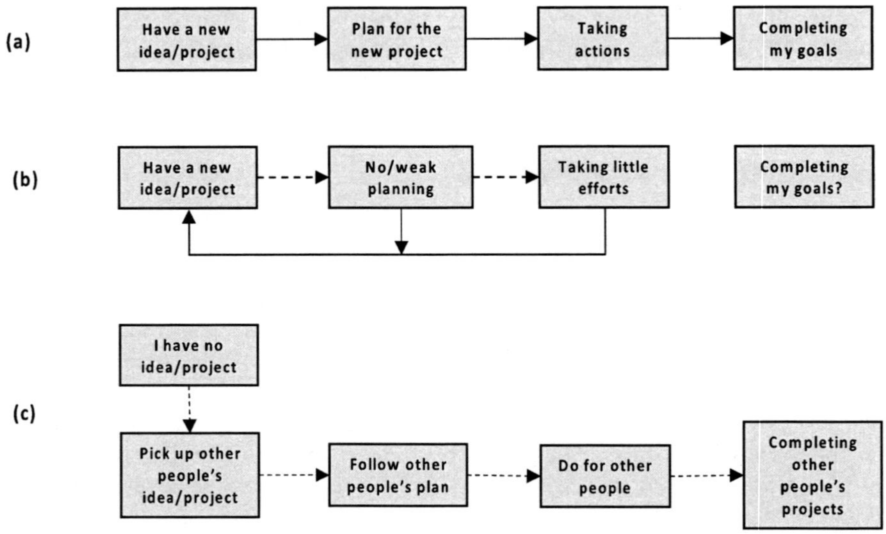

Fig. 4.1 The three types of researchers

important to remember that people in category (a) have bosses as well, who give them tonnes of tasks to do as well. A single research organization or department may have an imbalance in the ratio between categories (a), (b), and (c). It is likely that the majority of people in your workplace do not belong to category (a). There are some people who are able to control their fate (a) and there are others who are unable to (categories b and c). Under normal circumstances, PhD holders should have the power to decide their own fate, if their institutions allow it. Find out what the traps are in (b) and (c) and avoid them if you wish to get into category (a). Probably someone from your group (a) or a mentor will be able to offer you advice if you have no clue what these traps are.

There may have been academicians who spend all of their time writing grant proposals, taking new PhD students, conducting research, and writing and publishing. And do nothing else beyond that. It is completely fine if your tenure is only limited to that. However, in reality, we are required to do a wide range of tasks. To me, it is inappropriate if we spend 99% of our time doing research and do not bother to teach well, to attend meetings, to conduct audits, or to do anything else that will benefit the institution. By taking control of our research destiny, it does not imply that we surrender our other responsibilities to the poor colleagues, resulting in their failure as researchers. We must still complete tasks that fall under our responsibilities, even if research is our first priority. Only then will your colleagues and peers regard us with respect. As long as everyone does their own job, and controls their own destiny in research, the whole institution will progress together.

4.9 Lifelong Learning Tools

4.9.1 Publishers Upskill Tools

Recent decades have seen rapid advances in science and rapid changes in technology. Because of this, everything seems to grow so rapidly and information is exchanged so quickly. Continual improvement is necessary. A lot of talented and smart people are around in the academia and research worlds, but, as long as they remain comfortable in their comfort zones and do not make any significant changes to improve, it is likely that they will be left behind. A systematic lifelong plan is the only way to nurture oneself in order to achieve higher levels of success. Continue to dig. The learning process does not end after we receive our master's or PhD degree (Fig. 4.2).

The Cambridge Dictionary defines lifelong learning as "the process of gaining knowledge and skills throughout your life, often to help you do your job properly", while upskill simply means to learn new skills. In this digital era, we are now able to access many excellent online resources and many of these are either free or come with minimum cost. You may consider the following tools that may enhance your knowledge and skills.

The owner of Nature Masterclasses is Nature Research, a division of Springer Nature. Initially, it only included training in scientific writing, and then Nature Masterclasses added other modules, such as peer review, collaboration, and data management, and currently there are at least ten distinct major areas. The module on Persuasive Grant Writing, for example, has three online classes moderated by nine experts and a total of 3.5 h of learning time. There are some free classes; however, complete training requires either a laboratory subscription (up to 15 participants), departmental subscription (up to 50 participants), or institutional subscription, with all subscriptions taking effect on a yearly basis. You can check if your institution is a subscriber to the service.

Fig. 4.2 Keep digging and strive to acquire new skills

Researchers Academy is an online learning platform offered by Elsevier. It is something I recommend even to researchers who have been active for many years. This webpage consists of five main sections: (1) research preparation, (2) writing for research, (3) publication process, (4) navigating peer review, and (5) communicating your research. The first part contains dozens of lessons covering multiple aspects of securing funding and the importance of metrics in boosting funding and building networks. We often write articles using scientific jargon, but in one of the online lectures, we are taught how to write lay summaries so that non-academics can understand them. If you are interested in Open Science or an open-source model, or if you think about creating a social media network, or if you are looking for the basics of managing mental health, then this Researchers Academy is useful for you. Researchers Academy is free to use. On the other hand, Research4life is one of the partners of Researchers Academy. Research4life has the advantage of providing free access to low-income countries and low-cost access to low-middle income countries. Its website indicates that there are over 194,000 online materials aimed at improving teaching, research, health, agricultural, environmental, and social activities. Likewise, AuthorAID is a free global network team that provides training and mentoring to researchers in low- and middle-income countries.

4.9.2 Private Entities Upskill Tools

Editage is a company that has been providing proofreading services for more than 20 years. I have personally found that their proofreading services are efficient, accurate, and thorough. You can find many excellent free online articles, explanations, and advice about a wide range of subjects. Editage Insights periodically conducts webinars, either for free or with a small fee (i.e. few USD). Aside from the core topics covered by Nature Masterclasses and Researchers Academy, Editage Insights writes about researchers and their experiences. As an example, one article, entitled 'How COVID-19 has caused anxiety and sleeplessness in PhD supervisors in the UK', summarized the results of a survey of nearly 3500 supervisors. The supervisors as well as the students were suffering during the pandemic. There are also many other great articles about PhD life, motivational stories, mental health, stories of struggle, as well as many other topics. Many of these stories are inspiring. If you are inspired by some of the articles, I highly recommend you share them with your peers.

One other excellent online platform is Enago Academy, which offers courses covering a variety of topics similar to Editage. Additional resources include BrightTALK and Upskill by Researcher.life (paid yearly learning program). For microbiologists, especially those who work with prokaryotes, BISMiS Live online lectures are an excellent option. BISMiS stands for Bergey's International Society for Microbial Systematics. It hosts leading scientists from the field of microbiology. Videos of the earlier talks are available on YouTube. The free registration and participation in the live sessions provide a great opportunity to ask questions that have been bothering us for a while.

4.9.3 Massive Open Online Courses

If we are seeking technical courses for totally new skills, such as courses in artificial intelligence; big data; bioinformatics; data mining; R programming; MATHLAB; Bayesian statistics; genomics; data science; approaches in analysing microbiomes; Python; or professional development courses such as business skills, networking, or critical thinking, we may wish to look for free or paid courses available through MOOCs platforms. MOOCs stands for Massive Open Online Courses, and some examples of the platforms include MOOCs.org, OpenLearn, edX, Coursera, and many others.

4.9.4 Reverse Mentoring

We will discuss mentorship and collaboration in chap. 7. Typically, mentors are older or have far more experience than us. Have you heard of reverse mentoring? The concept is gaining increasing attention in the business world. In reverse mentoring, the more junior employee mentors the more senior employee. Technology, digital media, social media, software, apps, etc., are all developing too rapidly. Generally, younger generations are more skilled than older generations, but this is not always true. Would you agree that the most effective way for seniors to learn about social media is to ask those who are younger about it? We may be experts in a specific field in which we completed our PhD, but as time passes, new technologies, tools, software, and other means will become available. Do not be afraid to practice reverse mentoring when conducting scientific research. Learn from the juniors.

4.10 Be Visible

Increasing the visibility of our research is important. Societal visibility enhances your superiors' understanding of what you are capable of, and individuals from outside your institution may contact you to be appointed as a journal reviewer or associate editor, collaborator, thesis examiner, research proposal reviewer, or for giving a talk. What you can do to make yourself more visible is outlined below:

- It is advisable to publish in open access journals (if financially feasible) so that your articles are more likely to be downloaded, read, and cited.
- Make your ORCID account public.
- Set up a Google Scholar profile.
- Create a profile for your research on Publons, Scopus, and Web of Science. Merging different researcher IDs is only possible if you own an account.
- Keep a relatively active presence on Research Gate. However, only add your published journal if it is open access. The pdf file can still be saved in the system if the article is not open assessment. Sometimes, you may wish to share the file when someone requests it.

- Create a personal webpage with links to other major web pages.
- Consider posting your CV on the Internet, such as your institution's webpage or Academia.edu.
- Register a Loop account with the Frontiers.
- Make your Kudos Hub profile public. Kudos can also be used as a blog.
- Ensure that your basic information in the publishers' database is updated, especially your research interests.
- Keep an active presence on Altmetric.
- Twitter.
- LinkedIn.
- Deliver some good scientific lectures.
- Attending quality conferences and talking to people.
- Join a reputable scientific society and be active.

4.11 Summary

If you would read this chapter again, I would urge you to pay attention to the word 'habit' that kept popping up in the text. Changing a habit is much more challenging than conducting an experiment and collecting data. Are you in agreement with this statement? Research is not limited to proposals, data collection, and outputs. The right mindset, habits, and research practices make us better biologists and biotechnologists. A successful experiment design starts with a proper sampling strategy and labelling of samples in order to make the project more logical and easier to follow. The correct storage methods are essential since we work with biological samples. The importance of reproducibility cannot be overstated.

In this era, scientific protocols and methods have advanced so much compared to the earlier decades, which has led to higher expectations from examiners, reviewers, and readers. Remember that a single method of collecting data may not be as effective or as strong in generating a conclusion as it once was. Interdisciplinary (or multidisciplinary) approaches enhance the collection of data, which leads to more accurate conclusions and better findings. Some laboratories are reluctant to change and will find that getting the green light for publication is getting harder and harder. It is hard for us to change the thinking of editors and reviewers, but we can change the way we think and the way we design experiments. Readers may consider incorporating -omics, bioinformatics, machine learning, social sciences, data science, and other new areas into their current or future research.

In order to be a successful researcher, digital literacy is essential. The Google Workspace Marketplace is a good place to begin. The use of artwork or visual aids is essential for scientific communication. Producing high-quality figures is an art and a science that cannot be separated. In order to generate excellent figures more quickly, we will need to gain an understanding of digital literacy approaches. In the beginning, try something simpler, as suggested in this chapter, and when the time is ripe, learn something more challenging, such as a new programming language such as R.

I believe that reading is the gateway to knowledge and that reading materials serve as mirrors and quiet places for reflection. Is it hypocritical of us to not read, but to ask others (i.e. your students or postdoc) to do so instead? Poor science, poor publication, low grant approval rates, and poor KPIs are most likely the result of poor reading and planning. It has been suggested that we should stay away from our smartphones (or at least reduce the time) and begin reading instead. Take part in upskilling courses, such as Nature Masterclasses, Researchers Academy, Editage Insights, and Enago Academy. Learn new skills in MOOCs, if we are interested in artificial intelligence, big data, bioinformatics, data mining, R programming, MATHLAB, Bayesian statistics, genomic data science, approaches to analysing microbiomes, Python, or professional development courses. Be like a seed, grow every day.

It is important to develop good habits and habits of behaviour. We also discuss the power of calendars in this chapter, as stated by Rick Pastoor in his book *GRIP*. We can use it to plan our actions, publications, meetings, experiments, and holidays. Learn and practice "Think win-win" and "begin with the end in mind", as Stephen Covey advised. The book *Blue Ocean Strategy*, authored by W Chan Kim and Renée Mauborgne, advised us to think differently. Within these unexplored blue ocean fields, lies a plethora of research gaps, untapped applications, and potential opportunities for commercially motivated endeavors. Try avoiding red ocean fields because someone in another part of the world may have the same idea and may be able to publish it faster and more efficiently than you do simply due to their greater resources. Would it be possible to explore a different research field that is relatively new? Do you have the courage to take the first step?

We ended this chapter with the suggestion that it is important to be visible. Be as visible as possible so that students will know whom to contact when they wish to pursue a PhD. By being visible, you will be able to enhance your research, collaboration, and development. In the event that your visibility is causing you too many unnecessary appointments that are taking up your time, that is completely acceptable as long as you do not mind. Nevertheless, if your high profile is detracting from your research and your inner alarm bells are ringing, perhaps you should consider whether you need to reduce your visibility. Here is the message: Control your own destiny (fate).

References

Alexandra George and Toby Walsh (2022) Artificial intelligence is breaking patent law. Nature 605:616–618

Baker M (2016) 1,500 scientists lift the lid on reproducibility. Nature 533(7604):452–454

Chun J, Rainey FA (2014) Integrating genomics into the taxonomy and systematics of the bacteria and archaea. Int J Syst Evol Microbiol 64(Part 2):316–324

Flier JS (2017) Irreproducibility of published bioscience research: diagnosis, pathogenesis and therapy. Mol Metabol 6(1):2–9

Hasin Y, Seldin M, Lusis A (2017) Multi-omics approaches to disease. Genome Biol 18(1):83

Howell CR, Su W, Nassel AF, Agne AA, Cherrington AL (2020) Area based stratified random sampling using geospatial technology in a community-based survey. BMC Public Health 20(1): 1678
Pastoor R (2022) GRIP: the art of working smart. HarperCollins Publishers, UK
Plesser HE (2018) Reproducibility vs. replicability: a brief history of a confused terminology. Frontiers. Neuroinformatics 11:76
Schleifer KH (2009) Classification of bacteria and archaea: past, present and future. Syst Appl Microbiol 32(8):533–542
Shang Y, Kumar S, Oakley B, Kim WK (2018) Chicken gut microbiota: importance and detection technology. Frontiers in veterinary. Science 5. p. Article 254
Shiffrin RM, Börner K, Stigler SM (2018) Scientific progress despite irreproducibility: a seeming paradox. Proc Natl Acad Sci U S A 115(11):2632–2639

Publication 101

5

Abstract

Chapter 5 discusses A-to-Z on publications, with special emphasis on journal publications. Throughout the chapter, we discuss the reasons for publishing and how key performance indicators (KPIs) may lead us to have a distorted view of the process. Further, this chapter discusses Scopus and Web of Science, their content, and why indexation is necessary to combat predatory journals. There is a thorough discussion of ways in which to pick the right journal for us to submit our manuscript. Moreover, the author explains the purpose, tips, and mistakes of writing a review article. The author hopes that readers will develop good publication plans, write well, and reduce rejection rates.

5.1 Publication Purposes and Correct Mindsets

5.1.1 Reasons for Students to Publish

Let us begin with an end in mind. What is your expected number of publications if you were a PhD (or a master's) student at the beginning of the study? Publications are required in certain countries and universities before a student can submit their thesis for evaluation (viva voce). In the case of a student registered at the university where I serve as a lecturer, the PhD candidate must have a Web of Science (WOS) indexed article or two publications in Scopus. MPhil candidate KPI is much simpler—a single publication in a journal, conference (full paper), or book chapter is sufficient. Students reading this part of the article should consult their faculty or university for written regulations, as the requirements may differ depending on the country. What motivates postgraduate students to write and publish articles?

K. M. Goh, *Research Methodology in Bioscience and Biotechnology*,
https://doi.org/10.1007/978-981-99-2812-5_5

5.1.2 Reasons for Staff to Publish

How do you motivate yourself to write if you are a member of the academic staff, a full-time researcher, or a postdoctoral fellow? A publication is regarded as a key performance indicator (KPI) for a person's performance on an annual basis and for his or her promotion to a higher position. What are your publication milestones for promotions and before retirement? The Times Higher Education (THE) World Reputation Rankings and QS World University Rankings, for example, are also driving the higher education institute to pay attention to publications.

What is the appropriate number of articles (KPI) for an academic staff member? The quick answer is it depends. It is important to remember that quantity and quality are heavily influenced by the university policy, the position of the faculty member (the KPI for a full professor differs from a junior lecturer), or the country.

5.1.3 Publish for the Sake of Key Performance Indicators?

Have you heard the phrase 'publish or perish'? According to some online information, this phrase can be dated back to 1932. Although this may be true to some extent, it is not recommended to publish solely for the sake of publishing or to publish in an unscrupulous manner. While publishing may seem like a daunting task for someone who does not have much experience, keep in mind that 'no pain, no gain'. As practising makes a talented athlete even better, a researcher will also become more proficient at the process of writing and publishing.

Key performance indicators create a lot of pressure. Some academicians may decide to undertake simpler projects by reinventing the wheel and do publications of lesser quality rather than publications of greater prominence. I refer to such low-quality research strategy as 'instant-noodle science' and 'instant-noodle publication' because it requires relatively little time and effort. An excellent science project requires a larger grant, an appropriate experimental design, skill, knowledge, and more time to collect data and is more stressful. Is it rational that academicians publish a large quantity of low-quality publications (in order to survive), but none of these publications are read nor cited by other researchers in the field? While KPI may be driving factors, there are other reasons why we publish:

1. Sharing new knowledge with the scientific community
2. Enhancement of science
3. Influence the field or practice in order to improve it
4. Acquire recognition
5. Continual improvement and satisfaction
6. Dissemination of new protocols
7. Promotion to a higher position (or graduation for students)
8. Crediting those who share their data
9. And of course KPIs as well
10. Others

I rank the reasons in accordance with my genuine reasons for publishing. The current chapter is intended to provide sufficient guidelines on good practices in publishing. Do not chase after publication KPI set by your bosses, because that KPI will be the limit.

5.2 Types of Publishers and Documents

Table 5.1 summarizes the three most common categories of publishers and main publication types. Understanding the common features can assist you in choosing the right journal to submit your manuscript to.

Below is a list of the top 20 international journal publishers according to Dimension. The data has been retrieved based on the number of published articles in 2021 (not limited to bioscience and biotechnology).

- Elsevier
- Springer Nature
- Wiley
- MDPI
- Taylor & Francis
- IOP Publishing
- Wolters Kluwer
- Oxford University Press (OUP)
- Institute of Electrical and Electronics (IEEE)
- SAGE Publications
- American Chemical Society (ACS)
- Royal Society of Chemistry (RSC)
- Cambridge University Press (CUP)
- Hindawi
- BMJ
- EDP Science
- Emerald
- De Gruyter
- Thieme

'Paper' is often used by laymen and even professionals as a generic term, but technically, the word 'paper' is used to describe an article, a review, or any other type of documentation. There is a slight difference in the way Scopus and WOS classify documents, but they are generally quite consistent in their classifications. Scopus categorizes documents as Article, Review, Conference Paper, Book Chapter, Note, Editorial, Short Survey, Letter, Data Paper, and Erratum (Table 5.2, Fig. 5.1). A long review or short (mini) review is referred to by the Scopus as simply Review. The term 'Data Paper' refers to journals such as *Scientific Data* and *Data in Brief*. The journal *Microbial Resource Announcement* is classified as an Article, not as a

Table 5.1 Categories of publishers

Types of publishers	Characteristics
Institutional University- or faculty-based publisher Examples: *Jurnal Teknologi*, Sains Malaysiana, Chiang Mai University Journal of Natural Science, Taiwania, etc.	• Website parked under the institute when started. • Most of the editors are from the same institute. • Most of the submissions come from the same institute. • Number of readers is relatively low. • Less complicated submission and review process. • Some journals allow submissions in the language of the home country. • Some journal names contain elements of the name of a local institute, place, or country. • Journals and publishers that started as institutional can evolve into business entities.
Society or agency Examples: *Malaysian Journal of Microbiology*, *The ISME Journal*, *Applied and Environmental Microbiology*, *Infection and Immunity*, *Extremophiles*, *mBio*, *Environmental Microbiology*, *Microbiology Resource Announcements*, Algal Resources, *Korean Society for Biotechnology and Bioengineering Journal* (KSBB Journal), Society for Applied Microbiology, etc.	• Journals were initiated by a group of people sharing a similar interest from the same country, a region, or an international region. • The scope of the journal is specific or topical. • This website is parked under society, or any business entity listed below. • The submission and management system are good or excellent. • Excellent reputation. • Excellent review process. • Members usually pay a lower publication fee than non-members. • Read by researchers and experts in the fields.
Business entity Examples: Journals publish under Elsevier, Springer Nature, Blackwell, Taylor & Francis, Frontiers, MDPI, Hindawi, PLOS, Bentham Science, Oxford Academic, Oxford University Press, etc.	• An interdisciplinary team of experts with a business and science background initiated the journals. • Excellent system and procedures for submission and review. • The editorial office is staffed full-time. • Well-established authors and reviewers database. • Dominate the total number of publications. • Broad range of readers.

Data Paper. I consider *MethodsX* to be a method paper (based on the title of the journal), but I am wrong because Scopus classifies it as an Article.

In another system, WOS classifies documents as Articles, Review Articles, Proceedings Papers, Book Chapters, Early Access, Meeting Abstracts, Data Papers, Editorial Materials, Corrections, Letters, News Items, Retractions, Database Reviews, Retracted Publications, Software Reviews, Reprints, and others. In the Retraction category, an article is removed from a journal but the title of the retracted article is still listed in Scopus and WOS.

Table 5.2 Types of common publications

Types	Descriptions
Book	An example would be a textbook that is solely for educational purposes. There are only a few authors who contributed to the book. Other examples include biographies, fiction, and non-fiction.
Book chapter	A book that is edited by a small group of people. Individual chapters are written by different authors from different institutions.
Proceeding	This publication consists of a selection of academic papers published during a conference or symposium.
Article	This is the most common type of journal manuscript used for presenting data from a research project. Depending on the journal, it may be called a Research Article, an Original Article, or even a plain Article. This document includes a complete introduction, methods, results, discussion, summary, and reference section. Article usually has a word limit of 4000 to 6000 words. Examples: *Science of Total Environment*, *Bioinformatics*, *Biofilm and Microbiome*, *Nature Microbiology*, *Biotechnology for Biofuels and Bioproducts*, etc.
Short article	A strict length limit is often imposed on this format. Some refer to these documents as brief communications, short communications, or letters. Examples of journals with options of short communication are *ISME Journal*, *Annals of Microbiology*, *Journal of Basic Microbiology*, *Bioresource Technology*, *Microbial Resource Announcement*, etc.
Data descriptor	Scientific datasets and research that advances the sharing and reuse of scientific data are described in such articles. Examples: *Scientific Data*, *Data in Brief*, *MDPI Data*, etc.
Methods or protocols	These articles report on new experimental methods, tests, or procedures. Examples: *MethodsX*, *Methods*, *Nature Methods*, *Methods and Protocols*, *Current Protocols in Bioinformatics*, *Briefing in Bioinformatics*, etc.
Review	Review articles offer a comprehensive overview of the research on a specific topic, as well as a perspective on the current status of the field. Typically, these papers are written by leading experts from a particular discipline upon invitation by the editors of a particular journal. Authors may also nominate themselves. Review articles can be categorized into three types: literature reviews, systematic reviews, and meta-analyses. Examples: *PloS ONE*, *Biotechnology Advances*, *Biotechnology and Genetic Engineering Reviews*, *Current Opinion in Cell Biology*, *Annual Review of Physiology*, *Archives of Biochemistry and Biophysics*, etc.
Mini review	There are also journals that publish shorter reviews. Examples: *Applied Microbiology and Biotechnology*, *AMB Express*, *Cancer Letters*, *FEMS Journals*, etc.
Case studies or clinical trials	Case studies are descriptive accounts of real-life patient cases from medical or clinical practice. The purpose of case studies is to inform other researchers about the possibility that a particular phenomenon may occur. Besides, the purpose of this type of study is to frequently report previously unknown or emerging pathologies in medicine. Examples: *Frontiers* journal, *Trials* journal, *Contemporary Clinical Trials*, etc.
Editorial note	A brief article written by the chief editor or editor of a special volume due to a special purpose.

(continued)

Table 5.2 (continued)

Types	Descriptions
Opinion or commentaries	In opinion articles, the author presents his or her view on the interpretation, analysis, and methods used in a particular study. The author is able to comment on the strength and weaknesses of the theory or hypothesis. Commentaries are typically short pieces of writing, usually around 1000 to 1500 words in length.
Preprint	Preprints are versions of scholarly or scientific papers that have not yet been peer-reviewed and published in peer-reviewed publications. Examples: arXiv, bioRxiv, F1000 Research, Authorea, Research Square, etc.

(a) Scopus	(b) WOS		
Article	Articles	Data Papers	Bibliographies
Review	Review Articles	Retracted Publications	Poetry
Conference Paper	Proceedings Papers	Reprints	Database Reviews
Book Chapter	Meeting Abstracts	Book Reviews	Software Reviews
Note	Book Chapters	Corrections, Additions	Main Cite
Editorial	Editorial Materials	Books	Publication With Expression of Concern
Short Survey	Notes	Biographical-Items	Chronologies
Letter	Early Access	Discussions	Hardware Reviews
Data Paper	Letters	Retractions	Record Reviews
Erratum	News Items	Abstract of Published Items	Withdrawn Publication
	Corrections	Items About an Individual	

Fig. 5.1 Types of documents classified in (**a**) Scopus and (**b**) WOS

5.3 Authorship: COPE and CRediT

When it comes to publishing, authorship is a very important issue. Basically, we can follow the guidelines set by the Contributor Roles Taxonomy (CRediT) and the Committee on Publication Ethics (COPE). To avoid authorship disputes, it is a good idea to discuss authorship before drafting a manuscript. The readers are recommended to read further details on authorship in Chap. 3.

5.4 Indexation and Metrics

The journal indexes (also known as bibliographic databases) contain a list of journals arranged according to discipline, subject, or publication type. Journal articles must be indexed to maintain their reputation and ultimate impact. Scholars also tend to prioritize citing to and submitting to journals included in leading indexes, because indexing is regarded as an indicator of the quality of a journal. A journal may be indexed in more than one database, including EBSCO, ERIC, Ovid, ProQuest, DOAJ, SCIMago, Google Scholar, ERA, Scopus, WOS, Chemical Abstract Service, PubMed Central, etc.

5.4.1 Scopus and Web of Science

For biological, bioscience, and biotechnology journals, Scopus and WOS serve as the gold standard of indexation. Young researchers may find it difficult to comprehend why Scopus and WOS are important. Here are some illustrations to assist you with understanding. Patients receiving drugs, biologics, or vaccines prefer products approved by the FDA or WHO or your ministry of health in your home country. Halal certification of reputable bodies is an important factor for Muslims seeking processed foods and products. Additionally, consumers prefer products that adhere to GMP, HACCP, and ISO standards. What is the reason for this? The answer is likely to be quality and trust. Scopus and WOS are more than merely a bibliographic database. In webinars hosted by Clarivate, we have learned from speakers from Clarivate that WOS evaluates journals based on more than 40 different criteria. Every year, several reputable and prestige journals were withdrawn from the WOS JIF (journal impact factor) system for a variety of reasons. We can rest assured that good indexation parties such as Scopus and WOS are dedicated to quality. Do not misunderstand me, I am not implying that other indexing services are inferior.

In my university, publishing was not a mandatory requirement for a postgraduate student to graduate more than a decade ago. Over time, I have noticed a progression of publishing requirements from not compulsory, to encouraging it, to publishing in any indexed journal, and to publishing in Scopus/WOS. Many reputable universities require their students to have published in Scopus or WOS before they can submit the thesis or participate in the viva voce phase. Even so, some universities permit viva voce before publication.

For many years, I have been serving as the program coordinator for the PhD program. One of my responsibilities is to ensure that students fulfil their publication requirements prior to submitting their thesis. Occasionally, I discovered candidates publish in predatory (pseudo-) journals that the team initially thought had been indexed in Scopus. In this case, there is a lack of familiarity with journal indexation. We strongly recommend that you verify the indexation information that appears on Scopus/WOS webpages in preference to what appears on the website of journals or publishers or the email correspondence from editors and publishers.

5.4.2 Predatory Journals

Many readers may think it is rather cumbersome to check Scopus and WOS. Would it be possible to provide a list of journals that we should avoid? Several years ago, our university received a list of such journals; however, I think it is more desirable to become familiar with legitimate journals rather than predatory journals. You have probably heard about Beall's List. Jeffrey Beall was an American librarian and library scientist. By utilizing the key phrase 'Beall's List' in the Internet search engine, it is still possible to assess a list of possible predatory journals. Are the results reliable and should they be used as references? There are great discussions on the list, and if you are interested in the topic, you can read some more information on

ResearchGate and a few related publications (Silva 2017; Watson 2017; Krawczyk and Kulczycki 2021). Are we in need of another list of blacklisted predatory journals? In my opinion, no. It would suffice to focus on indexed authentic journals in Scopus and WOS.

Please do not misunderstand me once again. I am not suggesting that journals that have not yet been indexed in Scopus or WOS are inferior. Almost all indexation providers will not capture new journals when they are first introduced (see also the subsequent section for more description). Over time, editors and publishers can request Scopus and WOS to inspect their journals' quality. In order to be accepted for indexing, it must undergo a rigorous evaluation and selection process. Scopus and WOS have a comprehensive listing of quality criteria checklists. The collection of items includes editorial board members, journal policies, regularity, ethical issues, peer review process, academic contribution to the field, self-citations, and quality of content, including clarity of abstracts, readability of articles, iThenticate checking (plagiarism), article scope, and other components.

It is important to note that some of the journals on the Beall's List as well as those on the list of self-selected bad journals of some researchers are indexed in Scopus as well as in WOS. The concept of predatory journals and poorly managed journals should not be confused. In principle, Scopus and WOS do not allow predatory journals to be listed in their database. Journals with poor quality might be removed after some sort of auditing is conducted on a regular basis. However, there may still be journals under their monitoring that are poorly managed by editors or editorial officers. Thus, predatory journals are not the same as poorly managed journals.

Here is a brief summary of inferior predatory journals that should be avoided.

- A journal that declares itself to be indexed and with an impact factor (or misleads us into believing it is a WOS impact factor).
- Suspicious editorial team members (with no or weak publication records) whose names you've never heard before. A number of fake journals include the names of prominent researchers without obtaining their consent. This was done by stealing their names.
- Accepts everything we submit (you pay, they publish). These are the black sheep who caused APC (article processing fee) to be misunderstood.
- Poor quality of article content, including poor data, poor editing, poor language, poor diagram, and poor science.
- Self-declared rapid publication rate. Some promise less than a week.
- There is little or no manuscript review before publication.
- Responses are quick (conventional publishers or editors may take some time to respond to queries or emails).
- There is no proper submission platform. Submission of manuscripts generally occurs by email to suspect email addresses (e.g. Gmail, Yahoo mail, etc.).
- The title of the journal is very misleading or similar to the title of another established journal, or the website appears to be similar.

5.4.3 Non-indexed = Predatory Journals?

The ISME Journal is one of the most renowned journals in microbial ecology, microbial interactions, evolutionary genetics, and related fields. According to WOS Journal Citation Reports (JCR), the impact factor at the time of writing was greater than ten. In 2021, a new journal *ISME Communications* was launched by the society (publisher). *ISME Communications* was not indexed by Scopus or WOS for at least the first few years (probably at least 3 years). This journal should not be considered as predatory because the webpages of both journals clearly articulate the differences between the two journals; in fact, *ISME Communications* is a good journal with a reputable editorial board and a good publication record and is run by a reputable academic society. It will be indexed soon and receive its first impact factor score shortly.

Let us take a look at another example. It is well known that *Proceedings of the National Academy of Sciences* (PNAS journal) is a good journal. A wide range of cutting-edge research reports, commentaries, reviews, perspectives, and colloquia papers are published in this journal. In March 2022, the National Academy of Sciences published the first issue of its new journal, *PNAS Nexus*, under the publishing imprint of Oxford Academic. A comprehensive scientific coverage is provided by *PNAS Nexus* through timely dissemination of original research, both in engineering and medicine, as well as studies that have been conducted in the social, political, and economic sciences, as well as in the biological, physical, and mathematical sciences. As *PNAS* and *PNAS Nexus* are two separate journals, we should treat them separately. Since *PNAS Nexus* is a new journal, as of this writing, it is not yet indexed by Scopus or WOS.

Microbiome refers to the total microorganisms in the gut, soil, plants, or any other environment. The study of microbiomes involves high-throughput sequencing and bioinformatics. Wiley collaborated with the iMeta Science Society to launch a new journal called *iMeta* in February 2022. As a new journal, it has not yet been indexed. APC is waived for the first 2 years of publication. In light of the name of the publisher, the members of the editorial board, the national academy that is responsible for the publication, and the quality of the articles in the first few volumes, it appears that this newly published journal will provide a good platform for the communication of microbiomes and bioinformatics. Similarly, *Frontiers in Microbiomes* was launched in 2022. It should be viewed as a separate publication from *Frontiers in Microbiology*, which is well established.

I have introduced a few new journals to you that appear to be reputable. However, some new journals often have relatively low submission rates. KPI stress is a real issue, and some authors prefer to submit their work to other indexed journals because of the rules set forth by KPIs. Each month, new journals are created by departments, faculties, universities, societies, and many other organizations around the world. There are some new journals that just die out after only a few volumes have been published. This could be because there are not many submissions. Additionally, there are not enough readers, resulting in low citations. Do due diligence before submitting and it may be more appropriate to submit to a new journal run by a reputable publisher or editorial board.

5.4.4 Scopus' CiteScore and WOS' Impact Factor and Quartile

Different indexation providers may create their own impact factor (IF); however, in this discussion, we will focus on IF calculated by WOS Clarivate Analytics Journal Citation Reports (JCR). Clarivate Analytics was previously known as the Institute for Scientific Information (ISI); therefore, you might hear older researchers refer to it as the ISI impact factor.

A journal impact factor is determined by dividing the number of citations received in the current year by the number of citations received in the previous 2 years. The five-year impact factor represents the average number of times articles from the journal have been cited in the WOS JCR over the past 5 years. Impact factors change every year and are usually updated by June.

What does impact factor really mean? As an example, Journal-A has IF 1.0, indicating that, on average, articles published 1 or 2 years ago have been cited one time. Let's take one example. The IF of *Frontiers in Microbiology* for 2021 is 6.064 (released in June 2022). Thus the IF number is calculated by dividing the number of citations received in 2019 + 2020 by the number of cited items received in 2019 + 2020 (37,970/6262 = 6.064) (WOS JCR data). IF is a journal-based measure, not publisher-based, and not individual article-based measurement. Any journal that indicates an impact factor for the publisher or the article is likely to be a predatory journal.

In a given year, there are four quarters: Q1 from 1 January to 31 March, Q2 from 1 April to 30 June, etc. In statistics, the first quartile (Q1) corresponds to 25% of the smallest to the largest number, while the fourth quartile (Q4) corresponds to 25% of the largest number. Journal quartiles are associated with research fields, known as categories in WOS. Thousands of journals are indexed by WOS and are grouped into over 250 categories. In the WOS JCR, journals are categorized according to their topical relevance. Among these categories are Biochemistry & Molecular Biology, Biotechnology & Applied Microbiology, Cell Biology, Ecology, Environmental Science, Genetics & Heredity, Immunology, Plant Science, Marine & Freshwater Biology, Microbiology, Mycology, Multidisciplinary Sciences, Virology, Infectious Diseases, Toxicology, Biodiversity Conservation, Parasitology, and Mycology, among others. There are approximately 40 categories associated with biosciences and biotechnology.

Assume that there are 100 journals with impact factor (not a real number) in the category Plant Science. The journals are arranged in accordance with their impact factors. The top 25 journals with the highest impact factor are known as the Q1 (first quartile) journals in that category, whereas the bottom 25 journals are referred to as Q4. It is easier to understand this example if we use 100 journals as an example. A total of 238 journals are currently indexed in the Plant Science category (WOS-JCR-Science Citation Index Expanded index, SCIE) at the time of writing. *The Annual*

Table 5.3 List of journals (partial) in the Plant Science category

Rank n/ 238 in SCIE	Journal name	Total citations	2021 JIF	IF quartile
1	*Annual Review of Plant Biology*	25,803	28.31	Q1
2	*Trends In Plant Science*	32,492	22.012	Q1
3	*Molecular Plant*	20,242	21.949	Q1
4	*Nature Plants*	13,007	17.352	Q1
5	*Plant Biotechnology Journal*	16,345	13.263	Q1
...	...	...	...	...
58	*Plant Biology*	7206	3.877	Q1
59	*Mycorrhiza*	5021	3.856	Q1
60	*Perspectives in Plant Ecology Evolution and Systematics*	3073	3.842	Q2
61	*Journal of Plant Physiology*	17,683	3.686	Q2
62	*Journal of Soil Science and Plant Nutrition*	4107	3.61	Q2
...	...	...	...	...
119	*Euphytica*	13,742	2.185	Q2
120	*Journal of Berry Research*	747	2.156	Q3
121	*Weed Research*	3954	2.117	Q3
...	...	...	...	...
177	*Czech Journal of Genetics and Plant Breeding*	462	1.304	Q3
178	*Plant Genetic Resources-Characterization and Utilization*	1020	1.279	Q3
179	*Nova Hedwigia*	2065	1.254	Q4
180	*Notulae Botanicae Horti Agrobotanici Cluj-Napoca*	1982	1.249	Q4
...	...	...	...	...
235	*Rhodora*	357	0.323	Q4
236	*Iheringia Serie Botanica*	213	0.314	Q4
237	*Zeitschrift Fur Arznei- & Gewurzpflanzen*	85	0.262	Q4
238	*Castanea*	409	0.255	Q4

*Arrange according to journal impact factor (Source: 2021 Journal Impact Factor, JIF)

Review of Plant Biology with IF 28.31 is ranked first (1/238) in this category and is followed by *Trends in Plant Biology* (2/238) with IF 22.012 (Table 5.3). *Journal Perspectives in Plant Ecology Evolution and Systematics* is the first Q2 journal with an IF of 3.842, at the time of writing.

Journals may belong to more than one category. One such journal is *Systematic and Applied Microbiology*, published by Elsevier GMBH, which is grouped into two categories. As of this writing, the IF stands at 4.064. This journal is rated Q1 in the category of Biotechnology & Applied Microbiology (rank no. 38/167 using JCI index), and Q2 in the category of JCI Microbiology (rank no. 46/157). This journal is Q2 according to SCIE (66/158 Biotechnology & Applied Microbiology; 66/136 Microbiology).

As a general rule, the more citations a journal receives, the higher its impact factor score will be. Citations are influenced by a variety of factors. One important factor is the quality of the work. Poorly written descriptions may cause readers to doubt the credibility of the published work. Whenever a better alternative is available, we tend to cite a reputable journal or credible scholar.

Citations are affected by publicity or the number of readers. An academic journal that is run and maintained by a faculty of a particular institution may have a smaller audience than journals published by large international professional bodies, for example, ISME. Often, the Q1 journals are the leading locomotive that guides the direction of relevant research topics. A Q1 is an impressive journal not only because of its quantitative data but also because of its quality science and knowledge. Therefore, Q1 journals are pole settlers. We should read more Q1 papers.

Below is a list of the top ten publishers with the most citations; this information was taken from on a presentation slide presented by the chief editor of a journal where I served as an associate editor. The ranking of the articles is based on the average number of citations in 2020 to articles published in 2017–2019. I suggest you exercise due diligence since the ranking is not solely taken into consideration when deciding whether or not to submit your precious manuscript.

- American Chemical Society (ACS)
- Royal Society of Chemistry (RSC)
- Frontiers
- American Physical Society (APS)
- Public Library of Science (PloS)
- IEEE
- MDPI
- Elsevier
- Springer Nature
- Wiley

Instead of IF, Scopus uses CiteScore to calculate the average number of citations a document receives over a three-year period. If a journal is indexed in both Scopus and WoS, the journal webpage may display two numbers: CiteScore and impact factor.

The two-year IF and CiteScore are two of the most commonly used metrics, but they are not the only ones. It is important for readers to be aware that there are other metrics, such as the five-year impact factor, the Immediacy index, Eigenfactor Score, the Article Influence Score, the two-year Median, etc.

5.4.5 H-Index

The final topic that will be discussed in this subsection is the h-index (or Hirsch index). The h-index is a metric that measures the citation impact and productivity of an individual researcher or scientist. The Scopus h-index is often one of the gold standards. Have you ever wondered what the h-index value is for the world's top scientists? In case you are interested, you may visit the website Research.com and check out the first edition, which was published in 2022. In some online information, h-index scores between 3 and 5 are common for new assistant professors (senior lecturers), scores between 8 and 12 for promotion to associate professor, and scores between 15 and 20 for advancement to professor. The requirements may differ depending on the university. As attentive readers may have noticed, Scopus' h-index refers to author achievement and not an index for a journal or publisher. There have been predatory journals that provide h-indexes for their publications.

H-index values are also provided by other indexation services, such as Google Scholar and WOS. To my knowledge, Google Scholar and Scopus h-indexes include self-citations. The practice of self-citation occurs when we include references to our own earlier publications as part of our published work.

Let us look at two situations below. Figure 5.2 shows h-index of two researchers before and after excluding self-citations. The Scopus h-index for researcher-A is 21. When we clicked the box on the Scopus website to exclude self-citations, the number decreased to 16. Let us say that researcher-B has a Scopus h-index of 18 and after removing self-citations, the score is reduced to 17. Do you think Researcher-A is better than Researcher-B, or vice versa? Are they equally good? As a researcher, how would you like to be classified: Researcher-A or Researcher-B?

We may cite our own publications if the earlier work is directly related to the current submission. The practice of inducing self-citation to inflate a h-index is not

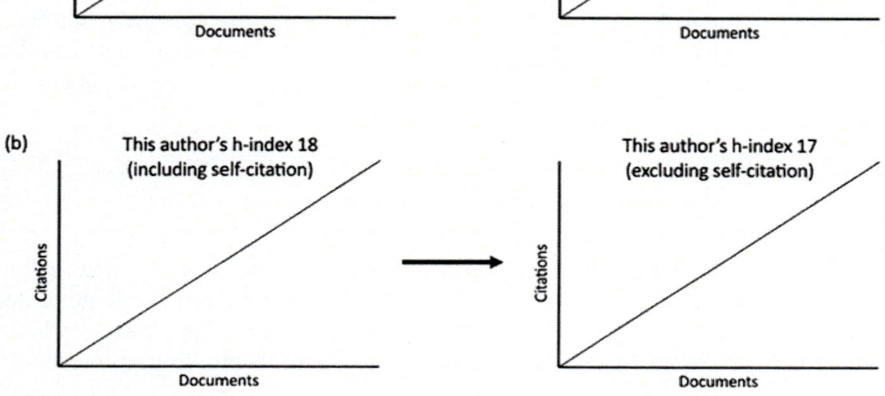

Fig. 5.2 Illustration of h-index plots

acceptable, particularly when it involves citing irrelevant documents with no relation to the subject matter of the research. We should avoid making thick sugar coatings on purpose since this is a violation of research integrity and ethics. A way to boost one's own h-index score is to keep focused on producing good science, writing excellent manuscripts, publishing in reputable journals, making yourself and the paper visible, making oral presentations in conferences, and patiently waiting for incoming citations. Years ago, I received an email from a person I did not know offering me a paid service to increase my citations. Basically, we pay them and they will cite our papers in their publication. I am not certain, perhaps this is part of the illegal scheme employed by paper mills. Make sure you do not fall into a trap.

5.5 Advice in Choosing the Right Journal

After months and years of collecting experimental data, it is time to write and submit a manuscript to a journal. Among the many submission systems available are Editorial Manager, which is used by Elsevier and PLOS, Manuscript-tracking-system by Springer Nature, SuSy by MDPI, ScholarOne by Wiley and Taylor & Francis, and Frontiers' collaborative peer review platform. Other journal managing platforms include eJManager, Open Journal Systems, OpusTech, and others.

As a general procedure during manuscript submission, we key or insert the title, authorship names, affiliations, keywords, abstracts, tables, and figures separately into the system before uploading the entire manuscript. Sometimes, it is tedious to copy and paste this information one by one. Some modern submission systems have evolved significantly as many smart systems are capable of capturing the essential information (title, abstract, texts, authors, etc.) directly from the .docx file we upload.

When the corresponding authors click the submit button or endorse the built PDF, the next step would be a preliminary check by the in-house publisher staff (or artificial intelligence) before the handling editor is invited and the review process will be initiated. The main purpose of the review process is to eliminate low-quality manuscripts or submissions that do not fit the scope of this journal. A quality review process is essential to ensuring that only high-quality submissions are published. How do you select which journal to submit to? I am often asked this question. In making a decision, readers may want to consider the polyphasic approach described below.

5.5.1 Consideration-1: Book Chapter or Indexed Journal?

According to my home country, Scopus, WOS, and ERA are appropriate index providers. The best place to submit simple data collected by good undergraduate students or short-term master's dissertations is probably proceedings, book chapters, or journals indexed by Scopus. Scopus indexes some proceedings from conferences. Getting started with something simple is a good way for young researchers to gain experience. A publication is better than none at all. Choose a WOS journal for a properly designed project.

It Is becoming more and more common for universities to publish their own book chapters using an in-house publisher. Typically, a small number of people form an editorial committee and search for a few authors. The editors may also contribute to a few chapters. This approach is, in my view, feasible from a KPI perspective. We should, however, ask ourselves, are we diluting our time by editing and writing at the same time? Another point is, who are the readers? Who will buy the final compiled book? Why not consider reserving a draft manuscript or good idea for a more appropriate book chapter project, often offered by an international publisher, with greater readership and marketing potential? For me as a writer, I prefer my articles to be read by others rather than placed on a bookshelf.

5.5.2 Consideration-2: Impact Factor

Considering your current stage, which option would you choose?

Option A: Short-term goal (low impact factor, IF). My immediate objective is to survive. KPI is pressing me. There is no time for long-term planning (too busy, no long-term interest, no grant, no good students, no research interest, not confident, no idea, or no plan).

Option B: Long-term goal (high IF). In order to improve my citations, I should put more weight on h-index and quality journals. I should make plans for promotion. I should produce scientific content of the highest standard.

Funding for research is subject to economic considerations. Research funding may be inadequate in your lab, and you may not have a very strong team, or you are at the beginning of the journey, so setting a short-term objective may be the only option. In the case of young academics, a decent number of simpler publications should be included on their CV before they are considered for larger research grants. But a few top-notch publications will enhance its appeal. Collaboration with a few established researchers is highly encouraged.

In my opinion, it would be best to balance the two options presented above—plan for both short-term and long-term projects. In the short term, microbiologists interested in bacteria isolation can publish their genome sequencing data in *Data in Brief* or *Microbial Resource Announcements*. Comparative genomics and transcriptomic analyses of these bacteria will be published in a better journal and that is considered as a longer-term objective.

Generally, young researchers at the beginning of their careers have limited funding available for their research. The more experienced and more confident we become, the more likely we should transition to journals with a higher impact factor, particularly those with Q1 or Q2. Plan or strategy is the key message.

5.5.3 Consideration-3: Building a List of Journals That We Are Familiar with

It is possible to make a list of journals based on the articles we frequently read and cite. If our experiment is comparable to the ones we read, then it is likely that our scope would be appropriate for that journal. Many outstanding students have good reading habits and may be able to offer some suggestions. We may also benefit from some excellent suggestions from an academician in our department or team. Anyhow, you should make sure that you read the 'Instructions to Authors' for the publication in question to determine whether or not your manuscript fits well.

5.5.4 Consideration-4: Building Journals List from the WOS

It was stated in an earlier section that WOS Journal Citation Reports group journals according to their categories. We can sort all journals associated with a category based on impact factors, alphabetically, total citations, IF, quartile, or any other criteria we choose.

As an example, we have two manuscripts where one is a study of a genome and the other is a study of the biochemical properties of an enzyme. We may find suitable journals like *Genomics* that is ranked Q1 (JCI) and *Preparative Biochemistry and Biotechnology* (Q3) in the category Biotechnology & Applied Microbiology. We will find other names of journals as well. If we have not previously considered these options, we may include them in the master list if we deem them appropriate.

It is unlikely that we know the names of all journals. In my early years as a faculty member, one undergraduate student completed a simple project, and we intended to write a simple article. We have the option of choosing a non-index journal or a Scopus-indexed journal. We were thinking about Q4-level journals at that time. We went through the list of journals within the category of Biotechnology & Applied Microbiology. We chose to submit this manuscript to a journal that has the lowest impact factor because the data collected is not very large and the manuscript is quite straightforward. Depending on the level of our appetite, we might consider this approach when identifying alternative lists that may have occupied the top position in Q1 or at the end of Q4 to look for some journals that we may have missed.

There may be other WOS Journal Citation Reports functions that are useful. For instance, we are able to see which organizations (institutions or universities) from which countries/regions have contributed the most articles in the past for a particular journal. We can then roughly estimate the readership.

Furthermore, we can observe whether the journal impact factor trend is upward or downward based on the journal impact factor plot. Let us consider the scenario in which we submit a manuscript to Journal-ABC during the month of March. Although the impact factor is decent, ABC has had a downward impact factor over the years and sits near the bottom of Q1. Assume that our manuscript is accepted for publication (say in early June). Usually, WOS releases information on Journal Impact Factor (JIR) in late June. There is a possibility that the impact factor for ABC may vary very little from the previous year, but it is now located in Q2 rather than Q1 because other journals in the same category have performed better in terms of total citations per publication. This will result in the loss of a Q1 journal on our CV. It might be advisable to choose another suitable candidate journal or wait for the most recent JIR release if the WOS JCI or SCIE or other functions are being examined carefully before making a submission.

5.5.5 Consideration-5: Specific or Multidisciplinary Category

Journal of Virology, *Journal of Medical Virology*, and *Journal of General Virology* are journals that are dedicated to the study of viruses and phages. These three journals fall under the category of Virology in the WOS JCI. The majority of JCI categories relate to certain topics. Some people are unaware of the Multidisciplinary Sciences category. Examples include journals such as *Nature*, *Science*, *PNAS*, *Scientific Reports*, *PLOS ONE*, *GigaScience*, *Heliyon*, *PeerJ*, *Current Science*, *Sains Malaysiana*, and many others.

For instance, *Scientific Reports* covers a broad range of topics in the fields of biological sciences, Earth and environmental science, and health science. *Heliyon*, another multidisciplinary science journal, contains at least 30 sections covering neuroscience, agriculture, clinical research, food science and nutrition, biology, education, and others. Should you consider multidisciplinary or specific journals? I would recommend both.

5.5.6 Consideration-6: 'Free' or Open Access

Those familiar with Elsevier will be under the impression that no charges are imposed on journals associated with Elsevier. Publication does not come free of charge. Let's take Elsevier as an example. Elsevier is a company that is a member of the RELX Group, a publicly traded company. Despite the fact that we do not pay Elsevier ourselves, many libraries or universities throughout the world do so. By way of example, if you are able to download articles through the university's Internet on the ScienceDirect, this means that either your library or your country has funded the subscription fees. Therefore, it is not totally free to submit an article to Elsevier or other publishers. In addition, it is a biased statement to claim that conventional 'free' journals are better, more ethical, and that acceptance is not by 'purchase'.

Open access is a different business model for publications. When it was first introduced, many people criticized it. Researchers who are unfamiliar with the publisher's business model tend to discriminate against open access journals and journals that charge a fee. Are you in agreement that *Nature*, *Nature Communications*, and *PNAS* have excellent reputations? Do you know that article processing charges (APC) are imposed for these articles? However, APC has been the subject of a lengthy debate because of its high cost. Many researchers, particularly those working in small laboratories in developing countries, find it difficult to publish in certain journals due to APC. Several journals and publishers recognize their limitations and provide fee waivers, either in full or in part. During the period of the COVID-19 pandemic, some publishers temporarily waived the APC for publications related to coronavirus.

It is my understanding that, despite the fact that APC is involved, most responsible publishers, editors, and reviewers do not compromise the quality of the journal. The APC should not be an obstacle to the publication of good articles if we (or the grants or funders) are able to cover the costs. It is important to note that APC is not directly linked with predatory journals. It is those predatory journals with APC that we should avoid. There are some publishers or journals that only offer open access, but most Elsevier journals allow authors to choose either the conventional publication mode or the open access option.

The time from submission to publication is shorter for certain open access journals based on my own experience but I am unable to speak on behalf of the publisher. The process is more rapid because publishers employ more full-time employees. The pace of publication is becoming increasingly competitive between publishers. I think no one would submit to the same journal again if they had to pay the APC from their own pocket or from a research grant, but the process was slow. The other advantage of publishing in open access journals is that more readers can access our articles. If the quality of our work is good, then the probability of getting higher citations increases. According to a publisher's statement, articles published in open access journals receive almost four times as many citations on average.

In conclusion, if you have a solid manuscript and are willing to pay an article processing charge, and you are short on time for any reason, consider open access. Should you lack funds but are willing to wait, consider conventional publishing. Hopefully this is helpful.

5.5.7 Consideration-7: Normal Track or Special Volume

The following is the procedure for journal publication. Submission, editorial office appoints a handling editor, review, correction (repeated review and correction), acceptance, minor proofreading and typesetting, indexing in volume and page, publication online and inclusion in Scopus, WOS, Google Scholar, and other databases.

Publishers can be creative by offering special volumes, special issues, or by branding this as a research topic. This approach serves more than one purpose. Publishers or editors-in-chief often request proposals for special topics from prominent scientists who are in line with the general scope of a journal. Researchers are invited to serve as guest editors or associate editors of a short-term editorial project until the project is completed. A guest editor will search for contributing authors, look for reviewers, and decide whether a submission will be accepted or rejected, though of course the official editor will make the final decision. It is the best time for a guest editor to become acquainted with the practice of the publisher. There is a possibility that they may be appointed to the editorial board if the collaboration goes well. *Scientific Reports*, Frontiers series, for example *Frontiers in Microbiology*, and MDPI series, for example *Microorganisms*, are examples of journals with special topics.

Do you think that our manuscript should be considered for a special issue, special volume, or research topic? If the journal is reputable, I would say yes. Rarely but sometimes, special issues may also offer additional benefits, such as reduced article processing charges. Authors will appreciate that their submission is handled by a real expert in the field, and the process may be expedited since the editor is likely to handle fewer submissions than the editor in the journal board. In some cases, certain publishers, such as *Frontiers in Microbiology* and their sister journals, produced an e-book with an ISBN based on the compilation of published articles. In such cases, the readability is higher.

5.5.8 Consideration-8: Format of Article

There is a distinct editing format for each journal and publisher. Using the template prepared by the publisher is recommended. As a rule, we refer to a template in terms of Latex or Word. A majority of people use Microsoft Word, and please learn to utilize its 'Style' function to its fullest extent.

A major focus of this subsection is on the format of either type-1, combining Results and Discussion, or type-2, which separates Results from Discussion. Certain journals are also highly particular about this matter, and it is clearly spelled out in their Instructions to Authors which writing style is most commonly used. The format of the report has nothing to do with the scientific viewpoint but selecting the right format will make our journey easier. Some writers tend to combine Results and Discussion in their writing style, while others prefer to separate them.

Young writers often experience higher rejection rates. If we keep switching between type-1 and type-2 while re-editing a manuscript, the editing process will be time-consuming. What is the appropriate strategy? Following our review of the six considerations guideline discussed earlier, we probably have a list of possible journal names in mind. Rearrange the lists by grouping them into either type-1 or type-2. In the event that a red light is given, revise the manuscript and submit to another journal following the same editing style. Try not to switch between the two styles, unless there is no other better choice.

5.5.9 Consideration-9: Using Journal Selection Tools

It is highly recommended that you consider all, if not most, of the suggestions made above. In case you think considerations 1–7 to be too complicated, or if you have already used such techniques before and wish there was a faster solution, here it is. With the manuscript title, abstract, and keywords in place, you can use some of these tools to obtain a list of possible journals. It is worth noting that some of the journal-finding tools were developed by the publisher; therefore, only journals associated with this publisher will be displayed. We will save a great deal of time by using these tools. Examples are:

- Elsevier JournalFinder
- Web of Science Master Journal List (Manuscript Matcher)
- Wiley Journal Finder Beta
- Springer Nature Journal Suggester
- American Journal Expert JournalGuide
- Jane (Journal/Author Name Estimator)
- Others

Please keep in mind that these tools only provide suggestions. You must make sure that your manuscript fits within the scope of the journal. I had already developed a short list of journals that I would try to submit to as part of a recent manuscript. I wanted to see if there were any other options that I had missed. With the title and the abstract as input text, I conducted a series of searches using the various tools listed above. My favourite manuscript matching tool is the Web of Science Manuscript Matcher. It is very similar to the list I manually compiled. Feel free to try yours.

Some searching tools, for instance Elsevier JournalFinder, provide some additional information such as impact factor, acceptance rate, average time to first decision, and the estimated publication time once accepted. But it is important that you undertake your own due diligence on the suggested list. I have noticed that some search engine engineers are less accurate. Additionally, there are some services that are paid to do this by reading our manuscript and suggesting a few possible journals for publication. Unfortunately, I have not yet used their service due to the cost. After reading the information in this chapter, researchers should now be able to make their own choice instead of buying suggestions.

5.6 Review Articles

A review article may be regarded as a critical evaluation of a previously published document in a systematic manner. It represents an important milestone in the systematic gathering of research findings, outstanding issues and matters over time, and represents a possible direction for the future for a narrow field of research. New results of experiments (unpublished data) are not usually included in review articles; however, some journals permit at least a brief mention of them. A well-written review can be used as a gateway to students learning the ins and outs of research and to understand new topics by novice researchers. A review article can help busy prominent researchers save time by capturing the most important messages from a single document instead of having to read dozens of articles.

Several years ago, some authors proposed more than ten types and methodologies of reviews (Grant and Booth 2009), which are listed below:

- Critical review
- Literature review
- Mapping review
- Meta-analysis
- Mixed-studies review
- Overview
- Qualitative systematic review
- Rapid review
- Scoping review
- State-of-the-art review
- Systematic review
- Systematic search and review
- Systematized review
- Umbrella review

Besides the jargon mentioned above, you might also encounter domain-based reviews, theory-based reviews, method-based reviews, etc. A student once asked me, "Was the review you published last year a critical review, scoping review, or systematic review? I'm writing one with my supervisor, so which should I choose?". My answer was, it was a review. In case you are not able to distinguish between the different types, do not worry as I am equally incapable of doing so without consulting the online guided definition. There are a number of overlaps in the concepts and strategies among the categories listed and, in my opinion, if you are a newbie in writing a review, it may be helpful to read and grasp these strategies.

We are already familiar with how Scopus and WOS categorize documents; however, these indexation providers only refer to such documents as Reviews or Review Articles. The majority of publishers use the term 'short- or mini-review', or a 'normal or length review'. As a journal editor and reviewer, I pay little attention to the types and methodologies listed above, and I am more concerned about the content and message of the manuscript.

5.6.1 Who Should Be Writing a Review?

A review article should be written by someone who has extensive experience in the field. There may be a misconception among some researchers that a review article is a summary of what they have read. It has come to my attention several times that some individuals are urging their work colleagues to write as much as possible review articles in order to meet the KPI. To fulfil the requirement for graduation, students may be asked to write a review. In my opinion, a review draft prepared by an expert is not problematic; however, it concerns me when academicians write on topics about which they are not familiar.

Are you confident that first-year postgraduate students are capable of preparing a quality review article? It was mentioned to me by multiple sources that supervisors require their inexperienced students to draft most of the text for a review. This is a good training strategy for students, as they are required to comprehend reading materials. In the event that the supervisor is a notable researcher and an expert in the subject, sufficient training from the supervisor may enable the student to become an equally competent writer and researcher. On the other hand, a strategy such as this may result in hundreds of mediocre review articles if none of the authors, including the supervisor, has any experience in the specialized field.

Many PhD (or master's) students may be exceptionally talented and hardworking; however, since they have limited experiences in experimental aspects (technical aspects) and depth of knowledge, they often lack the critical thinking skills required to make detailed assessments. High-quality review articles often contain recent information, critical evaluations, opinions, or judgements from the authors. As a side note, some of these students may fall into the plagiarism trap, that is stealing an idea from another review, even though iThenticate may have missed it in the first place.

It is important that high-quality review articles do not contain information that is outdated, or that is not pertinent to the main context or purpose of the review. For example, *Taq* DNA polymerase from *Thermus aquaticus* is well known to us. It is without doubt that the discovery and development of *Taq* between 1976 and 1983 was an important milestone in biotechnology. Almost all biotechnologists learn that from lectures and textbooks. I was reading some recent review articles that discuss microbial mat. In reading this review paper, I was surprised to see that *Taq* polymerase was extensively discussed by the authors. Perhaps I am mistaken, but I have the impression that this particular section was written by a novice writer, probably a young student unfamiliar with the field of microbial mat. Let me make it clear that I acknowledge the discovery and development of *Taq* polymerase. Nevertheless, *Taq* polymerase is an old story that does not relate to the entire content of the review. Obviously, a scientist and writer with experience would notice that, but why was such a mistake not corrected? One of the reasons could be a writer's inability to distinguish essential information from non-essential information. Their tendency is to take information that is easy to understand and read, but miss the main message. When compared with more recent and engineered DNA polymerase in the recent decade, *Taq* polymerase does not show the same level of performance. A new student may not have enough experience working with enough DNA polymerases, so they might not understand the development of various DNA polymerases such as *Pfu*, *Pfx*, Vent, Phusion, Q5, Kapa, and other more recent developments. Besides, students will never be able to determine the real situation without a thorough understanding of various PCR approaches (normal PCR, mega-primer PCR, overlap extension PCR, inverse PCR, PCR-mediated DNA shuffling, etc.).

You may observe some interesting things if you do a bibliography analysis of the leading experts in your field. Most of the world's top experts dedicate themselves to their research, and they publish primarily in article journals. When you have the opportunity to speak with them, they will tell you that they do not have time to write a review. Even so, a small number of top experts are occasionally able to contribute to outstanding review articles. You may have noticed that some of the reviews you have been reading in the past weeks were actually written by someone outside the field of expertise. The next time you download a review article, please do consider checking the background of the authors, especially the corresponding author, and by doing so, you will ensure that the document is of the highest quality and your due diligence will assist you in ensuring it is as accurate as possible.

After reading approximately a page, you are likely to be discouraged from writing a review. No, I do not mean that. Having a quality review article is more important than having a low-quality article just to satisfy a KPI. The habit of writing a review for KPIs is something I am against. In order to be able to write a review, take a moment to do a self-assessment below first before you get started. Are you knowledgeable about the research area you have been working on? If the answer is yes, you may begin the journey of writing a review as soon as possible. As a general rule, you are eligible to begin the first tiny step if the majority of the following lists are checked. Following is a checklist and some advice:

- You have been active on a particular topic for more than 5 years.
- Reading articles pertaining to the progress of the subject on a regular basis.
- You have made publication of relevant articles with good citations.
- Your articles have attracted the attention of prominent researchers or editors.
- A review of a particular topic is needed at this time.
- There is no relevant review article for the topic or scope, or the last review was published many years ago. There are some issues currently being debated by multiple groups.
- And are you able to contribute something?

In closing, let me reiterate the point that I made earlier. Most reviewer articles are still worth reading as long as the authors have done their jobs well, and the editor and reviewers have done their jobs as well. You should not write a review in order to meet a KPI. There is no way someone can write a review every year on a wide variety of topics in biological science, covering topics such as plants, soil, yeast, human, virus, energy, bioinformatics, metagenome, waste management, etc., where the topics are so diverse and require a high level of expertise. Does such a person or group exist? There is, indeed. Certainly, they will have a long list of publications, but if we think again, are they the real experts in the field? If you are not yet qualified, do not write a review. It was more appropriate for me to use the phrase 'not yet qualified' rather than 'unqualified' because, as we work hard, one day we will be able to write something when we are ready. You can do it, I believe.

5.6.2 Benefits of Writing a Review

Leaving aside the KPIs, there are two key benefits that an author can gain after publishing a review article. The first benefit is that the writing process is a Stop Checkpoint. A typical academic researcher's life cycle resembles a roller coaster ride. There is time with a high number of grants, students, and publications. At the bottom of the roller coaster ride, everything seems to be working at the minimum gear and this easily demotivates, losing momentum and making researchers feel incompetent. Stop Checkpoints are necessary to unlearn and relearn the latest developments in our fields of interest. In preparation for a review, we read and re-read more than a hundred articles, reviews, book chapters, and conference proceedings. Reading allows us to realign our knowledge, skills, and decision making in accordance with current developments. Lifelong learning is what makes us better researchers and writers. Furthermore, writing the review critically helps us to plan our next projects several years in advance.

A second benefit is that it enhances our career development. A well-written review can serve as a reference for other publications. As the citation numbers accumulate, our h-index will increase. We will also experience an increase in reputation. Thanks to two excellent reviews that I read many years ago, I became acquainted with the authors of that papers and later became good friends and collaborators. There are many other benefits, which we will not discuss here.

5.6.3 Getting Started and Keep Rocking

Writing the first review is challenging and you may need a good mentor. Be persistent and ensure that the process will continue until the article is published. Once we have acquired enough experience, our next review will occur after a few years, when we have an idea or when there is a need in the scientific community for a particular review. The following is a short description that may be of assistance. Here are the steps roughly involved in writing a review starting from scratch.

- Search different databases with different keywords. Mendeley is one excellent program for keeping track of references.
- It is important not to overlook articles with high citations. Furthermore, recent articles published within the last month or a few weeks should be included.
- Creating an alert or notification in Google Scholar and Mendeley. Any new related publications will be brought to our attention.
- Read at least 100 documents including articles, other reviews, book chapters, blogs, or whatever we can find. While working on such a project, I read between 100 and 150 articles, book chapters, conference papers, white papers, and reviews. My definition of reading is reading the entire document from beginning to end, line by line. It is never a good idea to be lazy, just read the abstract and take it blindly for your review. It is not a review, but a summary, which may contain plagiarized material.
- Make notes and summaries as we go. When reading keystone articles, be sure to read them multiple times to ensure we understand the message fully.
- During the long break between two academic calendars is the best time for me to begin reading and writing.
- After we have read through more than a quarter of the total collection, it is time to draft a backbone of the review that includes the main keyword or phrase. The first step for me is to write the potential titles of the sections and subsections of my

review. Throughout the process of reading and writing, some titles in the tentative backbone are merged or collapsed, or new titles are introduced as subsections.

- Having established the backbone, it is now necessary to categorize each article into different bins based on sections and subsections. Reading printed journals works better for me. I like to use stainless steel card ring binders (also known as loose binder, ring clip, etc., that look like key chain loops, but are operated by pulling on the loop). Hard copies of journal articles are grouped according to a theme or section. In the digital folder, I name the PDF file according to the title of the article. This step is critical to making the reading process more efficient and the writing process easier.
- It is recommended to put the main message of each article into its respective section of the backbone—one at a time. At this point, we should compromise with the linguistic, lexical mistakes, or the flow of the messages.
- From the beginning of our first written sentence, use Endnote, Mendeley, or another reference manager.
- Once approximately three-quarters of the references have been included in a specific subsection, it is time to connect the dots to create a complete narrative. Some rearranging, omitting the redundant message, and rewriting will take place in a logical flow. Continue reading to add additional information.
- Once we have more than 75% content written for all the subsections, we will have a first draft of the main body. First drafts are equivalent to a summary since no critical analysis is provided. I have spent approximately three months on the entire process, starting with step 1. Then, spend substantial time to add critical evaluations, opinions, or judgements.
- My practice is to write Abstract and Conclusion after the critical evaluations. In the next few paragraphs, we will discuss critical evaluations.
- If we are unfamiliar with certain subsection topics, it may be better to entrust the writing to another co-author that is more qualified.
- Take a fresh look at the subject matter and avoid plagiarism traps.

Consider using existing literature reviews (thesis Chap. 2) and selected discussions (thesis Chaps. 4 and 5, etc.) from theses or dissertations written by our PhD or master's students. The process will be easier. Compile, reorganize, and rewrite so that the draft does not look like a student's thesis, and add the latest articles. I ask my students to include a long comprehensive table that summarizes important data or information in their theses. Information of this kind is crucial to a review article. If I have theses and a few dozens of recent publications to work with, my first draft will take approximately a month or two. My opinion is that not all theses can be converted into reviews. Please bear this in mind.

5.6.4 Review Is Not a Summary

One of the definitions of the word 'review' in the Oxford Dictionary is "to describe the most important facts in a piece of academic research (= detailed study of a subject) that you have read, and give your opinion of the research". Among the words that are synonyms of 'review' listed at online Thesaurus.com are analysis, audit, check, inspection, revision, scrutiny, survey, reassessment, reflection, etc.

Summary provides key information. The journal review article is more than just a summary of multiple research articles. Reviews are expected to provide some form of evaluation, judgement, or opinion. Authors may choose to take a stand, either negative, positive, or mixed. Let's see an example. There is controversy regarding the status of Candidatus or the uncultured taxa of prokaryotes, for instance. In a review article (Pallen 2021), the author used SWOT analysis to analyse the earlier reported articles rather than just summarizing them all. A SWOT analysis is a tool used to identify strengths, weaknesses, opportunities, and threats. It is possible that some people feel SWOT is an outdated analysis tool. Perhaps you believe SOAR (strengths, opportunities, aspirations, results) is a more appropriate one. However, the focus here is not on comparing SWOT or SOAR; instead, we must analyse and review existing publications rather than simply stating their summaries. It is not necessary to use SWOT or SOAR in order to analyse. If you are a prominent researcher in that field, all the analysis can be generated by your brain without the use of SWOT or SOAR.

It is the responsibility of authors to read, analyse, and interpret the information in a review. On the Taylor & Francis website, the publisher suggested the following points: "It is important to include a discussion of both sides of the argument if contradictory research has been done in your field of study. The review can also help resolve conflicts between contradictory studies. Reviews should describe new ideas that have not already been explored".

It is necessary to include future research on the topic as part of the review article. Journals such as *Biotechnology Advances* and *Trends in Biotechnology* present excellent analyses, interpretations, and perspectives. Readers should read some of these journals' most recent reviews as part of their homework before starting their own review writing.

To sum up, review papers are not merely summaries of the material we read. Analysing and interpreting information is the responsibility of the authors. Several months ago, I was informed that some groups are making summaries from hundreds of journal articles using AI-driven technology with very little human intervention, intellectual effort, evaluation, or judgement. AI-computer-generated summaries are compiled and arranged into a review paper, again without much human intervention. In my opinion, such action may save a lot of time; however, what lessons have been learned as a result of the procedure? Do we learn knowledge about science and become wiser philosophical scientists as a result of this process? From these AI-generated summaries, do we really know what each word means? Although we might have a long list of publications listed on our CV, are we really experts in one (or a few) specific fields? It does not seem ethical to use a shortcut of this nature in this application, and it is not the intention of the AI developers to make us cheat. Active cooperation with AI can aid in the refinement of compositions. However, a concerning tendency has evolved in which some authors rely excessively on AI to produce manuscript material, ignoring their own originality, critical thinking, active participation in the writing process, and, most importantly, scientific correctness. This dependence on AI as a crutch can lead to indiscriminate copying and pasting of AI-generated material without careful consideration.

5.7 Best Practices to Avoid Being Rejected

It is the responsibility of reviewers and editors to ensure that only scientifically sound pieces of work with high-quality writing and illustrations are accepted for publication. I can think of three possible thinking styles (mindsets) among authors, so let me know which one you belong to.

Group 1. Let us assume that the project is only half-cooked, but the researchers may feel that they have collected enough information (minimum in fact) and decide to write something and submit it. From the point of view of more experienced scientists, the project appears incomplete, and no proper conclusions can be drawn. Nonetheless, the author is hoping that if he or she finds a reviewer that will close one eye or even both eyes, then the author may be able to get a favourable review. An example of a reviewer might be a former student, a supervisor, a friend, or someone from the same department. They hope to get some easy-going reviewer and pray that luck is on their side. It should be noted that such submissions are generally considered low quality by editors and reviewers, which eventually results in rejections.

Group 2 consists of a group of researchers and students who have done the best they could in terms of performing the experiments, analysing data, and writing up their findings. They appear to be very confident. Their submissions are sometimes handled and reviewed by very strict editors and reviewers. Sometimes, they may have a different background and have a different point of view when compared to what the authors may have. While researchers may have invested a lot of time and effort into their work and believe it to be flawless, when criticism is brought to their attention, they may be unable to accept it. It is their belief that criticism is equivalent to judging their abilities. Some perceive criticism as a judgment on their abilities, indicative of a fixed mindset. When faced with negative feedback from reviewers, their reaction is often one of anger and resistance, resulting in a reluctance to make the necessary revisions.

The third group consists of researchers or students who have a growth-mindset. They understand reviewers make comments (or criticism) because it's their job. Reviewers do not know you personally, so we should not take it personally. Comments are intended to highlight weaknesses in the design, data analysis, writing, or illustration of the work. If these comments are true, accept them and improve. In the event that these comments are not true as the reviewers misinterpreted them, you should do a proper rebuttal in a respectful manner.

As a general rule, our manuscript should be of the highest quality in terms of science, visual aids, formatting, and language. Some individuals may think that adding the name of a famous researcher could help increase their chances of acceptance. The claim may be true if the distinguished author has done his or her job by providing sufficient guidance from the beginning and ensuring that the data collection and final report are of the highest standard. As a matter of fact, if the overall design of the experiment and the quality of the writing are poor, I believe the majority of reviewers and editors will not compromise on that issue, even if a famous researcher is listed as an author. The following is a checklist. There are several online resources that address the same concerns.

- Plan our publication properly when applying for grants or starting a project.
- Collect all data that are enough to tell a complete story.
- Avoid reinventing the wheel.
- Read Instructions to Authors on the journal's webpage and follow it.
- To cover all the data and science in one article. Avoid dividing the story among multiple students and slicing a single project into multiple small articles. Some PIs utilize this strategy in order to maximize the number of enrolled students and ensure that the project can continue even if one of the students quits. The disadvantages of this strategy are: (1) Students compete against one another. (2) Once the first article has been published, it will be more challenging for the second and third manuscripts to be published due to the high degree of similarity between them. (3) Salami slicing will result in more articles with a lower impact factor. It is not advisable to use the concept of 'slice-by-slice', also called salami slicing (salami science), to develop several publications based on the same study.
- Consider using a native speaker proofreading service or a quality AI-based proofreading tool. We may use an internal or external proofreading service listed by the publisher. Examples include Editage, AJE, Enago, and others. Fiverr offers affordable freelance proofreading services. An effective proofreading service provides suggestions or acts as a reviewer. Editage is doing a good job on that. We will cover this part in the subsequent Chap. 6.
- If it is available, we can make full use of the template. Familiarizing yourself with the *style* function in Word. Format (arrangement, references, nomenclature, units and dimension, etc.) should be correct.
- High-quality keywords that are not too general or repetitions of the title of the abstract. Editors can locate reviewers more quickly by using proper keywords.
- High-quality (resolution, usually at least 300 dpi) and meaningful figures and tables.
- Never fabricate data or any part of it. Experts are capable of detecting it.
- Never use the services of a paper mill. It can be detected by publishers using tools.
- Do not attempt to commit image fraud, such as faking western blots. There are tools available to publishers for spotting it.
- It is important to highlight the entire story and novelty in the cover letter.
- Cover letter addressed to the chief editor. If possible, select the appropriate/ suitable handling editor. Avoid selecting an editor with a conflict of interest, such as someone we know.

- The following are best practices for nominating reviewers:
  - Nominate three to five reviewers.
  - Combination of senior, intermediate, and junior levels. Avoid a list that includes all senior professors. Avoid lists with all junior researchers. Note that some senior researchers have too many review invitations and are too busy with their own research. Assuming your suggested list consists entirely of senior researchers with heavy administrative duties, the editor will have some difficulty in finding reviewers for the manuscript.
  - Reviewers who have published in that journal, if possible (or who are listed in the publisher database/system). This will save the handling editor time when it comes to finding reviewers.
  - A mix of developed and developing countries.
  - Avoid conflict of interests.
- The following are best practices for correcting a manuscript:
  - Prepare a detailed list of the changes made.
  - Taking all comments seriously and adding experiments if necessary.
  - Investigate and systematically resolve 'odd/unclear comments' using inference (detective Conan's step-by-step reasoning). It is essential that if all of these alternatives are false, and if there is only one alternative left, that is the one that must be true.
  - Use scientific facts and proper arguments, it is possible to refute. You can use excellent published articles to support your claim.
  - It is possible to contact the handling editor via the system or directly by email.
  - Be polite and professional.

5.8 Setting Publication Plan

Students are the best people to learn how to write journals. Students have more time on their side, and they have greater control over the schedule. However, it is important to acknowledge the vital role of mentors and supervisors in shaping their journey. If a postgraduate student is provided with proper training and hands-on experience from the writing, deciding on a target journal, submission, rejection, improvement, resubmission, and so on, eventually when they become academic staff or research scientists, they are going to be very productive in publication.

5.8.1 What Is Your Publication Plan?

Publications are the bread and butter of lecturers, research scientists, and postdoc fellows. Readers, what is your plan for publication? In the first few pages of this chapter, I pose a similar question. 'Let us begin with and end with in mind. What is your expected number of publications if you were a PhD student at the beginning of the study? If you are a member of a faculty, what are your publication milestones for promotions and retirement?'

5.8.2 The Green and Red Octopus

Readers might find it interesting to read a Comment Article published in 2018 in *Nature* with the title 'The scientists who publish a paper every five days' (Ioannidis and Klavans 2018). This article came with a cartoon drawing where a green octopus wearing a lab coat performs various tasks such as meeting people, writing, discussing, teaching, preparing reports, etc. By the way, an octopus has arms and not tentacles.

The ability to publish a paper every 5 days is possessed by a few exceptional researchers and I am not one of them. I can, however, publish five papers every year. Each lab, project, student, and you are unique. It is relatively easy to collect data for some projects and publish them, but the rest may take some time. A person can publish well if he or she is honest, hardworking, and has the correct mindset. If we cannot publish a paper every 5 days, are we able to write five papers per year? Could you possibly lower it a bit, completing one or two papers each year, according to your current abilities? On the other hand, it may be an indication that something is amiss when a researcher has not published anything in 5 years. As an illustration, consider a red octopus with eight arms, in pyjamas instead of a lab coat, that produces nothing whatsoever except being a consumer (Fig. 5.3). There is something wrong if you are a red octopus. You should talk to your mentor about this. I recommend reading Chapters 4 and 8 of this book if you need more assistance.

Fig. 5.3 Two types of 'octopus'

5.8.3 Publication Planning

Setting a plan is essential for consistent output generation. It is my custom to sit with each student to discuss the publication plan together. We can, for instance, come up with a preliminary title, outline the content, suggest which data the student should collect, and most importantly, suggest which journal (with the expected impact factor or quartile) as well as the projected submission date. I believe this creates some pressure on students (and ourselves) and at least keeps that task in our minds.

I have been developing a habit to write a publication plan for over ten years. This does not require sophisticated software, merely a Word document or a Google Doc (Table 5.4). I print it out and attach it to my office wall. On a quarterly basis, the list is often updated by changing the date or name of the journal. By maintaining this list, we will remain motivated to achieve our goals. Please give it a try.

We may have heard that setting the goal higher is a good strategy; for instance, submit to a Q1 journal and then if that is rejected, a Q2, etc. It seems to me that this strategy is akin to 'trying my luck'. However, I would prefer something more specific. In general, tier one qualified manuscripts should not be published in tier three journals and tier four qualified manuscripts should not be accepted by tier one journals. When we create a research project for my student, we should have an idea that the publication output will fall into the Q1-Q2 or Q3-Q4 categories. If we have

Table 5.4 An example of publication goals

No.	Projects	Main writers	Journal	Submission date
	Project-A			
1.	Thermophiles and CA Zymes in	Goh and Sue	*Scientific Reports*	Jan
2.	Compendium of high- and medium-quality....	Ping and Goh	*Frontiers in Microbiology*	Jun
	Project-B (PhD Matthias)			
1.	Comparison of	Matthias	*STOTEN*	July
2.	Elucidation of.....	Matthias, Isaac	Q1/Q2. To be decided	Jan
	Project-C (MPhil Ben)			
1.	Characterization of....	Benjamin	Q3/Q4 maybe *J. of Basic Microbiol.* Or *3Biotech*	March next year
	Project-C review			
1.	A systematic review of....	Kuan, Goh	*AMB Express*? Or something similar	Aug
	Mini project			
1.	Draft genome of....	Fid	*Data in Brief*	April
2.	Complete genome of....	Liz	*Microbial Resource Announcement*	Feb

developed this ability through years of reading articles, please be as specific as possible when naming the target journals. The point I am making is if our intention is to publish an article in a Q1 journal, then our backup should also be a Q1 journal.

After making a good plan, ensure that you execute it until it is achieved. The process of writing can be tiring and requires a considerable amount of time and attention. Identify the time of day when you are most productive and take advantage of it. As an example, if your mind is more alert and productive in the morning, then schedule your manuscript writing for the morning and your meetings for the afternoon. The time after lunch may not be productive for most people. If your focus waned after lunch, sitting in front of your laptop would be extremely stressful since a 60-minute scope of writing could take up your entire afternoon and evening. You should read Pomodoro Technique if you find it difficult to sit down and get distracted easily. You might also be interested in the books *Deep Work* by Cal Newport and *The Power of Full Engagement* by Jim Loehr.

5.9 Summary

Let us conclude. Most people make a direct link between publications and key performance indicators (graduation requirement, annual appraisal, promotion, etc.). We should not publish solely for the purpose of meeting KPIs. It is important to recall that the main purpose of publications is to share new knowledge with the scientific community and to enhance science. Many journals and publishers are available; however, most researchers prefer journals that are indexed by Scopus or WOS. It is important to avoid inferior predatory journals. Journal Citation Reports (JCR) are updated on a yearly basis by WOS Clarivate Analytics. In order to be ranked as Q1, a journal must place within the top 25% of the top journals in a given category. CiteScore is used by Scopus instead. H-indices are metrics that measure the citation impact and productivity of individual researchers or scientists. When new journals are introduced, almost all indexation providers do not capture them. Please feel free to submit your manuscript to newly formed journals that have a decent editorial board and a good publishing background.

We provide several recommendations in this chapter as well regarding the choice of journals for your research when your manuscript is ready. You might consider using the approaches described in this chapter when trying to identify alternative lists that may have occupied the top spot in Q1 or at the end of Q4 to identify a journal that you might have overlooked. It is advisable to maintain a balance between the publication plans for short-term and long-term projects, especially for young researchers. When you gain more experience and self-confidence and learn the tricks, publishing in Q1 journals becomes easier.

The chapter also provides some advice on how to write a research article and a review. It is important to keep in mind that a journal review article involves more than just summarizing several different research articles. Students often lack the skills necessary to make critical assessments because of a lack of experience with experimental aspects (technical aspects) and a lack of depth of knowledge, but they can be trained. Therefore, the main author of a review article should possess extensive experience in the field and not be an inexperienced writer unfamiliar with the subject matter. Frequently, high-quality review articles contain critical evaluations, opinions, and judgements. A student can serve as a co-author and assist their supervisors or principal investigators during the writing process. This can be a valuable training experience for them.

For lecturers, researchers, and fellows, publications are the lifeblood of their careers. A proper plan will assist us in monitoring our progress at milestones. A section of the book illustrated two figures—a productive octopus and a sleepy octopus. If an individual is honest, hardworking, and has the right mindset, he or she may be able to publish well. The following quotation is attributed to Benjamin Franklin: "If you fail to plan, you are planning to fail". John J. Beckley said something similar: "Most people don't plan to fail, they fail to plan". It is well said by Ahmad Fauzi Ismail, the Professor and Vice Chancellor of UTM, that "when there is no vision, people will have no passion".

It is unlikely that we will accomplish our publication plans, if we do not translate them into concrete actions. All in all, when there is no plan, no vision, no goal, no love for science, and no actions, publications become a stressful KPI. Keep in mind that publication should be related to knowledge sharing and enhancement of science. Although it may sound naive, it is the innocent heart which keeps us in search of knowledge, wisdom, and truth. Blessed are the pure in heart, for they shall discover the science and truth behind the living organisms through their research.

References

Grant MJ, Booth A (2009) A typology of reviews: an analysis of 14 review types and associated methodologies. Health Inf Libr J 26(2):91–108

Ioannidis JPA, Klavans R (2018) The scientists who publish a paper every five days. Nature 561 (September):167–169

Krawczyk F, Kulczycki E (2021) How is open access accused of being predatory? The impact of Beall's lists of predatory journals on academic publishing. J Acad Librariansh 47(2):102271

Pallen MJ (2021) The status candidatus for uncultured taxa of bacteria and archaea: SWOT analysis. Int J Syst Evol Microbiol 71(9):005000

da Silva JAT (2017) Jeffrey Beall's "predatory" lists must not be used: they are biased, flawed, opaque and inaccurate. Bibliothecaeit 6(1):426–436

Watson R (2017) Beall's list of predatory open access journals: RIP. Nurs Open 4(2):60

Check for updates

6 Improving Writing and Presentation Skills

Abstract

Many non-native speakers learn English in primary and secondary schools, and even a few credit hours at university. Although understanding what one reads is not a major challenge, writing clearly or presenting a scientific talk (or a proposal presentation) is a major hurdle. Many non-native English writers feel more comfortable submitting their manuscripts to journals with a lower JIF owing to their limited English proficiency. Authors with a weak writing style may believe that publishers, editors, and reviewers are uninterested in their manuscripts, yet wonder why similar work can be published in a journal with a higher impact factor.

But it is incorrect to make this assumption. Having high-quality scientific content is crucial, but a well-organized writing style and clear wording help the evaluators and reviewers comprehend the work. If we have conducted experiments and collected sufficient data to generate conclusions that fill a research gap, language barriers do not pose a significant problem. Language barriers can be overcome in several ways. This chapter deals with communication skills. Many scientists devote their entire lives to the advancement of their fields, without spending sufficient time learning how to write and communicate in English. The purpose of this chapter is not to serve as an English textbook or a teacher that instructs us on tenses, sentence structure, or common errors. The author explains the importance of having a human proofreader as well as an artificial intelligence proofreader, especially for non-native writers. In addition to this, the chapter discusses how modern and carefully designed slides will capture the attention of attendees.

K. M. Goh, *Research Methodology in Bioscience and Biotechnology*,
https://doi.org/10.1007/978-981-99-2812-5_6

6.1 Communication Skills

Traditional Chinese kung fu drama is my favourite form of entertainment. In the novel published by Jin Yong in 1961, the Heavenly Sword and Dragon Slaying Sabre are the two most powerful and sharpest weapons. With its strength and sharpness, the Sabre could penetrate any common heavy metal weapons and break them into pieces. The novel shows that without a good piece of weaponry, the characters (actors) in the story will not survive.

A good command of English is imperative for effective communication of scientific information. If the document cannot be read easily, it is comparable to an unpolished and blunt sword that lacks power. Word and language are our instruments. It is common for scientists to devote their entire lives and efforts to the advancement of their respective fields while spending insufficient time learning how to write and communicate in English.

As a non-native English speaker, English has not been my main medium of instruction in my primary, secondary, and tertiary education. My first job was in an entirely English-speaking environment which caused me a great deal of difficulty. It is at this point that I realize how essential it is for me to re-learn English. It is never too late to learn a language properly. We must continue to sharpen and polish our sword (written and spoken English), and hopefully one day it will be as strong as Heavenly Sword and Dragon Slaying Sabre.

One day, I read a commentary in an online newspaper *The Star* in my home country. I found it extremely amusing. The journalist noticed funny signs or menus in restaurants or shops. The first funny sign is the restaurant serves 'lamp chop' instead of lamb chops. We all realize that a lamp is a device used to produce light, and no one enjoys eating an LED lamp. This is another humorous statement. Some of us prefer wholemeal bread, but the label on a hotel's breakfast bread corner reads 'whole male bread'. Here are some funny signs that made my food almost come out of my mouth. 'The rambutan fruit is cooked, exceptionally sweet, and ready for collection (ripe, not cooked)', 'Shoplifters will be prostituted (prosecuted)', 'Please pay your parking fee before existing (exiting)', and 'Employees must wash their hands before living (leaving)'.

Are you laughing? Could you imagine what would happen if such an error were to be discovered in our scientific writing? Biologists do not conduct research on lamp chop, so we are safe. If we mistakenly used the word 'prostituted' rather than prosecuted, we need to dig a hole to hide. The incident occurred to one of my classmates during a secondary school debate when the wrong word of prostitution was used. (I apologize to my former classmate. It was more than 30 years ago, but I still remember it.) Now for a quick quiz. Do you know the difference between the following words? Affect and Effect. Lose and Loose. Is it a HPLC stationary phase or stationery phase?

The first point I would like to emphasize is to pay attention to the spelling. Microsoft Word has a built-in tool for checking spelling. Do not disable it. Have a look at the Editor Score in Word. Second, use appropriate jargon when writing. Terms such as purified protein, DNA, and RNA, for example, are frequently used. A manuscript submission for which I served as the reviewer used 'cleaned protein'. Normally, purified means cleaned, but in biological science, we do not refer to clean proteins. Make sure you are extra cautious when it comes to nomenclature; for example, *Escherichia* can be easily spelled incorrectly despite being a common microorganism.

When writing scientific documents, we should be as accurate as possible. It is a common practice for students to look for alternative words on Thesaurus.com. One of my students liked to use words that were rare and difficult to pronounce. However, not all words would be appropriate for a given situation. The chosen term, verb, or noun must be as accurate as possible; it is as if, in archery, we shoot an arrow at the centre of the ring that scores 10 points. A weakly chosen word will shoot at the 8 point (not too bad); however, to pin it at 1 point or miss the ring would be problematic.

This chapter is not meant to act as an English textbook or to serve as a teacher who instructs us on tenses, sentence structure, or common errors. The three books in the reference are recommended if readers need more detailed reading materials (Day and Gastel 2006; Alley 2013, 2018). What should we do if we cannot write properly? Should we attend a night class to learn English? Should we purchase a grammar book and read it? (I purchased one, but only read ten pages). If we already learned English in primary school and secondary school, that will suffice. Nevertheless, the most effective solution involves writing and improving with the support of both a human proofreader and an artificial intelligence proofreader. By recognizing and accepting our writing weaknesses and actively striving to avoid them in the future, we can make progress. With that in mind, let us continue to the next part.

6.2 Proofreader

We have discussed research integrity and ethics in Chap. 3. One of the main purposes of viva voce or oral examinations is to ensure that the candidate conducted the research and wrote the thesis. If a postgraduate student pays someone to collect data or uses other people's effort to collect data (e.g. an undergraduate student working in the same lab) then the act is of low integrity. It is non-ethical to have another person helping us to write part of the thesis or the entire thesis. The student himself or herself is responsible for the content of the thesis, with some assistance and advice from the supervisors. Using a writing service company is unethical.

We understand that some writers are not proficient in English and may require assistance with language skills, tenses, or clarity. Supervisors typically find it a headache to read a draft thesis or a draft manuscript full of English errors that hinder the evaluation of the scientific content. Students who write their dissertations or theses in poor English are likely to encounter difficulties during the viva voce process because the examiners are unable to understand their writing. Additionally, a poorly written manuscript is easily rejected by the reviewers or editors because they are not able to understand the scientific points as the message is not effectively conveyed. There are two effective ways to resolve the problem.

6.2.1 Human Proofreader

Proofreading scientific writing or theses can be done by a number of individuals or companies. Publishers have in-house paid proofreaders where authors may consider taking advantage of their services. Some academicians, postdoctoral fellows, high school English teachers, or retired professors work part-time as proofreaders.

Obviously, the quality of human proofreading is highly dependent on the individual's skill level and knowledge of the subject matter. A number of proofreaders only focus on sentence structure, punctuation, grammar, and spelling whereas really excellent services also give suggestions regarding scientific content and the correct use of terminology.

A selection of proofreaders will depend on the type of document (manuscript or thesis), the stage of the document (before or after peer review), the length, the difficulty of the scientific content, and the cost (money). If the document is a long dissertation report that has been evaluated in viva voce, it may be appropriate to ask a friend who teaches English to check the document. If we intend to submit the document to a Q1 journal, we probably need someone with more experience and knowledge of the subject matter.

Generally, most proofreading services charge according to the number of words. The following list provides some suggestions regarding human proofreaders. There are some companies that employ English editors with PhD degrees who have the relevant expertise for our documents. They are aware of our needs and know how to provide us with advice.

- Elsevier Language Editing Services
- Springer Nature Research Editing Service
- Wiley Editing Service
- MDPI Language Editing Services
- Taylor & Francis Editing Service
- American Journal Expert
- Charlesworth Author Services
- Editage
- Edanz
- Enago
- International Science Editing
- LetPub
- Bioedit
- Fiverr

Disclaimer: (1) Please conduct your due diligence. (2) I am independent from any proofreading agency. (3) Although I have experience with some of the services listed, I do not have experience with all of them. Above are a number of companies, but their inclusion in the list does not imply they are excellent, while their absence does not imply they are inadequate.

The quality of the service is heavily influenced by the cost of the service and who handles your manuscript. Service providers recommended by publishers or journals are usually reliable. Fiverr offers affordable proofreading services for those on a tight budget.

Typically, service providers offer two main types of services. The first type is named basic editing, while some companies name it silver service. Some companies refer to their services as English language editing rather than scientific editing. Typically, basic service providers will retain most of the writing style and check for spelling, grammar, and ensuring readability.

The second type is advanced editing, golden services, premium editing services, or some name it scientific editing. The more advanced services include structural editing, re-structuring, relocation, suggesting correct terminology, and maybe even providing free re-editing within a set period of time or free checking once we have made any corrections based on the comments from the reviewers.

A skilled human proofreader will identify our writing mistakes and suggest ways to improve our writing. If we have been utilizing the services of a native speaker as a proofreader for an extended period and have not observed any improvement in our writing, it is likely that we have only clicked the 'accept all' track changes function in the Word program without understanding the rationale behind these suggestions. Every time I receive a well-proofread document, the hidden proofreader becomes my English tutor, imparting invaluable lessons to improve my writing skills. This is how I gained my writing proficiency.

Purchasing fruits (i.e. pineapple) is normally a straightforward process, and the result is typically either juicy and sweet or the opposite. From my experience, seeking out a relatively cheaper human proofreader service is comparable to purchasing a durian, and readers from Asian countries should understand this. We can only realize if we like a durian picked from the shelf by opening its shell and eating its fruit. The Musang King and Black Thorn cultivars, among the most costly durian cultivars, are simply too expensive if we have a large family with many people who enjoy durians. Consequently, alternative cultivars are considered because they are cheaper. After spending time in search of cheaper durians, we are disappointed when we open them because they do not taste good. I am only using this as a metaphor or parable, so please do not take it literally.

A good but reasonably priced proofreader can also be difficult to locate, especially if they are not referred by a publisher or journal. The process can be somewhat risky. We may find some companies to be cunning, whereafter we pay for the minimum cost, they may allege that the quality of our writing is too low and demand that we pay for additional proofreading. The service I received from such a company was not satisfactory and I will never use their services again. Service providers recommended by publishers or journals are generally reliable, especially if you do not have much experience and are uncertain about which company to choose.

6.2.2 Artificial Intelligence Proofreader

I appreciate human proofreaders who are detail-oriented and helpful. Engaging their services has been a valuable learning experience, especially when they take the time to explain the reasoning behind their changes. When we have a large volume of documents, paying on a regular basis is not cost-effective. We may benefit from AI proofreading that is relatively cheaper. AI proofreading works best for the following prerequisites:

- We have the ability to distinguish between clear and ambiguous sentences.
- Our writing is generally okay (error-free or nearly error-free), but simply does not read well (or is a bit awkward).
- We have the ability to spot mistakes or inaccuracies suggested by the programs.

An artificial intelligence proofreader is a great tool. However, it is worth noting that some suggestions provided by these programs may be unintelligible or employ incorrect terminology. There is a concept known as garbage in, garbage out in computer science, bioinformatics, or mathematics. The AI proofreader will not be able to read our minds, and its accuracy is heavily dependent upon the sentence we have written. Currently, there are a number of AI-based tools that are user-friendly. I will mention a few here (Appendix C).

For the start, consider Grammarly. We can obtain a demo version or a free version for Gmail, Outlook, Apple Mail, Slack, Word, Evernote, and even our mobile devices. For basic writing tasks or when our writing ability is not too poor, the free version may suffice. Grammarly Premium, on the other hand, is more powerful because it checks more features, such as style, tone, clarity, plagiarism, formality, fluency, punctuation, and advanced suggestions such as suggesting the best vocabulary terms. Check to see if your university has a subscription to Grammarly for Education. Usually after using Grammarly basic for several months, we will receive an email offering us a special discount on Grammarly Premium. Grammarly Premium is well worth the price if we are frequent writers.

I also recommend Wordtune. We may begin by trying the basic features for free, which include rewriting ten sentences a day. The basic Wordtune generator paraphrase sentences using a casual tone; however, sometimes it may be too casual for scientific writing. In the basic Wordtune version, formal tones, shortening, or expanding sentences are not supported. If Wordtune suits your needs, you can probably obtain an education discount and opt for the premium version, which can be accessed on a monthly or annual basis. Those who write regularly may find the premium version well worth it. QuillBot is another good tool you may want to try. Again, please keep in mind that QuillBot Premium is far more powerful than the free version. Another AI I've tried and highly recommend is InstaText. The web interface is user-friendly, ensuring ease of use and navigation for users. Additionally, InstaText focuses on improving styling, word choice, correcting grammatical errors, and enriching the content to enhance readability and understanding. InstaText is very well suited for academic writing, and readers are strongly encouraged to try the free programme and decide if you like it before subscribing. Most paid AI proofreading programmes allow you to subscribe on a monthly or annual basis, the latter often being much cheaper. ProWritingAid, another tool, has a lifetime premium package. Anyway, each of the tools suggested here and listed in the appendix has its own strengths and weaknesses.

Natural language processing, GTP, Codex, etc., are the basis of these tools. Technology is advancing rapidly and getting better and better. I have not tried other AI-driven tools; however, if we do an internet search, we will be able to find other such tools, such as the following: Ref-N-Write, Writefull, Wordvice AI, Pcanpi, PerfectIt, Ginger, WhiteSmoke, PaperRater, 1Checker, Scibendi AI, Jasper, Rytr, CloserScopy, PaperPal, and others. Some of these tools or applications are only compatible with online browsers, whilst others may have a Word plugin, or extension. There are some tools that allow dictation by speech-to-text. If you are planning to purchase one, please do your due diligence. It is always advisable to try the free version for at least a month before considering purchasing the premium version. Always remember that the premium version is far preferable to the free version.

Artificial intelligence technology is helpful, but they are still machines. I believe that if I put in a complex sentence with many errors and ambiguities as the input, these AI platforms will be very faithful in giving me a messy recommendation due to the idea of garbage in garbage out.

We need a legitimate license to subscribe to the premium version of the service, which comes with a fee. I have discovered that several proofreading programs have been hacked or pirated and are being offered in online shopping platforms. A common practice is for vendors to create a shared email address and password, which they then sell to dozens, if not hundreds, of users. I strongly advise you to obtain a genuine license. By using a shared online account, you may encounter the following problems: Firstly, proofreading companies can detect unusual login activity and block the account. Secondly, what you write and proofread in a web browser may be recorded, and other users who log in using the same ID may have access to your document. Thirdly, it is not ethical to engage in such practices. For students on a limited budget, there are many free online proofreaders available. For instance, you can try ChatGPT. Provide the chatbot with the right prompt (direction or guideline), and it will perform a fantastic job of assisting you in ensuring the wording of your content is precise, consistent, and acceptable. If you are inexperienced with prompts, you should definitely try ChatGTP-AIPRM.

It is necessary to read the AI-corrected documents again and again. Make sure that the selected term is consistent and suitable. The artificial intelligence may not be able to recognize the consistency of using commas (i.e. Oxford comma or serial comma), for example, "this enzyme hydrolyses starch to produce A, B, and C". It is also correct in English to omit the comma between B and C. At this point, I do not believe artificial intelligence can detect inconsistencies in our original sentences if we proofread the document part by part. Lets see another example. "The animals used in toxicity studies by researchers include fish, rabbits, rats, mice, hamsters, guinea pigs, and other small animals". There is a possibility that an artificial intelligence proofreader will recommend that we use the word 'animal' in certain sentences in place of rabbit. Let us assume that our experiments will involve both mice and rabbits. Readers may be confused whether this 'animal' refers to mice or rabbits if there are two or more models included in the manuscript. Besides, the AI may propose a tense that is inappropriate. For example, for the experimental portion, we often use the past tense, but the machine may use the present tense. When it comes to lamb chops and lamp chops, or the like, some AI may not be as intelligent as it appears on the surface. Sometimes, a weakness of AI-suggested paraphrasing is that it may not sound official or academic. Choosing the right tone is crucial. More importantly, ensure that the rewritten sentences are scientifically sound and match your original thought and the data you have, or accurately reflect the original meaning of the article you are citing. It is important not to exaggerate information when paraphrasing with AI. It is also essential for us to verify scientific accuracy.

Finally, do not abuse the rewriting function of AI tools. The act of stealing ideas, sentences, or entire paragraphs is unethical. There is a possibility that plagiarism detection software may be fooled by AI rewriting functions. Do not exploit the technology to do monkey tricks; rather, utilize it to polish our language only. Please do not try to be clever by stealing an article from another author and assuming that Crossref (or iThenticate or Turnitin) will not notice it. Machines can be cheated, but thousands of eyes will examine our published articles on online platforms. The published articles will be retracted if caught by the publisher and may stain our record. If a student performs the monkey trick and is caught during viva voce, then there will be many bad days to come.

6.3 Effective Presentation

Once a typical presentation style is established and becomes a habit, it is likely to be difficult to break the habit unless we are courageous enough to make a significant change. If our supervisor or students have commented that we need to polish our oral presentation skills, I suggest reading this book in the reference list (Alley 2013; Youknovsky and Bowers 2020). Also, if you observe that people start to doze off or look at their handphones within a few minutes after your presentation, you may need to modify the style of your presentation. There are many online videos available on YouTube that provide guidance on how to design a presentation. The majority of people, however, do not bother to learn it.

When buying tissue paper boxes from the supermarket we are likely to choose the ones that offer the cheapest price (or the ones with special discounts) or those that have aesthetically pleasing designs. When the price gap of two brands is small, it is likely that we will put the one that looks nicer in the shopping cart despite it being slightly more costly. Please do not be concerned if your response differs from mine. It is only an analogy. My point is that, like it or not, our eyes and brain are more likely to focus on something that appears pleasing. I will discuss two aspects here, slide design and the way we communicate.

6.3.1 Modern Slide Design

In the context of a presentation, particularly scientific communication such as a proposal presentation, viva voce presentation, or conference presentation, is it possible to make our slide more pleasing to the eye and brain? In my opinion, the following presentation style is outdated or ineffective:

- Self-designed style by starting with an empty or default PowerPoint (ppt) file.
- Choose and manually change the colour of each text box in the ppt.
- Like to use text boxes and add text boxes here and there.
- Using some very old-fashioned or odd bullet points (e.g. box or diamond).
- Slides with an entire paragraph.
- Slides with complete sentences.
- Adding a decoration by inserting a scanned photo or a photo taken from Google. However, they do not blend well. I will take as an example a picture with a white background on a slide that is black in colour.

A scientific presentation is not the same as delivering a lecture. Are the lists stated above similar to one of your most boring lecture notes, if we examine them closely? Do they resemble how presentations were made 20 years ago? Why do people continue to use a PowerPoint style designed 20 years ago?

Old-style presentations are not engaging the audience effectively. In no way am I implying that the content is not important. On the contrary, it is essential. Since I am unable to discuss the presentation content in this book, I can only provide advice regarding the slide design. The content of the message is important, but catching the attention of the audience is equally important.

Have you ever listened to TED talks or other excellent scientific presentations at conferences? Even though most TED speakers do not use slides whatsoever, we pay attention to what they have to say. This is exactly what the power of oral communication is all about. However, most audiences prefer both visually appealing aids (slides) and well-organized presentations. How does good slide design compare to other normal slides? Most good slides seem to be modern, without a lot of words, and much more focused on figures, am I right? I think we can learn how they design their slides—one word, catchy. Below are my suggestions:

- Do not begin ppt with a stretch. The process of manually editing the textbox, font, colour, etc., can result in inconsistency, poor coordination, or appearance issues unless we are very talented artists. The textbox can sometimes be the source of the problem, as it causes the text to appear incorrectly when the file is opened on a different computer or operating system.
- Use a proper PowerPoint (or Keynote for Mac) template. These can be downloaded from M365 or many other online platforms. I would recommend using an infographic style PowerPoint template. For example, AllPPT.com and fttp.com have nice infographic templates. It is safe and free to download the template at the time of writing. Canva is another good resource.

- Colour coordination for infographic templates is often good.
- Infographic template does not allow us to insert lengthy sentences. (That's my main point in suggesting an infographic ppt template.)
- Do not use lengthy text. (As a very bad habit, if I notice lengthy text on a presentation slide, my handphone or something else may take my attention instead. My apologies to the speaker.)
- Use the Slide Master function instead of making individual changes to a slide (under View, Master, Slide Master). For those who are unfamiliar with Slide Master, I recommend that we spend 30 mins on YouTube learning about it.
- Use high-resolution figures and avoid inserting unnecessary images. If my student adds an image or embeds an animated clip to the slide, I will ask them the reason for doing so. In the case that the answer is, they look good, nice, or I like this cartoon, the inclusion of them is not necessary. Generally, the human eye is drawn to images or motion rather than text. The unnecessary images disrupt the delivery and absorption of messages to the listeners' brains. Do not let the visual aid take over the attention.
- Make sure that the font size and type are appropriate. The vision quality of the elderly will decline over time. Avoid using indented text and all capitalization. The online Piktochart essay suggests 14 best fonts, such as Helvetica, Futura, Rockwell, Verdana, Raleway, etc.; of course, other webpages may have other preferences. In terms of font size, I prefer not to specify one since the colour of the background affects its clarity. However, overfilled text appears unattractive. To test if the font size is sufficient, I recommend standing approximately 3 m away from the computer screen. If you can read the text clearly, it will usually display properly on an LCD projector or a smart monitor display. An online presentation will be straightforward, but we should also consider users who are accessing it via a smartphone. But, usually, 24 points or slightly larger should be enough.
- It is recommended that you use at most three sizes of font. Use only one type of font and one colour. The use of bold or another suitable colour to make an emphasis on certain words is a technique that can be used at times. When you want to emphasize a word, bold typefaces may appear better than italics, underlines, or larger sizes.
- There are some presentation gurus (non-researchers) who advise against the use of bullet points altogether. If you are not comfortable removing all bullet points from your slides, consider reducing them.
- Slides should be arranged logically, not necessarily numbered or organized according to the flow of the proposal or thesis. As an example, a thesis consists of the introduction, problem statement, scope, literature review, etc. If we follow it strictly, background knowledge may appear twice in the Introduction and Literature Review.
- Consider arranging the slides in a creative manner. When we (evaluators) are required to examine several students in one day, we may become bored if we hear: (1) "Today, I will present my proposal entitled ___" or (2) a slide that contains the outline of the presentation (which is identical to the state of the thesis as a whole: introduction, problem statement, scope, literature review, method, results,

discussion, conclusion, etc.). A bit of creativity will catch the audience's attention. We can start with a scenario, a real-life example, a story, a metaphor, and a few wows. As an example, on the first slide we discuss the cases on dengue before describing our project title that relates to dengue. Likewise, you may wish to begin by mentioning a family member or a friend who had a COVID-19 infection, followed by the importance of research in coronavirus, and then continue the project title that is related to coronavirus. The creativity will attract the curiosity or the attention of the audience.

- Weigh the need for visual aids, such as animation, figures, charts, tables, sound clips, etc. Whenever possible, avoid visual aids that distract and detract from the content. Suppose all your audiences are from the same institute. If your slide template contains a large institution logo on each page, it will consume a lot of space and become distracting. When presenting at an international event, you may wish to include the logo at the beginning and at the end of the slide.
- Each slide represents one idea. I think that it is okay to have a ppt file with 50 slides with one message per slide, rather than a file with only 10 slides with lots of information on them.
- It is recommended to avoid slide transitions. A heavy transition between slides is in conflict with your presentation's efforts to attract the audience's attention.
- Make sure your slide is 16:9 since the majority of widescreen televisions and projectors now use this ratio instead of 4:3.
- Provide credit or references for photos or images that are not your own. Citations should not distract listeners. You may consider writing the reference in greyscale with a very small font size.

6.3.2 Oral Presentation

If you are not certain whether your voice, pronunciation, or tone is appropriate or not, you can present your slides (5–10 min) to a mentor or friend and ask for their opinions. Let us compare the features of good and weak presentations. In a nutshell, a good oral presentation consists of explaining the information in a slide in simple terms, rather than reading it. Remember that an oral presentation is similar to telling a story. Storytelling draws attention and helps attendees retain information. In order to improve communication effectiveness, three main factors must be considered: language, tone of voice, and body language. In Table 6.1, I summarize the characteristics of good and not-so-good presentations.

According to Daniel Kahneman's book *Thinking, Fast and Slow*, the brain works in a certain way. System 1 functions automatically and quickly, requiring little or no effort or control on the part of the user. In system 2, attention is paid to mental activities requiring exertion, such as complex computations or complex information.

Let us analyse this concept in relation to this section on effective presentation (slide design and oral). When we give a talk using slides, do we want the audience to use System 1 or System 2? I believe a more catchy presentation would emphasize System 1 and a portion (minimum or none at all) of System 2. Please use the System

Table 6.1 Comparison of good and weak presentations

Good presentation	Weak presentation
Explaining and give reasons	Reading the slides
Easy to follow	Jumping points
Logical and convincing	Creating confusion for listeners
Correct pronunciation	Strong accent and hard to listen
Complicated project explained in a simple way and simple language	Simple project explained in a complicated way, full of jargon, and unexplained abbreviations
Storytelling style	Robotic and monotone
Stay calm and control your nerves	Overly nervous and incoherent
Plan your speech in advance, but customize it depending on how the audience responds	Remember and memorize script
Adequate and suitable physical act onstage and making eye contact	Staring the screen
Leave a lasting impression on your audience. Let them remember something, especially the emotion	Focus too much on experimental results (individual numbers, decimal points, up-and-down, etc.)

1 concept in our proposal or viva voce presentation. The audience will enjoy the presentation since it is easy to follow. We should present a slide or presentation that is simple to understand, so the audience is not forced to read text or figures that are not needed or guess what we are trying to say. That would make for a nice presentation. System 2 should be used, however, if one wishes to motivate or inspire the audience to think deeper or make a decision. By the way, if you are not familiar with System 1 and System 2, there are some YouTube videos or online articles that discuss the concepts.

It would be helpful if you could search online using the name of this person, Albert Mehrabian, for his 7-38-55% communication rule. A point he made in his argument was that the words we speak, the voice we use, and the face (body language) we present affect the delivery of communication. Approximately 7% of the weight goes to our spoken words, 38% to our tone and voice, and 55% to our body language. The number should not be taken literally. Students, academics, and researchers often focus too much on the content of the talk and the design of the slides, and how we present it within the time limit. These contributions probably make up less than 10% of the maximum the listener can receive from our talk. Make sure to use your voice, tone, accent, facial expression, and body language in a way that will make your audience remember more of what you say.

I recommend a book by Chris Anderson entitled *TED Talks* (Anderson 2016). He served as the head of TED. Several suggestions have been made that will change our perception of public speaking. Presentation literacy is a skill that must be learned; it is not a natural gift. As a researcher, you can be famous and publish a large number of papers. Nevertheless, if you lack presentation skills, the moment you begin giving a public speech, nobody understands you, oops, that's a pity.

6.3.3 Artificial Intelligence-Voice Presentation

In the past, we would physically attend conferences, sit on a chair, and listen to presentations. Conferences during the COVID-19 pandemic were held on virtual platforms, either live or pre-recorded. Besides, it is also becoming more common to hear scientific podcasts. We have probably all tried recording some lectures or presentations and found that sometimes our pronunciation was less accurate, incorrect tenses were used, too much errs or quiet moments, or simply we did not like certain parts of the recording. Re-recording is a tedious process, and video-audio editing may not be something we are familiar with.

Artificial intelligence tools could be used to assist us in resolving the issues stated above. A few examples are Speechelo, Descript, Synthesia, Azure Text to Speech API, Murf.ai, Altered.ai, Amazon Polly, Yepic Studio, and others. For example, with Speechelo, we can choose from over 30 voices across 23 languages, and we can set the tone, speed, or pitch of the voice. A text file or script can be used as the input file, and Speechelo comes equipped with an AI-text editor that allows the user to check punctuation marks to ensure that the speech is grammatically correct. Unlike other software programs, Descript converts our audio into text, which can then be edited or deleted. The software converts the text back to audio after it has been edited or deleted. If this application is suitable for you, there are some video clips available on YouTube for readers to view. It might be a good idea to consider this tip if you want to become a scientist sharing on YouTube, but do not want to show your face or your voice.

6.4 Summary

For those of us whose native language is not English, do not worry. Let us focus on the scientific content and the scientific questions when we are preparing proposals or manuscripts. Utilize a human or artificial intelligence proofreader to maximize the clarity of the written message. A tool like AI proofreaders and paraphrasing can be a double-edged sword. Be ethical when using it. Original main ideas or sentences should always come from the author, while proofreaders should only serve to improve clarity.

We are comfortable with the method (style) of preparing slides and presenting a presentation that has been established for many years. It is always possible to organize and explain information on the slide in a better way. Are you up for making a change for betterment? The main takeaway is to use modern slide designs, for example infographics, and to avoid long wording, such as complete sentences and paragraphs. As I prepare this book, Canva is becoming increasingly popular. Additionally, there are several AI technologies that may assist you in creating presentations. It is not my intention to promote infographics, Canva, or any other trend or commercially available template. It is my objective to try to persuade you of the logic of making a change in the way we prepare slides and presentations, and the need for that change to be made. Moreover, remember that the goal of making a presentation is to describe the message clearly and catch the attention of the audience rather than reading the script or words displayed on the screen.

There is a special note I would like to make to all scientists who are interested in being impactful. It is not enough to polish your scientific knowledge, lab skill, or bioinformatics skill. It is important to polish your oral skills and your writing skills as well. This is something that needs to be done.

References

Alley M (2013) The craft of scientific presentations, 2nd edn. Springer, New York, NY

Alley M (2018) The craft of scientific writing, 4th edn. Springer, New York, NY

Anderson C (2016) Ted talks: the official TED guide to public speaking. Mariner Books

Day RA, Gastel B (2006) How to write and publish a scientific paper, 9th edn. GreenWood Publishing Group

Youknovsky A, Bowers J (2020) Sell your research. Public speaking for scientists, Springer Cham

7 Lessons Learned from Nature's Reports

Abstract

The journal *Nature* is one of the most influential journals. The journal is divided into several sections, including Research Articles (where most prominent scientists dream of publishing one in this section), News, Opinion, Research Analysis, Careers, Book & Culture, Podcasts, Videos, etc. On most weekdays, Nature Briefing will send us an e-mail recommending that we read their science news, opinions, analysis, and other informative articles. The author has handpicked a few interesting news and analysis. This chapter presents some key topics that young researchers and students should know about and some lessons we can learn from these non-biotechnology reports that will assist us in embracing change. In this chapter, we will discuss four major themes, including stress and mental health, good lab culture, mentors and collaborators, and possible quitting.

7.1 Good Lab Culture

It is unlikely that most of us would expect to find an article with the following title in *Nature* journal: 'How laughter in the lab can improve your science' (Dance 2018). Oxford University's Saïd Business School suggested that workers are 13% more productive when happy (Bellet et al. 2019). In my opinion, a happy lab culture is a good lab culture. I enjoyed the lab culture while I was pursuing my PhD because it was both fun and funny (of course it was stressful too). We had a few funny students who were humorous and some were really silly funny.

Having someone who is a ray of sunshine or cheerful will be a great asset to our lab. Usually, someone who is jovial is very talkative and enjoys socializing with others. It is important for us to have a few students in our lab who are jocular—

K. M. Goh, *Research Methodology in Bioscience and Biotechnology*,
https://doi.org/10.1007/978-981-99-2812-5_7

someone who is playful or likes to joke. Laughter strengthens the bonds of teamwork (Dance 2018). Do look for this reference and read it.

To shed more light on the question of whether students who study abroad are happy, it is worth referring to the online Career Feature in *Nature* dated 14 November 2022. The article titled 'Obstacle race: the barriers facing graduates who study abroad'. In a *Nature* survey, respondents were asked about their motivations for studying outside their home country. About 49% of respondents said the opportunity to experience another culture, 45% said funding opportunities were limited in my home country, and 44% said there were more job opportunities overseas after graduation. For international students, the first few months or even years can be challenging. Some of the problems they face are food, language, culture, homesickness, among other issues. According to that survey, 53% of respondents report that they have been treated unfairly or have felt unwelcome because of their race.

While it may be challenging to control the behavior of the general public, it is crucial that within our institute, international students are treated with equality and made to feel genuinely welcome. It takes foreign students some time to adjust to a new culture, language, and community. Fostering a sense of inclusion through shared laughter can be a powerful way to make them feel accepted. Laughter may result from jokes, provided they do not hurt or harass anyone. Laughter helps newcomers, especially those from foreign countries, adjust more quickly and feel more comfortable (Dance 2018). There is wisdom in the proverb that says, "A cheerful heart is good medicine, but a crushed spirit dries up the bones". We can aspire to create an environment in our lab that embraces this wisdom. Would you be interested in working together to make your lab a more fun and enjoyable environment? This is not a slogan; we can work towards achieving this goal.

7.2 Mentors and Collaborators

PhD students need a competent mentor (Woolston 2019). In the same way, all researchers, including principal investigators, need mentors too. Most people who perform well have a mentor or a group of mentors who have guided them throughout their careers. How does mentorship work? A mentor is defined by the Cambridge Dictionary as a trusted and experienced individual who provides advice, guidance, and support to another to assist them in developing. In certain universities, supervisors are the default mentors for doctoral or master's candidates. There are some institutes that match the senior academicians (mentors) with the younger members of staff (mentees). Do such default match-making systems work? In my opinion, there are indeed some success stories; however, the success rate of mentor-mentee relationships is low, for a variety of reasons, including lack of chemistry and other possible factors.

Mentors and mentees indeed play a crucial role in research, provided that the relationship is genuine and based on mutual trust and respect. I have had mentors all my life—a pastor mentor during my secondary schooling, a qualified-counsellor mentor during my undergraduate studies, and a professor mentor after I became a

Fig. 7.1 Experts vs. mentors vs. collaborators

principal investigator. The impact of these mentorship experiences has been invaluable. However, understanding what makes a mentor-mentee relationship effective can sometimes be puzzling. In my quest to explore this further, I came across two insightful resources: Rick Pastoor's book *GRIP* (Pastoor 2022) and Ruth Gotian's *The Success Factor* (Gotian 2022), I have found good answers.

Rick Pastoor wrote that expert and mentor are two distinct concepts. A supervisor of a project is an expert but is not necessarily a mentor to the student. Even though a prominent professor is an expert in a particular field, if he is not familiar with a mentee, therefore the professor is not performing the role of a mentor. We can refer to the professor as an expert, but not as a mentor (Fig. 7.1). The book *GRIP* defines a mentor as a person with whom we speak regularly, who is familiar with your situation, has a good rapport with you, and offers objective advice in addition to being knowledgeable and experienced. The institution has requested that I mentor a few staff in my department. I have been mentoring a colleague, who is a friend, for more than 10 years. We discuss a wide range of subjects, including life, family, health, mindsets, research, and publication on a regular basis. This ongoing dialogue allows us to provide support and guidance to each other in a comprehensive manner.

If a supervisor does not meet the criteria of a mentor, can a student find a real mentor elsewhere? Certainly. The supervisor should remain the supervisor of the project, but the student can receive support or advice from someone he/she trusts and has a rapport with. Generally, a mentor would be an experienced researcher in the lab, a co-supervisor, a postdoc, a senior student, or another lecturer who is directly or not directly involved with your project; however, the mentor must be a person of high integrity and ethics who will not steal the project's ideas or data. Young academics who are reading this, if you cannot find a mentor within your department, consider finding someone from outside. I think when the rapport is right, there is always someone who is willing to be your mentor.

Ruth Gotian, the author of *The Success Factor*, offers further insights in an article *Nature*'s Career Column titled 'My lesson from successful scientists: success can be learned'. In this article, she emphasizes the role of mentors in teaching us not only the necessary skills and knowledge but also the correct mindsets for success. While mentees may be eager for quick advancement in their careers, mentors often prioritize passing on their legacy to the next generation. The legacy could be in the form of a correct mindset and strategy. Good mentors are successful people wanting others to be successful. There are several other very practical guidelines for us to consider in Chap. 10 of her book *The Success Factor*, and you are highly advised to read it (Gotian 2022).

Mentors who are genuine will never attempt to mould you into a miniature version of themselves. Mentors believe in you more than you believe in yourself. A good mentor is someone who has achieved success and wants to make others successful as well. Mentors are individuals who wish to ensure that you do not make the same mistakes that they have made in the past. Get your mentor today, not just one person, you can have multiple mentors with different skill sets and expertise. Having a team of mentors in place provides you with a diversity of ideas, information, social capital, and network. By establishing a team of mentors, you will be able to minimize time spent on trial and error in areas such as research strategy, publications, grant management, etc. Mentors can easily see our blind spots. If you are selecting a mentor, consider whether you feel comfortable when the person is present. After speaking with the person, do you feel inspired? Are you in agreement with his or her philosophies or methods of doing things? You should first develop a strong friendship with the individual before asking him or her to mentor you. There may be a retiree who would make an excellent mentor for you. Mentors and mentees are not necessarily involved in a joint research project. Mentors who are genuine do not expect any benefit from their mentees, such as having their name on the paper.

My belief is that mentors can make excellent collaborators if the two of you are involved in similar research fields. But, if the mentor is a sleeping partner and expects the mentee to publish in return, I do not believe that there is a sincere relationship between mentor and mentee. There are some mentors who behave like bosses or managers and give instructions rather than advice. The mentees are treated as free labour. There are some mentors who attempt to leverage their own territory and empire by exploiting the resources of their mentees. The senior may claim all credit and ownership for the mentee's ideas or outputs. The mentor may attempt to influence the mentee to create a miniature version of themselves. These are not the true spirit of the mentor-mentee. I am hoping that such a relationship will not become sour.

Let us now discuss an article that was published in *Nature* on 13 May 2022, entitled 'How to pick a great scientific collaborator'(Pedersen 2022). The author used an interesting analogy to explain collaborators must share similar goals. If one party is seeking publication in *Nature* or a Q1 journal while the other demands a simple book chapter, then clearly this collaborator does not suit you. In his article, the author suggested a few important points.

1. The collaborator must be easy to work with.
2. The collaborator must be willing to contribute to the project.
3. You and the collaborator must share similar goals.

I consider finding a collaborator or mentor to be equivalent to making good friends with good bonds. It is natural that we tend to form stronger friendships with individuals who have similar goals or mindsets to ours. *New Scientist* published an article entitled, 'You're more likely to become friends with someone who smells like you' (available online). One of the experiments in the aforementioned online article involved smelling our own hands after shaking someone's hand. Certain people instantly pair up (click) due to their similar chemistry. The 'smell alike' concept has a long history despite the lack of proper experimental evidence to support it. I heard about that concept even when I was a kid. Discuss more frequently with the individual with whom you intend to collaborate and 'smell' (not necessarily by the nose, using the gut feeling will help) if both parties have the same chemistry. Collaboration can be miserable if the wrong individual is chosen at the outset.

7.3 Stress and Mental Health

Some universities or organizations may think that the issue of mental health is not a pressing issue that warrants the investment of time, money, and effort. But is that true? Mental health is defined as a "state of mental well-being that enables people to cope with the stresses of life, realize their abilities, learn well and work well, and contribute to their community". This is the official definition given by the World Health Organization (WHO). World Federation for Mental Health, founded in 1948 in London, has been committed to advocating and educating about mental illness. World Mental Health Day is celebrated worldwide each year on 10 October—the result of a global advocacy and awareness program started in 1992 by the World Federation for Mental Health. The theme for 2022 and 2021 was 'Make mental health & well-being a global priority' and 'Mental Health in an Unequal World', respectively. The theme for 2022 was 'Make mental health for all a global priority'. Having stress and mental health is not the end of the journey. The majority of people with mental illness can still achieve success if they receive appropriate support. This holds true for students, postdocs, academic staff, and researchers alike.

7.3.1 Large-Scale Survey

Since 2011, the Nature publisher's careers team has conducted a large-scale PhD survey. More than 6300 graduate students participated in the 2019 survey (before the COVID-19 pandemic hit) (Woolston 2019). The issue of students struggling, depression, and anxiety was raised in the 2017 survey, and it was raised again in 2019 (Woolston 2017). Sometimes, the source of the accumulated bad feeling may come from the project, flawed system, the supervisor, the peers, thesis writing, manuscript writing, or KPIs. There could be some relationship between personality factors (i.e. self-esteem, perfectionist, etc.), family history, stressful event (i.e. loss of a family member), and chemical imbalance in the brain (i.e. serotonin), with anxiety and depression. Stress is caused by a variety of factors. Some people respond positively to stress. A few accumulated stresses may, however, be the straw that breaks the camel's back.

The CACTUS Foundation has analysed a separate survey on mental health issues and published a report on their website. The report is entitled 'How to Foster a

Supportive Research Environment'. A total of 12,000 respondents from 169 countries participated in the survey. It is definitely a good read, covering topics such as harassment, discrimination, bullying, better pay for researchers, work-life balance, and move away from a publish or perish culture. Students and researchers put a lot of effort into the hard part of research, for example, finding grants, writing proposals, writing theses, submitting manuscripts for publication, developing networks, etc. However, not many people are aware of the importance of mental health. I recommend that readers consider the importance of maintaining healthy mental health during their careers. A mental health issue has emerged in the society as a whole. Academicians, students, and postdoctoral fellows also suffer from mental health problems that require our attention.

Mental illness can also lead to depression and anxiety. There is a short article in *Nature* entitled, 'Mental Health - Out of the Darkness' (Amanda C. Niehaus 2018). An individual said she was first diagnosed with schizophrenia when she was in her first semester in 1982, and it took her 10 years to become better and she struggled for 20 years before she was able to fully understand and accept her illness. Currently, she is a lawyer and professor at an academic institution. She has authored a few books that shared her experience and insights. She has given several talks, including a TED talk, which can be found on YouTube. Her name is Elyn Saks.

We may not understand the suffering that mental health patients endure. They may feel as if everything is wrong and that everything looks bad, withdrawal from social interactions, cannot sleep, are moody, cannot concentrate, or even attempt to harm themselves. I like this sentence written in an article in the reference, "Depression and related disorders can be devastating, but there are ways to fight them" (Amanda C. Niehaus 2018). This statement acknowledges the profound impact of these conditions while offering hope and emphasizing the importance of seeking appropriate support and treatment.

7.3.2 Stress Caused by Money

It was reported in an issue of *Nature* that the salary (stipend) of a biological science PhD student is below the basic cost of living in the USA (Woolston 2022). Such data may be scarce in other countries, but I believe they are approximately the same. Increasing inflation, increasing fees to be paid each semester, high gas and petroleum prices, costly food, high room rental fees (or mortgage) and other financial obligations impose a significant burden on academics, researchers, postdocs, and students. The stipend or salary of students is the lowest of these groups of people. There may be students who work part time and are pursuing a PhD at the same time. Given that biological experiment is typically time-consuming and labour-intensive, and students who have to work part-time will face an additional stress.

This book was written in 2022, a year that was marked by post-pandemic challenges and problems caused by the war(s). According to a *Nature* article published online on 31 October 2022 in the Career Feature column, postgraduate students were adversely impacted by the high cost of living crisis. As reported in

Nature's 2022 survey, 85% of students worry that they will not be able to afford food, rent, and other necessities. Because students are concerned about money for food, they have difficulty focusing on their responsibilities as students (reading, researching, writing papers, etc.).

Despite the challenges of limited of research funding, it is observed that some supervisors accept students even if they do not have sufficient funding. It is my understanding that some PIs indulge in 'gambling' by enrolling first a candidate in a program in the hope that the proposal for fundings will be accepted in a few months' time. However, what will happen if the proposal for applying grant is rejected? With rising costs, students without a proper stipend are likely to face shortfalls, and buying food or renting an apartment becomes a major burden. In the absence of grants, some lecturers are purchasing chemicals or kits for students from their wallets. The strategy, however, cannot be sustained. Do not be surprised that some students purchase chemicals or kits with their own funds (or their parents' funds). However, I do not believe the PI or supervisor is acting appropriately, even though this situation may not be regarded as research misconduct unless the student is being cheated by the supervisor with empty promises.

The consequences of insufficient funding in research programs can indeed be significant, leading to challenges and difficulties for students. Many of them may have ended up dropping the program or experiencing a great deal of difficulty at a later stage (i.e. not enough data due to an inadequate budget for buying kit supplies, reagents, cells, etc.). It is too expensive and unaffordable to carry out wet-lab biological science experiments on a personal basis. The lack of adequate funding restricts their ability to fully explore their research questions and limits the scope of their experiments. This situation not only hinders their academic progress but also adds to the stress and pressure they experience. I have been reflecting on the motivations behind supervisors accepting students without adequate grant funding, and it appears that the pressure to fulfill key performance indicators might be a driving factor in such decisions.

7.3.3 Stressful Postdocs and Principal Investigators

Are students in research the only ones who are stressed? Certainly not. Postdocs and principal investigators also face stress from time to time. The supervisors of PhD students are experiencing sleepless nights due to the COVID-19 pandemic. The resignation of workers, including researchers, has been occurring internationally, particularly in the USA and the UK, as a result of the pandemic, stagnant wages, pension cuts, bias, bullying, and hostile research funding environments. The following two short reports are available in *Nature*'s News and Work column for those interested in learning more: 'Mid-career mass exodus' and 'Sleeplessness and anxiety: PhD supervisors on toll of COVID'.

7.3.4 The Way Out Where There Seems to Be No Way

The terms 'stress', 'anxiety', and 'depression' have distinct meanings and are not synonymous. The WHO displayed some messages regarding depression on its Facebook page in Sept–Oct 2022. According to the WHO, depression is not an indication of weakness, but rather an illness that is characterized by a persistent sense of sadness and a loss of interest in things normally enjoyed by the individual. As a result of depression, they are unable to carry out their daily tasks. Symptoms of depression include lack of energy, changes in appetite, sleeping more or less, anxiety, reduced concentration, feelings of indecision, restlessness, feeling worthless, feelings of guilt, thoughts of self-harm, and feelings of worthlessness. If you or someone you know is experiencing depression, it is crucial to seek help and support. The World Health Organization (WHO) provides the following recommendations:

- It is important to talk to someone we can trust about our concerns.
- If we need professional assistance, it is a good idea to seek it.
- The third thing we can do is engage in some activities that we usually enjoy doing.
- Stay in contact with family and friends as much as possible. It is a great idea to go on walks with your family and friends.
- Regular exercise is recommended.
- Maintain a regular sleeping and eating schedule.
- The use of alcohol and illicit drugs should be avoided or restricted at all costs. Maintain a healthy diet.
- Have a talk with a trusted friend about your feelings.
- Whenever we feel suicidal, we should reach out for help from a trusted person or call the emergency services if needed.

Participation in a purpose-driven group is one method of coping with symptoms of anxiety or minor depression. One example of such a group is highlighted in the experience of Karra Harrington, a PhD candidate who organized regular meet-ups with six to ten students. Each meet-ups had a clear goal and purpose. Whether it was dedicated to writing a thesis quietly, discussing how to find a mentor, or just for a cup of coffee, the participants had a specific focus for their gathering (Harrington 2018). As I recall, peer support was very important to me when I was doing my PhD. Everyone was busy with their own projects when it was 3 pm and it was time for a few of us to leave for a short break. It is during this 15–30 minute break that we drank tea or coffee, and we discussed certain unclear points in our experiment or writing with a friend present for advice and support. Sitting quietly alone is not an effective method of releasing stress. Talking to someone can help.

From reading the article 'Why mental health matters' (Woolston 2018), we can learn the following. Support is crucial. Organizations like the Cheeky Scientist Association provides advice and support to students worldwide. In addition to providing mental-health support, campus counsellors should remain visible so that students and staffs will not be hesitant to contact them. There may be some students who are low in emotion, but they are afraid to see a counsellor for many reasons: the cost of the session (it is free at our campus), the long queue and their inability to wait, the language barrier, or other obstacles. Many university counsellors are doing a fantastic job. Student Minds in the UK is one of the organizations responsible for providing mental health and support to students. More organizations like this are needed in each country. In Malaysia, there are multiple support teams, for instance, Befrienders, Agape Counselling Centre, Malaysian Mental Health Association, Thrive well, MIASA, Life Line Association Malaysia, All Women's Action Society, Women's Aid Organization, etc. These organizations play a crucial role in offering mental health assistance to individuals in need. Additionally, researchers should not take lightly the stress caused by threats, harassment, and intimidation. Reporting such incidents is important. Researchers in the Netherlands can report threats and harassment through SafeScience.

The A2-size posters caught my attention in one of the universities I recently visited. The posters were placed on a notice board near the entrance where the research students are housed in a wing of the building. The first poster was produced by the university with the following words: "Please write to us if you have witnessed or experienced something that distresses or concerns you". Kudos to the university for caring about its students. A second poster was created by Befrienders, a non-profit counselling centre. This poster depicts a squatting individual wearing blue clothing under a grey rain cloud. It carries the message, "Feeling depressed and overwhelmed? You are not alone. Helps is just a phone call away. You can reach us by phone at ___, by WhatsApp at ___, or by email at ___. Mon-Sun 3pm to 12 midnight". The Befrienders team understands the importance of providing assistance in the night-time. Helping hands and listeners are always available (Fig. 7.2). Barnabas is present around us. Helping one person may not change the world, but it may change the world for that individual. (Barnabas, in Greek means son of encouragement. In the bible, Barnabas was very good in mentoring and supporting other people.)

When you are experiencing stress, anxiety, or minor depression, do not isolate yourself from the world. Instead, go out and converse with someone, go for a walk,

Fig. 7.2 Barnabas is present around us

go to the gym, seek out a good friend, consult a counsellor, speak to a mentor, or return to your religious community. One day, I was taking an after-lunch coffee at the canteen during the semester break, which is located within walking distance of my office. As I approached the bench, I noticed a student sitting alone with her computer on the bench. As she explained to me, she was preparing a thesis and needed a space that offered fresh air and an open view. It was a wise decision for her not to stay in her own room alone during this challenging time. It was by chance that I met her and encouraged her to move forward.

A helping hand is definitely needed for those in need. During challenging times, turning to one's religious beliefs, if they have them, can also provide solace and support. Mr. Don Moen sang a song in 1992, 'God will make a way', which was very encouraging to me and many of my friends during times of stress and difficulty. If you are unfamiliar with the lyrics, you can find them on the internet.

7.3.5 Is Stress Really Bad?

The CEO of an enzyme company I know is older than 80 years of age. He is over the retirement age for most people, but he is still working and willing to share his knowledge with the younger generation. As he stated, "productive stress is when I achieve something, and non-productive stress is when nothing is achieved". Whenever we encounter obstacle and seek to overcome it, we are experiencing productive stress. Ultimately, this type of stress results in happiness and satisfaction. We become better and resilient when we overcome these obstacles. In contrast, if a person has the habit of running away or avoiding facing the problem, non-productive stress generates no results at the end.

In a recent posting on a social media site, a picture of a signboard was posted with this message: "FEAR has two meanings: Forget everything and run or Face everything and rise. It is your choice". Someone I know sent this message to me: "If you get tired, learn to rest, not to quit".

Due to a variety of factors, we may sometimes become stressed and unable to do anything. It is important to note that most of our worries in research can often be very small and can be resolved easily, but our brain and feelings exaggerate it. Sometimes we are unsure of what we are worrying about. Those who tend to worry too much should simply convince themselves not to worry about tomorrow, for tomorrow will take care of itself. You might want to consider reading an old book titled *Stop Worrying and Start Living* by Dale Carnegie, whose content remains relevant today. As Dale Carnegie wrote, if we take good care of ourselves today, tomorrow will take care of itself. You can follow Dale Carnegie's advice by writing down the following if you are worried about something: (1) 'What am I worried about?' When we write down things we are worrying about, about half of our worries are gone. (2) The second question is 'Is there anything I can do about it?' Another 40% of your worries are gone when you write down all the possible options. The remaining worries will reduce once you start taking practical steps in resolving the problem.

Is it possible for a person to complete a PhD program or a research project without experiencing any stress or fear? Unfortunately, I do not believe it is possible. Dealing with stress is an integral part of completing a PhD program. Each student is an individual with his or her own strengths and weaknesses. During the PhD journey, strengths will continue to shine; however, weaknesses may contribute to the stress factor or prevent the student from reaching the next stage.

In writing the thesis, some students fall into a pothole of anxiety and find it extremely difficult to climb out. The other students may have relatively weaker skills in critical thinking, computer skills, or even a pair of good hands when it comes to experimentation. In some cases, students are too timid to speak up and others have social anxiety. A perfectionist may dig his own pothole and either jump into it or accidentally fall into it without noticing it. Occasionally, I encounter students who are too comfortable with their comfort zones or are accustomed to habits developed during their schooling and undergraduate studies; unfortunately, these students are unwilling or struggling to take on new challenges, overcome frustration, or overcome setbacks. Researchers and students sometimes fall into the trap of thinking in a way that leads them down a dead end. As a result, students will fall into their own potholes.

Occasionally, I may give my students a slightly different path for their PhD journey so that they will be exposed to weak points more frequently. In my opinion, PhD training involves one very important aspect, which is the ability to overcome one's own weaknesses. As a mentor, we may be able to recognize their weaknesses better than they do. Students are allowed to fall into the pothole and climb up slowly, and of course, we should give them a hand. Having gone through this process myself, I understand the difficulty and struggle, similar to the pain we experience as parents when our toddlers take their first steps and fall. Research is not a battle against data collection, a thesis, or a publication; it is a battle against ourselves. Do you think so?

It is likely that some readers are interested in marathons or half-marathons. My favourite outdoor activity is hill climbing. Engaging in hill climbing not only tests our stamina, resulting in exhaustion and thorn cuts, but it also exposes us to the risk of insect bites, muscular strains, and potential injuries from falls, leech silently sucking the blood from our feet, and other unexpected incidents such as monkeys stealing your drinks or food. Even though the journey is difficult, there are plenty of surprises along the way, such as fresher air, beautiful views, fauna and flora, waterfalls, breathtaking sunrises, and enchanting encounters with wildlife, making each moment of the experience truly memorable. Marathoners and climbers are aware that their greatest challenge lies within themselves and their determination. PhD is like climbing a hill (not a mountain) full of hardships, and once we have overcome all of these obstacles we will reach the top and know how well that sweat was rewarded (Fig. 7.3). Having an intrinsic motivation and passion, perseverance, and a bit of stubbornness that believes we are doing the right thing, we can get over, through, under, or around any obstacle.

Life is full of challenges, including PhDs. There will be hardships along the way. Patience is the key to success. Being tired is part of the job. Only if you do, you will

Fig. 7.3 The PhD journey is depicted as a climb up a hill

get it. Do not envy people with PhD degree if you really want it, strive for it. Those of you who are still feeling stressed and lost with your way, should try to read Chap. 8, as there are many helpful advice that will be of great help to you.

7.4 Quitting the PhD Program, an Option for Students?

An American motivational speaker named Eric Thomas once made this statement: "Pain is temporary. It may last for a minute, or an hour or a day, or even a year. But eventually, it will subside. And something else take its place. If I quit, however, it will last forever".

During the course of pursuing a PhD, many candidates will encounter a number of difficulties. For example, an experiment that fails and no data is collected for months, the breakup with a girlfriend or boyfriend, getting married and getting pregnant, caring for a newborn (or newborns), the untimely death of a close relative, a financial crisis or health crisis in the family, conflict with the lab members or supervisor, losing interest, depression, or other situations may pose challenges. These circumstances may at times make students contemplate quitting their PhD research.

Among the events that have taken place in the previous paragraph, I have encountered a few difficult ones. Honestly, I was questioning myself why I am doing a PhD and having such a challenging time right now. I struggled for more than a semester trying to purify my protein, which made me very frustrated because I considered myself an expert in protein purification because I had been doing this for 2 years in another research institute before beginning my Master, and subsequently converted to a PhD program. It was a frustrating and stressful time. Additionally, I faced other problems, particularly financial concerns and the responsibility of caring for a newborn baby, which often required late-night attention.

During a time of deep emotional distress, the poem *Don't Quit* was very motivating for me. We can find it online. It served as a reminder that challenges are a natural part of any journey, including pursuing a postgraduate degree. Having already established a clear vision of pursuing a postgraduate degree when I was a young adult, I was determined enough to remain on the PhD training journey. I witnessed instances where fellow students abandoned their PhD work midway, unable to find their answer to the question of why they chose this path, and succumbing to the difficulties they faced. However, I made a conscious decision not to run away from these obstacles. Instead, I faced them head-on, remaining committed to my lab work and writing, and ultimately completed my PhD within a span of three and a half years. And I published one Scopus and two WOS papers. In the absence of a strong reason and perseverance, the idea of quitting can become a persistent temptation. It is during these moments of doubt and difficulty that we must remind ourselves of our initial motivations and the long-term goals we set for ourselves.

A News-Blog published in *Nature* also expresses the opinion that it is normal for a candidate to consider quitting their studies at certain points. The blog 'Three questions to ask yourself before quitting your PhD' suggested these points:

(1) If the problem can be solved, it is likely not necessary to quit. If the supervisor is the problem, switching to another supervisor may be an option. (2) Before making a decision, consult with someone. There are times when peers may look at the problem from a different perspective, and their judgement may be more accurate than the student's. (3) Does the PhD fit with the student's long-term goals? One of my PhD students quit during the first semester of the program. Prior to enrolling in the program, her father passed away. Following months of consideration, she decided to resign and return to her home to take care of her mother and family business. As doing a metagenome project does not contribute significantly to the long-term goals of the family business, a PhD does not seem a good fit.

Gemma Conroy reported that approximately half of the total number of PhD applicants in North America fail to complete their doctoral studies due to a lack of support from their supervisor, in an article published in *Nature*'s 'How to choose the right PhD supervisor' (available online). Four points were raised by the author: (1) interview the supervisor before registering for the program; (2) speak with someone who is familiar with the potential supervisor, such as a former student or collaborator; (3) find a supervisor that can enable them to develop their career and skills (not limited to their thesis or project) after completing their PhD; and (4) take into account the supervisor's lab culture, research interests, meeting frequency, and working style.

The following is a testimony published in *Science* under the title 'My PhD adviser fired me. Here's how I moved on' (available online). The author of the article left her native country and began PhD studies abroad. The supervisor asked her to discontinue the program after a few months. There is no indication of the real reason why the supervisor made the decision, but in any case, this is not the main point here.

Although it was a difficult time for her, she is determined to obtain a PhD. In another country, she found a more appropriate project and supervisor. She took it positively. Whenever one door closes, another one opens. Do not stare too long at the closed door as we may miss the window opening.

I would like to remind readers that failing a study or quitting a study is not the end of the world. It may not always be our fault and sometimes, we may not be in control of the situation. On the Editage website, I found a document written by a student who failed a program: 'I failed my dissertation defence. But I am not a failure'. This article is available on the Internet for free and do read if you can. Stay positive and move on.

7.5 Summary

Readers are encouraged to read all the references provided here as well as the online materials suggested in this chapter. The main purpose of this chapter is to remind all researchers that they must take care of their physical and psychological well-being. In the event that you plan to dedicate your entire career to research, ensure that you maintain your mental health and remain happy despite facing all forms of stress. Running a marathon requires a high level of determination and stamina. If you are in the academic-research line, the journey of conducting research is also a lifetime endeavour. It is essential to have good mental health, a good research philosophy, a strong mind, and a solid set of skills and knowledge during our journey. Any difficulties we encounter during the research process can always be overcome. When we feel blue, talk to someone—spouse, parents, friend, lab mate, supervisor, mentor, counsellor, collaborator, religion leader, or personal God. This is written in Matthew 11:28: "Come to me, all you who are weary and burdened, and I will give you rest".

We all have experienced failures, whether they were large or small. It is time to let go of the past and to look forward to the future. Don't let failures keep you from moving forward; however, make use of your bad experiences to create a stairway. Whenever we are faced with a challenge, think positively and keep in mind that productive stress makes us stronger. The more obstacles we overcome, the better and more resilient we become. I want to borrow this idea from Philippians 3:13–15: "...Forgetting what is behind and straining toward what is ahead,... (NIV version), or ...I forget what is in the past and try as hard as I can to reach the goal before me (ERV translation version)". Some readers may be familiar with these verses. Many times we say that we have forgotten, but in reality we are still holding onto those setbacks. The word 'forget' in this verse does not mean erasing it from memory and trying not to recall it again. 'Forget' here does not mean that we should forget about our mistakes without reflecting on them. Take note of the past, but move forward. Do not let mistakes discourage you, but learn from them. Without learning from the past, we will make the same mistakes again and again. Would you agree with me?

Are you still feeling blue after reading this chapter? Take a jog or go to the gym and get wet and sweaty. I think you will feel a lot better after that.

References

Niehaus AC (2018) Mental health - Out of the darkness. Nature 557:267–269
Bellet CS et al (2019) Does employee happiness have an impact on productivity? In: Saïd Business School Research Papers Saïd Business School WP, pp 1–39
Dance A (2018) How laughter in the lab can improve your science. Nature 555(March 29):689–691
Gotian R (2022) The Success Factor, 1st edn. Kogan Page
Harrington K (2018) Harness the power of groups to beat the "PhD blues". Nature 559(7712): 143–144
Pastoor R (2022) GRIP: The art of working smart. HarperCollins Publishers
Pedersen CL (2022) How to pick a great scientific collaborator. Nature. https://doi.org/10.1038/d41586-022-01323-9
Woolston C (2017) Graduate survey: A love–hurt relationship. Nature 550(7677):549–552
Woolston C (2018) Why mental health matters. Nature 557(7703):129–131
Woolston C (2019) Just a minute PhD students voice concerns on mentoring. Nature 575:551–552
Woolston C (2022) PhD students face cash crisis. Nature 605:775–777

8 Success Can Be Learned

Abstract

The title for Chap. 8 is ‘Success Can Be Learned’. Do you believe that you can be successful? Are more successful people different from others? This is one of the most significant and interesting chapters in the book and should not be skipped. In order to provide readers with valuable insights, I reached out to several experts who generously shared their advice, opinions, suggestions, and best practices. They have been through many hardships and learned many lessons the difficult way. We hope that the next generation of research scientists will adopt the proper mindset, habit, strategy, etc., so that they can be better biologists and biotechnologists in the future. There are many tips and topics that are not discussed elsewhere in the book, making it an essential read for those seeking success in their scientific pursuits.

8.1 Can Success Be Learned?

In your opinion, what is the definition of a successful researcher or an impactful researcher? I have a friend who plans to publish a paper in *ISME* journal and that is her measurement for success. Another friend of mine aspires to become a full professor, and that is how success is determined. Do the large number of students, grants, and publications in our record indicate our success? Can impactful research be measured using metrics such as citation counts, H-indices, percentages of publications in top journals, etc.? It is also possible to measure success by receiving a particular award, such as for the greatest amount of research grants, supervising, publishing, exhibiting, or being named to the Clarivate Highly Cited Researchers list.

Do we measure our success based on the size of our research empire or our impact on society or the next generation?

K. M. Goh, *Research Methodology in Bioscience and Biotechnology*,
https://doi.org/10.1007/978-981-99-2812-5_8

The definition of success varies from person to person. As each of us has a different set of measuring sticks, it may be difficult to define what constitutes a successful researcher. As your career progresses, your definition of success changes. As defined by the Cambridge Dictionary, success refers to "achieving the results wanted or hoped for". Success is not a destination, it is not the end of the journey, but rather a process. I do not believe that success and luck are related. We are striving to obtain success, not sitting back and waiting for success to fall on us from above like a raindrop.

A Career Column in *Nature* published an article in 2022 entitled 'My lesson from successful scientists: success can be learned' by Ruth Gotian. In particular, she is interested in exploring the habits, mindsets, lifestyles, and other characteristics of Nobel Prize winners, award-winning physicians and scientists, and other extreme high achievers. The hypothesis of the research was that if there were similar mindsets across disciplines, it may be possible to learn success. She published part of her findings in the book *The Success Factor* (Gotian 2022). The hypothesis made by Ruth Gotian inspired this chapter.

There are some researchers who enjoy copying or mimicking the ideas and publications of other researchers. A copyist may mimic in detail the scope of a published work and adapt that scope for their own use. When we reflect carefully, the publication success of another individual is only the tip of the iceberg, and copying their work is not a wise decision. It is the mindsets, practices, and beliefs of high achievers that are the key to their success, so we should study how they think and work. Mindsets have a significant influence on our practices, habits, integrity, and publication achievements. How we define success is influenced by our mindsets.

The following text contains the written or oral statements provided by the respective interviewee. As discussed in this chapter, we can learn a great deal from the comments and advice provided by prominent scientists, institutional management teams, industries, and advisors to policymakers. Let's dive in.

8.2 Comments and Advice

8.2.1 What Advice Would You Give to a Young Scientist Just Beginning His or Her Career in Science?

Kok-Gan Chan

I will break down this question into many aspects. Although it is described as a simple piece of advice, I believe the most important step is to discover what we call the passion for the research topic, particularly in the area you intend to pursue. Do not subscribe to a project just because it will generate high impact papers, attract funding, etc., because the rule of thumb is that we work on topics we are passionate about. The second thing I would recommend to young scientists would be commitment. I will just add a condition to this advice; the commitment is not from 9 to 5; it is actually from 9 to 5 and from 5 to 9 until the project is completed, or until your last breath. This is what I refer to as a commitment at all costs. We have heard enough stories about impressive achievements. Nobel Prize winners, such as Barry Marshall and Robin Warren, have demonstrated their commitment to the fight against *Helicobacter pylori.* That is the type of commitment I am referring to. Another example would be Marie Curie, who was very dedicated to her work despite suffering from cancer while working on the radioisotopes radium and polonium. It is important to have that kind of commitment. Scientists' position is more than just a job. It is a calling.

To the young scientists, I would suggest that it is always a good idea to think outside the box. Consider being a bit more bold and taking a risk on an uncharted project. Working within comfort zones may not always be the best idea, as most of the science in comfort zones has already been discovered, whereas diving into unchartered waters will provide more scientific discoveries. If you work on a topic that has no known answers, you will be able to make a discovery. We conduct research not because of the rewards or awards. The satisfaction that a scientist receives from solving a mystery is much greater than any other reward. So I think these three pieces of advice will be the most helpful to young researchers.

8.2.2 Do You Agree that the Primary Objective of a PhD Program Is to Collect and Publish Data?

No, I disagree. These are administrative rules. This is not the true spirit of a PhD. I am looking at it from an ivory tower vantage point in medieval times when people came to universities as undergraduate students for years. They reside in the college. That is why the oldest universities in the world still maintain the colleges system where they live with their mentor, so-called the academic staff nowadays. It is through the mentor's example and actions that students learn and gain knowledge of the way in which they live and behave. The knowledge was passed on in this manner to younger generations, whatever was beneficial to them. That, I believe, is the essence of medieval unity. It is my belief that the spirit should remain the same. Although centuries have gone by, the essence of a PhD remains the same: you are embarking on a highly focused area of research as you undertake your PhD. It does not matter what the topic is. All disciplines share this common objective of solving the mystery in their projects in order to push the frontiers of knowledge in mankind. It is important to keep in mind that the whole idea of a PhD goes beyond simply collecting and processing data. When we have AI technology available, for instance, what is the need for a PhD when there is an AI that can do the data mining and data analysis and there is an AI that can write the research paper as well? Obviously, these practices are unethical. The purpose of doing a PhD is to investigate an area where there is no answer as to how it will be solved. I believe that will be the spirit of the PhD, rather than simply collecting data.

8.2.3 If You Were to Give Advice to Readers Considering Pursuing a PhD or Young Lecturers Supervising PhD Students, What Would You Recommend?

I would like to offer the following advice to readers who are considering pursuing a PhD. The PhD degree is very different from an undergraduate degree or a master's degree. PhD journeys can be very lonely. Unlike undergraduates, you always work in groups, with peers from similar colleges or universities, classmates. Nonetheless, as a PhD student in science, you are almost alone in the lab, since you are hardly able to speak to anyone in the lab while you are working. You are focusing on a topic that nobody has the answer to, even your supervisor does not know the answer to. In fact, you are not even aware that there are materials or methods available to you that can be used in your project. Thus, you may have to find new materials and new methods to solve the problem that you have fixed. All of these require a passion for the topic. You should think carefully before pursuing a PhD if you are considering it. Here are some analogies or comparisons that I would like to provide.

Relationships between boy and girl or husband and wife are usually not completely romantic, although they may seem romantic at first when you focus on the appearance, the beauty, the body lines, etc. But, believe me, no matter how appealing to you all these so-called phenotypes may appear to be, they will eventually fade away as he or she grows older. Believe me, human beings age in an unattractive manner. I believe that the passion is what binds the two together. We refer to this journey as a journey of love. In the journey of pursuing a PhD, I call it passion in the topic.

Here is the first piece of advice I would give to young lecturers supervising PhD students. This is the way I used to be advised by my supervisor or senior. They called it the green doctor, which is now called the young lecturer when you left your PhD not too long ago. Since you yourself are a fresh graduate, it would not be wise for you to supervise PhD students. You may not be doing the students and yourself justice if you do that. Start with the master's student. When I was a young lecturer, I did the same thing.

Regardless of the KPIs your institution may impose, as academics, we have our own industrial practices. We are quite different from other industries in that being a mentor for a PhD requires a great deal of effort, a lot of collaboration, a lot of energy, time and resources. It is possible that a young lecturer does not yet have all these things in place to provide him or her with appropriate supervision. That is what I stated. The supervisor as well as the PhD student will not be able to do justice if they refrain from doing so.

However, if you are extremely interested in doing that in the early stage of your academic career, then make sure you are well equipped, especially with regard to training PhD students in science and engineering. Now let's get practical. PhD candidates are going to need a great deal of equipment to accomplish their tasks. All of these lab facilities are required, including consumables and small items. It is very helpful to have a very decent lab where the PhD student can work in a comfortable manner. I would say that not all of the things are ready for the PhD students, but most of the things should be ready for them.

So again, once you have all this ready, financially equipped, infrastructurally possible, and you have the knowledge to do so, and then mind you the need to publish papers at the end of the PhD training. Depending on the institution, some students must publish their papers during their studies, while others may do so after they have been graduated. Whatever the case may be, publication is almost a necessity. As such, since this is another calling for academics, including young PhD students and young PhD lecturers, so publication is another journey, I wouldn't call it a journey of hell, but it is a journey of adversity. This is going to be a journey that you will need to take. It is not a simple journey for the faint-hearted.

8.2.4 How Do You Define Good Mentoring?

In my personal opinion, a mentor should be more experienced, wiser, and more knowledgeable than you are. This criterion is essential for qualifying as a mentor. Mentors and mentees must get along well. As the saying goes, it takes two to tango. An effective mentor must be willing to share his/her knowledge, wisdom, and experience with the younger generation.

Mentees should possess a highly receptive mind, humble themselves, and empty themselves in order to receive the wisdom from their mentors, who are often not necessarily older than them. Throughout my career, I have had the opportunity to speak with researchers from a variety of age groups and it has been my experience that the younger generation tends to view older generation researchers as outdated. To put it another way, the older researchers may have outdated techniques, but their minds are very, very fresh and youthful. It is true that the body may be weak, but the mind is indeed very incredibly sharp. It is the mindset of the mentors that we need to learn from in order to succeed. While the technique may be a bit outdated, he/she has the wisdom to see things that the younger faculty may not be able to see. That makes it a very helpful mentor. To make it even more clear, I will put it in the form of a simpler statement: as the mentor goes through the journey of his research, he gathers a lot of no-entry signs. It means that he made mistakes and learned from them. Once you have been provided with the no-entry signs, you will save yourself a great deal of time so you will not repeat the same mistakes again.

From the perspective of my own personal journey. I greatly appreciate their sharing and remember that in their old age, every minute counts to them, so if they open their mouths to share their wisdom with you, you better be attentive, keep your mouth shut, and learn from them. I have seen a lot of good mentors like this. It is a genuine intention. They are not expanding or extending their empire. Their goal is not to leave a legacy of themselves, but rather a good legacy of spirit, for example, a good spirit of science, a good spirit of research, and a good spirit of publication. In the context of someone expanding their empire, they might express it to junior staff by saying, "I'm in a senior position, and you work for me." This approach differs significantly from the earlier concept, where a good mentor focuses on imparting wisdom and avoids leveraging others' efforts for personal gain or leaving behind a legacy centered around themselves. Unfortunately, brain and knowledge transplantation is not viable in the medical field, and we cannot directly hook up a cable and download the wisdom from their brain to ours (just kidding), which is why the mentorship program is extremely important. As you look around, you will see angels all around you.

8.2.5 What Are Your Strategies or Tips for Generating Publications?

There are no tips, and it is not because I do not want to share them with you. Authentic publications require research, collection of data, writing, submission, and peer review; the force of peer review will crash most faint-hearted researchers. If there are any tips that we can share, I will share one. Working hard, working hard, and working hard.

I would like to encourage you to remain patience when someone attacks your publication, and do not take it so personally. The attacks should be viewed as constructive criticism that you can learn from. Criticism, whether constructive or destructive, is criticism. Consider it as a free gift that will help you to improve your research. Then you do not feel hatred, you do not feel anger, you do have hard feelings, you accept it, you improve your research, and then your publication journey may be smoother. Again, you should speak with your mentor who has gathered a large number of no-entry signs. They may be able to tell you whether you are digging your own graveyard or if you are taking the gold mine, which will help you know when you are there. While you go through the publication process, a bystander may provide you with additional tips. Consequently, I would like to reiterate the three answers I gave earlier: hardworking, hardworking, and hardworking.

8.2.6 How Do You Envision the Future of Bioscience and Biotechnology?

The last two centuries have been the century of physics, and the last century has been the century of chemistry. What else is there in science? I think it must be the century for the biosciences and the biotechnology, where we have been waiting for so long. Because physics was in the lead at the beginning, I wouldn't say there is much to discover, but when you're near the tip of the mountain, every step is like running along the Himalayan peaks. A breakthrough in the very established area of science will require a great deal of effort.

Biosciences and biotechnology, however, remain largely unknown due to the fact that there have been few breakthroughs during the last century. So there is much to be discovered in the field of biosciences, since it encompasses life sciences, diversity, and biodiversity, as well as flora and fauna. What about the microflora and microfauna? We have not yet seen all of them. I always think that a marine ecosystem is like Mars to me because it is so different from a terrestrial ecosystem. Therefore, while we are discussing things that are large, such as flora and fauna, it is important to consider the microorganisms underneath those things, as well as life at the extremes, such as in hot springs and at the poles. Biotechnology and bioscience will continue to advance with the advancement of other sciences, such as nanotechnology and the emergence of omics technologies. All of these are dependent on the chemistry and the physical sciences, and if they reach the breakthrough, then we can actually apply this to the research in the field of biosciences. The future looks very promising with the advancement of sequencing techniques and the development of omics technologies.

Artificial intelligence is advancing rapidly in this era but I personally do not believe that it can replace humans. Brains are still considered the world's best organs. If we consider AI as an organic organ, it pales in comparison to the human brain. AI applications are certainly useful, but they can never replace human beings. In big data mining, for instance, machine learning, AI is a very helpful tool, because humans would have a difficult time doing it manually. Let the machine do the data mining for you. Rather than having humans handle repetitive, boring tasks, let the machines do it. We should happily let the AI handle that part. The AI can be used to find patterns in big data, for example, or to detect certain phenomena in the bioscience, and then the computers would do the AI job for you, saying to you, "master, I've found it", but you cannot say you found it, you can only say that the LCD panel or monitor found it. Verifying AI-generated results still requires the human brain and wisdom. The AI is a good servant to us, and we are the master to identify the mysteries of science, research questions and gaps in science, and then verify what the AI suggests.

There are many bioscience and biotechnology topics that are emerging or that are expected to become hot in the near future. Some may represent the future, while others may not. If you are passionate about these topics, go for it. Should we jump into the bandwagon? It depends, again, whether you have passion in that topic, as I explained above.

It is up to you whether you wish to join the bandwagon. You may have several reasons for pursuing this, that is, you would like to engage in the new emerging topics because they can attract funding, get attention, attract mass media attention, and easily attract PhD students. It does not matter; however, the pre-requisite is that you must have a passion for that emerging research area. Please do not jump on the bandwagon without a passion for it. Assume that tonight there will be an announcement about the Miss Universe competition and that everyone will be rushing to get married to the winner of the competition. That sounds a bit odd, doesn't it? Instead, let your passion lead you to do research in areas you are passionate about.

8.2.7 Could You Please Briefly Explain the Importance of Social Science Research and What the Common Practices Are?

Lindsay C. Stringer

Social science essentially involves studying people—what they do, how they are organized, what they think, how they behave and interact (with both each other and non-humans), where, when, how, and why. Understanding people is important because as well as depending on each other both directly and indirectly, we also depend on external factors such as the environment for our survival. We need to know about these relationships and behaviours so that we can take decisions that support, reinforce, and improve the fundamental pillars of our existence, as well as making sure that the ways in which we are organized as a society are appropriate to that.

Disciplines that come under the social sciences include sociology, psychology, human geography, criminology, law, economics, anthropology, political sciences, linguistics, development studies, cultural studies, education, and more. In terms of how we do social science research, we use a range of qualitative and quantitative approaches. These include reviewing and analysing literature, policies, and other documents and archives; collecting new data for analysis (sometimes as input to models) through field-based quantitative methods such as questionnaires and experiments; and in-depth qualitative methods such as direct observations, communications, interviews, and so on.

8.2.8 How Can Social Science Elements Be Incorporated into Conventional Biotechnological Research? Can You Share with Us Some Examples?

One of the major drivers of biotechnological research is to advance knowledge to ultimately benefit people. Therefore, I think it is vital to incorporate social science elements into biotech research. This is important to understand how people might respond to advances in the field, but also to help identify where biotech could offer potential to meet people's outstanding needs.

It's all very well using biotechnology to improve, for example, the drought tolerance of a particular kind of crop, but if it then changes the crop in terms of, for example, the time taken to cook it, its texture or flavour, it could mean that people opt not to buy those so-called improved varieties. In that case, there would be little uptake and the economic cost of developing the improved variety will far outweigh the societal benefits that are delivered. If biotechnology and social science researchers can work together with those people in society who could influence or be influenced by the improved crop—these groups are known as stakeholders—it can provide useful insights into the demand side and allow the supporting societal structures and processes for the improved crop (including its storage, transportation, markets, supply chains, etc.) to be designed to facilitate its adoption.

Ideally stakeholder engagement should take place from the very beginning of a research project to ensure all relevant voices are included, but that is not always possible, as it can add to the project costs and extend the research timeframe. In many cases, the research funders also have their own agendas and priorities, so they are not necessarily fully open to stakeholder engagement from the research design stage.

It is nevertheless possible to involve social science approaches at other points in the research cycle, for instance, once a technology has been developed and researchers are looking to whether it can be exported to other places. Social science research can help analyse the policy context, to understand whether there are potential challenges and how they can be overcome, or whether the intended new market could be viable.

Another example comes from biotechnology that is looking to make food systems more sustainable. Technologies such as lab-grown meat or the increasing use of insects in food can often lead to a lot of assumptions being made within society about their safety. Identifying and understanding people's perceptions can help to identify where additional efforts might be needed in awareness raising and education about a particular technology. Opening these kinds of doors for dialogue between scientists and society can lead to much better outcomes over the longer term in relation to uptake and understanding.

8.2.9 What Message Would You Like to Convey to Young Women Who Are Considering a Career in Scientific Research?

Women have played a fundamental role in developing many of the world's most important scientific advances. As just one well-known example, the research of Marie Sklodowska Curie provided the foundation for understanding radioactive decay. Looking back in history, women have been advancing scientific knowledge for a long time. However, they have generally been under-acknowledged for their contribution. Across the Science, Technology, Engineering, and Medicine (STEM) fields, women have also been under-represented. While there are many systemic reasons for this, the world is gradually starting to change and it's an exciting time to be part of that change.

I would encourage young women who are considering a career in STEM to talk to female scientists, find out first-hand what the job involves and what the possible career pathways are. Ask them difficult questions, and if you are still keen, then go for it! Given the types of challenges the world faces today, science has a key role to play in developing solutions. Those solutions will be more equitable if the workforce developing them itself is more gender balanced. A career in science can help you to fulfil your intellectual curiosity while also helping to make the world a better place.

8.2.10 How Would You Describe an Impactful Researcher from Your Perspective?

Zhang Tong

We can answer this question from two perspectives. As a first step, it is important that a good project is designed and completed. I would define a good project as one that can provide answers to excellent scientific questions, and how the projects can offer scientists a new perspective or understanding of a topic they have never considered before. When we apply a better technology or a new approach to revalorizing wastes or adding values to products, some good projects can make a significant impact on society. As a whole, my first point focuses on the research or scientific outcome of the project.

As a second step, I believe that education outcomes are of utmost importance. From the beginning of my career as an academician, I have dedicated a substantial amount of time and effort to human training. In the course of time, our students will graduate or our postdocs will leave our group and may decide to work in another institute. There is a need for senior researchers or supervisors to guide and train the younger generation of researchers, with the hope that one day they might become independent researchers. It basically consists of five main elements in terms of how it works:

(a) Establish good communication between the Principal Investigator and the members of the team. Taking the time to discuss projects with young researchers is very important. Peer communication within a research group is essential. The PI or supervisor must be available at all times.
(b) Provide adequate guidance. In particular, new students often lack a clear understanding of how to handle a project. Communication is the key to guiding them in the right direction. A PhD candidate should be provided with enough guidance during their first one–two years of the program, as that period is usually sufficient to build their skills, understand, collect a small amount of data, and train the way they approach a scientific problem in a scientific context.
(c) I also believe that an impactful researcher is also an impactful mentor. It is important to give them a helping hand and provide them with support when they need it.
(d) We must let our hands go. As soon as I feel the time is right, I allow students to handle a more challenging task on their own. In order to achieve something worthwhile, there must be some pain. In order to become a better researcher, you must confront problems and difficulties and find out how to overcome them.
(e) Possess a philosophy mindset. When I meet with students, there are many times when we do not focus only on the project, the data, or the methods that are related to it. My emphasis is on the philosophy mindset, which refers to the way in which we view a given topic from a variety of perspectives.

This is a set of principles that I have been practising with heart for many years. My hard work has been well appreciated by the university as I have been awarded with the Best Supervisor Award as well as other awards. One of the things that makes me most proud is to see my well-trained students awarded the Outstanding Postgraduate Students Award, after they have graduated from our program, appointed in excellent universities, and excelling in their respective fields. In conclusion, if we can condense all of the information into a slogan and a logo, it will resemble this. Tagline: Empowering Researchers, Guiding Excellence. Slogan: Nurturing Impactful Researchers for a Transformative Future.

8.2.11 Would You Be Willing to Share with Us Your Secrets to Success in Science?

Humbly, I do not consider myself to be the most qualified person to answer this question. Let me share my own experience. My team is working on three main research fields, that is, environmental microbiome, environmental antibiotic resistance genes, and wastewater-based epidemiology. High throughput sequencing is our main tool. We have a good track record, and we are one of the few leading groups in environmental antibiotic resistance genes research.

Having studied and worked at the same university for many years, I had a shorter learning curve. I have adapted very well and have a successful start-up. After I joined the department as an assistant professor, I was given the opportunity to take over a lab due to the retirement of my PhD supervisor. In retrospect, I believe that it is important to have a good starting point. I had a short lag phase, which was a great advantage for me. As we know, microorganisms take a longer lag phase when they are placed in a new medium or under a new condition. In the same manner, any individual who is new to a work environment will require some time to adjust. My journey was shorter.

Owning a lab is one thing, but I believe it is more important to prepare and put effort into reading journal articles, thinking, planning, and making suitable decisions. In the beginning, I got my hands dirty with 454 sequencing, and then I progressed to multiple types of high-throughput sequencing. In my case, I was fortunate to have a keen interest in microbiome research. Our research coincided with the development of sequencing tools at a time that was just right. My achievements have been significantly enhanced by the advancement of technology.

I would like to acknowledge my team, particularly my students. Our assets are these individuals. A major reason for my success as an academic is the fact that my first batch of postgraduate students were very committed and willing to give themselves extra effort in collaboration with me to make their studies successful. My appreciation goes out to them. Many times, I have told people that the first PhD student of a new lecturer can provide a good indication of the future success of that supervisor. As long as the supervisor is willing to devote time and effort to overseeing the first batch of students, finding ways to overcome challenges, and maintaining excellent mentoring and communication skills, the fruits of success will follow. As a general rule, such supervisors have a positive reputation. Collectively, if the right time, right project, the right people (students), and opportunities are available to us, let's put them all together, and excel along the way.

8.2.12 In Your Opinion, What Are the Biggest Challenges Facing Biotech Scientists Today?

Probably most people, especially young researchers, find it difficult to obtain research funding, utilize new technology and tools, and publish their findings; however, these challenges can be overcome quite easily with some efforts.

Conducting research has more than just the goal of obtaining funding or simply utilizing the latest technology as its ultimate goal. Although funding, machines, and technology are important, they are merely tools. One of the most challenging aspects of being a scientist is identifying scientific questions, or good and sustainable scientific questions. Finding and solving the real scientific problems that deserve research is always the most difficult task.

The development of technology and research scopes in the field of microorganisms and microbiomes is just too rapid these days. If we are able to jump aboard an emerging research topic and new technology, we can progress to another level. Quite frankly, the pace of development in biotechnology is too fast for us to keep up with all of it. If we decide to take a short break from our research journey, or if we are without sufficient funding for some time, we will miss the boat and possibly be left behind. We can only grow and improve if we keep improving ourselves, keeping updated with the latest developments, and staying active and moving forward. As there are too many boats out there, we must take due diligence and make a good decision, and on occasion, we should predict which boat will be the best to board. Remember that there may be some new research topics and new technology that will only be popular for a few years at a time. As a supervisor, it is not wise to assign a student with a project that is relatively popular now for a period of one or two years, but after the student has graduated, the skills and knowledge they have gained may suddenly no longer be relevant or useful for the development of their career in the future.

With close to three decades of experience in the academic field, I have seen several changes in the characteristics of students from different generations. In the past decades, the global society has changed significantly; as the people and the society have become wealthier, the quality of living is increasing, and people are able to live more comfortably in the world. Parents will always do what they can to ensure that their children grow up happy and comfortable, and there is nothing wrong with that in principle. However, the drawback is that youngsters born and raised in such an environment stand a lower chance of facing setbacks, failures, or trails that can strengthen human character and make them stronger. I believe that this shapes the characteristics of each family over time and ultimately shapes the younger students we receive at universities. It may be that some of us place the blame on young students or young researchers. For my part, I believe that we cannot impose any negative tags on them, as that is simply the nature of our society. According to my opinion, this is not a problem of the younger generation, but it is a problem of the senior generations. The time has changed, and so has the generation. As supervisors, educators, and PIs, it is our responsibility to accept the characteristics of young people. Senior generation members have a duty as mentors, as role models, and as educators to guide the young generation in the correct way of doing things, for example, the correct philosophy mindset in research.

8.2.13 Is There Any Advice You Would Give to Young Academicians Who Are Struggling in Their Careers to Help Them Control Their Destiny? What Should They Do to Manage Their Work, Time, Career, Etc.?

Mohd Nazalan Mohd Najimudin
Set a clear goal, and the right goal is to keep on uncovering more new knowledge. From the first day you step into your office, you should establish your purpose in academia and research. In order to achieve your destiny, you must set a strategy and determine the direction to which you will move. It may be a lonely journey, but you are not really alone as you can set an excellent example for your students. Let them accompany you. Please keep in mind that your students will be watching you at all times. They will be watching what you eat, what you drink, what you do and what you read, as well as how you act and think. Educate them on how scientists think and work. Keep them thinking all the time.

In general, I had a very clear mission. I hope you have yours too. My mission was to change the entire research climate in the country. A number of unhealthy systems or KPIs exist today that cause young researchers to give up their initial dreams. Don't be afraid to stand up and hold strongly to the correct mindset. Keep your KPIs in check but don't let them choke you. My late colleague Dr Paul Dear (MRC Laboratory of Molecular Biology, LMB) and I believed that a researcher should be judged on how well his field of study and research are progressing based on wisdom. Never give up on your dream of having wild ideas that can become a reality. This requires good thinking. We are not a research paper factory that produces research papers using the same moulding pressing machine to generate very similar outputs with just a few tweaks here and there. I encourage readers to establish their own barometers for measuring progress. It has nothing to do with the number of papers you publish each year. Specifically, I am referring to how much wisdom and knowledge you have contributed to the advancement of science on the topic you are researching. Our profession as academicians and researchers is a category of thinker with a philosophical mind that contributes to the progress of science in a whole number of ways. It is my hope that the spirit of doing things right will continue to influence more individuals.

Full-time research, especially as an academic lecturer with a researcher's cap on, can be quite challenging. In addition to teaching, supervising, academic advising, etc., there is also a lot of paperwork, auditing, meetings, and other activities. Over the course of history, the system has become increasingly complex and demanding with a great deal of unnecessary work. In my opinion, KPIs are only guides; however, the whole environment has turned them into determinants. Some of us might have heard stories where certain work scopes that we spend hours working on are not captured in the KPI mark despite spending hours on them. Slowly, no one will be interested in such tasks since they will not be rewarded with a grade. It is possible to hear someone only working on KPI points in which the requirement is n-numbers of this and that, and they will never exceed the required number. Does this represent what academics and researchers should be doing?

Since many tasks are not taken up, someone has to do them, and eventually these responsibilities fall on the young staff. The problem is that they have too many things on their plates which results in them having less time for research. It is impossible for us to spend our time on everything, and therefore, we should be selective and focus on jobs that are meaningful and minimize the time spent on unnecessary tasks. There are many young academicians struggling with this issue, and the sad thing is that they are treated in a very punitive manner. We all make unintentional mistakes from time to time, don't we? We are more likely to make mistakes the more active we are, but sadly, there are also constant threats of punishment. You can avoid punishment by not acting, or by acting only as directed by your superior. However, there will be no actual outputs in either scenario, and once more, if KPIs are not met, a different form of punishment will be applied. Young people are under a great deal of stress as a result of this. The elders in the top management should reflect on this matter and take appropriate action.

Once you have determined your goals, I recommend staying focused. Stay focus on doing fundamental research to gain knowledge. If you are doing well, you probably have a greater sense of control. Please communicate with the elders along the way and seek some solutions together. Keep the spirit going. If, however, the environment you are working in is not what you desire, and you are unable to make any progress with it, there might be a better opportunity waiting for you somewhere else. Please keep in mind that there are problems everywhere. In the end, you need to know and be aware of what you are looking for. I believe that if you are sincere about conducting research, having a bench and the facility to toy around and experiment will be sufficient. Please do not allow any problems or disturbances to discourage you. Take a look at your students and those who have graduated and are doing well. Doesn't it make all the effort worth it?

8.2.14 You Have Been a Part of the Ups and Downs in Research and Publication, in Administration, in Teaching, in Making Great Decisions, and an Advisor to Decision Makers. Reflecting Upon Your Journey, Are There Any Distinct Segments Along the Way, or Are There Different Purposes or Aims or Focuses for Each Segment, as Well as Any Mindset-Changing Experiences that You Would Like to Share with Young People?

My career in academia spans almost 40 years. Over such a long period of time, I have witnessed and experienced significant changes in terms of facilities, science, systems, research culture, technology, etc., not only at my home university but also at universities in my country and other countries. Looking back over my entire academic and research career, I can see some important messages I would like to share with the readers.

I came from a biochemistry background, but I did my PhD in genetics at Cornell University in the US. If we do our Google-search online, you will see that Ezra

Cornell, 1868, said this: "I would found an institution where any person can find instruction in any study". The whole philosophy behind this is that the university's founder wanted an inclusive environment where people could be led to believe that they could conduct whatever kind of study and research they desired with excellence, integrity, and purpose. It was in this environment and with this kind of belief that I initially became an adventurous researcher; it was how I was trained in my professor's lab. My PhD training served as a crucial starting phase for my journey, during which I learned the importance of having an analytical and a research mindset, as well as being brave and adventurous in science. My subsequent years were filled with adventures. During my PhD study, I started doing research that people have not done before, for instance, trying to clone a repressor gene from Gram-positive bacteria when others were focusing on Gram-negative bacteria. This is only one example. During my career, I have jumped from one project to another, trying to explore things that have not been tried by others.

The second segment of my journey, which took place in 1989, was extremely challenging. Molecular biology and biotechnology were still at an infantile stage at that time. My expertise in recombinant DNA was among the first three in the nation. The majority of young researchers in this era will never have the opportunity to see how simple molecular biology protocols such as genome extraction and gene cloning were not so simple at that time. There was no molecular biology lab in the department, as well as no building and no facilities—an extremely demanding first challenge. The choice I made was to talk to people, and to find solutions. Along with a group of individuals with similar ambitions, we put together detailed proposals to obtain funding for the construction of a building, lab, and basic equipment. The task of persuading non-biologist decision makers to take a long-term view of what we wish to study was extremely challenging. Having worked hard for six years, I was finally able to take research students.

I want to make sure the reader understands my point—having facilities is important for us, but setting up the lab or building was not the ultimate objective. Our goal was not to build an empire for ourselves, but to raise the level of science of my team, of my university, and of other researchers and universities throughout the country. Aside from that, the last message or instruction my supervisor gave me before I left him and Cornell University was to be active in scientific societies in my home country when I returned. Probably because my professor forecasted well, or because he convinced or trained me well, or perhaps because he knew me well, I began to engage with a variety of research societies, individuals, and stakeholders. I met many individuals who shared the same values and goals of becoming more successful in the field of research that we had in common. We not only collaborated on research projects, co-supervision, and other common research activities, but we also encouraged others to adopt the right mindset. We have even established a national research laboratory hub and institute that was designed to foster and pursue fundamental research at a higher level.

Fundamental research, in my opinion, is the way to go, and anything applied from its findings will produce fantastic and impactful results. My beliefs remain the same even after I retired—we should focus on fundamental studies. It is unfortunate that most developing countries do not have an environment conducive to fundamental studies. In general, funding for fundamental research is typically less than that allocated to so-called applied science research. The system is looking for quick a tangible product, something that can be seen, touched, or felt. For example, people are more likely to acknowledge research if the researcher is able to generate new bone materials. There is no doubt that such new materials are important. However, if we discovered 50,000-year-old bones in a valley in our country, our society and funders would question why we should conduct DNA research on them. You might think that what I say about an ancient bone DNA is just a fabricated example, but that is not the case at all. An archaeologist and I tried to conduct such research, but I think such projects do not receive enough support. In contrast, we have seen some real-life examples. There is a research project that was published in *Nature* in 2021 by a team of scientists that looked at the DNA of extinct mastodons found in Greenland 2 million years ago and published an article about it. I watch BBC. A discovery of such an ancient bone in the United Kingdom will be very exciting, and BBC will cover it extensively, and the society will be intrigued and wish to learn more about it. The acceptance of such new knowledge is different between their society and ours. There is a shortage of funding for fundamental research in most developing countries. As a result, the system forces researchers to focus their attention away from their genuine enthusiasm for fundamental science and move towards product development instead. It is not possible to commercialize every research project. Do not be discouraged if you find yourself in such a dilemma, as I have been through it myself. Maintain your focus and excitement by surrounding yourself with people who share your interest and primary reason for being in academia. Try to juggle between satisfying the funder by fulfilling their requirements and demands while fulfilling your passion for science. Do not give up on your passion and pursue it very cleverly.

During my tenure, I have also been involved in administrative positions. My desire to become an administrator was never motivated by a desire for fame or a desire to climb the corporate ladder in order to build an empire. It was a privilege to serve as deputy dean for my faculty and dean of research my university for 12 years. I was offered an important and higher administrative post that was purely administrative, but I declined the offer. Because I am only interested in knowledge and the development of knowledge, I only wanted an administrative position that involved research. While I cannot make your decision for you, if possible, avoid administrative positions that can divert your attention away from developing your knowledge. As an administrator, I wanted to contribute primarily to the advancement of knowledge, influencing people to do science correctly, and fostering a positive attitude towards research. My position allowed me to support various types of research as well as the establishment of research centres and clusters, some of which are quite unique. My conclusion is the same as what I said earlier: as long as you are dedicated to excellence and purpose, you can research and study anything you like.

8.2.15 What Message Would You Like to Convey to Young Women Who Are Considering a Career in Scientific Research?

Maria Isabel Gonzalez Siso
My main message is that, if scientific research is what a young woman would like to dedicate herself to, she should not consider the fact of being a woman as a drawback or an obstacle. If you dream of it, work for it, regardless of your gender.

Scientific research is a beautiful job, but it is also very demanding. My experience leads me to give two pieces of advice to budding scientists. First, it is crucial to select a good research group for both the PhD and the postdoctoral—groups with relevant lines and publications, with continued financial support, and with demonstrated training ability. Second, women must not allow commitments like family care and housework to fall solely or mostly on them, to the point of impeding their professional development. The support and encouragement to move up the career ladder that one member of a couple finds in the other are essential in this regard. Serenity in personal life is a key to success, in science like in any other professional field. Moreover, it is important to separate time for work and time for leisure.

I am a mature woman and full professor in a Spanish university. I have seen a lot of changes about the role of women both in society and in scientific research. Despite the undeniable progress made in the last years, today in many scientific organizations, the distribution of the different steps in careers between men and women is still represented by scissor graphics. Thus, there are more women than men studying science degrees, but there are more men than women occupying the highest responsibility positions. The cross-point is around the postdoc. One reason for these scissor graphics could be the scarcity of close referents. It is true that there are more and more important women in science, but their numbers continue to be less than those of men. Another plausible reason for scissor graphics may be that the age of the postdoc usually matches the age of having children and starting a family. There are more women than men that choose family over a professional career. The governments and academic authorities have a great responsibility to promote conditions that facilitate the reconciliation of both things, for example, regulating that the inherent decrease in scientific production during parental leave does not penalize in job competitions, preventing gender discrimination in juries, providing free schools and nurseries from zero years, and so on. In universities, a proportional reduction in teaching hours is appropriate for staff who carry out research activities. There should be no need to choose between family and profession, both should be compatible and even complementary. A gender-balanced situation among leading scientists should be achieved and we must strive for it.

I fully encourage young women to work hard to get a leadership position in scientific research and benefit from their talent and training.

8.2.16 There Is a Perception Among Young Researchers and Student Authors that Journal Publication is Difficult. As an Author, Reviewer, and Editor of Renown, What Advice Would You Be Able to Give to Them?

There is a chapter in this book focused on how to write good research articles, choose a suitable journal, etc. I consider this chapter as a great help, and I have no more technical tips to give about this subject. My advice is, first, prior to writing, read a lot, not only scientific papers but also the classics of literature, philosophy, novels, etc. This is a good way to learn to organize the ideas and to express them in a reader-friendly manner. Secondly, to practice, learning by doing. Making mistakes and correcting them is the way to improve.

8.2.17 Could You Give Us Some Tips on How to Handle a Project with Researchers, Students, and Postdocs from Different Countries and Cultures?

Neil C. Bruce
Our team is continually seeking talented individuals from around the world, most of whom come from a well-established lab and will be able to integrate very quickly into our group. I have found that it is extremely important to have new people come in because they usually bring with them new skills, techniques, or different ways of thinking due to the fact that they have been trained by different people. I believe that it is healthy for the lab to include individuals from all over the world.

Throughout my career, I have always felt it is important to have two things in the group that are really significant to me. First, there must be some continuity. Having people who have been with you for some time and can train new people or students who arrive is a good thing so that the methods you have developed within your group can be taught to new members. But, I also like to have a healthy turnover of people, bringing in new people with their own backgrounds, who have a lot of different ideas and new techniques that they have developed during their time in other labs. Bringing in new people can help you spot new ideas and offer better solutions. You know that if you do not change your group members, you will all start thinking in the same manner. As for me, I think coming to work should be really enjoyable. Labs, departments, or institutes should be places where you enjoy interacting with friends, supportive colleagues, and do good science together. You should feel like part of a big family and feel a sense of belonging. It's really important to have a positive, enjoyable, and supportive environment. After a long career, I have managed over a hundred students and postdocs. I am fortunate in that most of them are friendly and easy to manage because we are a cohesive group. I treat them with the utmost respect.

Throughout the years, I have had many overseas researchers from around the world join our group, and I enjoy that very much. As I enjoy the diversity of the group created by overseas researchers who bring in their culture, I enjoy having this mix within the group. However, it's important to be mindful and understanding of their backgrounds. We should not expect them to assimilate immediately and conform to the norms of others, especially if they come from a country that may not be as developed as the United Kingdom. Sometimes, they are not very self-confident at the beginning, and you have to be very sensitive to that, and don't expect them to hit the ground running, and you need to support them, teach some techniques they do not know, or advise them on deepening their knowledge if they are unfamiliar with certain aspects of science.

Whenever the student is assigned to a larger project, I usually have the postdocs and other students working on it as well. The newcomer will certainly receive good support from the entire team. If they are working on a different project, I will spend more time with them and try to find them someone within the team that can offer them support. Later in my career, I realized that students from certain countries may have unique cultural characteristics. I am referring to the relationship between their superiors and themselves. In some cases, students expect to just be told what to do each day, probably because that was how they were taught during their undergraduate studies in their home country. In such cases, they may end up sitting at their desk, unsure of what to do, and lacking the confidence to ask for assistance. As soon as I am made aware of such circumstances, I will probably provide them with some support and direction, letting them know that things are different here, and they can knock on my door at any time to discuss anything they wish. In my group, I have always avoided any sort of hierarchical structure. All individuals are treated equally regardless of their background or position. Students may find that very awkward when they arrive, and some still wish to address me as Professor, which is fine, but they may address me by name instead. However, in their culture, calling a supervisor by name is considered impolite. It is not a problem for me. It is necessary for us to understand other issues that students may encounter. If you are aware that there are other students from their country, it would probably be beneficial for them to become acquainted and to support each other. The experience of being alone in another country can be quite challenging. Another problem I believe is the level of English. Having a good command of English makes life easier. It is, however, possible to unlearn and relearn that skill.

It is no secret that I am a strong believer in the power of a team. In my institute, I typically have about equal numbers of postdocs and PhD students. The number and current status of postdocs and research technicians available during that time period will determine the number of PhD students that I will accept. Throughout my career, I have been very fortunate to have had many PhD students and enough funding from them. There might be a number of reasons for this, including the track record of our groups. Students with government scholarships tend to seek out groups with significant track records rather than those who are new and unproven. I have had around 50 students, and all of them have completed their PhD within four years on average. I have never had a student fail their PhD in my lab. Nevertheless, we have had some extensions with COVID, but no one has failed their PhD due to the pandemic. There have been reports that the pandemic in 2020–2022 disrupted research activities and some students are experiencing mental health issues because of this and that. My group was able to sail the coarse journey during the pandemic because of the support from the entire team during the difficult time. Furthermore, two things have been done at the university and national level, especially for students nearing the end of their PhD projects. For students with scholarships, the UK research councils provide funding extensions. Students who self-fund their PhD can receive additional support and funding from our university.

With multiple funding sources, projects, students and postdocs, meetings, administrative responsibilities, teaching, etc., you need to demonstrate a high level of discipline. It is my practice to enter the timetable into the calendar. I schedule a meeting or mentoring time with each student, either one-on-one or in group, depending on the needs of the student. For students that need extra support, I will schedule meetings more frequently in order to ensure that they have the support they need. I find that scheduling works best for me. If we do not do that, if we say, "Let's try and meet up every couple of weeks", and we fail to schedule it, before you know it, two weeks have passed, and then it is three weeks later. Perhaps it will be a month before you meet with the student.

8.2.18 Why Is International Networking and Industrial Engagement Important in Your Opinion?

In general, networking is extremely important. It is very difficult to conduct good science research if all you do is work within the four walls of your laboratory. Networking is essential, especially with the best scientists, you must see what they are doing, hear them speak, see what resources they have, and understand how they solve a problem. As a researcher, networking is critical for the development of new research projects, for exposure to novel ideas, advanced techniques, for discovering scientific gaps, etc. In order to find new ideas, you could read journals, but it is not the same as attending a conference, meeting, or workshop with industrial representatives. Academic networking and networking with industry are equally important. By establishing those relationships, you can spark ideas that lead to new projects that you may not have otherwise gotten if you hadn't spoken to those people.

Please allow me to share my experience. After receiving my PhD, I was considering working in the industry or taking up a postdoctoral position. I had interviews in companies and had a postdoctoral interview at Cambridge University. A project at Cambridge aimed at developing a biosensor for illicit drugs in collaboration with the industry caught my attention. Despite the fact that it was a challenging project, I took up the offer, and I enjoyed it very much. The project was multidisciplinary in that it incorporated aspects of both fundamental study and applied science as well as industrial activity, resulting in a project that had a good mix of the three. At the centre where I worked, there were ongoing research projects that involved industrial interaction, engagement, and entrepreneurship in the form of spin-off businesses. Although I am capable of doing really good fundamental science, I believe that industrial engagement is very important. I realized at a very early age that at some point of my career the interaction and networking helped me move forward and to explore new areas of my own research, and industrial engagement often results in new sources of funding and new opportunities for me.

A piece of advice I always give to young academics is to obtain a broad portfolio of funding. Be careful not to rely solely on one research council for funding, as you might be popular and what you are doing might be popular at the moment. A few years from now you may have a hard time getting funding, so if you can balance the funding so that the funding is spread across all the different sources that it is, then you can keep your group at a relatively constant size and fund your core research interests. The situation is the same for a number of my colleagues. They rely on our Research Council for funding, and they'll get a grant for a postdoc or student, then two years later, another grant, but, once they do not receive funding for their project, postdoc will stop. Because the PI does not have any resources, and if there is no one else in your group, it will stop completely. Starting over from scratch with no one in the lab will be a significant challenge. Besides, in the absence of research income, you cannot generate data to support your next grant application, which creates a vicious circle, where you end up struggling and being under pressure. To avoid this trap, I recommend diversifying the portfolio of research funding, and that is why funding from industry is so important.

There may be some universities or institutes that do better than others when it comes to attracting funding from industry. There are several possible reasons why that might be the case. There should be a team at the university that can assist you, for instance, those with business knowledge, legal knowledge to prepare the legal agreement, etc., and who will do the right thing and ask the right questions on your behalf.

8.2.19 What Advice Would You Give to a Young Scientist Just Beginning His or Her Career in Science?

When you are just appointed as a young researcher or a lecturer, it is not a good idea to sit around for too long and not be very active. For a scientist to be successful, it is not necessary to be exceptionally brilliant; all you need is motivation, dedication, and a willingness to put in a lot of effort. There's no point in coasting and doing the bare minimum. You have to change that behaviour before it becomes a habit. An individual who is a good researcher should possess the following qualities: patience, perseverance, accepting knockbacks, and seizing opportunities when they arise.

I would also like to pass on another piece of advice that I was given. In the event that you are invited to give a talk, no matter where it is located, you should always take advantage of the opportunity since it benefits you. While attending a conference, you have the opportunity to network and meet new people. It is obvious that when you stand up there and give a talk, you have the whole audience sitting there, every single one of them knowing who you are, and there is a high probability that you will have someone approach you afterward in order to talk with you about your research, and that is how collaborations are naturally formed. Give a talk even if it is just ten people, but one of those people might be really influential, the one who would be the most important collaborator in the future, so don't hesitate to give a talk.

If your university has a policy that permits young lecturers to take on a reduced teaching load, you should take advantage of this and devote more time to networking and research. Once your teaching load and other duties increase, it becomes more challenging. Consider ways in which you can improve your track record. It is okay to start with something smaller, and when people know you, you may be able to unlock other opportunities.

Funding is difficult to obtain, especially from the industry. If you knock on people's doors, say hello, tell them who you are and that you think you are good, it will not work. There is nothing wrong with explaining to the industry what kind of science you do and asking if there is any way we might be able to collaborate, but the truth is that this is never going to happen. In order to be successful at networking, you must develop a relationship with other people on a genuine level. It is unrealistic to expect that just because you know someone, you will be granted benefits such as fundings immediately upon meeting them. There are times, for example, when I have been in contact and have kept in touch with a group of industry people or other researchers in another university for years, and we have become good friends, and we have continued to do so. The moment an opportunity arises, you will be the first person they contact since you are well known and they know what you have achieved so far.

In summary, here are my recommendations. Get out of the office and lab and meet more people. Get to know different people and find out what problems they are experiencing so you may be able to help them. Accept invitations to speak at conferences and other events. Get a broad portfolio of funding. Participate in grant panels, as being a member of the panel helps you develop a better understanding of what is required to craft a grant proposal that is really competitive. Establish connections with world-renowned scientists who can collaborate with you on your existing projects or new grant proposals. Push yourself forward and be dynamic. Take a risk and step outside your comfort zone.

8.2.20 Would You Be Able to Explain How Data Science and Biology Will Intersect in the Future? Would You be Able to Give Any Advice to Conventional Wet-Lab Researchers Who Are Comfortable with Their Existing Skills and Approaches but Who Would Like to Explore Bioinformatics and Data Science?

Aik Choon Tan

Data Science is an interdisciplinary field, where the three main domains are: computational science (including machine learning and artificial intelligence), statistical science, and domain knowledge (here will be referring to biological sciences). According to the National Institute of Health (NIH) definition, data science is "the interdisciplinary field of inquiry in which quantitative and analytical approaches, processes, and systems are developed and used to extract knowledge and insights from increasingly large and/or complex sets of data" (NIH Strategic Plan for Data Science, 2018, https://datascience.nih.gov/nih-strategic-plan-data-science).

It is now easier to learn programming skills as various massive open online courses (MOOCs) and tutorials are readily available for wet-lab researchers to attend and learn these skills at their own pace. In addition, some of the artificial intelligence approaches such as ChatGPT could generate example codes in various programming languages to solve some problems.

8.2.21 Describe What Makes You a Preserver. What Drives You to Keep Moving forward?

Mission, vision, and passion are what motivate me to continue to pursue my goals. As a data scientist or bioinformatician in the field of cancer research, my role is essentially to bridge the gap between basic science and clinical research. As a bioinformatics researcher, my role is to help the researchers to translate all these complex and big data into simple models or rules to be able to predict the next experiments or identify some biomarkers that can distinguish between responding and non-responding in clinical trials. When thinking about this, it is not a matter of attempting to become famous on your own or making your own name famous. Rather than focusing solely on publications and KPIs, my mission goes far beyond that. It is basically my mission, vision, and passion to help cancer patients who are most in need of this type of treatment at some point in the future. That is basically what motivates me to conduct my daily research.

For younger generations, I encourage them to establish a mission, and that will motivate them to become involved in research that can impact or benefit many people, or the society. As a fresh PhD holder, he or she needs to decide what research path to follow in the future. There may be some researchers who choose easy-to-do projects, while others opt for a path closer to fame, while others decide to follow other missions or calls.

Data scientists or bioinformaticians such as myself can work in a variety of research areas because almost all aspects of biomedical research require people in my profession. But my mission lies in cancer research since patients need it. As we all know, in order to train a machine learning model, a large number of samples must be collected as the training set. In other bioscience fields, soil, water, DNA, plants, etc., are used, and we would not have a strong sense or feeling about the training samples. However, in cancer research, the samples I used for training a machine learning model are from human patients, hundreds and thousands of them. These patients suffer from cancer and ultimately were defeated by this devastating disease. Unlike environmental samples, mine datasets are the result of the suffering of a human being. These datasets cannot be treated like environmental samples. I chose not to pursue a research area with a very narrow focus where the results will only be beneficial to a small group of individuals. To make a wise career choice at the beginning of your career path, you need to consider this factor. If possible, consider a value-driven path. This is the same reason why I chose to contribute my time and energy, along with our efforts, to doing good science with the ultimate goal of affecting the research of cancer in the future.

The following is a personal story that I would like to share with you. My postdoctoral experience began at Johns Hopkins University, School of Engineering, one of the top biomedical research universities in the United States. During that time, I was working on developing new machine learning tools in order to predict cancer from normal samples using gene expression profiles. The entire team worked very hard, and I was fortunate enough to have my manuscript accepted and published by the *Journal of Bioinformatics*, one of the most respected journals in my field at that time. Somehow, I felt that success in publication, some high-quality publications, was sufficient for me as a young researcher. I am glad that one event happened to me and changed my career trajectory.

One Thanksgiving night about 18 years ago, my pastor was called to give a prayer for a patient in the hospital, and he needed some assistance in translation of his prayer. My wife and I decided to accompany the pastor to the Johns Hopkins School of Medicine Hospital. Initially, we had no idea what was going to happen, we thought that our job consisted of merely translating. Later, my wife and I were surprised to learn that the patient was a four-year-old girl when we arrived at the hospital. It was her very last moment of life as she battled leukaemia. It was also my first encounter with a cancer patient, a very young patient, at the end of her life. We had to wait outside the room for a few hours until her passing. That is the pivot point in my life. The cancer samples for my machine learning dataset are no longer simply cold and meaningless numbers or data. Each piece of information we receive from the hospitals belongs to an individual who is battling cancer. As a scientist, I sometimes find myself in a dilemma since we need more patients to train the machine learning algorithm effectively. A very accurate model is what we are seeking. However, we require a large number of human patients. The sad thing is that this data pertains to an individual cancer patient. I completely changed my perspective on doing science as a mission-driven endeavour after I encountered the four-year-old leukaemia patient. I need to work harder and faster so that all stakeholders in our research will eventually be able to benefit from the work we are doing. That is basically the kind of mission, value-driven mission, I am referring to. A great mission requires a team consisting of individuals with diverse expertise who share the same vision and passion.

8.2.22 Are There Any Human Behaviours or Attitudes that May Prevent PhD-Holders from Progressing as Researchers?

In my opinion, there are three main types of behaviours. The first is the imposter syndrome, which refers to a psychological phenomenon in which a person doubts his or her abilities, talents, or accomplishments. Some researchers with the imposter syndrome lack confidence despite having obtained a PhD in his or her domain. One of the possible causes of this issue is a lack of proper training and experience during their PhD program. PhD projects can be very specialized or very narrow and focused. However, if one has a philosophical spirit, one should be able to absorb new concepts from literature, even if they are outside the scientific scope of their domains. Lacking that ability may lead to a lack of confidence. As time progresses, one begins to construct a wall in order to protect their existing comfort zone and is reluctant to move out of it. Researchers who are comfortable in their own surroundings tend to be afraid of taking on new challenges and risks. Therefore, being stagnant and to stop learning at any stage are risks to the researcher's career.

Secondly, losing curiosity hinders the progress of a researcher to the next stage. Science and technology are constantly evolving, especially in the field of genomics, for example, the landmark Human Genome Project, which was launched in 1990 and completed in 2003. During this timeframe, new sequencing technologies were invented to accelerate the complete sequencing of the human genome. At the same time, new computational and statistical methods were developed to align and construct the genome. Moreover, modern computational infrastructures were developed to facilitate data sharing across international collaborators. Similarly, for the past 20 years, we observed new technologies were developed to allow researchers to sequence with high-resolution at single cell level. Accompanying these technologies are the generation of a wealth of multi-modalities of data types in characterizing diseases. One of the current key challenges is how to process, analyse, integrate, and interpret these multi-omics data for understanding the biology. Therefore, it is imperative that there will always be new set of problems that arise in scientific research, and that will never change. The moment your curiosity begins to wane, that is the time that you need to reassess whether research is still a good field for you at this stage of your career. In the absence of curiosity, we are going to be left out from this exciting and fast-moving field. Thus, in order to be relevant and contribute to science, it is important to keep learning and stay curious.

The third thing is being overconfident. In my first point above, I discussed imposter syndrome as a result of a lack of confidence. A trap exists at the other end of the spectrum when we think that we are at the pinnacle of our academic and research careers. The moment that we start thinking that way, we will get into trouble. Self-reflection and reassessment are key to prevent overconfidence. Try requesting 360-degree feedback from your peers, including those who are junior to you, which is one of the best ways to stay sharp. It is not shameful to ask feedback, regardless of your level of seniority. Once an individual starts to think that he or she is the smartest person in the room, it will prevent the growth of that individual. I put up three possible human behaviours that prevent us from learning new information, expanding, and becoming better researchers.

8.2.23 Are You Afraid of Failure, or How Do You Deal with Failure and Rejection?

An individual researcher is likely to encounter multiple rejections every year, for example, our proposal or manuscript may be rejected multiple times. It may be uncomfortable when people criticize your proposal or reject your paper; however, this is part of academic research. It would be a big rejection if we submitted ten proposals and none were funded. In the eyes of some researchers, such rejections are equivalent to failure in scientific research and they feel extremely sad about it. Obviously, everyone will be upset.

There is no reason for me to be afraid of failure. There is a lesson in every rejection. I view every rejection as a lesson. I will illustrate what this lesson means to me by taking the rejection of a research proposal as an example. We do not know the answers to the research questions when we submit a proposal. When you already know the answer, that is not considered research. We may have a good idea, but the reviewers may not understand it fully if the idea is not communicated and presented well in the proposal. It is common for researchers to focus exclusively on the negative feeling or on the red flag of not receiving funding. Most likely, if you read through your comments and calm down, you will find that they contain some valid points which are basically your weaknesses in the writing, formatting, or way of expressing your ideas. Possibly, we did not put enough effort into preparing the proposal due to the lack of time. Many writers work until the very last minute, resulting in many typos or sentences that are not complete or do not make sense. Rejection can stem from various factors, and as a writer, you may not always notice certain errors. As a general principle, I believe that behind every failure there is a lesson to be learned. Focus your attention on the reason for the rejection rather than the rejection itself.

When I receive a rejection, I usually take a few days off or at least keep my eyes off of it. I do not give it much thought. Once I return with fresh eyes, I will be able to work on it again. I continue this cycle to learn and improve in my next proposal or manuscript.

8.2.24 What Are Your Thoughts on Students Who Fail Their Thesis?

Failure to complete a PhD program does not necessarily indicate that a person is a failure in life. Simply put, this is not the right path at the present time for that individual. Some universities, academicians, or professors may try to be inclusive or be compassionate to all PhD applicants. The process of determining whether an applicant is a suitable candidate for a PhD degree is extremely difficult at the outset. The purpose of pursuing a PhD should be clear to the student before considering it. Obtaining a PhD is very different from getting a bachelor's or a master's degree. PhD students should have a strong motivation, skill sets, and ability to pursue a PhD degree and pick up new topics along the process. Many students failed the thesis or dropped halfway in the journey because they are unclear of what they are looking for in their career paths. Some of them probably pursue postgraduate studies in order to avoid finding a job. In some cases, students were not prepared or ill-advised for the journey. Obtaining a PhD requires three to five years of commitment. PhDs are difficult to obtain, and they are earned by hard work and grit, which means consistently putting in enough or more than enough effort to achieve the certification and recognition.

We all need to know what we want in life, who we are, and what our life is all about. You need to know your interests, strengths, weaknesses, and abilities. Many students failed their theses because of the kinds of mistakes they made when they decided to enrol into the graduate programs. Both the student and supervisor are going to suffer if they are unaware that the decision was a bad one. It makes no sense to send a person who does not know how to swim and does not have any skills or talent in swimming to compete in a swimming competition, does it? A number of students are unaware of how to swim, but their parents, or themselves, or supervisors are trying to push them to participate in swimming competitions (PhDs in this case), which I do not think is feasible or a good idea. As academicians or thesis evaluators, sometimes we observe students who have reached the final stages of the examination, but we are aware that this individual is not qualified, but sometimes there are not enough options available to us.

Having failed a PhD program does not mean that a person is a failure in life. The person could be extremely skilled at something else, not necessarily in bioscience or biotechnology.

8.2.25 Writing Papers have been a Part of Your Professional Life for Many Years. Can You Share Your Experience on That?

There is a saying that goes 'publish or perish'. Basically, that is what a typical academic researcher's life is like. A career as an academic researcher is very competitive, especially in the USA. It is important that we have a good track record of publications in order to prove that our research proposal is feasible and we have the capability to accomplish the requirements outlined in the proposal. Along the way, there will be many failures, such as rejections by reviewers for your particular manuscripts or grant proposals. It is crucial to remain optimistic and humble while taking criticism from reviewers in order to improve the next version of either a manuscript or a grant proposal.

Taking responsibility for the quality of research publications is important for authors. I believe that those publications that undergo a rigorous peer review will provide researchers with a valuable resource that can be rigorous and replicated by other researchers, which will form the foundation for a good scientific research process. It is not a good idea to take your sweet time when it comes to publishing. Either you publish it or you will be the one reading an article that has been written by someone who has written similar things to what you intended to write.

8.2.26 Having Worked in a Variety of Academic/Research Institutions and countries, What Are Your Recommendations for a Young Researcher Who Wishes to Leave Their Comfort Zone and Move to a Different Country to Study or Work?

Do not hesitate to take action. Please do not worry too much. If you travel to other places, especially to a good research university, you will gain a great deal of knowledge. There will usually be journal club meetings, discussions, constant seminars, weekly seminars, lab meetings, mentoring, etc., and you will learn a lot. In addition, you can learn how to conduct research in the rigorous and reproducible manner. Try going to another country to get the exposure to these kinds of research environments if you cannot find them in your existing lab. It is obvious that you will gain new knowledge.

8.2.27 Are There Any Strategies or Tips You Can Share with Researchers at the Beginning of Their Careers in Order to Generate Good Publications or Become a Good Researcher?

A young researcher should read extensively. There are too many publications, and some are more important than others. How do you choose which papers to read? I will probably make a suggestion. For example, identify the 'leaders' in your field of research, perhaps between five and ten. These groups are essentially the pioneers or the poll-setters. They are at the forefront of research in the field of research you are interested in. You should follow the publications of those research groups. There is also a need to stay abreast of the most recent research and technological advancements. It is also recommended that you read some of the basic concept papers within your field, and then to look at the highly cited papers, and to follow the different subdomains or subfields within your field. It is important to do this systematically so that you may see the progress that has been made from the past to the present. Examine what is the latest knowledge as of today and then attempt to predict what will be the future in five years. By doing so, you can establish your own research directions instead of repeating research that has already been conducted by others. It is my opinion that this is one of the best strategies for young researchers or those who are new to a field of study. This is a good self-assessment too.

Once you are on a track or you are accumulating some achievements, it is a good idea to set a higher milestone and work towards it. For us to progress, we need a goal. In order to set a higher goal, one method is to engage in healthy and positive competition with other groups. Let's call that an 'imaginary competitor'. In most cases, they are not your enemies, and you may have never personally met them. The competition is a driving force that helps you and your team to maintain the right momentum in order to reach the next level. The idea here is that you want to see what your peers are accomplishing, and then you want to push yourself to reach those levels. If, for example, your 'imaginary competitors' publish new papers every six months or come out with new ideas, then you will have to keep up with them. In the absence of healthy competition, some researchers will take their sweet time in writing their papers and doing their research. Our true competitor is not with other groups, we are not competing with other researchers, we are actually competing with ourselves, and pushing ourselves to the next level. In summary, I hope this interview will be useful for the next generation of researchers in building their career paths and contributing to a better society!

8.2.28 You Have Developed Strong Collaborations and Networks with Researchers from a Variety of Universities or Countries. It Is Often Said that if You Want to Go Fast, Go Alone, Whereas if You Want to Go Far, Go with a Group. If You Believe this Phrase, to What Extent Do You Practice It?

Anna-Louise Reysenbach
In my career and due to the nature of my research, I have always worked with strong interdisciplinary teams of collaborators. We can address broad scientific questions from different perspectives and gain a deeper understanding of systems across multiple scientific disciplines. I also think that some of the biggest discoveries in science were made across disciplinary boundaries. Sure, it makes it harder, as you need to learn to communicate across disciplines, but I get a lot of personal satisfaction from such interdisciplinary international team research.

8.2.29 Taking into Account Your Experience and Observation, What Motivates High Achievers in the Field of Bioscience and Biotechnology to Achieve High Levels of Success? What Makes them Different from Those at Average Levels?

There are many ways to measure success. Is it happiness, is it being a leader in a research field, is it a very good publication record? What is success to you? For me, I feel so fortunate I have been able to follow my research trajectory throughout my career; when curiosity took me one way, I was able to follow that. So, for me that is success. I have been very happy doing the research I do, and having the people, both students and collaborators, and the funding that has enabled us to investigate some very interesting scientific questions.

What motivates for this definition of success? Definitely, insatiable curiosity. For me that is in the natural world, and how little we still know and understand about how it is all interconnected. This scientific curiosity is often called ‘passion for science’. But ultimately, it’s what my postdoc mentor Norman Pace called ‘the WOW factor’. Once you have that, everything else seems to just fall into place. What is needed too though is a strong careful work ethic, a critical thinker that can step outside and self-criticize. While it does take working hard and long hours, work should not be drudgery but should be what you want to be doing; even to the extent you wake up at night pondering and often finding that solution to a result that had so befuddled you. You have to be content to work through the boring routine of some experiments which will ultimately give you the explanation for a scientific question.

8.2.30 Are Women Scientists Generally Faced with More Challenges Than Men? Is There Any Advice for Female Readers Who Feel So Stressed Up with Family Matters, Teaching, Supervising, Administrating, and Doing Research and Publications?

Yes, I do think women in science are faced with more challenges than men. Some of my advice for early career women scientists is to choose your mentors and collaborators carefully. Learn to set realistic goals and priorities. Ultimately you will gain respect for having these clear boundaries. This will also help you say 'no', which is often really hard to do as you will have lots of amazing opportunities.

8.2.31 You Are a Senior Editor of Many Good Journals. In Your Own Experience, How Have You Benefited from the Responsibility of this Position? Could You Please Tell Me How Good Editors Can Make a Difference to the Scientific Community? How Would You Advise Young Researchers Who Have Just Begun to Publish Their Work?

Being an editor is a weighty responsibility and is a way one can give back to your scientific community through service. Good editors can help foster rigorous, balanced, and careful scientific reporting and transparency. With this in mind, when you are starting to publish your work, spend a lot of time with your data and results. Find the story in the data, and then fine-tune the figures and tables to provide the narrative of your manuscript. Once you have done writing, step back for a few days, revisit the manuscript, check for clarity, and think critically about your conclusions. Consider if there are other explanations to the patterns you see in your data, and report on those alternatives too. Sometimes you will be surprised what your results reveal that you had not previously considered. When you get reviews back from your manuscript, address each reviewers' comments, in detail, point by point. If you disagree, say why. Do not get offended by something a reviewer may have overlooked. Stay kind, thoughtful and measured, and consider adding the different opinions into your manuscript. Ultimately, the review process will help improve the manuscript presentation.

8.2.32 As the Chair (Dean) of the School of Graduate Studies, You Have Witnessed the Joy and Achievement of Research Students, as Well as Their Tears Due to Failures and Rejections. Do You Consider Obtaining a PhD Certification to be the Highest Achievement in Life or Quitting a PhD a Failure? Are there Any Messages You Would Like to Convey to Students Who Have Passed (Graduated) or Failed Their Studies?

Noor Hazarina Hashim

It has been three years since I was appointed as the chair of the graduate school at my university. It is a job that I really love, but it is also a job that can be heart-breaking, as I have seen many excellent students drop out—both within my university and from other institutions. Many students come to me for advice, and I have kept a collection of tips I have learned based on successful candidates and outstanding supervisors.

Many students consider obtaining a doctoral certification as the greatest accomplishment in life. I am not denying that celebration is a must-have after three or four years of dedicated work. You ought to celebrate. The challenge lies on 'What's next?'. Many students have the intention to join the university after completing their doctorate, which is perfectly fine. Spending time at universities doing research, teaching, mentoring, and writing can't be more rewarding. What is more rewarding, however, is the opportunity to give back to society and prosper in community life.

I have witnessed the frustration of students and supervisors when students drop out, which is the part of my work where I despair most. Whenever I encounter this situation, inside of me, I want to say this: PhD is not your life, it is just a part of your life. It is heart-wrenching and dismal, but it is not the end of everything. Take a break, meditate, reflect, and do things that you like to do and makes you appreciative. PhD is not everyone's cup of tea, so buckle up and move on.

8.2.33 Would You be Willing to Share with Us Your Secrets to Success in Science, from an Industrial Point of View?

Teh Chee Keng

First of all, a deep understanding of the research problem is key. My observation is that many of us place too much emphasis on fancy technologies instead of understanding the real problems encountered by the industry. It is partly for this reason that the younger generation is moving away from classical sciences, such as zoology, physiology, plant breeding, and so on. Like many plantation companies, biotechnology alone cannot solve or expedite breeding progress without the foundation of germplasm collection and breeding trials. I notice that many researchers working on crops jump immediately into OMIC analyses, such as transcriptomes and others, to determine trait differences, and sometimes end up with negative or inconclusive results. Nevertheless, I learned that some research problems can be resolved through a relatively lower-cost and simple experiment, if we are well acquainted with the problem or the fundamental knowledge. While OMIC technology may represent the state-of-the-art, from an industrial perspective, it remains merely a tool. Indeed, many of us overlook fundamental questions such as 'have we identified genuine susceptible and tolerance samples at the first place?' 'Are these differences due to genetic contamination?'. When we specialize too much, we are likely to miss out on information, which is why multi-disciplinary collaboration becomes increasingly relevant. Besides, I have observed that academic literature surveys alone are not sufficient to understand the entire issue in agricultural science, simply because many details are not published in journals or books. Therefore, it is imperative that we engage with farmers and individuals associated with any industry who possess firsthand knowledge of the real problem requiring research. The task of convincing industry stakeholders to adopt cutting-edge technologies, methods, or processes is always a significant challenge for researchers. With considerable experience in the plantation and crop industries, I believe that some issues have remained unresolved for decades. This is due to the unwillingness of some stakeholders and investors to accept research findings due to multiple reasons. It is my opinion that success in research does not end with the generation of a thesis or a journal article. In order to close the loop, the research should also influence the stakeholders and investors in the industry as well. I believe that this is the first step to a successful research project, to be followed by the possibility of commercialization (if possible).

8.2.34 If You Were in Your Position, How Would You Recommend Universities in Order to Align Their Research to Meet the Needs of Industry and Society?

In my opinion, most students (or even some academicians) are unaware of the role and actual applications of a scientific subject. While visiting some universities, I noticed something very interesting. We have excellent and well-trained students in a specific area, such as bioinformatics, but they have no clue about their future prospects, and many of them have changed career paths immediately after graduation. It is a significant loss for our research. Academic institutions may consider inviting industrial researchers as guest lecturers to discuss their experience and the latest challenges they face. I do believe that both academia and the industry can work together to create a win-win ecosystem to train a strong talent pool which matches the industry's needs. In this way, the unemployment rate of recent graduates may be reduced. It may surprise you that despite the large number of graduates we have at each convocation, many companies are having difficulty employing local graduates and we have to look overseas for the right candidates.

8.2.35 What Are the Most Challenging Issues that Scientists in the Field of Bioscience and Biotechnology Face? What Is Your Advice?

Tam Heng Keat
I have observed that the challenges scientists face are largely related to local issues. In any case, I would like to address two common issues. Firstly, research funding. Many small labs, university departments, or even countries face financial difficulties, which may limit the scope of their research, the equipment they can own, and the technology they can use, and eventually affect their interest in science. Unless a drastic change is made in the research policy, this problem will continue to persist. As we know, gross domestic product (GDP) of each country influence total fundings.

A young scientist, especially one with a great idea and big dreams, will face many challenges and eventually feel disappointed if they are placed in a position with restricted research funding. Possible ways to walk away from the trap are, for instance, (1) to join or collaborate with big research groups or prominent senior researchers with whom you share the same interests or common goals within your home institute, or a nearby institute within your country and (2) in the case of a lack of essential facilities and funding, as well as a lack of a team, relocating to another country may be one option. As a rule, we do not have the power to influence other people or change the entire situation, therefore, we must learn to adapt and change on our own. It may be necessary to do what we can at a particular time when funding is limited. Store our great ideas somewhere so that when the time is right, we can turn them into a reality. Several developing countries lack access to advanced technologies. Collaboration is an effective short-term solution in scenarios where funding is limited, technology is not available at your home institute, or skills are lacking. Across countries and across research disciplines, collaboration is very common and essential for complex projects. Even though not all collaborations are fruitful, we should at least try, and it is beneficial to get to know more people. You can find collaborators through fellowships, professional networking events, online platform like ResearchGate, research consortia, conference, etc.

Lack of skills or ideas is the second most common problem. We must maintain the right mindset to learn both the conversional technique and newer skills as technology continues to evolve. Never be afraid to explore and learn something new. Many researchers place too much emphasis on high-end devices; however, they rarely use these devices even when they are available. In some laboratories, we may have seen that the equipment and funding are adequate and the publication outputs are above average, but in this situation, I feel that the issue is related to the student and/or principal investigator's skills and ideas. Certainly, no one is born with all the necessary skills needed to conduct research. The main problem is that people are afraid to try something new.

8.2.36 There Are Multiple Short-Term Fellowships That You Have Received. Why Did You Apply and Why Did You Travel to Different Laboratories in Different Countries?

Most fellowships are free to apply for, and they do not impose any processing fees. If we apply and are rejected, we have nothing to lose, at least we have tried and learned something from the process. I have personally experienced that attachments to other laboratories facilitate knowledge exchange and provide us with a great deal of opportunity to learn new technologies, new lab cultures, and new ways of thinking and solving scientific problems. Fellowships can enhance my career development. In some cases, getting a fellowship is a way for a person to be recognized for his or her work.

8.2.37 Having Worked in a Variety of Academic Institutions and Countries, What Are Your Recommendations for a Young Researcher Who Wishes to Leave Their Comfort Zone and Move to a Different Country to Study or Work?

Be courageous. As a young adult, you have nothing to lose. My future was uncertain when I had just completed my secondary education, as was the case with most young people. At that time, a school alumnus who worked in US NASA gave a motivational talk at our school. I remembered how he inspired me to be brave, and not let the condition of my family of origin or background limit my potential. Financial constraints within the family may be a problem, but they should not be used as an excuse for not working hard. Being poor in terms of finances is not a matter of shame, but being poor in terms of dreams and dedicated to work is something to be concerned about. When one is well prepared, there are always opportunities available to him or her. Since I was not born into a wealthy family, my parents were unable to finance my overseas study. Back when we were still using low-speed internet access, I spent months searching for scholarships, potential projects, and potential supervisors. It was a difficult time, and I had to put more effort into breaking through the barrier in order to reach another milestone. We now have access to high-speed internet, and there are a variety of opportunities if we sit in front of a computer and conduct a thorough search. You should utilize your phone, tablet, computer, and the Internet on the correct target that is aligned with your aim in order to become a better researcher. Sadly, I have observed that some young people are unwilling to take the initial step towards seeking such opportunities. In order to apply for a job or a position for postgraduate studies overseas, a significant amount of communication and submissions are required to different offices, institutes, or supervisors. Shortly stated, chances are for people who are prepared for them.

8.2.38 Do You Believe that Moving Abroad to Study or Work Is Only Suitable for Young Adults?

Individuals who are 20+ years old may find it relatively easy to go overseas for a fellowship attachment, to study, or to work. I myself, at the age of about 28, moved from my home country to another country for my PhD that was located thousands of miles away. Over the course of my postdoctoral career and my employment as a scientist, I moved to a new city with a totally new environment when I was over 30 years old. Eight years later, as I was about 40, I decided to relocate to another country in order to gain better research opportunities and contribute to scientific research. Generally, people believe that young people find it relatively easy to leave their comfort zone. It becomes more challenging as we age, with spouses and children being among the major concerns.

Taking a different perspective, since I have been moving and adapting fairly rapidly since I was in my 20s and 30s, relocating to another country is not a difficult decision to make. The process of navigating a turning point is never easy, but as young individuals, we should have the ability to resolve any problems that may arise. Young adults have a better ability to overturn a bad decision, even if it turns out poorly.

However, if I had not previously left multiple comfort zones, I will not accept the offer to relocate to another country at my current age. My point is that if I have been living in a country or working at one university for 20 to 30 years, moving to another country to work will be extremely challenging. We know some senior researchers who left their comfort zones in the 50s to pursue better employment in other countries. It is certainly possible for young people who have recently graduated from college or who have earned a master's or doctorate degree to do the same if they wish. Again, be courageous.

8.2.39 How Do You Set Research Goals and Maintain Momentum to Achieve Them?

Enketeswara Subudhi

For me, research goals are about understanding the 'what' and the 'why' of research. By answering these two basic questions, we can establish any research goals of any topic of global interest. A well-supported literature review should be undertaken before the establishment of the research goals, and research gap in the literature (knowledge) should be recognized in order to understand the research problem of the scientific community in particular and the society as a whole. Research is done in order to seek out answers to problems related to a particular topic. I think one of the most important aspects of setting research goals is to consider the relevance of those goals around the world so that it becomes the principal aspect to consider. Additionally, we must consider how we can use regional sources, methods practised in ethnology, and concepts of local origin around us to address goals with global implications. Basically, I am referring to the concept of using local resources, local facilities, local technologies, local methodologies, and using the modern techniques and tools which shall result in a sustainable solution. If a research goal is exclusively focused on local needs, it may have lesser impact or influence on the scientific community at a global scale and will have a worldwide identity.

Here is an example that will help readers understand the concept. Recently, we have explored thermophilic microbial diversity in a local hot spring located at Deulajhari in India. Using amplicon sequencing, we determined the microbial population. In some ways, this project may seem a bit localized since it focuses only on a local sample. It is interesting to note that this hot spring is surrounded by pandanus plants. There is a worldwide interest in finding thermostable enzymes that can be used to decompose and saccharify plant biomass. There are only a few hot springs that have plant litter or endogenous plants surrounding them. Our aim is to perform more detailed multi-omic analysis, such as metagenome-assembled genomes and transcriptomics, in order to discover new enzymes. Look at this example. The hot spring is a geothermal site in India, which is very localized, isn't it? Pandanus plants may also be a bit localized, since other hot springs may not have these plants. By using this example, you should be able to understand how local resources can be connected to global scientific research. If the project is too localized with no global interest, we may find it difficult to secure funding or even publish in a good journal. It does not matter what kind of research we are focusing on, there is always a way to look at it from a local and global perspective, that is how I believe we can narrow down and fix our research goals.

To maintain momentum in research, we should reflect on a few basic questions. Why do you do research? Are you really passionate about it? Is it merely a means to earn a living? Do you conduct research solely for the purpose of meeting certain due dates or meeting certain publication set by your employer? When it comes to doing research, there is a fundamental difference between those who have a passion for it and those who consider it a job only. Having a job as a researcher or academician, it is very different from having a job in other fields. Those who work in an office, for example, have to complete all their tasks within a specific time frame and then go home to relax. Researchers must think about their own expertise, building and improving their expertise, research goals, career development, and all these factors must be aligned properly using individual efforts. Generally, scientists design their own career goals, putting in the time and sweat to achieve them, and investing time and effort into them. In terms of the goals or plans that I pointed here, I would say that they are some kind of distant vision goals and not something that can be achieved within days or months. Distant vision goals or far-sighted aim should be seen as a universal response to the bigger research questions that are being asked across the globe. This means you are part of a global effort to find a solution that benefits the entire scientific community and society as a whole. By doing so, you will be able to find value in what you are doing or set a value for it. That is one way to maintain our momentum.

My opinion is that there are several reasons why some researchers do not have clear research goals and do not have momentum. Complacency is one of the causes. I am a full professor at a private university, here in Odisha, India. In order to acquire funding, buy equipment, and take on postdoc or PhD students, we compete with government-funded universities and institutions. Due to the fact that we are in a competitive environment, we have no time to idle around. Although I am sorry to say, I have come across academicians and researchers who are so complacent after they get a job at government universities or institutions, and I am not saying this for my country, but rather as an observation that I have observed when I have visited other universities and talked to many people. Research and academic work are not the same as ordinary jobs, as I said earlier. A person who treats their tenure as an ordinary job will probably feel smug or think that they do not have to do anything about a situation. "I only need to do the minimum, or to just do the almost fixed task scopes that are assigned to me, and after all, I can still go home on time. I get my pay. Extra work doesn't get me extra money. My job is so secured and almost no possible to lose my job, then, why should I work harder", this is often the slogan in their career goal. These researchers and academicians are kind of sad to see. They have the ability, they were well trained during their studies, and they have good facilities and infrastructure, yet they are complacent. In conclusion, passion is the key to success.

8.2.40 How Do Students Find Motivation in Research?

In the earlier part of the conversation, we discussed the scenario for full-time academicians and researchers. Let us now discuss students. Success still depends on passion. In reality, young students don't know what they like when they're inexperienced. Over the last decade or so, the program in my country has changed to prepare students for PhDs. Students are motivated to do some research and some publications in their master training. Basically, they are exposed to different aspects of research, and they get a feel for what it's like in the real world or they get a feel for how they'll handle a PhD program. A PhD is something you choose, not something you're forced to do. It should be by choice and not by chance. You should not be like, if you did not find any success in other disciplines or could not qualify some other national level tests for any technical graduate programs, you should resort back to research as a career. Students should pursue different research activities at graduate and post-graduation levels in a small scale to get inspired and keep themselves motivated to jump into this field. Universities can think along these lines by giving young minds the chance to try in several fields of research. During master's training, we offer scholarships to help students develop their skills and interests, and with that, they will be able to find a direction that is personal to them before pursuing a PhD. But the initiations taken should be in a bigger scale. Students must feel secured and supported by the system, if they're going to pursue this as a career. Students should dare to get their hands dirty by trying new things and explore their own interests. Once you find it, continue developing it. Continue exploring different things until you find one that interests you.

8.2.41 What Are the Major Differences Between Research in Developed and Developing Countries? Could You Please Explain What Causes This? Could You Please Provide Advice to Young Researchers Who Have Limited Resources but Wish to Achieve Excellence in Their Research?

In each country, there should be a government ministry or agency for managing research funding matters, such as granting funding or scholarships. We understand that developed countries will have better facilities, better equipment, and greater opportunities for funding. Again, there might be more competition over to attract these funds by the developing countries as a whole. Researchers in developing countries struggle more than their counterparts in developed countries. It seems that the overall percentage of researchers with relatively poor facilities and funding is higher than those with well-equipped facilities. The disadvantageous situation may cause us sadness or disappointment. Real passionate researchers must find a way out. The first thing that we should do is to gradually accumulate your funding and equipment as you go along. In the second step, you need to collaborate with partners who have better facilities and opportunities in the other parts of the world. There are various ways in which collaboration can take place, such as the creation of an official agreement, the sharing of equipment and analysis facilities, student and faculty exchange programs, or co-authorships in papers within the country as well as outside your country. In the third step, you should demonstrate your ability and convince anyone or any stakeholder who may be able to assist you in moving forward. Finally, because we have limited resources, we should be creating smarter projects: those that are precise, more creative, more innovative, as well as those that have more reasonable and achievable goals. It is always possible to achieve the goals in a cheaper and more effective manner. Even if we do not have the means to own a private airplane, we are still able to travel to the place we want to go if we are in the economy class.

I would like to reiterate the following point once again. Genuine researchers, scholars, and students have a strong passion for their work, and they do not treat their tenure as a job. The outcome will be different if you have the motivation and the passion. Young researchers from developing countries should not give up, and rather build your passion, develop your knowledge and skills, and make connections and you will also be able to excel.

8.2.42 Some Students Believe That Simply Focusing on Textbooks and Exams or Scoring an A Is Sufficient. Thus, They Decide Not to Strive for Any Further Goals. In Pursuing a Master's Degree or PhD, the 'Students' Focus Remains on Grades. What Are Your Thoughts on This Matter?

Lulie Melling
When a student pursues an A grade on a test, their goal is limited to just getting the grade. However, if you prioritize knowledge and understanding, the 'A' will follow

naturally, and your knowledge and understanding will actually go beyond the boundaries of the 'A'. While it is understandable for students to pursue good grades, I have witnessed situations where individuals face a lack of direction and purpose after graduation, especially when their primary motivation is solely to please their parents. These students often lack self-motivation and become lost after completing their first degree or master's degree, having spent their entire school and university lives following in the footsteps of their parents or peers. Similarly, there may be individuals who pursue master's or doctoral degrees, albeit without a clear sense of purpose or self-motivation.

Moreover, individuals who pursue advanced degrees solely for the sake of status may limit their potential. To prove this point, they can feel disoriented and lost even with a PhD degree in hand. When individuals aim solely for a status or a PhD certificate, their goals are limited to a piece of paper and a PhD (Dr) title. However, if they strive for knowledge, their understanding will expand as there is no limit to exploration and learning. The more one reads, the more they realize how much they do not know, which will then trigger them to explore further.

It is insufficient to solely focus on textbooks, exams, or journals; we must also possess the ability to critically evaluate existing scientific theories for their accuracy and reliability. In recent years, it has become increasingly evident that many university students are neglecting the importance of reading textbooks, which is crucial for developing a strong foundation in their field. This oversight can significantly hinder their acquisition of essential skills such as scientific exploration, critical thinking, evaluation, and summarization. However, please do not misconstrue my message. My emphasis lies not on the specific reading materials or subjects we engage with, but rather on the habit of reading itself. To put it more bluntly, it is the absence of a reading habit that concerns me. However, by cultivating a habit of consistent reading, we can nurture our aptitude for scientific exploration, critical thinking, evaluation, and summarization, empowering us to become well-rounded and knowledgeable individuals in our respective fields.

One prevalent issue I have observed is that many individuals nowadays prefer to engage in brief reading sessions. In libraries or bookstores, we have the opportunity to learn more and be exposed to new ideas not merely in science but in all other aspects. Even so, I have been told by certain people that they are almost lost in libraries or bookstores, which proves that they rarely, if not never, enter the ocean of books that are abundant with knowledge.

In today's digital age, people can quickly search for information online. While resources like Wikipedia can be a helpful starting point, they should not be relied upon as a comprehensive source of information. It's important to recognize that true expertise comes from dedicating significant time and effort to a particular subject. Simply reading a few sentences or pages or taking the brief information from ChatGTP doesn't make someone an expert. Only through in-depth study and practical experience can one acquire a deep understanding of a subject.

We should not treat acquiring knowledge in the same manner as having fast food; no shortcut should be taken. Besides, the action of acquiring science and information in an instant manner is similar to the practice of relying solely on lecture notes to prepare for exams, memorizing the text, and regurgitating it to the lecturer during the exam. This approach does not create advancement in science nor improve the students' level of understanding. In fact, acquiring knowledge is not merely about learning the subject, but it also entails developing exploratory and critical thinking skills and conducting in-depth research to uncover the underlying depths of a topic.

Unless university students alter their approach to learning science, significant progress and success will elude them. A few years ago, while seeking written resources on the essential skills necessary for thriving in universities and research institutes, I stumbled upon the 'Academic and Research Skills Handbook' compiled by Crawford Academic and Research Skills Advisors at the Australian National University. This comprehensive eBook is available as a downloadable PDF file online. I highly recommend all readers to explore this valuable resource.

8.2.43 How Would You Define PhD?

In my opinion, a PhD programme is not only about data collection. Rather, it presents an opportunity for personal growth and self-discovery. If you pursue knowledge and wisdom with the correct mindset, you will achieve success in your PhD and future research. Philosophical thinking requires the training of the mind, which is the core element in PhD training (learning). We are aware that the '5W & 1H' questions include the what, who, why, when, where, and how. However, I view the 'how' question as the most powerful question, followed by 'why'. In addition, the philosophy of science aims to explore the nature of science, including its underlying principles and the process by which scientific knowledge is constructed through logical reasoning. The term 'scire', which means 'to know' in Latin, reflects the goal of scientific inquiry. As a form of metacognition, the philosophy of science delves into the methods used to define science, its purpose, and the various approaches to building scientific knowledge. As a PhD candidate, you should possess a philosophical mindset and maintain the mindset even after receiving your certification.

8.2.44 In Your Opinion, What Soft Skills or Interpersonal Skills Should One Possess to Be Not Only Successful but Also Be Outstanding After Completing Their Studies and Entering the Workforce?

For readers to remain relevant in their field and successful in their workplace for the long term, I will share the opinions that I have discussed with my staff in my institute on a regular basis.

The equation is Success = IQ + EQ + 2SQ + AQ + LQ, where:

- IQ is merely the passport to enter the graduate club.
- EQ is how you would move forward.
- SQ is social and spiritual.

- AQ is agility—adaptation, perseverance.
- LQ is leadership and planning.

While certain keywords may seem self-explanatory, their true meaning is best understood through daily practice. It is important to dispel the common misconception that IQ is synonymous with academic achievements such as exam scores or CGPA. In reality, this notion is sometimes inaccurate. For instance, possessing a high IQ alone does not guarantee success in scientific pursuits, particularly for those aspiring to pursue careers in research or academia. A high CGPA also does not imply high intelligence; rather, some students are just good at memorizing.

I firmly believe that attitude plays a pivotal role in determining an individual's success, as even the most intelligent person can underperform without the right mindset. It is crucial to never become complacent with our achievements and continuously strive to learn and explore new possibilities. In his famous 2005 lecture at Stanford University, Steve Jobs urged us to 'stay hungry, stay foolish'—a message that emphasizes the need to remain curious, embrace new experiences, and continually work on our weaknesses. I strongly feel that obtaining a PhD does not define one's worth or intelligence; rather, it is the willingness to constantly improve and push beyond our limitations that truly sets us apart. In contrast, if we remain in our comfort zone, not being productive, and not even doing the bare minimum or improving despite holding a PhD, this situation should be considered an embarrassment.

Constant progress is crucial within the academic community and among researchers. Prolonged periods of stagnation, regardless of whether they occur at a higher or lower level, often indicate an underlying issue with attitude. It is therefore essential for researchers to possess a profound sense of vocation that encompasses respect, commitment, and responsibility towards their occupation. Furthermore, researchers must acknowledge that their profession is distinct from other occupations. Unlike many jobs, research involves a lifelong pursuit of knowledge, necessitating continuous problem-solving and exploration. This demanding endeavour requires exceptional dedication and diligence from researchers.

I would like to elaborate on two crucial attitudes: humility and curiosity, which are essential for establishing a strong foundation in any field, particularly in research. It is important to acknowledge that researchers do not possess all the answers, and they should be open to learning from anyone. The purpose of research is to expand knowledge and enhance understanding. To achieve this, researchers must approach their work with a deep sense of curiosity, continuously questioning and critically evaluating every piece of information they encounter.

While reading scientific journals is indeed important, the ability to assess and evaluate each article holds even greater significance. It is vital to recognize that not all information published in journals is inherently accurate and reliable. Therefore, it is necessary to seek clarifications whenever needed by posing the right questions to the appropriate individuals. By embracing humility and curiosity in our work, we can foster continuous learning and personal growth as researchers. These attitudes enable us to make meaningful contributions to our respective fields and advance knowledge in a meaningful way.

There should be a direct correlation between every penny earned from the salary (or scholarship received by students) and the effort placed into their work. Simply passing the time at universities or research institutes is unacceptable. Researchers must take responsibility for their work and be held accountable for it. By maintaining a high level of accountability, we can strive for excellence in our field.

Within an organization, there may exist a group of exceptional researchers who consistently produce a remarkable number of scientific papers. However, when compared to the average activity level of their colleagues, these individuals may be considered outliers. It is worth noting that many high achievers possess a sense of humility, allowing their work to speak for itself. Ironically, despite their exceptional output or publications, they may not necessarily enjoy high popularity. We do not conduct research to gain fame or popularity; we solely focus on conducting research that makes a difference to science and society. If you are one of the researchers whose good actions are not recognized by your colleagues, please do not be discouraged as the world will acknowledge and recognize you in due course. Luke 4:24 states, "no prophet is accepted in his hometown". You will be able to understand my point if you have heard the story and understand the history behind it.

In the equation of success that I previously formulated, two important components are Social and Spiritual. As a woman with children, I am acutely aware of the challenges faced by female researchers. As my responsibilities at home include supporting my husband and attending to family matters, I ensure that all the responsibilities are fulfilled. At the same time, I also ensure that my responsibility to society is fulfilled. Besides, the job as a researcher may be more suitable for women compared to a teacher or a few other occupations because you have more freedom and time flexibility.

Perseverance serves as the key to achieving success. To gain success, tremendous effort and being organized are crucial; we must practice good time management, a balance between family and careers, and properly manage the kitchen. Mothers who work as scientists can successfully balance motherhood and their careers, as they stay informed about the latest developments and can impart knowledge to their children.

Throughout the journey to success, perseverance and being purpose-driven are essential. For individuals who are religious, it becomes essential to include spirituality in their equation for success. As a Christian, I am deeply aware of my calling and responsibilities. Furthermore, the SQ affects our lives and actions and is directly connected to the concept of purpose-driven living. Aligning my daily life and responsibilities with my life's purpose is of great importance to me. This practice ensures that I am on the right path and enables me to make a positive impact on society and improve the lives of others. Incorporating this motivation into the equation for success becomes an integral part of my journey.

The LQ (Leadership Quotient), which encompasses leadership and planning skills, is another vital component of success. In the realm of research, two types of researchers can be identified. The first type is primarily dedicated to the research itself and may not have a strong inclination towards management or leadership roles.

However, it is important to recognize that effective management, leadership, and planning are integral to the success of any research project.

Tasks such as managing research grants, supervising students, engaging stakeholders, overseeing laboratory staff, and procuring necessary supplies are essential for the smooth progression of a research endeavour. In fact, leadership and project management are inherent in the very initiation of a research project. As the saying goes, 'if you fail to plan, you are planning to fail'. Furthermore, even in cases where there may be a lack of equipment or laboratory space, proper planning and resource management remain indispensable. To some extent, leadership and planning are inevitable in the field of research. They contribute significantly to the success and effective execution of research projects, ensuring the optimal utilization of resources and the achievement of desired outcomes.

The second type of researcher is open to combining research with management positions and leadership roles. From my perspective, management and leadership encompass qualities of a leader who mentors, motivates subordinates, leads by example, and supports others. True leadership involves actively engaging with the team and progressing together. Merely occupying a directorial position and giving orders does not define leadership in my opinion. Genuine leaders are those who possess substantial expertise and content knowledge. They are more likely to be recognized as effective leaders, as their competence and understanding enable them to guide and inspire others.

8.2.45 How Important Do You Think It Is to Have Good Communication and Planning Skills?

I have observed that students who prioritize chasing after 'A' grades often describe themselves as 'shy'. They tend to struggle with effective communication and frequently lack strong negotiation skills. In such cases, I encourage you to take initiative, be resourceful, and not let timidity or fear hinder your progress.

Additionally, it is crucial for young adults to recognize the importance of communication with individuals at all levels of the hierarchy. I have witnessed individuals who solely interact with their supervisors while disregarding those at the bottom of the hierarchy, such as cleaners. This approach is not appropriate, as there is no justification for ignoring the people around us. Whether working in a laboratory or any other setting, establishing good relationships with all individuals is imperative, as success often relies on effective teamwork and collaboration.

It is important to emphasize that these relationships should be genuine and built on mutual respect, rather than being solely driven by personal benefits. Building healthy relationships based on respect and genuine connection will contribute to a positive and supportive environment for everyone involved.

To share my experience, I served coffee six times a day to my sensei (my project supervisor) during my PhD. Although he did not ask me to brew the coffee, it was my way of showing respect, communicating, and building good relationships with him. In no way did I dream nor expect anything from this coffee; it was my way of meeting him, talking for a few minutes, and building rapport and trust on a sincere and respectful level.

Essentially, the keyword is communication. Good medical doctors possess good communication skills. Do we not prefer seeking doctors who have excellent communication skills and can explain to us in a more understanding manner? However, scientists often fail to realize that written and oral communication are equally crucial to the design of their experiments and the analysis of their results. Another recommendation is to learn to listen instead of merely hearing with your ears. Rather than simply hearing, make sure you listen to the message and try to understand the reasons behind the message. The ability of a researcher to listen and think is paramount to their ability to communicate.

Polite, effective, and structured communication is essential. It goes beyond language and tenses. Let me provide a few examples. It is considered rude and ineffective to submit a poorly edited report to your boss, expecting them to review and make necessary changes within a limited timeframe. The lack of clear instructions can hinder effective communication. For example, submitting documents without proper guidance on platforms like WhatsApp can be problematic. To ensure success, it is important to make extra efforts. For instance, when submitting a lengthy report, consider providing a separate summary page to help superiors grasp the main points quickly. Similarly, receiving an email attachment without a proper description can be disrespectful. When seeking a supervisor's signature, politely explain the purpose of the document. It is crucial to learn communication skills and show respect in facilitating others' work. By demonstrating your ability and attitude, you contribute to a culture of respect and responsibility among all individuals, including young researchers.

8.2.46 What Is the Funniest Behaviour That You Have Seen from Certain Researchers That You Think Should Be Changed?

If we don't take action on our dreams, success remains nothing more than an imaginary slogan. Some individuals find enjoyment in the cycle of 'reading, analysing, reading, analysing', which can be valuable for planning and decision-making processes. However, if this cycle is not followed by action or execution, it could lead to a lack of progress or even failure in completing tasks. 'Analysing until you are paralysed' refers to the inability to make a decision due to overthinking a problem. Overthinking can indeed lead to decision paralysis. However, it's important to recognize that sometimes we need to break free from this cycle and adopt the mindset of "just do it". By taking the necessary steps towards our goals, we can overcome analysis paralysis and make progress.

Besides, we can'3t know everything from A to Z before we begin a project, especially in research. Despite the importance of preparation, learning or understanding the topics is more effective if you act on them and learn along the way. Moreover, humans would be unable to think if they do not understand the content. Making mistakes is okay as long as we learn from them, improve our knowledge and skills, and do not repeat the same mistakes. It has been said that we do not know what we do not know. Are you of the same opinion? Ultimately, the ability to learn from mistakes and take risks is one of the most essential components of the path to success.

8.2.47 The Suggestions You Have Provided Cover a Wide Range of Topics. Is It Possible for You to Summarize Them in a Concise Manner?

As a conclusion, I have prepared a list of main key points.

1. Scientific Foundations (improving principles, basics, and theory).
2. Professional Mastery (polishing research skills and expertise, accuracy, and unbiased)
3. Organizational Skills (leadership, prudence, planning, and allocating and monitoring of resources)
4. Cognitive Skills (conceptualization, theory building, analysis, self-assessment, reflection, intellectual humility)
5. Integrity (trustworthiness, dependability, reliability, and accountability)
6. Curiosity (open-minded, interest, exploration, and experimentation)
7. Collaborative Skills (building and maintaining relationships/networks, and shared vision)
8. Perseverance (determination, decisive, and persistent)
9. Self-Reliant (self-confidence, working under pressure, and resourcefulness)
10. Self-Motivation (sense of purpose, personal vision and drive, pride in one's work, responsible, timely deliverables, and set achievement goals)

The list is still a bit lengthy, but if you really want to make a change, these are the three main steps you should take. First and foremost, I think you ought to have a correct mindset. Second, you need a solid foundation that is grounded in fundamental science and knowledge. Third, you need to create your network and PR with people around you from multiple layers and make contacts outside your small circle of influence. There is a lot of humanity and attitude interspersed among the three main things stated above.

8.2.48 What Does Stagnation in a Person's Research Achievement Mean to You? What Causes Stagnation and What Should Be Avoided?

Annarita Poli
The stagnation in a person's research achievement for me represents the lack of professional growth, goal development, and the achievement of specific objectives. The lack of enthusiasm to achieve the professional successes is the main cause of

stagnation. The main point is to understand whether stagnation can occur as a temporary condition or it is so ingrained that it becomes a permanent and irreversible condition.

There are several ways to avoid this stagnation, including comparing yourself to other researchers working in the same field, or in fields similar to it, participating in specific conferences and workshops, and continuing training on the most recent techniques in your field in order to keep on top of the latest developments in the field.

8.2.49 Do You Believe that Mentoring and Collaboration Are Important? Why? What Advice Would You Give to Young Researchers Searching for Mentors or Collaborators?

It is my belief that the presence of mentors and collaborators can help to stimulate new horizons in work and to foster openness to new challenges. Even healthy competition between colleagues in the same research group can promote success and overcome stagnation. My advice for young researchers is to look around and choose the working group and the mentor for which he/she has the greatest affinity. You should seek out mentors and colleagues who are able to inspire and transmit enthusiasm for new challenges.

8.2.50 Could You Please Provide Some Advice to Candidates Who Are Considering Doing a PhD in the Field of Bioscience or Biotechnology?

Consider microbiology if you are unsure what area to study. The area is so huge and full of mysteries that you will fall in love with it and spend your whole life exploring it. The majority of life on our planet is microorganisms, both in number and biomass. They are ubiquitous in the biosphere and fundamental for the functioning of ecosystems, and in fact they regulate biogeochemical cycles and the recycling of organic matter; they are the main producers and absorbers of greenhouse gases; they play an essential role in the soil (structure and fertility) and in the quality and productivity of soils, seas, lakes, and rivers; and they cover the surfaces of all other organisms, influencing their physiology and well-being. Microbial biodiversity is considered a valuable resource for science and industry, with implications for the economy and finance, since microorganisms are the engine of the bioeconomy and environmental sustainability to promote global development and human well-being. I strongly recommend a course of study based on biotechnologies which for me represent the future for eco-sustainable and competitive development.

8.2.51 Based on Your Experience and Observations, Do You Believe that Master's and PhD Degrees are Only Suitable for Students with Excellent Academic Records, Such as Top Students in High School or First-Class Students in Undergraduate Studies? Do You Agree that Only Brilliant Individuals Will Be Successful in PhD Research?

Stephen Brian Pointing
I would say that based on my personal experience in academia the response to this question is a resounding no! Whilst I have had some excellent graduate students that have come to my lab from top-tier universities, the most successful students were all graduates from what are perceived as mid-tier universities. These highly successful individuals shared some common traits: they were all highly motivated, were capable of recognizing opportunity where others might see only difficulty in research, and they all had a fastidious attention to detail. I don't think that attending an elite university or achieving the top honours at undergraduate level necessarily reflects or nurtures these qualities. This is why the selection process for postgraduate students needs to be holistic.

For future postgraduate students, I would like to offer the following advice. My advice would be to commit wholeheartedly to your aspirations. If you are an undergraduate and you have decided that postgraduate research is your goal then the most important thing to do is to get some experience as a laboratory or field intern at your university, or contribute as a volunteer to an organization working in your chosen scientific field, for example, a conservation charity, local hospital, etc. Grades matter of course, but this experiential learning will give a strong indicator to any selection panel that you have a passion for your field of study.

8.2.52 Would You Also Agree that Only Lecturers and Researchers Who have Obtained Training and Graduated from Top Universities Will Have a Better Chance of Succeeding as Researchers and academics? Some Young Researchers Feel that They Are Already Lost at the Starting Line Because They Lack a Relatively Strong History. Does the Future Hold Any Hope for Them?

I think that attending an elite university brings certain advantages related to networking and perceived pedigree. However, there are many academics who have reached the top of their field without this advantage. My personal view is that there is far too much academic imperialism involved in determining who is 'successful' in academia, and this is pervasive in decision making from faculty hiring to grant awards and peer review of research publications. I hope that in the future greater effort will be placed on truly decolonizing science as I believe it will be beneficial overall.

8.2.53 A Successful Faculty Member Must Take into Account a Number of Factors. It Could Include Things Such as (i) 'Strategy', Things that Can be Seen or Measured, Like Lab Facilities, Funding, a Winning Proposal (Idea), and (ii) 'Attitude', for Example, Mindsets, Behaviours, Effort, and Personality. If You Were to Choose One, Which Would You Pick—Strategy or Attitude?

I consider both important. If I had to choose only one, I would choose strategy, provided the individual does not have a major attitude issue. It is vital to have a plan for how your research will address an important issue and to have a roadmap for how you will evolve this into a coherent long-term research plan. However, I also feel that being kind, supportive, and recognizing that as a supervisor you have a wider duty to prepare your postgraduates for a career in science is an essential hallmark of a good supervisor.

8.2.54 Researchers from Less Developed Countries, Especially PhD Graduates Who Have Just Joined as Academics, May Feel That They Are Unable to Compete with Large Laboratories that Have Considerable Funding and Laboratory Resources. They Have Great Ideas and Wish to Conduct Good Research. What Would be Your Advice to Them?

Rajesh K. Sani
It is true that on the surface less developed countries may appear to have greater challenges because of a lack of funding; however, when viewed from a different point of view, the competition in developed countries is much greater. The idea that other labs are better than mine may always be ingrained in our minds, but that could very well be a myth. It is true that some people will think, well, you are working in the US, and you have large grants from federal agencies like National Science Foundation (NSF), but that kind of funding is not available in our country. Moreover, people assume that getting NSF funding is easy, but NSF has a 10–20% success rate. Translating that figure into the number of attempts simply means I may be funded only once or twice if I submit 20 proposals. In my opinion, most countries have budgets for research, and I am confident that if you prepare a good proposal, you will be funded. Here, the emphasis is on how much time and effort we have invested in preparing the proposal, not on the total amount of funding available in your country.

In the US, we usually have between three and five people to evaluate a proposal. Say there are four evaluators who read your proposal, you'll get plenty of feedbacks. Don't get discouraged if your proposal gets rejected. Learn from the comments you got, take them positively, and submit a better proposal next time. Your idea may be good, but you may have missed some details that the reviewers were looking for. In such a case, it would be useful if you included these details in a later version of your submission. On the other hand, if all the reviewers don't like your idea, remember that they are some one more senior and have read dozens or hundreds of proposals; they can tell if an idea is good or not. Nevertheless, if they do not like the idea, speak to your mentors or someone who can make some fair comments, and eventually, you may realize that the idea is not mature. In that case, it would be time to demolish the proposal and discard it. There is no point in persisting with a research idea that is not feasible. This is the learning phase. No proposal is ever guaranteed to be accepted, so don't get discouraged if yours is rejected. Formulate a better idea and move on.

No matter where they come from, fresh PhD graduates share some common characteristics. The follow my recommendations. When young adults have just graduated and become the junior faculty members, they tend to be narrow in their view of research and tend to focus only on their PhD work. I would suggest you think outside of the box a little bit, think a little bit larger, try not to focus only on a narrow discipline on what you were doing for your PhD, and instead try to think more about multidisciplinary or interdisciplinary research rather than focusing on one discipline. Obviously, at the beginning of your career, you have to spend a lot of time reading and learning about a new discipline, but I do think it is worth it in the long run. We may not be familiar with other disciplines well, so enlist the help of researchers with expertise in engineering, geology, chemistry, physics, medicine, computational science, and even social science. Try to engage in some transformative projects that are new and in risky projects that can help you develop your thinking skills and other skills at the same time. Give it a try and see if it works. Make sure you have read enough literature before writing the proposal. Collaboration with senior researchers you trust can help you do good research if you have great ideas but lack funding or resources.

8.2.55 Describe the Top Three Things That Young People (Researchers and Academicians) Should Prioritize and Put Effort Into, So They Can Enjoy the Benefits in the Future

The first thing you need to do is focus on research and give it priority. It is not my intention to suggest that you let go of any of your other responsibilities like teaching or some essential administrative activities, but, you must put priority on your research and give it your full attention. Keep your focus on the development of the field that you are going into and keep improving as you go along. The only way to come up with better ideas is to be focused. The second is travelling. You need to make time to attend conferences, present orally, put up a poster, visit another laboratory, and meet the industrial people. These are all activities that should be added to your calendar. The experience of travelling and meeting people allows you to learn about what is happening outside of your office or lab. From there, you will be able to analyse how people perceive your work and you will be able to improve accordingly. The third step is to focus on students. They are the ones who actually do the work in the lab. Ensure that you work closely with them, spend quality time with them, and make sure they are mentally not disturbed. These are the good students who will take your research to the next level. It is important to build good relationships with your students. You will not be able to succeed in your research unless they commit to it.

8.2.56 In Your Opinion, What Should the Head of the Department or a Senior Professor Do to Support the Professional Development of Young Faculty Members?

From the moment young faculty members join the department, the department can do a variety of things to assist them in their professional development. As an example, provide them with a workshop that teaches them how to write a persuasive proposal and how to generate good hypothesis, prepare a reasonable budget, etc., in principle, to make sure that the proposal quality is higher. There would be an even greater benefit if the coach or mentor already knew what the funders were looking for, for example, the scoring metric or rubric that they were looking for. Training like this is of utmost importance because when the students did their PhDs, they did not receive any training on how to write proper proposals for funding from the national and international agencies. Certainly, they do write proposals for their PhD projects, or a short proposal for a scholarship, but it is entirely different from how to write a proposal for securing fundings. In the US, each funding agency usually has one program manager in charge of its funding. It might be a good idea for the department to allocate some funds to support faculty members' travels to meet with program managers one-on-one, so they can get advice or feedback before they write their proposals. I think the department can provide basic resources for their lab start-up, provide them with a nice workforce, and provide them with good support.

8.2.57 In Your Opinion, What Are the Biggest Challenges Facing Biotech Scientists Today?

Edgardo Rubén Donati
Firstly, it might be important to distinguish between scientific challenges and those faced by scientists. Biotechnology, in my opinion, has the ultimate objective of improving human life quality; this implies a range of applications aimed at improving the health of individuals, human ecosystems, and terrestrial environments, as well as integrating them physically, philosophically, and sustainably. Biotechnology is a discipline of applied science that should have a positive impact on communities. The majority of my research and studies are focused on environmental biotechnology, which is usually referred to as grey biotechnology. A major challenge to the development of new 'clean' technologies as well as remediation and restoration of the environment is aiming to make these technologies asymptotically approximate natural processes in order to prevent "rebound effects", which are like "where a medication has more devastating effects than a disease itself". This requires more in-depth studies on how the various processes—including the microbial processes—

have been established and how they have evolved in order to build the complex network of interactions that define the functioning of terrestrial ecosystems and the adaptation of life to various environments and conditions. I am describing a general problem I have encountered in my research.

Other fields in biotechnology will undoubtedly present their own unique challenges, and I will not go into great detail about them. However, I observe that the key challenge lies in how we approach and address these individual challenges. Researchers often make the mistake of focusing on the wrong thing, such as spending too much time on trees that block the view of the forest. We often spending excessive time on minor details instead of seeing the bigger picture. For example, new tools or methodologies are just the vehicle that drive us to the real research objectives; however, people often spend too much efforts on the vehicle. Please keep in mind that scientific activity is inherently a human activity based on social contexts and societal demands. I am certain that we are both in agreement that the current KPI systems in academic institutions are exerting pressure on scientists, primarily through the quantification of publications. In spite of this, we should put more emphasis on the value or quality of research, especially in terms of its societal impact. We need to prioritize research that contributes to positive societal outcomes rather than solely focusing on publication metrics.

8.2.58 Why Is Networking and Industrial Engagement Important in Your Opinion?

My opinion is that the general aim of scientific research should extend beyond the publications or theses, although not all projects may be able to do so. We should develop the mindset that recognizes the potential application of our research projects in the future, whether by private or public organizations. There are countless examples of scientific or technological advances that have occurred without the involvement of business or industry. Part of the problem may be solved through the establishment of networks and industrial engagement. Industrial participation allows many projects to be reoriented and channelled by defining specific goals and deadlines aligned with industrial requirements.

It is imperative for industry and academia to work together to solve real-world problems. Participating research groups benefit financially from the funding provided by industrial partners. As a result of this collaboration, a number of benefits are available to the companies in many cases, for example, tax reduction (or in some cases even double tax reduction) in certain countries. However, there is a lack of business-industry-academic partnerships in developing countries, particularly in the field of environmental science research. University institutions should work harder on this issue, but I believe it may be difficult for individual institutions to accomplish this on their own. The creation of bridges between companies and academia may be initiated by scientific foundations or associations at a national level. Participating companies can share their difficulties and needs through such organizations.

8.2.59 Research and Publications have been a Part of Your Professional Life for Many Years. Describe What Makes You a Preserver. What Drives You to Keep Moving Forward?

It is likely that we all become preservers after engaging in an activity for a long period of time, regardless of how challenging or novel the activity may be. For me, I am still driven by the same goals I had when I began my scientific career when I was very young: curiosity about how things work and an interest in applying that knowledge in concrete ways. The process of imagining hypotheses, constructing appropriate experimental designs to test them, analysing the results, and putting them together in a harmonious way is also pleasurable to me; I enjoy working on this same sort of scientific construction process as I do when teaching undergraduate students. Choose a topic that interests you and conduct research on it. You will also serve as a preserver.

8.2.60 The Research You're Doing Now Aligns Very Well with the Sustainable Development Goals, and You've Been Putting a Lot of Effort into Studying Microbial Ecology, Biogeochemical Cycles, Greenhouse Gas Emissions, Etc., for a Long Time Now. What Is the Driving Force Behind Your Interest in This Research Area for Such a Long Period of Time?

Adrian Ho Kah Wye

My short answer is interest. Since an early age, I have always enjoyed being close to nature. I recall catching insects in my garden and fishes (mainly, guppies) in nearby streams for my aquarium to study them. Eventually, this shaped my current research, being actively involved in studies to restore and preserve pristine environments. I reckon it is important that one is passionate about their research to sustain long-term goals. At the same time, I am always open to new research ideas (e.g. applying concepts in other fields to my work); cross-disciplinary work usually leads to unexpected and interesting findings! Admittedly, nature preservation and conservation is often in conflict with land development to generate food, feed, and fuel to meet the demands of an ever-growing human population. To this end, my recent work involves looking into ways to increase crop yield, employing regenerative agricultural practices to limit further deforestation for agriculture. That is my genuine research goal that drives me for these years.

8.2.61 It Is Important to Have a Solid Foundation of Knowledge and to Publish in the Academic Field. In Your Capacity as a Researcher at a Research Institute and a Biotechnology Company, What Advice Would You Give to a Researcher When He or She Is Developing a New Research Proposal?

Although developing a research proposal implies work to be carried out in future, it is essential to have convincing preliminary results that support the formulated hypotheses. The term 'preliminary results' can be subjective and may vary in interpretation. In many cases, these results represent solid research findings that have either been published or are publishable. Based on personal experience, I reckon having strong preliminary results are crucial for a successful research proposal, at least when submitting a research grant proposal to the national funding agencies in Western European countries. It is also useful to incorporate technique development or improvement on existing methodologies, as added-value to the proposed work.

8.2.62 How Would You Recommend Universities in Order to Align Their Research to Meet the Needs of Industry and Society?

I have been conducting research at (semi-)public institutions (e.g. Max-Planck-Institute for Terrestrial Microbiology, the Netherlands Institute for Ecology, and universities in Belgium and Germany) that primarily address basic and curiosity-driven scientific questions. However, I noticed a trend in recent years where funding agencies are encouraging public-private research collaborations. Hence, there is incentive to align basic research to industrial and societal needs. To this end, the university could set up a dedicated office to link public and private entities. The success of this office can be evaluated based on the number of collaborative grant applications, as well as the success of these grants (approval and completion).

8.3 Reflection

This chapter is entitled 'Success Can Be Learned' and discusses various considerations and approaches to achieving success. There are many ways in which success can be measured. I would like to extend millions of thanks and appreciation to participated interviewees. In spite of their tight schedules, they were willing to squeeze in some time for interviews. The purpose of this endeavor is straightforward and sincere—to equip the next generation of research scientists with the right mindset so that they can be better biologists and biotechnologists in the future.

The earlier subsection is quite lengthy, but the dataset of interviews may appear small; therefore, it is not possible to draw a major statistical conclusion from it. In addition, the interview questions were not designed in such a manner that they would generate any trends or novel findings. It serves as a purely advisory service, not as a research tool or data. The interviewees' advice may be useful to them, but please be aware that you need to exercise due diligence and check whether it applies to your situation before taking everything at face value. I have attempted to contact a wide range of individuals from diverse backgrounds. Unfortunately, due to a heavy workload, other researchers and industrial players were not able to commit to the first edition of this book.

It is universally acknowledged that knowledge can be acquired, as well as skills. It is often forgotten that success and leadership can also be learned. I would recommend that you read Ruth Gotian's book *The Success Factor* or listen to some of her talks on YouTube or various podcast platforms if you have not already done so. Having reviewed all the comments and advice provided by the interviewees in this chapter, you will be able to identify several key points. Among these interviewees, only a few have read the book *The Success Factor*, but they have practised the habits and have the same concept that Ruth Gotian discussed in her book. To achieve success, Ruth Gotian said there are four key elements that need to be considered: motivation, perseverance, building a strong foundation, and constantly learning through informal means (upskilling) (Gotian 2022). Following the interviewees' responses, I attempted to capture the major keywords and concepts they mentioned and match them with Gotian's four major factors.

8.3.1 Motivation

According to some interviewees, they describe motivation as their passion, mission, purpose, enthusiasm, purpose, aspiration, or calling. Motivations can be classified into two general types, intrinsic motivations and extrinsic motivations. Even in the face of difficulties, intrinsic motivation is their greatest ally. You might want to refer back to Chap. 1 if you do not recall or are unsure what intrinsic motivation is. In the majority of the interviewees whom I interviewed, I found that they had a very clear sense of intrinsic motivation, which is something they derive from within themselves, such as their core values, interests, morals, religions, aspiration, or passions. The high achievers focus on true and genuine motivations when doing research and publishing their findings. Their motivations are not KPIs, fame, empire, or anything extrinsic. They do research because they want to. The aim of their work is to acquire knowledge and generate new knowledge as much as possible. They conduct research for societal impact. The societal impact includes students, their colleagues, patients, people, country, as well as the industry.

In the absence of proper motivation, people tend to stay in their comfort zones and become complacent because of it, and eventually end up stagnant in life. When it comes to doing research, there is a fundamental difference between those who have a passion for it and those who regard it as just another job that they need to do in order to make a living.

There is only one person that can really help you if you still cannot find your own intrinsic motivation, and that is yourself. Ask yourself if what you're currently engaged in is something you genuinely love. Listen to your inner heart. I am sure that if you are not able to hear the voice from your heart, then you probably are drained by tons of tasks that are keeping you busy, but they are not making you happier. Did you join the academic line or research institute because you are interested in the field or are you just interested in the pay? "Big payoffs far down the road just are not enough to keep us motivated", this sentence was written by Katy Milkman in her book, *How to Change*.

There are plenty of young PhD holders in some countries who have difficulty finding a job in a university or institute due to the lack of opportunities for them. If you know that the subject of science is not of interest to you, would it be possible for you to do something that you really enjoy, and let go your seat to someone who has a great passion for the subject? One of my lecturers (who was also the chairperson for my PhD viva voce) quit his job as an associate professor a few years ago. He was a very inspiring lecturer, as well as a very smart one. He was once the head of the department as well. He was a PhD student of a famous top scientist who wrote a book that is used around the world as a textbook. He had a good academic background. He published a lot. It was clear that he was not a lazy person. Unfortunately, I did not have a chance to interview him about his reasons for leaving the university. At the beginning, I was wondering whether his resignation was due to the fact that he had a more attractive offer at another university with a full professorial position for him. Someone I can trust told me he was leaving for his dream. He is not passionate about becoming a professor. Guess what, he has a passion for tour guiding and food seekers looking for vacation packages abroad. That's awesome. My respect goes out to his brave decision. If you are not passionate about what you do, working as a lecturer or researcher can be a nightmare for you, your director or dean, your institute, and your students. It results in a waste of time, both yours and that of others, as well as the taxpayer's money that has been allocated to support your research endeavors. Do you agree with this perspective? Now back to you, have you discovered your passion?

8.3.2 Perseverance

It is common that when we look at a person's publication list in Scopus, WOS, or in Google Scholar, we are like an audience that is only paying attention to the outlook of that person. Wow, this person has a lot of publications under his or her name. We often forget how much pain, struggles, setbacks, and even failures they have encountered and have endured along the way. Rejection and failure are no one's cup of tea, but based on the interviewees' responses, we can tell that they are not intimidated by failure and the challenges they face. Almost all of them have a growth-mindset, which entails the ability to see rejections, knockbacks, and failure as opportunities to grow as researchers and become better and more impactful over time. You have probably heard the saying that "a pessimist sees the difficulty in every opportunity; and optimist sees the opportunity in every difficulty" (Google wrote that Winston Churchill said this). A perseverant person is humble enough to recognize his/her limitations, but bold enough to push them as far as they can. For them, fear is nothing to be feared. The high-achievers have the persistence to try again and again, finding ways to overcome weaknesses, and just keep on trying until they achieve their goals, of course in an ethical manner. To them, hard work is part of good research ethics. There are no short cuts in perseverance.

In December 2022, Argentina and France were the finalists of the World Cup that was held in Qatar, and Argentina won the tournament. Argentina last won the World Cup 36 years ago in 1986. In 2022, Lionel Messi was 35 years old when he played for possibly the last time in a World Cup. Sometime ago, Messi stated, "You can overcome anything if and only if you love it enough". Some readers may not be aware that Lionel Messi was diagnosed with growth hormone deficiency when he was 11 years old. We would not be able to understand what he meant by 'overcome anything' if we did not know his background. Low achievers and average achievers tend to focus on problems and speculations. A high achiever focuses on how to overcome the challenges they face. Perseverance, perseverance, and perseverance.

8.3.3 Building a Strong Foundation

In the interviews, we noticed that one of the keys to success was to focus on and master the latest knowledge, skill, or technology. Curiosity is the key. In the absence of curiosity, we are going to be left out of the loop very soon. It is mandatory to read, and it is not just about reading journals, but also about reading other documents that can help train our brains. According to interviewees, if one wants to be able to read effectively, some sort of reading strategy is imperative. A participant in the interview said: "I feel so fortunate that my research trajectory has followed me throughout my career; when curiosity took me one way, I followed that direction. That, for me, is success".

To build a strong foundation based solely on journal articles and within your four walls in your office or lab is not enough. Jump up from your seat, attend conferences, meet with mentors, ask good questions to mentors, connect with someone from another university, and meet industry people—all are part of the homework. Networking, networking, and more networking. The networking should be diversified and the sources of research grants should also be diversified. Establishing a strong foundation also involves having a good team in place that you can count on to assist you in the process. With the help of a dynamic and good team, they are able to push themselves beyond their comfort zone and expand their horizons.

There is a lot of need for spending time with students in order to help them develop, and students will bring your research to the next level. Our success cannot be achieved on our own, we cannot do it independently, we cannot do it alone, we need help and cooperation from others. In the same way, you can be a part of someone else's success plans. There is no doubt that building a strong foundation takes time, but planning is the key, and some of the interviewees emphasize the importance of using a calendar to do your planning and execution. Strong foundation = improving, improving, and improving, for multiple aspects.

If you are a student or young researcher reading this book, please remember that your role is to learn, to keep learning, and to continue improving. Have you heard of the concept of clay and potter? Like wet clay, you will become something more valuable and valuable in the hands of the potter. It is necessary for you to be willing to be shaped by the potter and to be placed in the fire to keep that shape. The best time to unlearn and relearn is during the PhD training period, and those who are reluctant to change the habit will have a difficult time later on. If you already hold your PhD and graduated 10 years ago, I would think that the best time to make a change in habits was 10 years ago. I'm sorry, but we can't change the past. If you want to change a habit, the second best time to make the change is right now. There may be times when you think, I will just change my habit later, but I am sorry to say, as the time passes, and the moisture in the clay is drained away, it may become more and more difficult to make a significant change to the shape and value of your sculpture. If you are the clay, who is the potter? Supervisor, mentor, and who else?

In the history of badminton, Lin Dan (China) and Lee Chong Wei (Malaysia) were two of the most influential players. They were both among the best players in their respective countries. As far as skill, speed, strength, and accuracy are concerned, they have reached their own personal limits. Some Malaysians were of the opinion that if Chong Wei had not been born in the same generation as Lin Dan, then Chong Wei would have been the best player in the world. It is recommended that you view some YouTube video that explains the lives of Chong Wei and Lin Dan if you have the opportunity. A number of newspapers and magazines interviewed both of them separately, and both of them expressed their gratitude to have counterpart as the opponent. Since the opponent is so strong they have to work harder and harder to push their own limits. Healthy competition keeps us alert and drives us to strive for better results. Do you have a competitor or an imaginary rival? You should have one to maintain your momentum in research and publication.

As long as Chong Wei and Lin Dan stay in their comfort zones, they will never be able to grow. Whether we are a researcher, academician, postdoc, or student, the same idea still holds true. A warning alarm should sound if we think we are the smartest person in the room or nobody in our department can beat us. It is probably a worthwhile idea for us to read this book, *Ego Is the Enemy*, by Ryan Holiday. This book offers valuable insight into how our ego can limit our potential and how we can take steps to overcome its negative aspects. It also provides strategies to help us stay humble and focused on our goals, which can be invaluable in helping us achieve success. The ego is seen by Ryan as a threat to long-term success. The ego transforms confidence into arrogance. I personally believe that people with humble and pure hearts are very blessed. Those with pure hearts will discover the science and truth behind living organisms through their research. Arrogance and ego are the enemies of a pure heart.

Many people think that success depends on your upbringing, your education, or your family background, or where you got your PhD from. Although I do not have statistics to support this claim, I think that people with better upbringing may have a smoother route with fewer worries. A badminton player, Chong Wei, was born into a relatively poor family. He practised badminton using whatever he could find (a book, wood, or a pot, instead of a racket) according to a movie about him. If you come from a relatively less privileged background or have a lab with little resources, the journey may be more challenging. Be not discouraged. One of the interviewees said this: "An individual who is a good researcher should possess the following qualities: patience, perseverance, accepting knockbacks, and seizing opportunities when they arise". A good opportunity coupled with a strong work ethic can still lead to success. In my opinion, dedication and hard work are the keys to success for those coming from a relatively less privileged background.

Some researchers feel less confident due to factors such as not having a first-class degree, or not graduating from one of the top world universities, or English not being their mother tongue. But, throughout my professional career, I have seen examples of people who went to very renowned universities who have difficulties in adapting to their new roles as lecturers or researchers after returning to their native country; as a result, they have poor performance for the first few years. There are many others who can do well, of course, right from the start. My point is something like this. Say there are two people, one of whom graduated from a local university (not so famous university, for example) and the other from a renowned overseas university. Upon completing their PhDs, both worked in a developing country in the same university. They have similar knowledge and technical skills. The two people will now have the same career starting point. The PhD they earned may not determine their success in the future, but their mindset and habits will. It is likely that the individual who believes I should work harder will be more successful than the individual who does not. Of course work harder needs a roadmap. If you keep complaining and reminisce about all the good things that happened during the PhD training, you are likely not to perform well. We are influenced to some extent by our education and background. Nevertheless, your past history should not be a barrier to success. The past history refers to both the good and the bad histories.

Many people think that success depends on where you got your PhD from. In case you recall, one of the interviewees wrote the following: "Whilst I have had some excellent graduate students that have come to my lab from top tier universities, the most successful students were all graduates from what are perceived as mid-tier universities". It is possible that some of us have had the same experience as the interviewee. Even so, there is no statistical evidence to support the assertion. In my opinion, people with growth-mindsets, whether they graduate from a mid-tier university or not, will still have a fair chance of excelling in research.

I initially felt a lack of confidence after I obtained my PhD from a mid-tier university. The only thing I can do is to keep trying and trying and trying, improving and improving, and improving. My self-belief and confidence grew quickly. It took me a while to realize that I am not substandard and no one looks down on me since 'I am made in Malaysia'. Malaysia was the country in which I was born and where I received all my education. There is nothing I am more proud of than the country I am from and the education I received. How about you?

8.3.4 Upskilling

There is a separate part in this book that discusses how to unlearn and learn new things. Even the most prominent professors that we have interviewed say that learning is a never-ending process. There is no limit to what can be learned and learned about. High-achievers do not simply strive for a grade A, as that would set a limit on their growth. They do not chase after achieving full compliance with KPIs, because the KPI will be the limit if they do so. Their true goal is to acquire knowledge and wisdom. The sponge-like nature of their learning keeps them absorbing and learning new things. By pursuing knowledge and wisdom, they naturally achieve excellent results and meet the requirements of predetermined KPIs.

During the course of the interview, we found out that many interviewees mentioned that mentoring had helped them in the development of their careers. A mentor may not necessarily be an older person, since high achievers also learn from those who are younger (a concept known as reverse mentoring). Mentors provide advice to their mentees. Mentors have collected a lot of no-entry signs over the years. In other words, they have made mistakes and they have learned from them. It will save you a great deal of time if you are provided with the no-entry signs, so you won't make the same mistakes again.

Try to be a mentor too. The truth is that we tend to benefit from giving advice to others, for example, we can build up our confidence, self-belief, and remind ourselves what is the correct approach to a problem when we give advice to others. In a situation where you are mentoring a new lecturer just learning how to become a student's supervisor, and you say this, "well, you should communicate more with your students". Psychologically, it reminds us that we should follow suit, or we will be hypocritical.

Be aware of the imposter syndrome trap which makes you doubt what you are capable of, what you are talented at, or what accomplishments you have achieved. Even after obtaining a PhD, some researchers struggle with the imposter syndrome. Perform a self-audit and discover what causes you to lack belief. It would be awesome if you could spot it. If you are a member of the 'imposter syndrome society', please fill out a form to cancel your membership as soon as possible. Upskilling is one of the membership cancellation forms. Drop the idea of becoming the president of 'imposter syndrome society'. We cannot ignore confidence issues.

8.4 Conclusion

You might recognize as we draw to a close that the book's messages can all be divided into two main themes: strategy and mindset. On one hand, we are encouraged to take strategic actions such as gathering data, enhancing lab facilities, crafting winning proposals, staying updated with literature, improving our skills, networking, planning our career growth, mentoring students, and executing our plans to achieve our goals. These actions fall under the umbrella of strategy, which involves refining our core knowledge in a specific biological field and developing smart and sharp skills. Persuading you that factors like attitude, enthusiasm, perspective, mindset, habit, effort, personality, etc., are equally significant is the second major theme of this book. It highlights the value of cultivating a positive mindset, maintaining enthusiasm, embracing different perspectives, cultivating beneficial habits, putting in dedicated effort, and nurturing a strong personal identity. These elements contribute to our overall success and fulfillment.

Is it possible for anyone working in the research line to succeed? I would guess that it is not necessary. Even if you have read the entire book from page one to the end, it does not guarantee that you will be successful. We stated that goals and plans are important factors for success; however, James Clear, the author of the book *Atomic Habits*, said that both groups of people who succeed and fail to succeed may share identical goals. Habit is at the core of James Clear's philosophy, and he believes that if we work 1% better every day, we will see tremendous results (Clear 2018). If you have to choose between talent and effort, which one is more important to you and influences your success? A friend of mine told me she lacked research talent. She said to me, "Hey, you're crazy talented". In response, I said that I wasn't talented nor smart, and my degree was in engineering, not bioscience. Since I am not talented, I put in a lot of effort. It is not only talent that counts. Efforts, passion, and habits make a big difference.

I asked my colleagues and students what they thought was the most important factor to success: (a) a measurable strategic planning or (b) an unmeasurable attitude, mindset, habit, motivation, etc.? There is a good chance that you will have a different answer. In my opinion, both are equally important. According to the interviewees, both sides were mentioned in their responses. A person who is well versed in strategic planning, but does not possess a scientific and philosophical mindset and a growth-mindset may find themselves engaged in routine projects that lack innovation and are similar in nature. Publication became a product of a factory, all look alike. On the opposite side of a coin, without a plan, we are heading no way. Without a blueprint, no roadmap, and no strategy, a researcher will be lost. Strategy, attitude, mindset, habit, and motivation are essential.

It is unlikely that we will accomplish our publication plans, if we do not translate them into concrete actions. All in all, when there is no plan, no vision, no goal, no love for science, and no actions, publications become a stressful KPI. Keep in mind that publication should be related to knowledge sharing and enhancement of science. Although it may sound naive, it is the innocent heart which keeps us in search of knowledge, wisdom, and truth. Blessed are the pure in heart, for they shall discover the science and truth behind the living organisms through their research.

Make sure you remain focused on what you are trying to accomplish. I would like to emphasize that staying focused on your research and career development plan does not mean that you should let go of other responsibilities. You are responsible for your health. I know a few good researchers who suffer from health problems due to spending too much time at the computer or concentrating on their research. Parents, your kids need you. Kids are more important than your research. Whenever you have to teach, do it well. It is not an excuse that we do not have time to teach well because we wish to write for publications. Regardless of how small the matter is, be

Fig. 8.1 Ascending the pyramid of success: strategies and characteristics for a resilient researcher

responsible and committed. "Whoever can be trusted with very little can also be trusted with much, and whoever is dishonest with very little will also be dishonest with much" (Luke 15:10).

Success can be learned? The answer is definitely yes. It is not a wise idea to steal or copy ideas from someone who is better than you. You will not be as successful as them since you are focusing on the outlook and on a few traces of success. Focus on their attitude, mindset, perseverance, their strategies, and their habits, as well as how they handle challenges and how they continue to improve. Copy their strategy and habits. If you learn these things, you will be as good (wise) as they are. Remember, genius is 1% inspiration, 99% perspiration. Do walk extra miles. In a delightful culmination, the essence of all these chapters harmoniously unfolds in a captivating cartoon, vividly depicting the transformative path of research, publication, and personal growth, leaving a lasting impression on our pursuit of success (Fig. 8.1).

References

Clear J (2018) Atomic Habits: An easy & proven way to build good habits & break bad ones. Avery
Gotian R (2022) The Success Factor, 1st edn. Kogan Page

Appendix A. Tips to Teaching Lecturers

Do you teach courses on conducting research, writing proposals, writing dissertations, and publishing? (No, skip this part). Yes. In that case, congratulations, because it is a very important job, but it may be time-consuming if you are doing it well.

My advice is do not just focus on the strategy in preparing their dissertation or just to pass the proposal or viva voce defences. It is not enough to teach them tips, such as how to write objectives, methodology, how to prepare slides, etc. It is also not the focus of the lesson to teach them about the machine or methodologies they will use. It does not make sense to tell every research protocol to 20 students if each has a different project. Try to make your course practical to all students in the class. It is not sufficient to come and share your PhD journey with the class for hours, days, and weeks. Some students are not interested in hearing about your PhD experience since it doesn't interest them. This is not a class where you only tell motivational stories.

I would suggest that you seek out the subtitle of the book on the cover of the book and look at it. My choices were 'Research Mindset • Best Practices • Integrity • Publications • Societal Impact'. I define the Research Mindset as a combination of philosophia (love of wisdom and generate new knowledge) plus growth-mindset plus intrinsic motivation. Encourage the students to think about what is right and wrong in research and take part in the whole scientific community in order to improve the quality of science. In this context, the term 'Best Practices' refers to any activity that has been shown to be most effective for the intended purpose. The term 'Best Practices' refers to multiple aspects of research, including laboratory procedures, writing, presenting, networking, collaboration, etc. For example, in the case of making citations, show them how to do it in the most time-efficient manner. Best Practices also mean to focus on the quality of the research work by providing proper guidance to the students. This will ensure that the students have a thorough

K. M. Goh, *Research Methodology in Bioscience and Biotechnology*,
https://doi.org/10.1007/978-981-99-2812-5

understanding of the research process and will be able to apply the knowledge they have gained to their own work.

Research must be conducted with integrity and ethics. Please emphasize integrity, not just plagiarism. Wouldn't you like to prevent any of your students from going viral because of a lack of integrity and ethics? If one of the students I taught before ended up on the front page of a newspaper with the headline 'X was kicked out of university due to ethical problems', I will be devastated.

The next subtitle is Publications. Do not turn your students into paper factories, don't teach them to chase after KPIs, but rather to chase after knowledge. The publication of the article will follow automatically as a result. We must consider the Social Impact of our research, which means we do our research in order to benefit people in the long run. This can include SDGs, industrial benefits, environmental benefits, etc. Is it too difficult? Then, train your students to become better scientists, and I believe that type of impact cannot be ignored.

I believe there is no better course than yours for changing a student's habit, their thinking about research, and their motivation to do research in the future. Encourage them to ask questions in class or even by email. As soon as possible, you should respond to all questions and provide them with some suggestions for their consideration. Further, it is important to provide feedback on the students and to help them improve and refine their writing abilities.

The following is a summary (Table 1) of the content I taught my students. In general, I spend three hours lecturing and discussing each week, and I cover Topics 1-8 over the course of approximately ten weeks. Engage in active learning rather than lecturing in a single direction. Feel free to reach out to me if you would like to discuss further; just find my email address and drop me an email. I will be more than happy to hear from you. I hope you find this information useful.

Table 1 Suggestion of topics for lecturing

Topics	Contents
1	Introductory (Refer to Chaps. 1 and 8) • Course Information and course management. • Concept of 'Science' and 'Philosophy'. • Differences in research in undergraduate and postgraduate. • Program Standards (if available in your country). • Supervisor's expectations on research, thesis, publication, etc. • Examiners' expectations on research, thesis, publication, etc. • Avoiding failure in postgraduate. • Basic science vs. applied science research. • Intrinsic motivation vs. extrinsic motivation. • Growth-mindset vs. fixed-mindset. • Give some homework: (1) Why did you study postgraduate? (2) Ask them to find ten possible reasons for failing a PhD.
2	Chapter 2 of proposal/dissertation/thesis (Refer to Chap. 2) • Defining 'literature' and 'review'. • Describing the essentials of the literature review. • Construct a literature review using various sources and the limitation of a single database. Setting new article alert (Demonstration). • Using proper reference management programs/tools: Endnote, Mendeley, etc. (Demonstration). • Tips in using WOS and Scopus: Exporting data and bibliometric analysis (Demonstration). • Other bibliometric tools: Lens.org, Dimensions, VOS View, etc. (Demonstration). • Common mistakes/misunderstandings in preparing the literature review. • Finding research gap (literature gap). • SDG, Open Data, Patent search, etc.
3	Preparing Chap. 1 of proposal/thesis/dissertation (Refer to Chap. 2) • Prepare five to eight examples of poorly written problem statements. Give it to students and ask them to grade it. Discuss the characteristics of poorly vs. excellently prepared problem statements. • Differentiate SMART vs. BE-SMART objectives. • Ask students to draft problem statements and objectives for their project (Review and provide feedback to them).
4	Good Research Practices (Refer to Chap. 4) • Managing digital files • Managing and keeping biological samples • Reproducibility in experimental data • Thesis Manual and tips for using the template • Preparing high-resolution and effective figures • The power of Mendeley • Other minor topics: Maximize the purchasing power using research grants, stress management, create collaboration, rapport with supervisor, etc. • IP protection and potential commercialization
5	Ethics in performing research and reporting (Refer to Chap. 3) • Plagiarisms, Turnitin • Authorships conflicts • Research ethics issues • Conflict of interest • AI: friend or threat (refer to Chap. 7 and Appendix C)

(continued)

Table 1 (continued)

Topics	Contents
6	Publication 101 (Refer to Chap. 5) • Publication requirements. • Major types of publishers. • Defining authorship, corresponding author, first author, and first student author. • Indexation in publication—WOS, Scopus, ERA, etc. • Characteristics of predatory journals. • Paid vs. open access. • Category in WOS, impact factor, quartile, h-index, etc. • How to find a suitable journal to submit my manuscript? • Advice on writing the review article. • Correct mindset in publications.
7	Self-reading and reflection (refer to Chaps. 7 and 8) • Students are given non-scientific (non-technical) articles/comments/notes published in *Nature* or other sources. Students will do a reflection. Read their reflection report. Communicate with them and give them support and advice. Examples of topics can be: • Research culture, habits, and careers • Stress, dilemma, and mental health • Life after PhD • Gender equality • Publication and writing tips • International student • Handling failures and setbacks
8	Upskilling (Refer to Sect. 4.9 and Chap. 6) • Online upskilling tools • Improve writing using the help of AI paraphrasing • Slide design
9	Preparing a proposal • Ask the student to write a proposal • Using thesis format • Using Mendeley/Endnote • Prepare Chaps. 1, 2, and 3
10	Feedback on students' written proposals and presentation • Feedback on Chap. 1 (focus on problem statements and objectives) • Feedback on Chap. 2 (focus on coverage and content of literature review) • Feedback on Chap. 3 (focus on tenses, flow chart, and suitability of the procedures) • Feedback on writing style and presentation style • Feedback on the use of references

Appendix B. A Letter to the Young Generation

There seems to be some kind of tag attached to various generations by society. Gen X, for instance, is born between 1965 and 1980, Gen Y is born between 1981 and 1996, iGen is born between 1997 and 2025, and Gen Alpha is born between 2012 and 2025. The range of years is not a golden standard since different websites may have slightly different ranges. According to a variety of online sources, the following are some characteristics of each generation:

- Gen X is a workaholic generation that produces and works hard.
- Millennials were initially characterized as lazy, narcissistic, and spoilt. It is sometimes called the me-generation.
- Gen Z were born at a time when internet, smartphones, tablets, and the internet were already mature. Social media and technology influenced Gen Z, which sometimes leads to stereotypes such as tech-addicted, anti-social, or social justice warriors.

Taiwanese created the term 'strawberry generation' for people born in the 1990s. Strawberry is a very soft fruit that can be easily pressed. This is due to its high water content and relatively low cell wall strength. The combination of these two factors makes it much easier to mash or break down, compared to other fruits. According to Wikipedia, strawberry generation members are prone to bruises and unable to cope with social pressure or hard work since they have been overprotected by their parents and raised in unchanging environments.

Thereafter, you may hear the expressions *tang ping* and *bai lan* for a trend in China. *Tang ping* refers to lying flat, while *bai lan* refers to allowing it to rot. Both terms refer to resistance movements or countercultures. In essence, it is a form of minimizing social activity, saying no to existing systems, not cooperating, and not listening to advice, but instead following one's own style and one's own anti-social behaviour. In Chinese, the term *tang ping* plus *bai lan* means to actively embrace deteriorating circumstances rather than trying to reverse them. These terms are commonly used in the context of slackers, unmotivated people, those who put in minimum effort, give up easily, and those who do not care about anything. If you would like to learn more, you can probably start by searching these two words on

K. M. Goh, *Research Methodology in Bioscience and Biotechnology*,
https://doi.org/10.1007/978-981-99-2812-5

YouTube. Channel News Asia also has some very good commentary on the CNA Insider channel. Anyway, as we move forward, we are likely to see other tags.

Younger society is often tagged with negative tags. Are these tags 100% accurate and apply to everyone? Without a doubt, no. We cannot ignore the fact that the generation has changed, and the characteristics of students at university have changed as well. In terms of research, is the Gen X generation better than the Gen Y generation and better than the Gen Z generation? To my knowledge, there is no unbiased way to measure this. Strawberries, *tang ping*, and *bai lan* behaviour are beginning to appear on campus, affecting some students.

The world may be cruel, and the reality may be harsh, but there is still hope. If you are young and dislike the negative tags that are placed on you, then the right way to counteract this is to show the world that they are wrong and you are equally capable as the older generations. Encourage your peers to adopt the right mindset and habits. Show the world that you can make a positive difference by leading by example, proving that age is no barrier to success. In 1 Timothy 4:12, it is written "Don't let anyone look down on you because you are young, but set an example for the believers in speech, in conduct, in love, in faith and in purity".

Remember, Ruth Gotian has said there are four key elements which must be considered in order for you to succeed: motivation, perseverance, building a strong foundation, and continuously learning new skills through informal means. There are three additional pieces of advice that I would like to draw your attention to:

Advice #1. It is never a good idea to try *tang ping* and *bai lan*. If you are a PhD or a master's student, you should never consider them. *Tang ping* and *bai lan* do not provide students with the necessary skills and strategies to problem solve and make good decisions. This can lead to feelings of helplessness and lead to feelings of anxiety. This can have long-term consequences such as depression and other mental health issues. In case you see your friends *tang ping* and *bai lan*, wake them up and act as their Barnabas, so they can get back on track. It would be a great honour if you could help your generation in any way. Show them the right path and be the beacon of hope for them; it is the least you can do for those who are lost.

Advice #2 Communication. In this context, communication does not mean being able to write or present effectively. Effective communication involves more than simply sending an email, or sending a short text, or posting a status update on social media. It also involves taking the time to carefully consider the message you want to deliver, the tone in which it should be delivered, and the best medium for delivering it. It also involves taking the time to listen to the other person's response and engaging in a two-way conversation. Communication also includes being able to listen to, understand, and empathize with other people's perspectives. It also means being able to express oneself clearly and articulately and to understand the nuances of language and the implications of one's words.

Advice #3 Keep a long-term perspective. Do not focus solely on what you can earn now, but rather on what you will earn after the immediate effect occurs. As an example, if you select a research project that has no long-term scientific value, is easily accomplished with simple skills, and has an accommodating supervisor, well, that is your choice and strategy. By doing the minimum required, you can obtain a

PhD certificate in three years. There is no better time to acquire knowledge, skill, and wisdom than during a PhD. If you do not incorporate a longer-term strategy into your whole strategy planning, you will have a difficult time achieving your goals for the following year. It is important to have a plan in place that not only sets short-term goals but also has a vision for the future. You should begin planning your investments (time, effort, strengths, and money) at a young age. Do not be afraid to push the boundaries of your comfort zone.

Appendix C. List of Useful Tools and AI

Since 2020, a staggering number of AI tools and software have been developed and are expected to change the way students, professors, postdocs, principal investigators, and researchers do their daily work. This appendix provides a list of potentially helpful tools. However, by the time you read the print edition or the e-book, this list may already be out of date. Although you will encounter increasingly advanced and powerful tools, keep in mind that they are not your masters. Artificial intelligence is nothing more than a tool, a machine, or a toy in its most basic form. In the past, a biologist's lab bench consisted solely of instruments for data acquisition and analysis of analytes or macromolecules. In the last decade, both cloud computing and bioinformatics capabilities have grown significantly. With the development of AI and the many useful software, we now have an additional tool. You should always be the source of the content or main idea, and never the artificial intelligence. Use AI technologies as a resource, but remember that they are not a substitute for original thought. Stay ahead of the competition by always being the master of your own ideas.

Writing, Paraphrase, Translating, and Proofreading (Writing Assistant Tools)

InstaTest, Quillbot, Wordtune, JasperAI, Grammarly, Anyword, CopyAI, Copymatic, Rytr, Trinka, WordAI, Chibi AI, ProWritingAid, Writer, Writesonic, Ref-N-Write, Writefull, Wordvice AI, Pcanpi, Ginger, PaperPal, PerfectIt, Sincore, Jenni, CorrectEnglish, Notion AI, ChatGTPWritesonic, BioGTP, Caktus, Compose. AI, Hemingway Editor, Bard, ChatGTP (and Chrome extension, i.e. ChatGTP for Google, Merlin, Tactiq, AIPRM, ChatGTP Writer), and others.

K. M. Goh, *Research Methodology in Bioscience and Biotechnology*,
https://doi.org/10.1007/978-981-99-2812-5

Literature Review, Bibliography, Making Summaries, Taking Notes, Explain Articles, or Verifying Information

Sematic Scholar, Scite, Assistant by Scite, Perplexity.ai, Wordtune Read, Scholarcy, ChatGPT, BingGTP, Bard, TLDR, SMMRY, ChatGPT Prompts Splitter, ChatGTP Splitter, FlowGTP, Flowcite, Quiibot, Jasper, Genei, Study Crumb, Explainpaper, Iris.ai, SciSpace, Elicit, Consensus, OpenKnowledgeMap, Google Trends, Rayyan, System®Beta (System®Pro), Paper Digest, ResearchRabbit, ChatPDF, OpenRead, etc.

Personalize Learning

Brainly, StepWise Math, TutorAI, Rewind, etc.

Meetings Tool

Otter.ai, Fireflies, etc.

Writing Computer Code

Programming Helper, ChatGPT, Bard, AutoGPT, OpenAI Codex, CodeT5, etc.

Art, Music, Slide, Presentation, Speech, and Video

Adobe FireFly, Dall-E, Midjourney, Stable Diffusion, DiffusionBee, Sounddraw, Beautiful AI, Design Beast, Adobe Express, Descript, FlexClip, Synthesia, Toonly, Voicemaker, Speechify, ElevenLabs, Runway, Descript, Synthesia, Azure Text to Speech API, Murf.ai, Altered.ai, Amazon Polly, Yepic Studio, Speechelo, AssemblyAI, Canva, Gamma, Microsoft Designer, SlidesAI.io, Ecoute, etc.

Data Analysis

ChatGPT Code Interpreter, Power BI, etc

Website Builder

Kleap, Pineapple, Leia, Hocoos, Hostinger, Durable, Appy Pie, Wix ADI, 10Web, Bookmark, Jimbo, etc.

Daily Job and Productivity

Notion, Microsoft 365 Copilot, AgentGTP, ClickUp, Monday.com, Google Workspace (with AI), Reclaim.ai, etc.

Maximum Dissipation Non-Equilibrium Thermodynamics and its Geometric Structure

Henry W. Haslach Jr.

Maximum Dissipation Non-Equilibrium Thermodynamics and its Geometric Structure

Springer

Henry W. Haslach Jr.
Department of Mechanical Engineering
University of Maryland
Glenn L. Martin Hall
College Park, MD 20742-3035, USA
haslach@umd.edu

ISBN 978-1-4419-7764-9 e-ISBN 978-1-4419-7765-6
DOI 10.1007/978-1-4419-7765-6
Springer New York Dordrecht Heidelberg London

Printed on acid-free paper

Springer is part of Springer Science+Business Media (www.springer.com)

Preface

This work explicates a geometric model for non-equilibrium thermodynamics and a maximum dissipation criterion assumed to supplement the second law of thermodynamics. The evolution equations resulting from the author's maximum dissipation construction provide a constitutive model for non-equilibrium processes. The construction can account for the effect of loads, heat, electromagnetic effects, chemical effects, and transport processes on fluids and solids. In the process of producing a mathematical construction describing the evolution of non-equilibrium processes, some insight is also obtained about the foundations of thermodynamics. A further goal is to put this model in the context of the history of thermodynamics and constitutive modeling of solids.

A constitutive model is a mathematical construct that describes the inter-relations of a set of variables that are presumed to describe the physical state and behavior of the particular materials making up the body. The model might be a system of algebraic equations, a system of differential equations, a combination of algebraic and differential equations, a geometric structure, a statistical description, etc. Constitutive models may be based on a phenomenological characterization of experimental results or may be based on fundamental physical laws. A phenomenological (or empirical) model is one that simply describes data, but is not necessarily derived from physical principles. Such models are commonly used in solid mechanics to describe the equilibrium stress-strain response. An example is the Mooney-Rivlin stress-stretch relation for rubber. No one has yet successfully derived such a model for rubber from the laws governing the molecular response of polymers.

Alternatively, the mathematical model could be chosen by analogy. An electrical circuit could be used to describe the circulatory system in humans. The spring and dashpot models for the time-dependent behavior of solids are also models based on analogy. The constants in spring and dashpot models seldom have a meaningful physical interpretation. For example, the spring and dashpot based Prony series, commonly used in industry to model polymers, often require a large number of coefficients just to fit the data. The fact that a model fits a set of data does not imply that it can explain the cause of the material response. Because it is arises from established science, a model based on fundamental physical principles such as those of thermodynamics is more desirable than a model based on analogy.

A good mathematical model must fit the experimental data, but fitting the experimental data does not by itself prove or verify that the proposed mathematical model is good. A good model offers physical insight that helps explain why a phenomenon occurs and also how that phenomenon is connected to other phenomena. Finally, a good mathematical model must produce predictions that may be experimentally tested.

The maximum dissipation construction of non-equilibrium evolution equations starts with a phenomenological model for the equilibrium state of a material, which is relatively easy to produce from time-independent experimental data. A primary advantage of this non-equilibrium thermodynamic construction is that it gives a simple method to generate evolution equations from the thermostatic constitutive model and a thermodynamic relaxation modulus. A model for the thermostatic states is the foundation for the construction of the time-dependent description of non-equilibrium processes because the thermostatic states organize the dynamic non-equilibrium response at a speed that depends on the thermodynamic relaxation modulus. The mathematical model for the evolution of non-equilibrium processes is a system of differential equations.

The intended audience is researchers and graduate students in mechanics, mathematical physics, and mathematics, including those working on non-equilibrium thermodynamics and the mathematical modeling of material behavior. The book could be used as the basis for a graduate course. To make the work accessible to members of such disparate groups who bring different training to this book, background information is provided. For example, the history of thermodynamic models is briefly reviewed in Chapter 1 as is nonlinear dynamics, and Chapter 2 summarizes needed ideas about thermostatic functions and energy methods.

The subject is the thermodynamics of non-equilibrium processes in materials. Time-dependent processes such as the viscoelastic behavior of polymers, the viscoplastic behavior of metals, the behavior of biological soft tissue, transient thermoelectric effects, and the rate of crack growth in metals and polymers may be modeled by this non-equilibrium maximum dissipation construction. The construction distinguishes the equilibrium or long-term states of the system from the non-equilibrium states as the extrema of a generalized energy function; the philosophy is that of energy methods. The construction is presented first in Chapter 3 in terms of simple real valued generalized thermodynamic functions defined on the Euclidean space of thermodynamic variables and then in a differential topology setting in Chapter 7. Engineers and beginning graduate students can therefore understand the construction without knowing differential topology. The approach is inspired by the Gibbs thermostatic energy surface construction which is developed to define the maximum dissipation non-equilibrium evolution process in Chapter 3 for homogeneous systems. The general construction admits large deformations, and in such a case the models based on the current configuration must yield objective evolution equations, which are achieved using the Lie time derivative. An inequality on the Gibbs contact one-form acting on admissible non-equilibrium paths reproduces the Clausius-Duhem inequality; in this sense the one-form measures dissipation. Chapters 4 and 5 give examples of the application of the maximum dissipation construction to viscoelastic and viscoplastic behavior. Chapter 6 provides an application to the multi-scale modeling of the elastin-water system in a hydrated artery. Many of the examples provided are biomechanical problems. These three chapters give examples of the application of the maximum dissipation non-equilibrium evolution equations in which the relaxation modulus captures the microstructure of the material. The mathematical structure of the maximum dissipation non-equilibrium construction is given in terms of contact structures in Chapter 7. This chapter may be of particular interest to mathematicians and mathematical physicists. Chapter 8 provides applications when the generalized energy density has a singularity. Bifurcations in the generalized energy are analyzed by catastrophe theory. No variational techniques are used. The maximum dissipation non-equilibrium construction is extended, in Chapter 9, to non-homogeneous thermodynamic systems, those in which the thermodynamic variables differ from point to point in the body, based on a generalized thermodynamic function that accounts for entropy production. The focus is on solid bodies, but the construction is shown to apply to fluids as well. The construction accounts for the evolution of fluxes in non-equilibrium, non-homogeneous systems in a way that predicts a finite velocity of propagation of disturbances, rather than an infinite velocity. Chapter 10 applies this non-homogeneous maximum dissipation construction to thermoelectric systems. Chapter 11 takes a different view of the thermostatic states as critical states for the propagation of a crack to develop dynamic fracture mechanics in terms of the maximum dissipation evolution equations.

College Park, MD Henry W. Haslach Jr.

Contents

Chapter 1
Short History of Non-equilibrium Thermodynamics

1 Introduction

Thermodynamics is intended to be a universal description of the world we live in. Thermodynamics began in the attempt to understand the cycles of the steam engine in the mid-nineteenth century in which the cycles were essentially considered quasi-static processes, those consisting of a sequence of equilibrium states. Most real processes, however, are non-equilibrium processes and therefore are time-dependent. They are dynamic, not quasi-static, and generally involve dissipation. Dissipation is common in natural phenomena including friction, permanent deformation of solids, release of heat, etc. and cannot be described by quasi-static or time reversible physics. But even so, the earliest researchers in the subject identified the first and second laws of thermodynamics. In the twentieth century, researchers began to describe non-equilibrium thermodynamic behavior, an effort that required reformulation of the second law, which, even if vague, is an attempt to ensure that, in the thermodynamic description, heat can only flow from hot to cold regions unless the process is forced. This chapter does not review in detail other thermodynamics theories, but rather identifies the issues and problems to be dealt with by the maximum dissipation non-equilibrium thermodynamic construction.

2 Gibbs Thermodynamics

The non-equilibrium thermodynamics construction presented in this book is based on a generalization of the Gibbs energy surface for homogeneous system. A homogeneous system is described by thermodynamic variables, external or internal, which are functions of time only; no spatial gradients exist. In his 1873 paper (Gibbs, 1948b), Gibbs geometrically describes the equilibrium states of a homogeneous substance by the graph of the internal energy, ϵ, as a function of entropy, η, and volume, v, the control variables. Based on the general equation $d\epsilon = \theta d\eta - pdv$, the pressure, p, and temperature, θ, at an equilibrium state are interpreted as the angles of inclination of the tangent plane at that point on the surface in the planes perpendicular to the volume and entropy axes, respectively. The Gibbs

H.W. Haslach Jr., *Maximum Dissipation Non-Equilibrium Thermodynamics and its Geometric Structure*, DOI 10.1007/978-1-4419-7765-6_1,

relation used in later thermodynamic analyses is $\theta d\eta = d\epsilon + pdv - \sum_i \mu_i dw_i$, where μ_i and w_i are respectively the chemical potential and mass fraction of the ith substance.

The advantage of this geometric description is that stability of an equilibrium state may be defined. Gibbs defines an equilibrium state to be stable if the graph of the energy surface near the point lies above the tangent plane to the state except at the point of contact. This is just a statement that would be expressed today as that the energy surface is convex at a stable equilibrium point (see Chapter 2). The equilibrium point is defined to be unstable if any part of the graph near the point falls below this tangent plane. If the graph contacts this tangent plane at more than one point, the equilibrium is defined to be neutral. Gibbs points out that the function $\epsilon - \theta\eta + pv$ must take a minimum at a stable equilibrium.

The extensive variables, such as volume, energy and entropy, depend on the size of the system, while the intensive variables are bulk properties that do not, like temperature and pressure. In a solid body, stress is intensive, while strain is extensive. In this construction, the entropy is the variable conjugate to the temperature. The idea of entropy, with its units of energy per degree, apparently was invented to account for the fact that the amount of heat received at one temperature is not equivalent to the same amount of heat received at a different temperature, as pointed out by Gibbs (1873a) in his paper on graphical methods in thermodynamics. Both Carnot and Clausius seem to have been aware of this fact prior to the Gibbs paper.

Gibbs also was active in the development of statistical thermodynamics, but none of these ideas are used in the non-equilibrium construction presented in this book. Gibbs and Boltzmann produced different H-functions that led to different expressions for entropy. Boltzmann's H-theorem, suggests Truesdell (1984, p. 23), is the basis for the common idea that thermodynamic functions may only be defined near equilibrium. Truesdell argues that the work of Boltzmann on the kinetic theory of monatomic gases contradicts this belief.

3 Twentieth Century Thermodynamic Theories

Except for Carathéodory's attempt to axiomaticize equilibrium thermodynamics at the beginning of the century, the main thrust of twentieth century thermodynamics is the description of non-equilibrium behavior. Non-equilibrium thermoelectric processes were known prior to the work of Clausius in the mid-nineteenth century. Possibly, the first recognized non-equilibrium processes were the Seebeck-Peltier effects (Garcìa-Colìn, 1988) in thermoelectric systems. The effects describe the conversion of a voltage difference across a body into a temperature difference and conversely. Thermocouples involving the junction of two metals, and thus using the Peltier effect, were known in the middle of the nineteenth century. For example, Joule used thermocouples to measure the slight change in the temperature of rubber as it is stretched.

An important thread of thermodynamics in the twentieth century was linear irreversible thermodynamics (LIT) growing out of the work of Onsager and of Prigogine. This attempt was followed concurrently by extended irreversible thermodynamics (EIT) and continuum thermodynamics in the last half of the century. The main issues of concern were the meaning of entropy and the extension of thermodynamics to non-equilibrium behavior far from equilibrium. A long dispute about whether ideas such as entropy are only defined for equilibrium states persists today. Continuum thermodynamics attempted to correct some deficiencies of the Onsager analysis and was followed by the extended rational thermodynamics of Müller and Ruggeri (1998) based on the idea that the entropy flux should be given by a constitutive relation. However, the structure of non-equilibrium thermodynamics remains a controversial subject.

3.1 Carathéodory

The era in which Carathéodory (1909) produced his axioms for equilibrium thermodynamics included attempts to axiomaticize other mathematical systems such as the Peano postulates (1889) for the natural numbers and the various attempts to develop axioms for set theory. The perceived need to find axioms that imply a given scientific theory has greatly diminished today. A primary goal of Carathéodory's work was to prove, from his axioms, the existence of a global entropy and absolute temperature in equilibrium thermodynamics. Thermodynamics, influenced by the need to understand the steam engine, was originally the study of thermodynamic cycles. Carathéodory dropped the Carnot thermodynamic cycle as the fundamental building block of thermostatics.

The first axiom of Carathéodory says that during an adiabatic process of a multiphase system, the change in internal energy equals the work done. Truesdell (1984, pp. 49–57) critiques the attempt to define heat in terms of mechanical work since heat would be part of the internal energy. He argues that because the Carathéodory theory is formulated only in terms of mechanical energy, heat plays no role. The second axiom, used to derive the second law, is translated by Pogliani and Berberan-Santos (2000) as "In the neighborhood of any equilibrium state of a system (of any number of thermodynamic coordinates), there exists states that are inaccessible by reversible adiabatic processes". The Carathéodory axiom of inaccessibility states that given an equilibrium thermodynamic state, $\mathcal{E}$, some nearby equilibrium thermodynamic states cannot be reached by an adiabatic transformation, one that occurs without heat exchange, that has $\mathcal{E}$, as initial condition and the nearby equilibrium state as final condition. Any two equilibrium states can be connected by an adiabatic transformation, but one cannot always choose either as initial condition. In this way Carathéodory tries to capture the observed fact that heat flow from a hot to a cold body. However, many have found this postulate to be physically unreasonable (e.g. Truesdell, 1984). The axiom is also needed to mathematically establish his theorem on pfaffian forms.

Carathéodory's theorem for thermostatic systems states

Theorem 1 *Let a pfaffian equation* $dx_0 + X_1 dx_1 + \cdots + X_n dx_n$ *be given, where the* X_i *are finite, continuous, differentiable functions of the* x_i*, and it is known that in every neighborhood of an arbitrary point* P *in the space of the* x_i*, there are points which cannot be reached along curves which satisfy this equation, then there must be a multiplier of the expression which makes it an exact differential.*

The existence of the multiplier was to be a proof of the existence of absolute temperature; the multiplier itself was to be interpreted as absolute temperature. He also uses the existence of this integrating factor for his Pfaffian form to define entropy. Unfortunately, Carathéodory made mathematical errors in his proof. The proof was corrected by Bernstein (1960) by viewing the Carathéodory theorem to be local rather than global. But in a local version, the multiplier cannot be interpreted as temperature. Bernstein's new argument makes the result global to allow the multiplier to be interpreted as temperature. It requires that the coefficients are infinitely differentiable rather than just once differentiable. But the Bernstein mathematical work still assumes the Carathéodory axioms.

Carathéodory pioneered the application of differential forms to thermodynamic analysis by expressing his axiomatic thermostatics in terms of differential forms. However, his axiomatic approach has not been accepted because of the untenable machinations implied in some of his axioms. The Carathéodory use of differential forms is not that employed in this book as the foundation for a geometric structure of non-equilibrium thermodynamics.

3.2 Linear Irreversible Thermodynamics

Many of the ideas of linear irreversible thermodynamics (LIT) were developed either by Onsager or by Prigogine and can be approached from the point of view of fluctuations or of the Onsager relations. Four assumptions are listed as the foundation of LIT (Demirel and Sandler, 2004). A flux is usually the time derivative of an extensive parameter that describes the response of the system. In the linear theory, the components of the thermodynamic fluxes, J_i, are assumed to be a linear combination of the components of the thermodynamic forces, X_k, (sometimes called affinities) so that

$$J_i = \sum_{k=1}^{n} L_{ik} X_k, \tag{1}$$

where the L_{ik} are the phenomenological coefficients. The symmetry relations are

$$L_{ik} = L_{ki}. \tag{2}$$

Such relations are treated in this book as constitutive relations for the steady behavior of fluxes.

The entropy production is assumed to have the form $\sigma = \sum_i J_i X_i \geq 0$. Substitution of the linear expression into the entropy production produces a quadratic form in the X_i that must be positive definite if $\sigma > 0$. The entropy production is $\sigma = 0$ at equilibrium.

The Onsager reciprocity theorem is based on the idea that the equation of motion for each individual particle is time reversible as in classical dynamics or quantum mechanics. The macroscopic result obtained from this assumption by Onsager is that the relations $L_{ik} = L_{ki}$ hold in the linear expression for the fluxes. The objection usually raised is that microscopic time reversibility should not be able to produce macroscopic irreversibility. The Onsager relations have been severely criticized by Truesdell (1984, Chapter 7) for the linearity of the stationary state constitutive models deduced, for a weak mathematical foundation, and for their lack of physical foundation.

The remaining assumption for LIT involves a restriction on which fluxes and forces may be coupled in the linear relation for the flux. This idea was proposed by Curie in 1908 and later taken up by Prigogine. It says that flux and force tensors of differing order may only be coupled if their order differs by an odd number. Perhaps due to the strong criticism of the Curie principle by continuum thermodynamicists, the Curie restriction is replaced in (Garcìa-Colìn, 1988) by a local equilibrium assumption based on the Gibbs relation.

3.2.1 Applying the LIT

The method finds applications to mass flux problems, to chemical reactions, and to biological cellular processes. In the Prigogine non-equilibrium thermodynamics model as described by de Groot and Mazur (1984), a constitutive balance of entropy equation is used to identify the conjugate fluxes, $\mathcal{J}_i$, and gradients $\nabla \mathcal{X}_i$. de Groot and Mazur use a variational statement to show that their entropy production has a minimum at the stationary states. Stationary means that the state variables are independent of time. The so-called "linear approximations to non-equilibrium", as statistical thermodynamicists call the Onsager relations, are just constitutive models for stationary states.

An alternative procedure involves a dissipation function, which is defined using the Gibbs relation in terms of all required thermodynamic variables (Demirel and Sandler, 2001, 2004). The dissipation function is the product of the temperature and the rate of entropy production within the body (as opposed to entropy transport through the boundary of the body). The entropy production is written as the product $dS_i/dt = \sum_k J_k X_k$, where J_k is a flux and X_k is the associated force. To derive this product form, the Gibbs relation is differentiated with respect to time, multiplied by the density (assumed not to be time-dependent), and the mass, momentum, energy and entropy balance relations are used to obtain the local entropy production per volume as a product of fluxes and forces (Demirel and Sandler, 2001). The local equilibrium assumption for non-equilibrium states near equilibrium arises from the use of the Gibbs relation. Then the Onsager relations $L_{jk} = L_{kj}$ are assumed based on microscopic reversibility. The fluxes and forces produce linear phenomenological equations for non-equilibrium behavior of the form $J_j = \sum_{k=1}^{n} L_{jk} X_k$ for L_{jk} constants.

One of the problems of this construct is the identification of the thermodynamic fluxes and forces, J_k and X_k. If the selection is made based on entropy production, a different solution may arise than if they are selected using the dissipation function to identify the conjugate fluxes and forces, as promoted by Demirel and Sandler (2001) in their dissipation phenomenological equations (DPE) approach. Therefore, apparently no unique choice exists.

The Gouy-Stodola Theorem states that the entropy production due to irreversibility is proportional to the energy lost. Demirel and Sandler (2004) argue that the entropy production can be therefore used as quantitative measure of irreversibility.

Disadvantages of LIT are that it is restricted to linear systems, no general means of minimizing entropy production is available, and the phenomenological coefficients must be determined experimentally. An advantage is that it can relate the fluxes and their coupling in very fast processes (Demirel and Sandler, 2004). The Navier-Stokes equations for fluids may be obtained (de Groot and Mazur, 1984). Some phenomena cannot be modeled in this theory, such as the elastic response of rubber or of the effect of shear on density or on viscosity (Garcìa-Colìn, 1988).

3.2.2 Prigogine Minimum Entropy Production Theorem

To put linear irreversible thermodynamics on a foundation similar to other branches of physics, Onsager (1931) defined a variational principle, much as the least action principle gives dynamics. The Onsager variational principle operates in thermodynamic flux space and begins with the function $[\sigma(X_i, J_k) - \Psi(J_i, J_k)]$, where the dissipation is $\Psi(J_i, J_k) = \frac{1}{2}\sum_{i,k} R_{ik} J_i J_k$ and R_{ij} is the coefficient matrix. The principle says that if values of the irreversible forces X_i are assigned, the process fluxes J_i must maximize the expression, $[\sigma(X_i, J_k) - \Psi(J_i, J_k)]$. In terms of variations with respect to the fluxes for fixed X, $\delta_J[\sigma(X_i, J_k) - \Psi(J_i, J_k)]_X = 0$. The Onsager variational principle implies (1) and (2), the two key assumptions of LIT, by a simple computation.

One of Prigogine's contributions was the converse theorem that the relations (1) and (2) imply that the entropy production, $\sigma(J_i) = \sum_i X_i J_i$, takes a minimum on stationary states.

Theorem 2 *(Prigogine) The entropy production takes a minimum on the stationary state associated with fixed irreversible thermodynamic forces, X_i $i = 1, \ldots, j < n$, where n is the number of forces on the system. Fluxes J_i, $i = j + 1, \ldots, n$, which are conjugate to unfixed forces, are zero.*

In flux space, the problem is to determine the thermodynamic forces as a function of the fluxes, or conversely to predict the fluxes given the forces. The Ziegler maximum entropy production principle states that for given irreversible forces X_i, the fluxes must be those J_i that maximize the entropy production $\sigma(J_i) = \sum_i X_i J_i$. The Zeigler maximum entropy production principle implies the Prigogine variational principle as a special case, even though the Zeigler principle is based on maximum and the Prigogine on minimum entropy production. Further, in the case of LIT, the second law in the form $\sigma(J_i) > 0$ may be derived from the Ziegler principle.

Assuming the constraint making the system linear, $\sum_i X_i J_i = \sum_{i,k} R_{i,k} J_i J_k$, and setting equal to zero the variation of the entropy production using a Lagrange multiplier, λ, to produce the function $\sigma(J_k) - \lambda[\sigma(J_i) - \sum_i X_i J_i]$, gives a mathematical expression of the Ziegler principle. The Lagrange multiplier acts like the proportionality constant in classical plasticity theory to produce an orthogonality condition defining the J_k by a plane tangent to the graph of the entropy production function. The variational statement once the entropy production is written in terms of the fluxes immediately yields the Onsager variational principle and as a consequence the Prigogine minimum entropy production theorem. A clear derivation of these relations between the Onsager variation principle, the Prigogine theorem, and the Ziegler variational principle is given by Martyusheva and Seleznev (2006).

The proof of the Prigogine minimum entropy theorem may be viewed as taking entropy as a potential and showing that it reaches a minimum on the stationary states. This might be viewed as a thermodynamic stability problem (Lavenda, 1993). Lavenda points out that the first variation of entropy does not equal zero on the non-equilibrium stationary states so that this theorem is not truly a stability statement. He states (1993, p. 65) "Thermodynamic stability cannot be derived from any single non-equilibrium thermodynamic potential". A counterexample to this statement is given in the construction for non-homogeneous systems (Chapter 9).

Another important question is to identify the speed at which a non-equilibrium system moves to its final state. Martyusheva and Seleznev (2006) suggest that the Ziegler maximum entropy principle requires that the system moves at the largest possible rate to its state of maximum entropy because the system moves through a sequence of maximum entropy states. Such a conclusion does not seem to be required by the mathematics. The linear evolution equations derived, such as the Fourier law, predict instantaneous propagation.

3.2.3 Chemical Reactions Using LIT

Today, Onsager-Prigogine non-equilibrium thermodynamics seems to be more often applied to modeling chemical reactions. The extent of a chemical reaction parameter, ξ, is defined in terms of a measure of the amount of the substance involved in the reaction, the stoichiometric coefficient of a substance $\nu_i = dN_i/d\xi$, where N_i is the number of molecules of the substance (see Prigogine and Defay, 1954). Moroz (2008) defines an energy function and a dissipation function for a chemical reaction. Moroz defines the thermodynamic potential as $\Psi = 0.5\sum_{i,j}^{m} l_{ij}\xi_i\xi_j$ and dissipation $\Phi = \sum_{i,j}^{m} R_{ij}\dot{\xi}_i\dot{\xi}_j$. The Lagrangian $\Psi + \Phi$ is used to produce the Euler-Lagrange equations, and these functions are maximized at the highest speed possible, as an optimum control strategy. A Hamiltonian, $H = -\Psi - \Phi + \sum_{i=1}^{m} p_i\dot{\xi}_i$ such that $\dot{p}_i = \partial H/\partial \xi_i$, is defined that leads to a least action principle. Also see Martyusheva and Seleznev (2006).

3.3 Extended Irreversible Thermodynamics

One question that led to the development of extended irreversible thermodynamics (EIT), which arose in 1980's, is to what extent or distance from equilibrium is the LIT valid (Garcìa-Colìn and Uribe, 1991), although it is not clear how one measures the distance. The extended irreversible thermodynamics takes some gradients and fluxes as thermodynamic variables, as does rational thermodynamics (Müller, 1967), discussed below. This choice of variables allows the definition of a local entropy that is not required to satisfy the local equilibrium hypothesis.

Extended irreversible thermodynamics (EIT) avoids the local equilibrium assumption by not using the equilibrium Gibbs relation (Garcìa-Colìn, 1988). Further the thermodynamic variables are of two types: (1) fast or non-conserved variables and (2) conserved densities. A function, η, takes the place of entropy and an entropy balance equation is assumed as promoted by Müller (1967). Two approximations are made about the entropy function and the derived evolution equations. First, the coefficients are assumed to expand in power series of the non-conserved thermodynamic variables. This assumption means that phenomena involving singularities cannot be represented. Finally the coefficients have been taken as constants even though they can be space and time dependent, resulting in linear evolution equations with constant coefficients. The bulk of the results are similar to those of LIT. The proponents of EIT seem unhappy that η cannot be proven to be entropy, even though no definition of non-equilibrium entropy is offered that would allow a proof to be attempted.

3.3.1 The EIT Derivation of the Maxwell-Cattaneo Equation

One of the first points of emphasis of extended irreversible thermodynamics was the Maxwell-Cattaneo equation that yields finite velocity propagation rather than the instantaneous propagation implied by the Fourier law. In contrast to the derivation based on either the maximum dissipation construction defined in this book or on rational thermodynamics, the EIT derivation involves several additional assumptions. Assuming a rigid conductor, Garcìa-Colìn (1988) obtains a first order generalization of the Gibbs equation,

$$\frac{d\eta}{dt} = \frac{1}{\theta}\frac{du}{dt} + g(u)\mathbf{J}_q \cdot \frac{d\mathbf{J}_q}{dt}, \tag{3}$$

where $\mathbf{J}_q$ is heat flux, u is the specific internal energy, and η is the specific entropy function. An entropy balance is then assumed;

$$\rho\frac{d\eta}{dt} + \nabla \cdot \mathbf{J}_\eta = \sigma_\eta, \tag{4}$$

where σ_η is an unknown scalar. Then a computation leads to $\mathbf{J}_\eta \sim \beta(u)\mathbf{J}_q$. It is then argued that $\sigma_\eta = \mu(u)\mathbf{J}_q \cdot \mathbf{J}_q$. Combining these equations and making them consistent with the LIT produces

$$\frac{\rho g(u)}{\mu(u)} \frac{d\mathbf{J}_q}{dt} = -\frac{1}{\mu(u)} \nabla \theta^{-1} + \mathbf{J}_q. \tag{5}$$

When the left side is small, the stationary state is reached. To make the resulting equation agree with the Fourier relation, $\mu(u) = (\kappa\theta^2)^{-1}$. Then the equation can be written in the form of the Maxwell-Cattaneo equation

$$-\tau \frac{d\mathbf{J}_q}{dt} = \kappa \nabla \theta + \mathbf{J}_q, \tag{6}$$

where $\tau = -\rho g(u)\kappa\theta^2$. Since τ must be positive, this proof requires that $g(u) < 0$. See Chapter 9 for a derivation of the Maxwell-Cattaneo equation from the maximum dissipation construction.

3.4 Continuum Thermodynamics

Rational thermodynamics, as named by Truesdell (1984) because he thought other thermodynamic theories were not logically consistent, uses the balance laws and the second law in the form of the Clausius-Duhem inequality to derive constraints on admissible constitutive relations, but does not explicitly construct a geometric structure in which non-equilibrium thermodynamics can be placed and studied. No stability criterion is available. Continuum thermodynamicists exerted considerable effort to make the foundations of thermodynamics precise and rational.

One major contribution from continuum thermodynamics is the Clausius-Duhem inequality introduced by Truesdell around 1952 to include the influence of all thermodynamic variables that can act as sources of heat throughout the body. In the time of Clausius around 1850, many still adhered to the Carnot choleric theory which states that heat is a substance. In contrast, Joule and later W. Thompson (Lord Kelvin) viewed heat as a dynamic behavior due to atomic vibration. Various formulations of the second law of thermodynamics have been proposed since the middle of the nineteenth century. The second law is intended to account for the fact that in a process, unless it is a forced process, heat only can move from hot to cold regions of the body. Clausius stated the second law as requiring that compensation is required to pass heat from cold to hot regions. Compensation apparently means that the process is forced in the sense used in this book. The Clausius inequality requires that over a cycle, $\oint (Q/\theta)dt \leq 0$ for heat Q. The process is reversible if the integral is zero and irreversible otherwise. Some take $d\eta/dt \geq 0$ as a statement of the second law. Perhaps this idea comes from Planck (1879), who made as a basic assumption that the sum of the entropies of all components of the system increases during a process. Planck proposed $\Delta\eta \geq \int dQ/\theta$. The Duhem (1906) version of the second law of thermodynamics uses the Stokes heat flux vector $\mathbf{q}$ to give the entropy flux though the boundary of a body as

$$\dot{\eta} \geq \int \frac{\mathbf{q} \cdot \mathbf{n}}{\theta} dA, \tag{7}$$

where **n** is the normal to the boundary area.

Truesdell and Toupin (1960) added the radiation term r to Truesdell's formulation of the Clausius-Duhem inequality (see Truesdell, 1984, p. 42). For example, the Clausius-Duhem inequality is usually stated in the uniaxial case as

$$\dot{\eta} \geq \frac{1}{\theta}\left(r + \frac{dq}{dx}\right) - \frac{q}{\theta^2}\frac{\partial \theta}{\partial x}. \tag{8}$$

If the displacement is $\mathbf{x} = \chi(\mathbf{X})$, then energy balance for internal energy density u and Cauchy stress, σ

$$\dot{u} = \sigma \frac{\partial \dot{x}}{\partial X} + r + \frac{\partial q}{\partial x} \tag{9}$$

is used to eliminate r. The resulting modified Clausius-Duhem inequality is

$$\dot{\Psi} \leq \sigma \frac{\partial \dot{x}}{\partial X} - \frac{q}{\theta}\frac{\partial \theta}{\partial x}, \tag{10}$$

where the Helmholtz energy density is $\Psi = u - \theta\eta$.

Rational thermodynamics proposed six requirements that constitutive equations must obey (Truesdell and Toupin, 1960). (1) consistency with the balance laws and the entropy inequality in the form of the Clausius-Duhem inequality; (2) coordinate invariance of the constitutive law in all fixed coordinate systems; (3) just setting in the sense that "there should exist unique solutions corresponding to appropriate initial and boundary data, and these solutions should depend continuously on this data"; (4) material frame indifference or independence of the constitutive model from the frame of reference (The term "objective" is sometimes used instead of "material frame indifference".); (5) material symmetry must be represented in the constitutive equation; (6) equipresence of the independent variables in all constitutive equations for a given material (also see Bowen, 1989, pp. 95–98).

The equipresence requirement seems to have lead continuum thermodynamics astray. The thermal flux and other time derivatives were given the status of a thermodynamic variable in continuum thermodynamics and the energy was allowed to be a function of these rates (e.g. Truesdell, 1984). For example, the Helmholtz energy function is assumed to depend on the heat flux, q, and on the time rate of change of the deformation gradient (e.g. Bowen, 1989, p. 20). Bogy and Naghdi (1970) to allow finite heat condition, while continuing to assume the Fourier relation, assumed that the specific entropy depends on the time derivative of temperature. Unfortunately, application of the Clausius-Duhem inequality to their example falsely predicts that addition of heat can lower the temperature (Green and Naghdi, 1977). In contrast, in the maximum dissipation construction, energy is not a function of the thermal flux, of the temperature gradient or any other rates.

A significant contribution of continuum thermodynamics to the mathematical modeling of non-equilibrium processes is the Coleman (1964) idea of a material with fading memory (see Chapter 4). Unfortunately the integral formulation requires time-dependent material functions in the integrand. The time-dependent fading memory integrals proposed for viscoelastic behavior (e.g. Coleman, 1964; Gurtin, 1968; Christensen, 1982) have lost favor because it is very difficult to experimentally measure the time-dependent kernels in the integral over long time periods.

Serrin (e.g. 1986) tried to account for the fact that heat at one temperature is qualitatively different than the same amount at another temperature. Serrin develops a theory based on cycles and hotness, represented in the hotness manifold, and represents the quality of heat through his accumulation function and the hotness manifold. But this construction does not seem to have attracted many practitioners.

3.5 Extended Rational Thermodynamics

A long-standing problem is to represent finite time transport, such as heat conduction, rather than instantaneous transport. In thermodynamically non-homogeneous non-equilibrium systems, fluxes such as the thermal flux must be given a constitutive model. However, many flux constitutive equations, such as the Fourier for the heat flux or Ohm's law, produce an instantaneous response rather than a physically realistic response requiring a finite time. A key motivation of the rational extended thermodynamics construction of Müller and Ruggeri (1998) is to produce evolution equations which, in contrast to the Fourier law, predict finite propagation speeds of processes. Müller (1967) proposed that the entropy flux should be given by a constitutive relation that extends the older idea that entropy flux is equal to the heat flux divided by temperature. Likewise internal entropy production is represented by a constitutive relation. If the thermodynamic state varies from point to point in the body, then thermodynamic fluxes such as the heat, mass, chemical reactions, and stress fluxes, become thermodynamic variables in the entropy production constitutive model. Müller and Ruggeri (1998) obtain a system of non-linear hyperbolic first order evolution equations by minimizing their entropy production subject to the constraints given by the balance laws.

Rational extended thermodynamics, as developed by Müller and Ruggeri (1998), has had its greatest success representing gases of particles, especially when some variables exhibit large gradients. However, this theory has not been developed to the point of producing viscoelastic evolution equations (Müller and Ruggeri, 1998, p. 372).

4 Maximum Dissipation Criteria

Dissipation is commonly represented in the constitutive models of the last fifty years for the non-equilibrium behavior of solids, but is rarely given a thermodynamic foundation. Traditional methods of modeling dissipation include the introduction of a viscous stress, assumption of the principle of fading memory, or the use of internal state variables whose evolution describes dissipation (Coleman and Gurtin, 1967). An internal state variable, α_i, is sometimes called a hidden state variable. Each time rate of change of the internal state variable is presumed to be a function of the internal state variables and the external state variables. Coleman and Gurtin wrote evolution equations for the internal state variables and applied the stability theory of dynamics. The role of the internal dissipation is identical to that in Coleman's exposition on materials with fading memory.

The evolution of both viscoelastic and viscoplastic solids is often expressed in terms of a dissipation potential, a function of the rates of change of the internal and other state variables $a_1, \ldots, a_n$, in analogy with thermodynamic energies and from which the evolution equations are derived (e.g. LeMaitre and Chaboche, 1994). Such a potential commonly has no a priori physical meaning. The choice of the dissipation potential is essentially an assumption of a constitutive model. The dissipation potential $\phi(\dot{a}_1, \ldots, \dot{a}_n)$ defines the dual variables by $A_i = \partial\phi/\partial\dot{a}_i$. The evolution of the state variables is obtained from the Legendre transform, $\hat{\phi}(A_1, \ldots, A_n)$, of ϕ as $\dot{a}_i = \partial\hat{\phi}/\partial A_i$. Such dissipation potentials are not usually derived from thermodynamic foundations. Nor do those using this construction always discuss objectivity, perhaps because small displacements are implicitly assumed. The dissipation is given as $\sum_{i=1}^{n} A_i\dot{a}_i$.

To explicitly capture finite viscoelastic long-term behavior, Holzapfel (2000, p. 282ff) split the free energy into a sum of the volumetric elastic energy, isochoric elastic energy, and a viscoelastic contribution. The latter term, which is a function of strain and other internal variables, accounts for the dissipation by a model like the Maxwell model that evolves to zero over time under constant control variables, leaving a hyperelastic expression that accounts for the long-term behavior. Essentially, this model is a superposition of viscoelastic and long-term hyperelastic terms. The thermodynamic force, an affinity, which drives the non-equilibrium process is the derivative of the dissipative portion of the energy by the strain. The dissipation is this driving force times the time derivative of the corresponding variable.

The imposition of a maximum dissipation criterion is not unusual and many different types have been employed to develop constitutive models. In finite strain time-independent plasticity, such a condition on the plastic dissipation gives kinematic plastic stability and the normality rule (Lubliner, 1984). Maximum dissipation models have been proposed for plasticity by Hill (e.g. 1948, 1958) and by Mandel (e.g. 1973) in which the dissipation function is maximized by requiring that the evolution of the variables conjugate to the generalized forces form a vector in the cone of normals to the convex space of generalized forces. Nguyen Quoc Son (1984) used a similar construction for anelastic "standard" solids. The maximum dissipation criterion in other works takes many distinct forms. For example, the dissipation may depend on a penalty function of the difference between the current value of the stress and the corresponding equilibrium stress (Deseri and Mares, 2000). In the rate independent plasticity model of Cermelli et al. (2001), the dissipation, which is relative to an intermediate configuration, depends on the difference of an internal couple density and the configurational stress. A constitutive isothermal dissipation potential depending on the deformation gradient from the reference state to the natural configuration and on the rate of change of this gradient has also been assumed to govern the evolution of the intermediate natural configuration (Rajagopal and Srinivana, 1998, 2000). In a system based on the evolution of the intermediate natural configuration defined by instantaneous unloading, the dissipation function depends on dissipative fluxes, and the overstress drives the evolution as determined by a maximum dissipation criterion (Hall, 2008). A maximum dissipation criterion has been developed by workers in the Onsager school of thermodynamics (e.g. de Groot and Mazur, 1984; Ziegler and Wehrli, 1987; Yang et al., 2005; Fischer and Svoboda, 2007).

5 Nonlinear Dynamical Systems

Thermodynamics must involve dynamics. The Introduction to Thermodynamics course taught in many engineering schools is typically a course in thermostatics. The dynamics can be put back into thermodynamics by using nonlinear dynamical systems to define evolution equations that describe non-equilibrium processes. To establish some of the possible types of evolution equations, some nonlinear dynamical systems are reviewed. An autonomous dynamical system is one in which time does not appear explicitly in the differential equation defining the system. The autonomous nonlinear dynamical systems include potential systems, Hamiltonian systems, and Hamiltonian systems with damping.

A gradient dynamical system is one for which the evolution of the state variables, $x_i;\ i = 1, \ldots, n$ is given by $\dot{x}_i = -\partial f(\mathbf{x}, \alpha)/\partial x_i$, where α is a set of parameters, and $f(\mathbf{x}, \alpha)$ is the potential function. In other notation $\dot{\mathbf{x}} = -grad_{\mathbf{x}} f(\mathbf{x}, \alpha)$. The equilibrium states are those for which $grad_{\mathbf{x}} f(\mathbf{x}, \alpha) = 0$. In other words the equilibria are determined by the simultaneous solution of the n equations, $\partial f(\mathbf{x}, \alpha)/\partial x_i = 0$. The function $f(\mathbf{x}, \alpha)$ decreases on trajectories of the system.

A Hamiltonian system is conservative and is described in terms of the total energy, H, of the system. A system of generalized coordinates is a minimum sized set of coordinates that describe the configuration of a system; it need not be Cartesian. For example, the motion of a rigid pendulum may be completely described in terms of the angle the pendulum makes with the vertical. Let $q_i,\ i = 1, \ldots, n$ be the generalized coordinates of the system, let $\mathbf{q}$ be the vector of generalized coordinates. The kinetic energy is $K = 0.5\langle\dot{\mathbf{q}}, \dot{\mathbf{q}}\rangle$ and the potential energy is $V(\mathbf{q})$. The potential energy exists since the system is conservative. Define the generalized velocities to be $\mathbf{p} = \dot{\mathbf{q}}$ and define the total energy as $H(\mathbf{q}, (\mathbf{p}) = K(\dot{\mathbf{q}}) + V(\mathbf{q})$. The action is

$$a = \int_{t_1}^{t_2} (K - V)dt, \tag{11}$$

where $L = K - V$ is the Lagrangian function. The dynamic response is that which minimizes the action.

The function minimizing an integral, $\int f(x, y, y', y'')dx$ is determined by the Euler-Lagrange equations,

$$\frac{\partial f}{\partial y} - \frac{d}{dx}\frac{\partial f}{\partial y'} + \frac{d^2}{dx^2}\frac{\partial f}{\partial y''} = 0.$$

The Euler-Lagrange equations that give the extrema of an integral function for the action yield the n equations of motion

$$\frac{d}{dt}\frac{\partial L}{\partial p_i} - \frac{\partial L}{\partial q_i} = 0, \quad i = 1, \ldots, n. \tag{12}$$

In other words, $\ddot{\mathbf{q}} + grad_{\mathbf{q}} V = 0$.

Since $\dot{\mathbf{p}} = \ddot{\mathbf{q}} = -grad_{\mathbf{q}} V$, the Hamiltonian equations defining the system are

$$\dot{\mathbf{q}} = \frac{\partial H}{\partial \mathbf{p}} \text{ and } \dot{\mathbf{p}} = -\frac{\partial H}{\partial \mathbf{q}}. \tag{13}$$

The time rate of change of H is zero on a trajectory; energy is conserved. In Chapter 7, a symplectic structure, often associated with a Hamiltonian system, appears in one technique to define the thermostatic states. An autonomous gradient system cannot have a periodic orbit, but a Hamiltonian system can. In a gradient system, the potential function may serve as the Liapunov function.

A damped Hamiltonian system is one with a damping term added so that the equation of motion is $\ddot{\mathbf{q}} + grad_{\mathbf{q}} V = -c\dot{\mathbf{q}}$, where c is a positive scalar. The rate of change of the total energy is now negative on trajectories.

5.1 Equilibrium States as Attractors

The set of equilibrium states may be an attractor or a repeller for a dynamical system. While many definitions have been given for stability of a dynamical system, the focus here is on Liapunov stability.

Recall that

Definition 3 A function is called C^o if it is continuous and C^n if its first n-derivatives exist and are continuous. A C^∞ function is infinitely differentiable and continuous.

Let $dx/dt = f(x)$ be an autonomous dynamical system that $f(0) = 0$. Denote the solution on $\Re^n$ by $g(x, t)$. Assume further that f is C^1 on $\Re^n - \{0\}$ and that f satisfies the Lipschitz condition that $|f(x_1) - f(x_2)| \leq C|x_1 - x_2|$ for some constant C.

Liapunov stability depends on the system behavior after a perturbation. Let $|| \cdot ||$ be the metric in $\Re^n$.

Definition 4 The equilibrium state $x = 0$ is Liapunov stable if given ϵ there exists a $\delta(\epsilon)$ such that if the perturbation $x(0)$ is within δ of the equilibrium, $||x(0) - 0|| < \delta(\epsilon)$, then $||x(t)|| < \epsilon$ for all t.

The equilibrium state $x = 0$ is Liapunov asymptotically stable if it is stable and if $\lim_{t\to\infty} x(t) = 0$.

A simple means of verifying asymptotic stability uses a Liapunov function. A weak result is based on the following definition.

Definition 5 A continuous function $V(x) : \Re^n \to \Re$ is a Liapunov function if

1. its first derivatives exist (C^1) in a region D about the origin;
2. $V(0) = 0$ but $V(x) > 0$ in D for $x \neq 0$;
3. $\dot{V} \leq 0$.

Lyapunov (1892) gave a proof of sufficiency under the assumption that V is C^1.

Theorem 6 *For an autonomous system, the trivial solution $x = 0$ is asymptotically stable if a Lyapunov function exists.*

Proof Let the initial condition for a path $x(t)$ be a small perturbation from the minimum, $x = 0$, of V. Then the condition $\dot{V} \leq 0$ implies that $x = 0$ is an isolated minimum for V and that the path must approach $x = 0$ as $t \to \infty$.

The converse was not proved until the 1960's and produces a C^∞ function V. The converse result requires a stronger definition of a Liapunov function by assuming in addition that V tends to a constant at the boundary of D. The stronger statement of the asymptotic stability theorem that gives the equivalence of asymptototic stability to the existence of a Liapunov function is

Theorem 7 *Assume $\dot{x} = f(x)$ such that $f(0) = 0$, f satisfies a Lipschitz condition at 0, and is C^1 on $\Re^n$. A necessary and sufficient condition that $x = 0$ be asymptotically stable in a domain D is that there exists a C^∞ Liapunov function for $\dot{x} = f(x)$ in D.*

One goal is to identify the level sets and the basin of attraction. The description requires some other mathematical ideas.

Definition 8 A continuous function $h : X \to Y$ is a homeomorphism if there is an inverse, $h^{-1} : Y \to X$ such that 1) $hh^{-1} = 1_Y$ and 2) $h^{-1}h = 1_X$. Then X and Y are called homeomorphic.

Definition 9 A homeomorphism is a diffeomorphism if it is C^∞.

Example $X = Y = \Re$; $y = h(x) = 3x$; $x = h^{-1}(y) = y/3$.

Definition 10 Functions $f, g : X \to Y$ are homotopic if, for the unit interval $I = [0, 1]$, there is a map $h : X \times I \to Y$ such that $h(x, 0) = f(x)$ and $h(x, 1) = g(x)$.

Definition 11 Spaces X and Y are homotopically equivalent if fg and gf are homotopic to the identity maps.

Example In the plane with the standard metric, let $D = \{x, y | x^2+y^2 < 1\}$. Then the constant map $D \to (0, 0)$ is homotopic to the identity map $D \to D$ by $h(x, y, t) : D \times I \to D$ where $h(x, y, t) = \big((1-t)x, (1-t)y\big)$.

Definition 12 A level surface for V is the set of all x satisfying $V(x) = k$ for a positive constant k.

The closed n-disk is $D^n\{(x_1, \ldots, x_{n+1}) \mid \sum_{i=1}^{n+1} x_i^2 \leq 1\}$. The n-sphere $S^n = \{(x_1, \ldots, x_{n+1}) \mid \sum_{i=1}^{n+1} x_i^2 = 1\}$ in $\Re^{n+1}$ bounds the closed n-disk.

The basin of attraction, $V^{-1}([0, k])$, is diffeomorphic to $\Re^n$ and the level surface, $V^{-1}([0, k])$, is homotopic to S^n (Wilson, 1966). The proof is based on the idea that the trajectory of the system can only pass through a level surface once.

A more recent application of these ideas is the Brockett (1983) theorem for controls that create a stable response.

Theorem 13 *A necessary condition for the existence of a C^1 feedback law, $u = u(x)$ making x_o locally asymptotically stable for the system $\dot{x} = f(x, u(x))$ is that $f(x, u) = y$ is solvable for all $||y||$ sufficiently small.*

Solvability of $f(x) = y$ for fixed y means that the vector field $\dot{f} = f(x) - y$ has an equilibrium, x, i.e. a point where $\dot{f} = 0$.

To produce a definition of dynamic stability for thermodynamics, Duhem (1911) defined a ballistic energy whose extrema subject to constraints give the equilibria, an idea taken up by Ericksen (1991). Duhem tried to build thermodynamics from this viewpoint by constructing a function whose minima correspond to the dynamically stable equilibria (Truesdell, 1984, p. 39). The ballistic energy, B, is a function of the spatial position and time which satisfies $\dot{K} \leq -\dot{B}$, where K is the kinetic energy and the dot indicates the material derivative. Duhem took the ballistic energy to be the sum of the potential energy of the external forces and the available energy. If a function is defined by $\Phi = K + B$, then $\dot{\Phi} = \dot{K} + \dot{B} < 0$ for all processes. Therefore, Φ decreases in time as a process proceeds, allowing a stability analysis. Φ resembles a Liapunov function so that the equilibrium point to which the process converges is globally asymptotically stable. The generalized energy is defined in Chapter 3 for similar reasons. The use of generalized thermodynamic functions for non-equilibrium thermodynamics allows the definition of evolution equations in which relaxation processes are represented by gradient systems. But forced thermodynamic processes must be defined by non-autonomous systems obtained by modifying this construction. The generalized thermodynamic function permits a simple definition of stability and also the study of systems in the neighborhood of singularities, such as phase changes, non-unique elastic deformations, plastic instability, etc., by singularity classification schemes such as catastrophe theory (see Chapter 8).

6 Goals for a Non-equilibrium Thermodynamic Construction

The above review of the state of non-equilibrium thermodynamics suggests several goals for a non-equilibrium thermodynamics construction. A system of evolution differential equations seems most appropriate to represent non-equilibrium processes because non-equilibrium processes are time-dependent. Further, many solution techniques, both in closed form and numerical, are available for differential equations. These evolution equations must be valid for processes that are far from equilibrium to improve on the Onsager type theories. Non-equilibrium thermodynamics should be represented by a nonlinear dynamical system in which the stable thermostatic states act as an attractor.

The construction must be capable of representing the time-dependent behavior of solids. Many current time-dependent models, in order to guarantee thermodynamic consistency, are formulated to ensure that the evolution process obeys the second law of thermodynamics in the form of the Clausius-Duhem inequality. However, the second law only determines those processes which are allowable. An additional criterion is required to select the particular process actually taken from a given non-equilibrium state. Such a criterion may involve a condition on dissipation. Therefore, the construction must include a means of representing dissipation, preferably without using an ad hoc separate dissipation function.

The Onsager-type theories develop non-equilibrium constitutive models largely for steady state processes and only predict instantaneous propagation of disturbances. A non-equilibrium construction should be capable of predicting general finite velocity transport.

In contrast to almost other thermodynamic theories other than that of Gibbs, the construction must be strongly embedded in a geometric construction that represents and relates both thermostatic and non-equilibrium processes. The geometric interpretation of the construction would facilitate the application of differential topology and nonlinear dynamics to the study of thermodynamic systems. The construction must be capable of, and amenable to, dealing with the bifurcations common to many important physical phenomena. Many thermodynamic theories cannot deal with energy bifurcations such as phase transitions.

The construction must solve technical matters such as whether the energy should depend on the rates of thermodynamic fluxes and other variables. If these variables cannot appear in the energy as suggested by the false conclusions drawn in continuum thermodynamics, they must be accounted for in another essential thermodynamic function. The construction must distinguish between fundamental principles, such as the balance laws, and constitutive statements. It must also guarantee frame invariance (objectivity) in systems involving large deformations of solids and clearly distinguish the reference and current states of the body.

Constitutive modeling is one practical application of non-equilibrium thermodynamics. The application to solids motivated the construction described in this book because much of thermodynamics focuses on fluids and because the time-dependent methods available in continuum thermodynamics such as the principle of fading memory are exceeding complex to apply. A main goal is a non-equilibrium thermodynamics construction that allows for a straightforward method for the constitutive modeling of the time-dependent behavior of solids.

References

B. Bernstein (1960). Proof of Carathéodory's local theorem and its global application to thermostatics. *Journal of Mathematical Physics* **1**(3), 222–224.

D. B. Bogy and P. M. Naghdi (1970). On heat conduction and wave propagation in rigid solids. *Journal of Mathematical Physics* **11**, 917–923.

R. M. Bowen (1989). *Introduction to Continuum Mechanics for Engineers*, Plenum Press, New York.

R. W. Brockett (1983). Asymptotic stability and feedback stabilization. *Differential Geometric Control Theory*, Birkäuser, Boston, pp 181–191.

C. Carathéodory (1909). Investigations into the Foundations of Thermodynamics. In *The Second Law of Thermodynamics*, ed. J. Kestin, Dowden, Hutchinson and Ross Inc., Stroudsberg, PA, 1976. Originally, (Untersuchungen über die Grundlagen der Thermodynamik, Math. Ann. **67**, 355–386).

P. Cermelli, E. Fried and S. Sellers (2001). Configurational stress, yield and flow in rate independent plasticity. *Proceedings of the Royal Society of London A* **457**, 1447–1467.

R. M. Christensen (1982). *Theory of Viscoelasticity; an introduction*. 2nd Ed., Academic Press, New York.

B. D. Coleman (1964). On thermodynamics, strain, impulses and viscoelasticity. *Archive for Rational Mechanics and Analysis* **17**, 230–254.

B. D. Coleman and M. E. Gurtin (1967). Thermodynamics with internal state variables. *Journal of Chemical Physics* **47**, 597–613.

Y. Demirel and S. I. Sandler (2001). Linear-nonequilibrium thermodynamics theory for coupled heat and mass transport. *International Journal of Heat and Mass Transfer* **44**, 2439–2451.

Y. Demirel and S. I. Sandler (2004). Nonequilibrium thermodynamics in engineering and science. *Journal of Physical Chemistry B* **108**, 31–43.

L. Deseri and R. Mares (2000). A class of viscoplastic constitutive models based on the maximum dissipation principle. *Mechanics of Materials* **32**, 389–403.

P. Duhem (1906). Recherches sur l'Elasticité. Gauthier-Villars, Paris.

P. Duhem (1911). *Traité d'Enérgétique ou Thermodynamique Générale*, Gauthier-illars, Paris.

J. L. Ericksen (1991). *Introduction to the Thermodynamics of Solids*, 1st Ed., Chapman-Hall, London.

F. D. Fischer and J. Svoboda (2007). A note on the principle of maximum dissipation rate. *Journal of Applied Mechanics* **74**, 923–926.

L. S. Garcìa-Colìn (1988). Extended non-equilibrium thermodynamics, scope and limitations. *Revista Mexicana de Fisica* **34**, 344–366.

L. S. Garcìa-Colìn and F. J. Uribe (1991). Extended irreversible thermodynamics, beyond the linear regime: a critical overview. *Journal of Non-Equilibrium Thermodynamics* **16**, 89–128.

J. W. Gibbs (1873a). Graphical methods in the thermodynamics of fluids. In *The Collected Works*, Vol. 1, Yale University Press, New Haven, 1948.

J. W. Gibbs (1873b). A method of geometrical representation of the thermodynamic properties of substances by means of surfaces. *Transactions of the Connecticut Academy* II, 382–404. Also in *The Collected Works*, Vol. 1, Yale University Press, New Haven 1948, 33–54.

S. R. de Groot and P. Mazur (1984). *Non-equilibrium Thermodynamics*, Dover, New York.

A. E. Green and P. M. Naghdi (1977). On thermodynamics and the nature of the second law. *Proceedings of the Royal Society of London, Series A* **357**, 253–270.

M. E. Gurtin (1968). On the thermodynamics of materials with memory. *Archive for Rational Mechanics and Analysis* **28**, 40–50.

R. B. Hall (2008). Combined thermodynamics approach for anisotropic, finite deformation overstress models of viscoplasticity. *International Journal of Engineering Science* **46**, 119–130.

R. Hill (1948). A variational principle of maximum plastic work in classical plasticity, *A. J. Mech. Applied Math.* **1**, 18–28.

R. Hill (1958). A general theory of uniqueness and stability in elastic-plastic solids. *Journal of the Mechanics and Physics of Solids* **6**, 236–249.

G. A. Holzapfel (2000). *Nonlinear Solid Mechanics*, 2005 reprinting. Wiley, Chichester.

B. H. Lavenda (1993). *Thermodynamics of Irreversible Processes*, Dover, New York.

J. Lubliner (1984). A maximum-dissipation principle in generalized plasticity. *Acta Mechanica* **52**, 225–237.

A. Lyapunov (1892). Problème general de la stabilité du mouvement. In *Annual of Mathematics Studies* **17**, Princeton University Press (1949).

J. Mandel (1973). Thermodynamics and Plasticity. In *Foundations of Continuum Thermodynamics*, ed. J. J. Delgado Domingos, M. N. R. Nina, J. H. Whitelaw, John Wiley and Sons, New York.

L. M. Martyusheva and V. D. Seleznev (2006). Maximum entropy production principle in physics, chemistry and biology. *Physics Reports* **426**, 1–45.

A. Moroz (2008). On a variational formulation of the maximum energy dissipation principle for non-equilibrium chemical thermodynamics. *Chemical Physics Letters* **457**, 448–452.

I. Müller (1967). On the entropy inequality. *Archive for Rational Mechanics and Analysis* **26**, 118–141.

I. Müller and T. Ruggeri (1998). *Rational Extended Thermodynamics*, 2nd Ed., Springer, New York.

Nguyen Quoc Son (1984). Bifurcation et stabilité des systèmes irréversibles obéissant au principe de dissipation maximale. *Journal de Méchanique théorique et appliqué* **3**, 41–61.

L. Onsager (1931). Reciprocal relations in irreversible processes I, *Physical Review* **37**, 405–426, and Reciprocal relations in irreversible processes II, *Physical Review* **38**, 2265–2279.

M. Planck (1879). Über den zweiten Hauptsatz der mechanischen Wärmetheorie (Phd dissertation presented to the University of Munich).

L. Pogliani and M. N. Berberan-Santos (2000). Constantin Carathéodory and the axiomatic thermodynamics. *Journal of Mathematical Chemistry* **28**(1–3), 313–324.

I. Prigogine and R. Defay (1954). *Chemical Thermodynamics*, Chapter IV, translated by D.H. Everett, Longmans, Green & Co.

J. Serrin (1986). An outline of thermodynamical structure. In *New Perspectives in Thermodynamics*, ed. J. Serrin, Springer-Verlag, Berlin.

C. Truesdell (1984). *Rational Thermodynamics*, Springer Verlag, New York.

C. Truesdell and R. A. Toupin (1960). The Classical Field Theories. *Handbuch der Physik* Vol. III/1, Ed. S. Flugge, Springer-Verlag, Berlin, Sec. 293.

F. W. Wilson (1966). The structure of the level surfaces of a Lyapunov function. *Journal of Differential Equations* **3**, 323–329.

Q. Yang, L. G. Tham, and G. Swoboda (2005). Normality structures with homogeneous kinetic rate laws. *Journal of Applied Mechanics* **72**, 322–329.

H. Ziegler and C. Wehrli (1987). On a principle of maximum rate of entropy production. *Journal of Non-equilibrium Thermodynamics* **12**, 229–243.

Chapter 2
Thermostatics and Energy Methods

1 Introduction

Energy methods analyses of mechanics, dating back to the eighteenth century, obtain equilibrium (thermostatic) states as the extrema of energy functions. By analogy, a goal of the construction given in Chapter 3 is to define generalized thermodynamic functions for which the thermodynamic equilibria or long-term states are extrema with respect to the state variables. The equilibrium states can also be defined in terms of state equations, as done by Gibbs for example.

Energy methods use techniques such as virtual work, the minimum total potential energy or complementary energy density to determine the equilibrium deformation of a body given the loads, or vice versa. These techniques can often quickly solve statically indeterminate problems. Calculus of variations is used to analyze, usually minimize, those systems in which the energy function is expressed in terms of an integral. While not discussed in this book, approximations are given by the Raleigh-Ritz, Galerkin, and finite element methods. The thermostatic behavior of bodies made of non-linear elastic materials can be more easily studied by energy methods. Energy methods can also be used to study abrupt changes in structures, such as buckling, to determine the post-abrupt change behavior from the energy function, including displacements and stability.

Definition 1 A variational principle is a statement that a certain response of a mechanical system is determined by minimizing a particular real valued function associated with the system and the type of response.

Variational methods are common in mechanics, but are not used in this book except as a model to define the generalized thermodynamic function. Some facts about these variational principles are briefly reviewed to give background for the non-equilibrium thermodynamics construction presented in Chapter 3.

2 The Principle of Virtual Work

The classical variational principle is that of virtual work, which gives a necessary condition for equilibrium of a system. Its essence was known in the sixteenth

H.W. Haslach Jr., *Maximum Dissipation Non-Equilibrium Thermodynamics and its Geometric Structure*, DOI 10.1007/978-1-4419-7765-6_2,

century. For example, Galileo used it to analyze simple machines. It was first formulated precisely in 1717 by Jean Bernoulli and developed further by J. L. Lagrange in a 1760 publication. Euler had used it as early as 1732 to analyze the behavior of simple structures. The principle was extended to kinetics by Hamilton in 1834. In the nineteenth and twentieth centuries, many other variational principles have been proposed and the methods have found important applications in determining the stability of the equilibrium states of a mechanical system.

If the position of a point mass is defined by a vector $\mathbf{r}$, the work done by a force $\mathbf{f} = f_x\mathbf{i} + f_y\mathbf{j} + f_z\mathbf{k}$ over a path is

$$\int_{x_1}^{x_2} \mathbf{f} \cdot d\mathbf{r} = \int_{x_1}^{x_2} f_x dx + f_y dy + f_z dz. \tag{1}$$

The force may change over the path in either magnitude or direction. In general, the work done depends on the path, as for example, when the force is friction.

The configuration or shape of a body is described by indicating the position of each point in the body in a coordinate system. A system of generalized coordinates for a system is a set containing the minimum number of coordinates required to completely describe the configuration of the system.

As the body goes through an imagined displacement, called a virtual displacement, the real fixed forces acting on the body do work, called the virtual work, δW, as they move through a distance during the virtual displacement. One restriction is that the only virtual displacements allowed are those which preserve the constraints on the body.

The mathematician and engineer Fourier (1768–1830) proposed that a body remains in equilibrium if the virtual work done by any virtual displacement is less than or equal to zero. The statement that $\delta W \leq 0$ is often called Fourier's inequality. Fourier's inequality is only a sufficient condition for equilibrium. For example, a ball balanced on the top of a hill is in equilibrium, but its weight does positive work under any virtual displacement consistent with the constraint that the ball must remain in contact with the hill. An alternative condition which is both necessary and sufficient is given by the principle of virtual work. This statement is a variational principle.

Principle of Virtual Work A body is in equilibrium if and only if the total virtual work, δW, done on the body by all actual forces during any virtual displacement is zero.

This condition is equivalent to the force equilibrium condition determined in Statics. Let a particle be subjected to forces $\mathbf{f}_1, \mathbf{f}_2, \ldots, \mathbf{f}_n$. For a virtual displacement $\delta\mathbf{s}$, the total virtual work is $\delta W = \sum_{i=1}^{n}(\mathbf{f}_i \cdot \delta\mathbf{s}) = (\sum_{i=1}^{n}\mathbf{f}_i) \cdot \delta\mathbf{s}$. So $\delta W = 0$ if $\delta\mathbf{s}$ is perpendicular to $\sum_{i=1}^{n}\mathbf{f}_i$ or if $\sum_{i=1}^{n}\mathbf{f}_i = 0$, and conversely. But the result must hold for all $\delta\mathbf{s}$, therefore $\delta W = 0$ iff $\sum_{i=1}^{n}\mathbf{f}_i = 0$. Therefore the principle of virtual work has the same status as Newton's Laws. It is an assumption whose validity is verified by the accuracy of the predictions it makes about the behavior of mechanical systems.

A condition for the virtual work to be zero can be computed easily if the work is a function of a finite number, n, of degrees of freedom, $x_1, \ldots, x_n$. Then the variation of the work is defined to be

$$\delta W = \frac{\partial W}{\partial x_1}\delta x_1 + \ldots + \frac{\partial W}{\partial x_n}\delta x_n \tag{2}$$

Since the coordinate variations, δx_i, are arbitrary and independent, the virtual work can only be zero, $\delta W = 0$, if

$$\frac{\partial W}{\partial x_i} = 0 \quad \text{for} \quad i = 1, 2, \ldots, n. \tag{3}$$

3 The Principle of Stationary Potential Energy

The integral (1) is exact if there is a function $g(x, y, z)$ such that $\partial g/\partial x = f_x$, $\partial g/\partial y = f_y$, $\partial g/\partial z = f_z$. An exact integral is independent of the path. In such a case, the force is said to be conservative. The work done on the system only depends on the endpoints of the displacement path. For example, gravitational force, spring forces, and loads which do not change direction or magnitude over the path are conservative.

Potential energy is only defined for conservative systems, those mechanical systems on which all forces, both internal and external, are conservative.

Definition 2 Suppose that the work done by the forces during the displacement of a system from configuration X_1 to X_2, denoted by $W(X_1, X_2)$ is conservative. Then the potential energy, $\Pi(X_2)$, at configuration, X_2, with respect to configuration, X_1, is $\Pi(X_2) = -W(X_1, X_2)$.

This defines the potential energy as the negative of the work if the work is conservative. For example if a body magically rises through a height h, the work done by gravity is $-mgh$ since the gravitational force opposes the motion. Therefore the potential energy is mgh.

By the principle of virtual work, the system attains an equilibrium state at those $x_1, \ldots, x_n$, where the virtual work is zero. This occurs if each $\partial W/\partial x_i = 0$ for $i = 1, 2, \ldots, n$. But since $W = -\Pi$, the principle of virtual work becomes the principal of stationary potential energy.

Principle of Stationary Potential Energy A conservative system described by coordinates, $x_1, \ldots, x_n$, and having a potential energy $\Pi(x_1, \ldots, x_n)$ reaches an equilibrium state at those points for which $\partial\Pi/\partial x_i = 0$ for $i = 1, 2, \ldots, n$.

This principle gives n equations in n unknowns to determine the coordinates of the equilibrium states. Since the equations $\partial\Pi/\partial x_i = 0$ for $i = 1, 2, \ldots, n$ may be nonlinear, there can be more than one equilibrium state.

Apparently the application of the principle of virtual work to conservative systems is called stationary because the derivatives are analogous to the derivative with respect to time. Then just as a system is stationary if its velocity is zero, the potential is called stationary at the extrema.

4 Stability of Equilibria in Conservative Systems

Energy methods such as the principle of stationary potential energy allow the determination of the stability of an equilibrium state.

Definition 3 An equilibrium state is said to be stable if its potential energy is a minimum. The equilibrium state is said to be unstable if its potential energy is a maximum.

In general, the behavior is stable if the Hessian is positive definite and unstable if the Hessian is negative definite. If the potential energy depends only on a single variable and if the second derivative is non-zero, the sign of its second derivative determines stability. The case in which the Hessian has zero determinant is discussed in Chapter 8.

Example A rigid L-shaped member whose vertical arm is of height $L = 50$ cm, and whose horizontal arm is of length $\ell = 10$ cm is supported by a frictionless pin at the base, O, of the vertical arm and by a horizontal spring at the intersection of the vertical and horizontal arms. The spring constant is $k = 100$ N/m. The spring is supported in a frictionless slot in the wall so that as the L-shaped member rotates, the spring remains horizontal. A weight, $W = 20$ N, is hung from the end of the horizontal arm. Determine the equilibrium positions and their stability. Ignore the mass of the L-shaped member.

The single generalized coordinate required to describe the configuration of the L-shaped member is the angle θ measured counterclockwise from the vertical to the equilibrium position of the vertical arm of the L-shaped member (Fig. 2.1). The weight moves down through a vertical distance $(L - L\cos\theta) + \ell\sin\theta$, and the spring

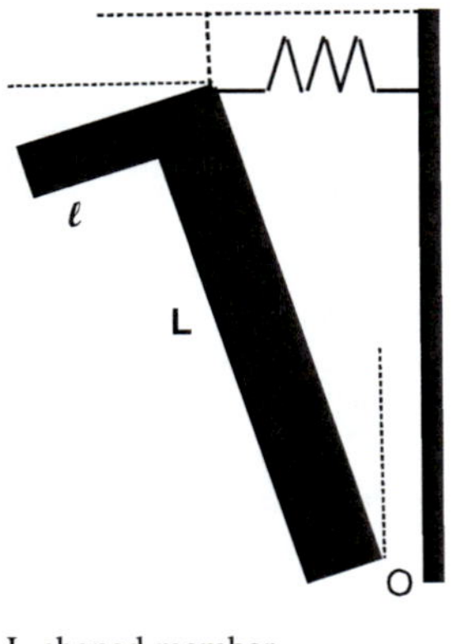

Fig. 2.1 The configuration of the L-shaped member

stretches a distance $L \sin\theta$. Since the motion of the weight is in the same direction as the gravitational force, the work done is positive. Therefore, the potential energy of the forces acting on the L-shaped member is

$$\Pi = -(L - L\cos\theta + \ell\sin\theta)W + \frac{k}{2}(L\sin\theta)^2. \tag{4}$$

The equilibrium positions are determined from the principle of stationary potential energy by

$$\frac{d\Pi}{d\theta} = -(L\sin\theta + \ell\cos\theta)W + \frac{k}{2}L^2\sin 2\theta = 0. \tag{5}$$

Substituting the values of the parameters and solving numerically yields an equilibrium configuration at $\theta = 0.1346$.

The stability is determined from the second derivative

$$\frac{d^2\Pi}{d\theta^2} = (-L\cos\theta + \ell\sin\theta)W + kL^2\cos 2\theta. \tag{6}$$

The equilibrium position $\theta = 0.1346$ is stable since the second derivative is positive there and equals 23.6.

Setting the moment about the frictionless pin O of the forces on the L-shaped member equal to zero recovers the equilibrium equation (5). Therefore if the system is conservative, the energy method gives more information than the vector statics method taught to beginning engineering and physics students because stability is not determined by vector statics.

5 Hyperelastic Thermostatic Energy Density Functions

If the work done by the internal forces in a body is conservative, i.e. if the body is elastically loaded, the potential energy of the internal work is called the strain energy, U. The strain energy is often written in terms of the strain energy density, U_o, so that $U = \int_V U_o dV$, where V is the volume.

Many finite deformation energy density functions have been proposed for finite deformations in addition to the linearly elastic strain energy density for small deformations. Hyperelastic models for isotropic, isothermal materials are local, current elasticity models describing the thermostatic states. No viscous behavior is considered, nor is any dependence on the past history of the loading. Many such functions have been defined to predict the equilibria of various solid materials. The idea of a strain energy dates at least to George Green around 1840. The idea of a hyperelastic material was proposed by Green, but the name was suggested about 1960 by C. Truesdell and coworkers.

Definition 4 A hyperelastic material is one that has a strain energy density function and whose state of stress is computed as the derivative of the strain energy density function by the strain.

This means that the strain energy density function determines the stress-strain relation.

5.1 Linear Elastic

The isotropic generalized Hooke's law is, for normal stress σ, shear stress τ, normal strain ϵ, shear strain γ, the elastic modulus E, and the shear modulus G,

$$\begin{aligned}\epsilon_x &= \frac{1}{E}[\sigma_x - \nu(\sigma_y + \sigma_z)];\\ \epsilon_y &= \frac{1}{E}[\sigma_y - \nu(\sigma_x + \sigma_z)];\\ \epsilon_z &= \frac{1}{E}[\sigma_z - \nu(\sigma_x + \sigma_y)];\\ \gamma_{xy} &= \frac{1}{G}\tau_{xy};\ \gamma_{xz} = \frac{1}{G}\tau_{xz};\ \gamma_{yz} = \frac{1}{G}\tau_{yz}.\end{aligned} \tag{7}$$

The Hookean strain energy density at a point for a linear elastic material, such as a metal, is

$$\begin{aligned}U_o &= \int_0^{\epsilon} \sigma_x d\epsilon_x + \sigma_y d\epsilon_y + \sigma_z d\epsilon_z + \tau_{xy} d\gamma_{xy} + \tau_{xz} d\gamma_{xz} + \tau_{yz} d\gamma_{yz}\\ &= \frac{1}{2}[\sigma_x\epsilon_x + \sigma_y\epsilon_y + \sigma_z\epsilon_z + \tau_{xy}\gamma_{xy} + \tau_{xz}\gamma_{xz} + \tau_{yz}\gamma_{yz}],\end{aligned} \tag{8}$$

where U_o is a function of just the strains. To compute U_o, the stresses must be written as a function of the strains. Solving (8) for the stresses produces

$$\begin{aligned}\sigma_x &= \frac{E}{1+\nu}\left[\epsilon_x + \frac{\nu}{1-2\nu}(\epsilon_x + \epsilon_y + \epsilon_z)\right];\\ \sigma_y &= \frac{E}{1+\nu}\left[\epsilon_y + \frac{\nu}{1-2\nu}(\epsilon_x + \epsilon_y + \epsilon_z)\right];\\ \sigma_z &= \frac{E}{1+\nu}\left[\epsilon_z + \frac{\nu}{1-2\nu}(\epsilon_x + \epsilon_y + \epsilon_z)\right].\end{aligned} \tag{9}$$

This Hookean model is given in terms of the elastic modulus, E, and the Poisson ratio, ν. An alternative is the equivalent Lamé expression for the stresses. The sum of the strains, $\epsilon_x+\epsilon_y+\epsilon_z$, the trace of the strain tensor, is the dilatation $e = \epsilon_x+\epsilon_y+\epsilon_z$. If one puts

$$\lambda = \left(\frac{E}{1+\nu}\right)\left(\frac{\nu}{1-2\nu}\right) = \frac{\nu E}{(1+\nu)(1-2\nu)},$$

then the stress equations can be written using $2G = E/(1+\nu)$ for the shear modulus G.

$$\begin{aligned}\sigma_x &= 2G\epsilon_x + \lambda e;\\ \sigma_y &= 2G\epsilon_y + \lambda e;\\ \sigma_z &= 2G\epsilon_z + \lambda e.\end{aligned} \tag{10}$$

The coefficients $\mu = 2G$ and λ are called the Lamé constants.

The equivalent Lamé expression for a the strain energy density of a linear elastic material is, using the Lamé equations (10) for the linear elastic constitutive relation,

$$U_o(\epsilon) = \frac{1}{2}\left[\lambda e^2 + 2G(\epsilon_x^2 + \epsilon_y^2 + \epsilon_z^2) + G(\gamma_{xy}^2 + \gamma_{xz}^2 + \gamma_{yz}^2)\right]. \tag{11}$$

The linear elastic model is hyperelastic since each stress component is a derivative of the strain energy density with respect to the corresponding strain. For example, from equation (11) and the Lamé equations,

$$\frac{\partial U_o(\epsilon)}{\partial \epsilon_x} = 2G\epsilon_x + \lambda e = \sigma_x.$$

The complementary energy density, Ψ_o, is the strain energy density written in terms of the stresses. An expression for the isotropic three-dimensional complementary energy is obtained by using Hooke's law to write each strain in terms of the stresses and the fact that $U_o = \Psi_o$.

$$\Psi_o(T) = \frac{1}{2E}(\sigma_x^2+\sigma_y^2+\sigma_z^2) - \frac{\nu}{E}(\sigma_x\sigma_y+\sigma_x\sigma_z+\sigma_y\sigma_z) + \frac{1}{2G}(\tau_{xy}^2+\tau_{xz}^2+\tau_{yz}^2). \tag{12}$$

The derivative of the complementary energy density with respect to stress yields the corresponding strain. For example, $\partial\Psi_o/\partial\sigma_x = \epsilon_x$.

A relation between load and displacement in a deformable body was obtained by Castigliano in his 1873 thesis and is also sometimes called the *principle of complementary energy.*

Theorem 5 (Castigliano's Theorem) *For a linear elastic structure, the partial derivative of the complementary energy, written in terms of the external loads, with respect to the magnitude of a point load,* **f***, is the deflection of the body at the point of application of* **f** *in the direction of the force,* **f**.

5.2 Nonlinear Elastic

Finite deformation nonlinear thermostatic behavior is usually described in terms of the stretch. Rubber is the classic case of such a hyperelastic material. The strain is usually measured by the stretch (not be confused with the Lamé constant) $\lambda = L_f/L_o = 1 + \epsilon$, where ϵ is the engineering strain. When the deformation is large, the stretch is defined in term of the displacement function $\mathbf{x} = \chi(\mathbf{X})$ where X_i are the reference Cartesian coordinates and x_i are the current Cartesian coordinates. Then $\lambda_i = \partial\chi/\partial X_i$ is the stretch in the X_i direction. The stretches may vary from point to point in the body. In the following, hyperelastic strain energy functions are described in terms of the three principal stretches, λ_i. Deformations of the form

$$\mathbf{x}(\mathbf{X}) = (\lambda_1 X_1, \lambda_2 X_2, \lambda_3 X_3), \tag{13}$$

for λ_i constant over the body are called purely homogeneous, where X_i are the reference Cartesian coordinates and x_i are the current Cartesian coordinates.

Definition 6 The deformation gradient is the differential of the deformation function at fixed time t, $D\chi(\boldsymbol{p}^0, t) \equiv F(\boldsymbol{p}^0, t)$, where $D\chi(\boldsymbol{p}^0, t) = (\partial\chi_i/\partial \boldsymbol{p}_j^0)$ and $\chi(\boldsymbol{p}^0, t) = \boldsymbol{p}$.

In principal stretch coordinates, the deformation gradient is diagonal with entries, λ_i. The left Cauchy-Green strain tensor, $B = FF^t$ is defined in the current configuration. In principal stretch coordinates, it is also diagonal with entries λ_i^2. A more refined definition of the Cauchy-Green strain tensor is given in Chapter 7.

Each tensor, T, has an associated set of three invariants obtained from the eigenvalues equation, $\det(T - \alpha I) = 0$. When this equation is expanded it yields the characteristic polynomial. For the tensor, B, the characteristic polynomial is $\alpha^3 - I_B\alpha^2 + II_B\alpha - III_B = 0$. The coefficients are independent of the coordinates and so are called the invariants of the tensor. In principal stretch coordinates,

$$\begin{aligned} I_B &= \mathrm{trace}(B) = \lambda_1^2 + \lambda_2^2 + \lambda_3^2; \\ II_B &= \lambda_1^2\lambda_2^2 + \lambda_1^2\lambda_3^2 + \lambda_2^2\lambda_3^2; \\ III_B &= \det(B) = \lambda_1^2\lambda_2^2\lambda_3^2. \end{aligned} \tag{14}$$

In the incompressible case, the determinant of B is 1 since it is a measure of the volume ratio in the current and reference configurations. Then $\lambda_1\lambda_2\lambda_3 = 1$. The incompressibility condition $\lambda_1\lambda_2\lambda_3 = 1$ implies that $II_B \equiv \lambda_1^2\lambda_2^2 + \lambda_2^2\lambda_3^2 + \lambda_1^2\lambda_3^2 = \lambda_1^{-2} + \lambda_2^{-2} + \lambda_3^{-2}$.

Various constitutive modelers of rubber-like materials have viewed the isothermal strain energy as either a function of the strain invariants or of the stretches, λ_i. The corresponding strain energy function, $\varphi(B)$, can be rewritten as a function of the strain invariants I_B, II_B, and III_B. Experimenters in the mid-twentieth century often tried to work in terms of the invariants. However, this approach is not followed here because the λ_i have thermodynamic conjugates (see Chapter 3), but the invariants do not.

A neo-Hookean incompressible material is one for which $\varphi_I = \partial\varphi/\partial I_B = c_1$, a positive constant, and $\varphi_{II} = \partial\varphi/\partial II_B = 0$ so that the strain energy density is

$$\varphi(\lambda_1, \lambda_2) = c_1(I_B - 3) = c_1(\lambda_1^2 + \lambda_2^2 + \lambda_3^2 - 3) \tag{15}$$

A Mooney-Rivlin incompressible material is one in which $\varphi_I = \partial\varphi/\partial I_B = c_1$ and $\varphi_{II} = \partial\varphi/\partial II_B = c_2$ are positive constants. The Mooney-Rivlin strain energy density for the isothermal equilibrium states is

$$\varphi(\lambda_1, \lambda_2) = c_1(\lambda_1^2 + \lambda_2^2 + \lambda_3^2 - 3) + c_2(\lambda_1^{-2} + \lambda_2^{-2} + \lambda_3^{-2} - 3) \tag{16}$$

Ogden (1972) generalized the first and second strain invariants of B to define strain invariants of the form

$$\bar{\varphi}(\alpha) = \lambda_1^\alpha + \lambda_2^\alpha + \lambda_3^\alpha - 3,$$

where α is any fixed real number. He built constitutive models for rubber from linear combinations of the $\bar{\varphi}(\alpha)$ so that the form of the strain energy density is

$$\varphi(\lambda_1, \lambda_2, \lambda_3) = \sum_{p=1}^{N} \frac{\mu_p}{\alpha_p}\bar{\varphi}(\alpha_p), \tag{17}$$

for empirical constants μ_p and α_p. The coefficients are constrained by $\sum_{p=1}^{N} \mu_p\alpha_p = 2\mu$, where μ is the shear modulus. A sufficient condition for the model to satisfy Hill's inequality for admissible constitutive models for incompressible materials is that in each summand $\mu_p\alpha_p$ be positive.

The thermostatic response of biological soft tissue has often been predicted using a hyperelastic model. An exponential strain energy density was defined by Fung. It has been applied to both arterial and lung membrane tissue by Fung and coworkers. The Fung (1990) strain energy density is

$$\varphi(E_1, E_2) = c\exp(a_1E_1^2 + a_2E_2^2 + a_3E_1E_2), \tag{18}$$

where c, a_1, a_2, a_3 are material constants and the principal Green strains are $E_i = (\lambda_i^2 - 1)/2$, for the principal stretch λ_i.

6 Stability of Classical Thermostatic Energy Functions

The classical thermostatic energy functions, such as internal energy, the Helmholtz function, and the Gibbs function, do not yield equilibrium by an energy methods criterion for extrema because they are only defined for equilibrium states. But a stability condition is still desirable. One cannot consider critical points as in energy methods, but the second derivative, the Hessian, may be required to be positive definite. As Ericksen (1991) has shown, this criterion for stability leads to the statement that the classical thermostatic energy function must be convex in order that the equilibrium be considered stable.

Definition 7 A function $g : \Re^n \rightarrow \Re$ is convex at the point $x_o \in \Re^n$ if $g(x) - g(x_o) \geq \frac{\partial g}{\partial x} ||x - x_o||$ for x near x_o.

The graph of a convex function near x_o therefore lies above the tangent plane to the graph at x_o. An alternative statement of convexity is that for $0 \leq \alpha \leq 1$, $g(\alpha x_1 + (1 - \alpha)x_2) \leq \alpha g(x_1) + (1 - \alpha)g(x_2)$.

If thermodynamics is to be given an energy methods form in order to take advantage of the mathematical machinery associated with energy methods, a generalized energy function must be defined so that the thermostatic equilibria occur at critical points of the generalized energy function for thermodynamic states.

References

J. L. Ericksen (1991). *Introduction to the Thermodynamics of Solids*, Chapman and Hall, London.

Y. C. Fung (1990). *Biomechanics: Motion, Flow, Stress, and Growth*, Springer, New York, NY.

R. W. Ogden (1972). Large Deformation Isotropic Elasticity - on the Correlation of Theory and Experiment for Incompressible Rubberlike Solids. *Proceedings of the Royal Society of London A*. **326**, 565–584.

Chapter 3
Evolution Construction for Homogeneous Thermodynamic Systems

1 Introduction

Evolution equations are constructed that predict the particular non-equilibrium process that a body follows from a given non-equilibrium state. This construction has several advantages over classical constructions for non-equilibrium processes described in Chapter 1 Most importantly, the nonlinear evolution equations constructed satisfy the second law of thermodynamics. The non-equilibrium construction generalizes that of the Gibbs geometric surface for an equilibrium thermostatic energy density function rather than assuming a linear flux evolution equation as in Linear Irreversible Thermodynamics (LIT) or rather than assuming the Clausius-Duhem inequality as in continuum thermodynamics. No local equilibrium assumption is required. Because the construction generalizes the Gibbs construction, the non-equilibrium processes lie on a simple geometric surface that is the graph of a generalized energy density function. A more complex geometric interpretation in terms of contact structures is given in Chapter 7.

The construction of the evolution equations begins with, rather than produces, a thermostatic energy density model for the long-term equilibrium behavior. The physical interpretation of the set of long-term states may differ from application to application. Often it is the equilibrium behavior. In other cases, it may represent the behavior that organizes the response. This set might be called the distinguished states in the abstract mathematical construction.

In this chapter, the behavior is assumed homogeneous, in the sense that the set of thermodynamic variables is the same at all points of the body so that no gradient of any thermodynamic variable is involved. Essentially the construction applied to a homogeneous system produces a constitutive model for time-dependent behavior, even though the construction also may be applied for arbitrary loadings.

A generalized energy density function built from the thermostatic energy density is the foundation for the fundamental construction of evolution equations for homogeneous non-equilibrium processes that is valid for solid materials. The generalized energy density function is defined on both equilibrium and non-equilibrium states which need not be near equilibrium. The equilibria, or distinguished states, are extrema of this function. In this sense, equilibria organize the material

H.W. Haslach Jr., *Maximum Dissipation Non-Equilibrium Thermodynamics and its Geometric Structure*, DOI 10.1007/978-1-4419-7765-6_3,

non-equilibrium response. This fact allows an investigation of the stability of the thermostatic states in terms of the generalized energy density function. The distinguished states are either attractors or repellers under a dynamic relaxation process, as opposed to a forced process. Non-equilibrium relaxation processes are defined in terms of new variables that are nonlinear generalizations of the affinities and fluxes used in classical linear non-equilibrium thermodynamics. A relaxation process is driven by the approach of these nonlinear affinities to zero at equilibrium. An analogous construction is applied to non-homogeneous non-equilibrium processes in which the steady states as a function of fluxes become the distinguished set and the non-equilibrium function is the generalized entropy production (Chapter 9). Therefore the chapter begins with a definition of a generalized thermodynamic function to account for both cases.

The second law of thermodynamics is incomplete in the sense that it only gives a criterion that every possible process must satisfy. An additional criterion is needed to determine which non-equilibrium process the system takes from a non-equilibrium state. The physical hypothesis assumed to supplement the second law of thermodynamics is that non-equilibrium relaxation processes proceed in state space in the direction of maximum dissipation of the generalized energy density function with respect to the affinities. The speed at which the response follows this path is determined by a thermodynamic relaxation modulus. This hypothesis produces a thermodynamically consistent system of evolution equations for the state variables, which is the constitutive model for the non-equilibrium behavior. Dissipation is measured as the rate of change of the generalized energy density function.

The evolution equations are most useful if they involve information that is relatively easy to obtain experimentally. The two criteria, a physical hypothesis based on maximum dissipation during relaxation and the organizing of the non-equilibrium response around the long-term equilibrium states, reduce the number of experiments required to determine a model. The only experiments required are those to determine the long-term thermostatic energy density function and those to determine the thermodynamic relaxation modulus. Because neither of these terms is explicitly time-dependent, the experimental analysis is much simpler than that required to determine the time-dependent kernals in constitutive models formulated in terms of integrals. All material parameters except the thermodynamic relaxation modulus appear in the long-term behavior model so that only a small number of time-independent material coefficients must be found experimentally.

2 Thermostatics

The construction is inspired by the 1873 paper of Gibbs (Gibbs, 1948) that states that the relationship between the total entropy, $\mathcal{S}$, volume, $\mathcal{V}$, and internal energy of a fluid in equilibrium can be represented by a surface in three dimensional space, the graph of the total internal energy, $\mathcal{U}$ (Chapter 1). The partial derivatives, $\partial\mathcal{U}/\partial\mathcal{S}$ and $-\partial\mathcal{U}/\partial\mathcal{V}$, are the temperature and pressure, respectively. To generalize Gibbs, the body is thought of as a subset $\mathcal{B}$ of three-dimensional space, $\Re^3$, and associated with each point are the thermodynamic variables. The physical systems to be represented are those whose distinguished thermostatic states are described by $2n$ thermodynamic variables divided into control variables $y_1, \dots, y_n$ and state variables $x_1, \dots, x_n$. The thermodynamic variables are, for example, components of the stress, strain and internal variable tensors, the components of flux vectors, as well as temperature and entropy. The distinguished states are described by a function of the controls $\hat{\varphi}(y_1, \dots, y_n)$ such that $x_i = \partial\hat{\varphi}/\partial y_i$ for $i = 1, \dots, n$. This relation defines how the $2n$ thermodynamic variables are divided into pairs of conjugate variables (y_i, x_i), one of each pair is a control and the other is a state variable. The state variable x_i is called the conjugate of y_i. For equilibrium, i.e. thermostatics, each point of the n-dimensional thermostatic surface is associated with a set of $y_1, \dots, y_n$ and the corresponding $x_i = \partial\hat{\varphi}/\partial y_i$ for $i = 1, \dots, n$. One could imagine an n-dimensional space composed of control variables attached to each point in the body. If the system is homogeneous, then the same space is associated with each point since any particular control variable is the same at all points of the body.

The classical energy density functions such as the Helmholtz, internal, and Gibbs energies, depending on the choice of control variables, define the thermostatic states, sometimes thought of as the equilibrium, quasi-static, or long-term states. These energies are assumed to be local and depend on the point in the body as well as the thermodynamic control variables, one member from each relevant conjugate pair. The equations, $x_i = \partial\hat{\varphi}/\partial y_i,\ i = 1, \dots n$, are the constitutive equations defining the thermostatic thermodynamic states at each point of the body, and the equations must be objective. For example, the long-term behavior may be hyperelastic so that the energy is a function, $\hat{\varphi}(F)$, of the deformation gradient, F (Chapter 2, Definition 6), say for a viscoelastic polymer. Or, the long-term behavior for a plastically deformed metal may be a function of internal variables, a_i, so that $\hat{\varphi}(F, a_1, \dots, a_m)$. Either may be a function of temperature.

A constitutive equation satisfies material frame indifference if it is invariant under rotations of the ambient space in which the body moves (Marsden and Hughes, 1994, p. 194). The energy function is required to satisfy objectivity with respect to isometries. An isometry is a distance preserving function mapping one metric space to another metric space. Any two observers are assumed to have chosen the same ambient coordinate space for the reference state. However at subsequent times, the view chosen by one observer may be rotated by $Q(t)$ from that chosen by the first observer. A vectorfield $\mathbf{u}$ transforms by $Q\mathbf{u}$. Material second order tensors are seen as the same by both observers. Two point second order tensors, such as the deformation gradient, F, are seen as $F' = QF$ by the second observer. Spatial second order tensors, such as the Cauchy stress σ, are seen as $\sigma' = Q\sigma Q^t$ by the second observer since the dyadic form (e.g. Malvern, 1969, p. 588) satisfies $Q\mathbf{u} \otimes Q\mathbf{v} = Q(\mathbf{u} \otimes \mathbf{v})Q^t$.

Definition 1 Suppose that the thermostatic energy density is given by a) $\hat{\varphi}(F)$, or b) $\hat{\varphi}(T)$, where T is a spatial second-order tensor. An energy function is objective if in case a) $\hat{\varphi}(F) = \hat{\varphi}(QF)$ since $F^* = QF$, in case b) $\hat{\varphi}(T) = \hat{\varphi}(QTQ^t)$, since the second observer sees $T^* = QTQ^t$ where Q is an orthogonal transformation.

Since the internal variables are not observable, objectivity for an energy density depending on internal variables requires $\hat{\varphi}(F, a_1, \ldots, a_m) = \hat{\varphi}(QF, a_1, \ldots, a_m)$.

2.1 Thermodynamic Variables

Denote the thermodynamic scalar variables as α_i, $i = 1, \ldots n_0$, the thermodynamic vector variables as $\mathbf{v}_i$, $i = 1, \ldots n_1$, and the thermodynamic second order tensor variables as T_i, $i = 1, \ldots n_2$. A thermodynamic conjugate to a thermodynamic variable, a, is a thermodynamic variable b of the same type having units that are energy density divided by the units of a. Tensor fields must be paired with tensor fields, vectorfields with vectorfields, and scalars with scalars. One of each conjugate pair of vector or tensor fields is covariant and the other is contravariant. A classical energy density function, $\hat{\varphi}$, defines the conjugates by $\partial\hat{\varphi}/\partial a = b$. The thermodynamic variables at $\boldsymbol{p} \in \mathcal{B}$ each have a conjugate, $\hat{\alpha}_i = \partial\hat{\varphi}/\partial\alpha_i$, $\hat{\mathbf{v}}_i = \partial\hat{\varphi}/\partial\mathbf{v}_i$, $\hat{T}_i = \partial\hat{\varphi}/\partial T_i$.

Conjugate pairs of scalars include the temperature, θ, and the specific entropy η. Possible conjugate pairs of strain and stress tensors are F and first Piola stress, P; the material tensors the Green strain, E, and second Piola stress, S; and the spatial tensors the Euler-Almansi strain e and the Cauchy stress, σ. Fracture growth may be modeled by the conjugate pair of scalar thermodynamic variables, the crack length and the crack driving force (or the energy release rate) (Haslach, 2009). Conjugate vectors include the classical electromagnetic fields. E is the electrical field in volt per meter; H is the magnetic field strength; D is the electric displacement field; and B measures the magnetic field. The two conjugate pairs of thermodynamic variables are E and D as well as B and H. For example, the enthalpy, ψ_e, satisfies $D = -\partial\psi_e/\partial E$ and $B = -\partial\psi_e/\partial H$ (Fabrizio and Morro, 2003, p. 252). Internal conjugate pairs of thermodynamic variables may also be defined. For example, the back stress and back strain may be defined to account for time-dependent and plastic behavior in metals (Chapter 5).

2.2 Construction of Thermostatic Energy Density Functions

Traditionally the relationship between the state and control variables at equilibrium has been given in terms of a set of equations of state. For example, the ideal gas law and the van der Waals relation are time-independent equations of state for fluids, and Hooke's law is an equation of state for a metal. If the Maxwell type relations hold, then the equations of state can be integrated to obtain a thermostatic energy density function. Recall that the choice of which of a pair of conjugate variables is the state and which is the control variable is up to the user. To define the generalized thermodynamic functions, the Legendre transform of the equilibrium energy density function, $\hat{\varphi}(y_1, \ldots, y_n)$, to depend on the state variables is more useful in the generalized energy density construction than the equilibrium energy density function in terms of the control variables.

Assume that equations of state for the control variables are given in terms of the state variables. The thermostatic energy density function, $\varphi : \Re^n \to \Re$, with domain the state variables, $\mathbf{x} = (x_1, \ldots, x_n)$, can be constructed from the set of equations of state defining equilibrium behavior. Suppose that equations of state are obtained from either theory or experiment:

$$y_i = g_i(x_1, \ldots, x_n) \qquad i = 1, \ldots, n,$$

with the Maxwell relations,

$$\frac{\partial g_i}{\partial x_j} = \frac{\partial g_j}{\partial x_i},$$

for all i and j. A function, $\varphi(x_1, \ldots, x_n)$, is obtained, up to a constant, by integrating the first order partial differential equations,

$$\frac{\partial \varphi}{\partial x_i} = -g_i(x_1, \ldots, x_n), \qquad i = 1, \ldots, n.$$

$\varphi(x_1, \ldots, x_n)$ is a Legendre transformation of the thermostatic energy density function of the controls, $\hat{\varphi}(y_1, \ldots, y_n)$.

3 Generalized Thermodynamic Functions

The conjugate relations, $x_i = \partial\hat{\varphi}/\partial y_i$, for $i = 1, \ldots, n$ define a distinguished subset, M_e, of $\Re^{2n}$ at each point of the system. The subset, M_e, is a submanifold (Chapter 7). For example, if $\hat{\varphi}$ is the hyperelastic strain energy density of a solid, then M_e is the set of possible equilibrium states. To model time-dependent non-equilibrium behavior, a generalized function, φ^*, of all thermodynamic variables is defined on all states rather than just those in M_e such that M_e is the manifold of zero gradient states of φ^* with respect to the state variables. The construction of the generalized function rests on viewing the derivatives of $\hat{\varphi}$ as variables so that the set of thermodynamic variables, $y_1, \ldots, y_n$ and $x_1, \ldots, x_n$ is the domain of the generalized thermodynamic function associated with each point of the body. The state variables are now explicit and are allowed to take values unequal to $\partial\hat{\varphi}/\partial y_i$ for a given set of controls, $y_1, \ldots, y_n$. Note that the generalized function achieves the same goal as Duhem's ballistic energy discussed in Chapter 1.

The body is viewed, as usual, as a subset of three dimensional Cartesian space, $\mathcal{B} \subset \Re^3$. Let $\varphi(x_1, \ldots, x_n)$ be the Legendre transform of $\hat{\varphi}(y_1, \ldots, y_n)$ such that $y_i = -\partial\varphi/\partial x_i$.

Definition 2 A local generalized thermodynamic function defining a thermodynamic system is a smooth function, for $(p_1, p_2, p_3) \in \mathcal{B} \subset \Re^3$, $x_1, \ldots, x_n$ the state and $y_1, \ldots, y_n$ the control variables,

$$\varphi^*(p_1, p_2, p_3, x_1, \ldots, x_n; y_1, \ldots, y_n) : \Re^3 \times \Re^{2n} \to \Re$$

such that at each $(p_1, p_2, p_3) \in \mathcal{B}$

$$\frac{\partial \varphi^*}{\partial y_i} = x_i \qquad \text{for} \qquad i = 1, \ldots, n, \tag{1}$$

and the distinguished submanifold, M_e, is described locally by

$$\frac{\partial \varphi^*}{\partial x_i} = 0 \qquad \text{for} \qquad i = 1, \ldots, n. \tag{2}$$

Also, the restriction of φ^* to M_e is the function, $\hat{\varphi}$, of the control variables up to a sign.

A distinguished manifold M_e is associated with each $\boldsymbol{p} = (p_1, p_2, p_3) \in \mathcal{B}$, but need not be the same at all points of $\mathcal{B}$ unless the thermodynamic system is homogeneous.

The simplest form of the generalized function at $\boldsymbol{p} \in \mathcal{B}$ is, in terms of tensors,

$$\varphi^* = \varphi(\boldsymbol{p}, \alpha_1, \ldots, \alpha_{n_0}, \mathbf{v}_1, \ldots, \mathbf{v}_{n_1}, T_1, \ldots, T_{n_2}) + \sum_{i=1}^{n_0} \alpha_i \hat{\alpha}_i + \sum_{i=1}^{n_1} \mathbf{v}_i \cdot \hat{\mathbf{v}}_i + \sum_{i=1}^{n_2} T_i : \hat{T}_i, \tag{3}$$

where the hat indicates the corresponding conjugate variable, the product of vectors is the scalar product and the product of tensors is the double dot, or contraction, scalar product. In terms of the components,

$$\varphi^*(x_1, \ldots, x_n; y_1, \ldots, y_n) = \varphi(x_1, \ldots, x_n) + \sum_{i=1}^{n} x_i y_i. \tag{4}$$

The local generalized thermodynamic function is always assumed to have the form of Eq. (4).

Proposition 3 *The generalized function (3) is objective if φ is objective.*

Proof The sum of objective functions is objective. The scalar product of vectors is objective. The tensor contraction, or double dot, requires one tensor to be covariant and one to be contravariant (See Chapter 7). The contraction of second order tensors is objective.

Example The generalized thermodynamic function, $\varphi^*(x; y) = x^3 + yx$, has y as control and x as state variable. The equilibrium manifold, M_e, is defined by the equation of state, $x = (-y/3)^{1/2}$, for y negative, from $\partial\varphi^*/\partial x = 0$.

The thermostatic function $\varphi(y)$ is obtained by inserting the equation of state into φ^* to obtain $\varphi(y) = (2y/3\sqrt{3})(-y)^{1/2}$. The same result can be obtained by integrating the equation of state, $d\hat{\varphi}/dy = x = (-y/3)^{1/2}$. The Legendre transform $\varphi(x) = x^3$ is obtained from $y = -d\varphi/dx = 3x^2$ in agreement with $\varphi^*(x; y) = \varphi(x) + xy$.

3.1 Stability in the Distinguished Manifold

Gibbs identified the stable and unstable equilibrium states in terms of the curvature of the thermostatic surface, the graph of $\mathcal{U}$. More recently, if φ in Eq. (4) is an energy density for the thermostatic states, it is required to be convex (Ogden, 1984, p. 523), in the sense that its graph always lies above its tangent hyperplanes, in order that a thermostatic state be stable.

Definition 4 A real valued function, ψ, on a vector space is strictly convex if for all $\mathbf{u}$ and $\mathbf{v}$ in the domain, it satisfies the inequality

$$(\psi(\mathbf{v}) - \psi(\mathbf{u})) - (\mathbf{v} - \mathbf{u}) \cdot \frac{\partial \psi}{\partial \mathbf{u}}(\mathbf{u}) > 0. \tag{5}$$

If the thermostatic energy function is convex, the thermostatic manifold is a set of minima for the generalized energy.

Proposition 5 *The generalized thermodynamic function* $\varphi^*(x_1, \ldots, x_n; y_1, \ldots, y_n) = \varphi(x_1, \ldots, x_n) + \sum x_i y_i$ *has zero-gradient manifold,* M_e*, composed of minima or singularities of* φ^* *iff the function* φ *is convex.*

Proof If φ is strictly convex, then the Hessian of φ is positive semi-definite (Bernstein and Toupin, 1962). But the Hessian of φ^* is the same as the Hessian of φ so that the Hessian of φ^* is positive semi-definite. Therefore the elements of the zero-gradient manifold for φ^* are either minima or singularities. Further the singularities are nowhere dense.

Conversely, assume that $(\mathbf{u}, \mathbf{y})$ is a minimum for φ^* for fixed control variables $\mathbf{y}$. Then for nearby $(\boldsymbol{v}, \boldsymbol{y})$, the inequality $0 < \varphi^*(\boldsymbol{v}, \boldsymbol{y}) - \varphi^*(\boldsymbol{u}, \boldsymbol{y}) = \varphi(\boldsymbol{v}) - \varphi(\boldsymbol{u}) - \mathbf{y} \cdot (\boldsymbol{v} - \boldsymbol{u})$ holds. But $\mathbf{y} = \partial\varphi/\partial\mathbf{u}$ if $\partial\varphi^*/\partial\mathbf{u} = 0$ by (2) and (4) so that φ is strictly convex by (5).

Geometric singularities occur at points where the Hessian, $(\partial^2\varphi^*/\partial x_i \partial x_j)$, is singular. For example, a phase transition may occur at such a point. An advantage of working in terms of the generalized thermodynamic function is that the local behavior near isolated degenerate singularities may be investigated using catastrophe theory (see Chapter 8). The distinguished manifold M_e may act as an attractor or repeller for the dynamical non-equilibrium processes, depending on whether M_e is composed of minima or maxima.

3.2 Examples of Generalized Thermodynamic Functions

The generalized thermodynamic function need not be an energy density function.

Example 1 Uniaxial Long-term Linear Elastic: Let ϵ and σ be the small displacement uniaxial strain and stress respectively, and let E_1 be the elastic modulus to distinguish it from the Green strain tensor. In strain control, $\varphi(\sigma) = \sigma^2/2E_1$, and $\varphi^*(\sigma;\epsilon) = \sigma^2/2E_1 - \sigma\epsilon$ yields the distinguished states defined by $\epsilon = \sigma/E_1$. In stress control, $\varphi(\epsilon) = E_1\epsilon^2/2$ and $\varphi^*(\epsilon;\sigma) = E_1\epsilon^2/2 - \sigma\epsilon$. In the first case, the evolution equations to be developed for relaxation processes represent stress relaxation and in the second case, creep.

Example 2 Deformation of a Biaxially Loaded Rubber Sheet: The equilibrium response of an in-plane biaxially loaded hyperelastic sheet is thought of as the distinguished states. Let the in-plane stretches λ_1 and λ_2 be the state variables and the in-plane principal stresses per reference area T_i, be the control variables. The potential energy function for a strain energy density $\varphi(\lambda_1, \lambda_2)$, since $T_3 = 0$, is also of the form of a generalized function.

$$\varphi^*(\lambda_1, \lambda_2; T_1, T_2) = \varphi(\lambda_1, \lambda_2) - T_1\lambda_1 - T_2\lambda_2. \tag{6}$$

The equilibrium stresses are calculated from $\partial\varphi^*/\partial\lambda_1 = 0$ and $\partial\varphi^*/\partial\lambda_2 = 0$. The thermostatic manifold M_e is defined by $\partial\varphi/\partial\lambda_1 = T_1$ and $\partial\varphi/\partial\lambda_2 = T_2$. An analysis of the role of bifurcations for various choices of $\varphi(\lambda_1, \lambda_2)$ is given in (Haslach, 2000).

Example 3 Viscoplasticity: The long-term response, rather than the equilibrium response, is the distinguished manifold, M_e. The simplest model is a quadratic energy density (Haslach, 2002).

$$\varphi^*(\epsilon, \boldsymbol{B}; \sigma, \boldsymbol{b}) = \frac{1}{2}E_1(\epsilon^e)^2 + \frac{1}{2}H\boldsymbol{B}^2 - \sigma\epsilon - \boldsymbol{B}\boldsymbol{b},$$

where $\boldsymbol{B}$ and $\boldsymbol{b}$ are the back stress and back strain respectively, H is a constant, ϵ^e is the elastic strain component, E_1 is the elastic modulus, and ϵ and $\boldsymbol{B}$ are the state variables. In the long-term manifold, the back stress and back strain are related by $\boldsymbol{b} \equiv HL$, where L is the value of the back stress on the long-term manifold. This is the value to which the back stress tends over time.

Example 4 The Fourier Law: The Fourier relation, $\boldsymbol{q} = -K\nabla\theta$, is stationary. Let the gradient, $\nabla(1/\theta)$, be the control variable, and the heat flux vector, $\boldsymbol{q}$, be the state variable. The local generalized entropy production function is

$$\Psi^*(\boldsymbol{q}; \nabla(1/\theta)) = \left(\frac{1}{2K\theta^2}\right)\boldsymbol{q}\cdot\boldsymbol{q} - \boldsymbol{q}\cdot\nabla\left(\frac{1}{\theta}\right). \tag{7}$$

The Fourier relation is recovered from the zero gradient condition on Ψ^*.

Example 5 Griffith-Irwin Fracture: Classical fracture theory can be presented in terms of generalized thermodynamic functions (see Chapter 11). The crack length, l, and the crack driving force at the crack tip, G, are the state and control thermodynamic variables respectively. Griffith described his quasi-static fracture theory in terms of the equilibria of a potential energy function, Π, which may be written as

$$\Pi = 4wl\gamma - \Delta U = 4wl\gamma - \frac{\pi l^2 \sigma^2 w}{E'}. \tag{8}$$

Here γ is the surface energy per unit area, w is the width of the crack face, σ is the applied stress, and in plane strain $E' = E_1/(1-\nu^2)$, where E_1 is the elastic modulus and ν is the Poisson ratio. The equilibria of Π obtained by setting $d\Pi/dl = 0$ determine the states for incipient crack propagation. This potential function is related to the generalized thermodynamic function, φ^*, by letting $G = 2\gamma$ (Haslach, 2010).

$$\varphi^*(l; G) = -\frac{(1-\nu^2)}{2E_1}\sigma^2 \pi l^2 + Gl, \tag{9}$$

where $w = 1$. This construction also represents the Irwin theory because $K_I^2 = E'G$. The distinguished manifold is the set of pairs (G, l) at which crack propagation is impending, where G is a function of other variables including l. The distinguished manifold composed of minima of φ^* is a dynamic attractor, in which case the crack growth is called stable. Unstable crack propagation is repelled from a distinguished manifold composed of maxima.

4 Evolution Equations for Non-equilibrium Processes in a Thermodynamic System Defined by a Generalized Function

The non-equilibrium states are represented by points in $\Re^{2n+1}$ which lie on the $2n$-dimensional graph of φ^* off the distinguished submanifold, M_e. A process is a time-dependent path on the graph of φ^*, $\gamma : \Re \to \Re^{2n+1}$ which is given in local coordinates by $\gamma(t) = \big(\boldsymbol{x}(t), \boldsymbol{y}(t), \varphi^*[\boldsymbol{x}(t), \boldsymbol{y}(t)]\big)$. This viewpoint generalizes the Gibbs analysis in terms of the graph of the thermostatic energy function and improves the Gibbs energy surface since it explicitly involves the equations of state, $x_i = \partial\hat{\varphi}/\partial y_i$, $i = 1, \ldots, n$, to define the submanifold of the graph of φ^* corresponding to M_e. Processes whose paths on the graph of φ^* are not subsets of this submanifold are non-equilibrium processes in this construction (Fig. 3.1).

A constitutive model for the non-equilibrium response of a system to the variation of controls $y_i(t)$ for $i = 1, \ldots, n$ is usually given by a system of evolution equations for $i = 1, \ldots, n$,

$$\frac{dx_i}{dt} = f(\boldsymbol{x}, \boldsymbol{y}). \tag{10}$$

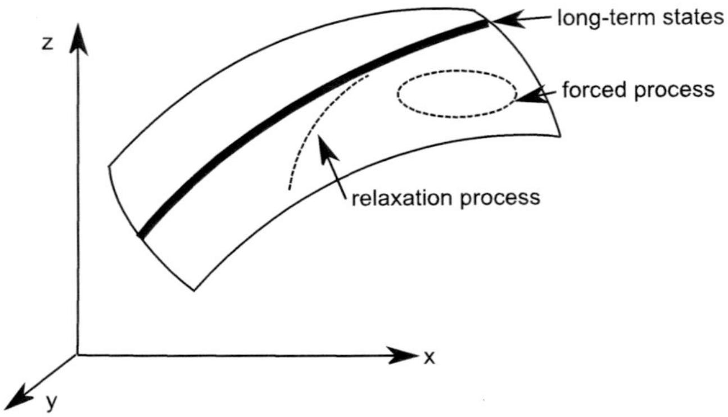

Fig. 3.1 Non-equilibrium processes on the graph of the generalized energy function

Such a system is just a nonlinear dynamical control system with control parameters, $\boldsymbol{y}$. The system is autonomous if the controls are fixed, but the control variables may be functions of time. In an engineering problem, the goal may be to adjust the controls to produce a fixed desired result such as stabilizing the system.

4.1 Affinities

In classical non-equilibrium thermodynamics, processes are driven by affinities, often called thermodynamic forces. Affinities act as generalized forces driving the system to the thermostatic states. The analogous affinities in this construction are defined from the generalized thermodynamic function, φ^*, and correspond to the scalar α_i, vector $\mathbf{v}_i$, and second order tensor T_i state variables.

$$\mathcal{A}_i = \frac{\partial \varphi^*}{\partial \alpha_i}; \qquad \mathcal{V}_i = \frac{\partial \varphi^*}{\partial \mathbf{v}_i}; \qquad \mathcal{T}_i = \frac{\partial \varphi^*}{\partial T_i}. \tag{11}$$

Recall that the local coordinates are an ordered list of the components of each tensor. In local coordinates,

Definition 6 For the generalized function, $\varphi^* = \varphi(\boldsymbol{x}) + \sum_i x_i y_i$, the affinities in local coordinates are

$$X_i = \frac{\partial \varphi^*}{\partial x_i} = \frac{\partial \varphi}{\partial x_i} + y_i, \qquad \text{for} \qquad i = 1, \ldots, n, \tag{12}$$

where x_i are the state variables.

In this point of view, the affinities arise naturally from the generalized thermodynamic function, and indicate how far the system is from thermostatic states. At a thermostatic state, each X_i is zero. The fluxes, $J_i = dX_i/dt, i = 1, \ldots, n$, then describe the evolution of the system towards the thermostatic distinguished states, if the controls are fixed, by a system of first order ordinary differential equations for dX_i/dt.

The generalized function, φ^*, can be rewritten as a function of the affinities and the control variables by a coordinate transformation in the domain, $\Re^{2n}$, in any neighborhood where the Hessian with respect to the state variables, $(H_{ij}) = (\partial^2\varphi^*/\partial x_i \partial x_j)$, is non-singular. Define a map $h : \Re^n \times \Re^n \to \Re^n \times \Re^n$ while holding the control variables fixed by

$$h(\boldsymbol{y}, \boldsymbol{x}) = \left(\boldsymbol{y}, \left.\frac{\partial \varphi^*}{\partial x_1}\right|_{(\boldsymbol{y},\boldsymbol{x})}, \ldots, \left.\frac{\partial \varphi^*}{\partial x_n}\right|_{(\boldsymbol{y},\boldsymbol{x})} \right),$$

and put $X_i = (\partial\varphi^*/\partial x_i)|_{(\boldsymbol{y},\boldsymbol{x})}$, $i = 1, \ldots, n$. The map h is a diffeomorphism since (H_{ij}) is non-singular.

The function $\varphi^* : \Re^{2n} \to \Re$ is expressed locally in terms of the new coordinates $(y_1, \ldots, y_n, X_1, \ldots, X_n)$ as a function $\varphi'^* : \Re^n \times \Re^n \to \Re$ so that $\varphi^* = \varphi'^* \cdot h$ (Fig. 3.2). By Definition (2), the affinities are zero on M_e. Further, $\partial\varphi'^*/\partial X_i = 0$, when $\mathbf{X} = 0$, by the chain rule and the non-singularity of the Hessian of φ^*.

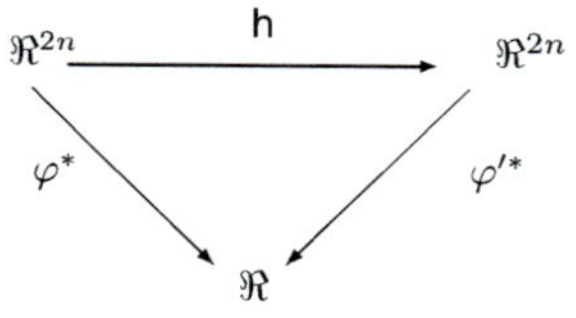

Fig. 3.2 Transformation to affinities

In the case of large displacements, the deformation gradient induces relations between the tensors in the reference and current configurations. The push-forward of a tensor in the reference configuration is an associated tensor in the state configuration. Conversely, the pull-back of a tensor in the current configuration is an associated tensor in the reference configuration. To prepare for the definition of the evolution equations below, consider how the affinities pull back under the deformation $\chi : \mathcal{B}_o \times \Re \to \mathcal{B} \subset \Re^3$ taking the reference state, $\mathcal{B}_o$, to the state, $\mathcal{B}$, at time t in ambient space. The covariant and contravariant tensors are pulled back in different ways (e.g. Holzapfel, 2000, p. 83) because they have different domains and ranges (see Chapter 7). If the second order tensor T is covariant, then its associated affinity, $\partial\varphi^*/\partial T$, is contravariant, and vice versa. Denote the deformation gradient by F and its transpose by F^t. See Chapter 7 for the differential topology interpretation of these tensors.

The function φ is defined per unit volume. The thermodynamic function must be a function either of reference state thermodynamic variables only or of current state thermodynamic variables only; variables from the reference configuration and the current configuration cannot be mixed in the domain of φ. Therefore the thermodynamic function in the reference configuration φ_r and that in the current configuration

φ_c are related by $\varphi_r = J\varphi_c$, where $J = \det(F)$ is the ratio of the current to the reference volume.

Proposition 7 *Let T be a second order spatial tensor and let the tensor $X = \partial\varphi_c^*/\partial T$ be the associated affinity. Let Z be the pull-back of X to the reference space. Let Y be the pull-back of $\partial\varphi_c'^*/\partial X$, then $Y = J^{-1}\partial\varphi_r'^*/\partial Z$.*

Proof The pull-back of a contravariant affinity X is $Z = F^{-1}XF^{-t}$. A chain rule computation yields

$$\frac{\partial\varphi_r'^*}{\partial Z} = \frac{\partial\varphi_r'^*}{\partial(F^{-1}XF^{-t})} = JF^t\frac{\partial\varphi_c'^*}{\partial X}F. \tag{13}$$

The result for covariant affinities is obtained by replacing F by F^{-t} in the above equations.

$$\frac{\partial\varphi_r'^*}{\partial Z} = \frac{\partial\varphi_r'^*}{\partial(F^tXF)} = JF^{-1}\frac{\partial\varphi_c'^*}{\partial X}F^{-t}. \tag{14}$$

Example The pull-back relation in Proposition (7) also holds in a generalized function, φ_c^*, that is a function of a spatial second order tensor. Let E be the Green strain tensor and e be the Euler-Almansi strain which is the push-forward of E under the deformation diffeomorphism χ so that $e = F^{-t}EF^{-1}$, since the tensors are contravariant. Then the relationship between the Cauchy stress, σ, and the second Piola stress, S, may be obtained as follows using the strain energies for which σ and e are conjugates in the current configuration and S and E are the corresponding conjugates in the reference configuration. Since $\varphi_r = J\varphi_c$,

$$\sigma = \frac{\partial\varphi_c}{\partial e} = \frac{\partial\varphi_c}{\partial(F^{-t}EF^{-1})} = J^{-1}F\frac{\partial\varphi_r}{\partial E}F^t = J^{-1}FSF^t. \tag{15}$$

Criteria for the objectivity of the affinities are needed to verify that the evolution equations defined below are objective. Let Q be a rotation in either the reference or spatial configuration.

Proposition 8 *Suppose that φ is either a long-term specific energy or a generalized thermodynamic function. (a) If $\varphi(F)$ is objective in the sense that $\varphi(QF) = \varphi(F)$, then $P = \partial\varphi/\partial F$ is objective. (b) If T is an objective spatial tensor and if $\varphi(T)$ is objective in the sense that $\varphi(Q^tTQ) = \varphi(T)$, then $\partial\varphi/\partial T$ is objective.*

Proof a) A simple calculation using the chain rule shows that, for $*$ indicating here a second observer,

$$P = \frac{\partial\varphi(F)}{\partial F} = \frac{\partial\varphi(QF)}{\partial F} = Q^tP^*.$$

Therefore $QP = P^*$ as required. An extended proof of case a) is given by Šilhavý (1997, 8.3.2).

b) A straightforward computation shows that

$$\frac{\partial \varphi}{\partial T} = Q^t \frac{\partial \varphi}{\partial T^*} Q.$$

Corollary 9 *If T is objective and if φ is objective then the affinity X associated to T is also objective.*

4.2 Objective Rates

The objectivity of a time-dependent thermodynamic process becomes a question when the thermodynamic variables are defined with respect to the current configuration because many spatial time derivatives are not objective. There is no problem when the rate is given by the material time derivative in the reference configuration.

The Jaumann rate is the simplest of the objective rates, but it can lead to physically unrealistic predictions (e.g. Simo and Phister, 1984; Haupt, 2002). The Oldroyd rate (Oldroyd, 1950) has the advantage that, since it is the Lie derivative in the direction of the velocity, it pulls back under χ to the material derivative in the reference space, where calculus can be more easily performed. This Lie derivative is called the Lie time derivative by Holzapfel (2000). Let f be a spatial tensor field of order one, two or three. Then the Lie derivative of f is the push-forward of the directional derivative of the pull-back of f to the reference configuration in the direction of the spatial velocity, $\mathbf{v}$, denoted $\mathcal{L}_{\mathbf{v}}(f) = \chi_*(\mathrm{D}_{\mathbf{v}}\chi_*^{-1}(f))$ (Holzapfel, 2000, p. 106), where $\mathrm{D}_{\mathbf{v}}$ denotes the material time derivative. Here the push-forward under χ is denoted χ_* and equals F and the pull-back χ_*^{-1}. (The function χ_* is between tangent bundles; see Chapter 7). The Lie time derivative of a scalar spatial field is the material time derivative.

4.3 Gradient Relaxation Processes

In classical non-equilibrium thermodynamics the second law does not identify the precise non-equilibrium response; it only determines a class of allowable responses. To define uniquely which of the non-equilibrium processes allowed under the second law is actually followed, an additional condition to supplement the second law must be assumed. A unique non-equilibrium response is selected in the construction here by imposing a maximum dissipation criterion on the generalized energy written as a function of the affinities, which forces the process to move in the direction of the gradient of φ' with respect to the affinities. The relaxation path in $\Re^{2n}$ induces a non-equilibrium relaxation process on the graph of φ^* in the local coordinates $\Re^{2n+1}$.

For example, a system originally in equilibrium has its control variables suddenly perturbed, so that the substance has the old state variables, but new control variables. A typical problem is to predict how this non-equilibrium state relaxes to a new equilibrium state if the control variables are held fixed. The states through which the system passes as it approaches equilibrium form a path, $\gamma(t)$, called the relaxation path. The fundamental non-equilibrium process that defines the class of constitutive relations considered here is called a gradient non-equilibrium process. The relaxation path $\gamma : \Re \rightarrow \Re^{2n}$ having coordinates $(\boldsymbol{y}, X_1(t), \ldots, X_n(t))$ is a gradient non-equilibrium process if the tangent to the relaxation path is proportional to the negative of the gradient of $\varphi'^*(\boldsymbol{y}, \boldsymbol{X})$. A gradient process in spatial coordinates is required to pull back to a gradient dynamical system in the reference configuration.

Definition 10 A thermodynamically homogeneous gradient relaxation process is locally defined by the system of equations, for spatial affinities $X_i,\ i = 1, \ldots, n$, that are tensors,

$$\mathcal{L}_{\mathbf{v}} X_i = -k_i \frac{\partial \varphi'^*}{\partial X_i}. \tag{16}$$

If the affinity, X_i, is in the reference configuration, then the associated equation is

$$\frac{DX}{Dt} = -k_i \frac{\partial \varphi'^*}{\partial X_i}. \tag{17}$$

All material parameters defining a gradient relaxation process except the relaxation modulus, k, appear in the thermostatic function, φ. This fact simplifies experiments required to determine the evolution equations in applications.

Lemma 11 *A thermodynamically homogeneous gradient relaxation process is objective.*

Proof The process is objective because the Lie time derivative is objective and because the derivative on the right is objective by Corollary (9). The tensor k is required to be objective by definition.

This description of non-equilibrium processes is given within a contact geometric structure in Chapter 7.

4.3.1 The Thermodynamic Relaxation Modulus

The speed of the gradient process is adjusted by a thermodynamic relaxation modulus k associated with each affinity; k is an objective, positive definite tensor, usually diagonal. The smaller each component, k_i, of k, the slower the relaxation. Note that the tensor k may be viewed as a metric tensor or as the inverse of a metric tensor. In principle, it need not be diagonal.

The components of the relaxation modulus, k_i, depend on the particular choice of the control variables. The relaxation modulus, k, contains information on the micro- or nano-structure of the material that determines the speed of relaxation in thermodynamic space. For example in the viscoplastic metal model (Haslach, 2002), $k = ||\dot{\epsilon}^p||$, the norm of the plastic strain rate that depends on variations in the dislocation structure in the metal. The plastic strain rate is obtained from a potential whose existence is proved by Rice (1971). The relaxation modulus is constant for viscoelasticity (Haslach and Zeng, 1999) when the long-term manifold is represented by any of the classical hyperelastic models for large or small deformations or by any of the hyperelastic models proposed for soft biological tissue (Haslach, 2005). Multi-scale modeling can be captured in the thermodynamic relaxation modulus. In Chapter 6, the thermodynamic modulus k is shown to depend on the molecular bond energy in a model of hydrated biopolymer behavior.

The magnitude of the relaxation modulus determines whether the response is instantaneous (elasticity), or takes time (viscoelastic). The interpretation is seen by considering a system with just one pair of conjugate thermodynamic variables, x and y, and in the associated version of (24) letting $k \to \infty$ so that

$$\frac{1}{k}\frac{dx}{dt} = \left(\frac{\partial^2\varphi}{\partial x^2}\right)^{-2}\left(-y(t) + \frac{\partial\varphi}{\partial x}\right) \to 0. \tag{18}$$

In this case, $\partial\varphi/\partial x = y$ so that the response lies on the long-term thermostatic manifold. Quasi-static means that the material relaxes immediately to an equilibrium state. In contrast, the smaller the value of $k > 0$, the more viscous and the slower is the uniaxial response.

Example Newton's Law of Cooling can be obtained as a gradient relaxation process in which the entropy is held constant. A rigid rod is placed in a heat bath. The temperature of the heat bath is θ_B and is unequal to the temperature, θ, of the rod. Therefore the rod is initially in a non-equilibrium state and relaxes to equilibrium at θ_B. Let θ, the temperature, be the state variable and the entropy, η, be the control variable. The rod is homogeneous so that the equilibrium Helmholtz energy is

$$\varphi(\theta) = -0.5\frac{c_0}{\theta_0}(\theta - \theta_0)^2,$$

where c_0 is the heat capacity at the reference temperature, θ_0. The generalized energy function for this system is, by Eq. (4),

$$\varphi^*(\theta, \eta) = 0.5\frac{c_0}{\theta_0}(\theta - \theta_0)^2 - \theta\eta. \tag{19}$$

At equilibrium, since $\partial\varphi^*/\partial\theta = 0$, the entropy, $\eta = c_0(\theta - \theta_0)/\theta_0$, is an increasing function of temperature, as expected. Also the equilibria are stable.

Imagine the entropy perturbed to its equilibrium value, $\eta_B = c_0(\theta_B - \theta_0)/\theta_0$, at θ_B and held constant. The resulting gradient relaxation process describes the relaxation to equilibrium. The affinity is

$$X = \frac{\partial\varphi^*}{\partial\theta} = \frac{c_0}{\theta_0}(\theta - \theta_0) - \eta.$$

The generalized energy under the coordinate change to X and η is

$$\varphi'^*(X,\eta) = 0.5\frac{\theta_0}{c_0}(X+\eta)^2 - \eta\left[\frac{\theta_0}{c_0}(X+\eta)+\theta_0\right], \tag{20}$$

with gradient $\partial\varphi'^*/\partial X = (\theta_0/c_0)X$. The gradient relaxation process for fixed η_B is defined by

$$\frac{dX}{dt} = -k\frac{\partial\varphi'^*}{\partial X} = -k\frac{\theta_0}{c_0}X.$$

If this is now rewritten in terms of the original state variable θ, using the fact that $dX/dt = (c_0/\theta_0)(d\theta/dt)$, and if the entropy is fixed at η_B, the process becomes,

$$\frac{d\theta}{dt} = \alpha(\theta_B - \theta). \tag{21}$$

where $\alpha = k\theta_0/c_0$ is a positive constant. This is exactly Newton's Law of Cooling. Therefore

$$\theta(t) = \theta_B - (\theta_B - \theta_0)\exp(-t/\alpha).$$

The rod relaxes towards, but never reaches, the equilibrium temperature as time tends to infinity.

4.3.2 Gradient Relaxation Processes in Terms of the State Variables

The system of evolution equations Eq. (17) is given in terms of affinities, X_i. However, in many practical nonlinear problems, Eq. (12) is not easily, or cannot be, inverted to write φ^* as a function of the affinities, X_i, and the control variables, y_i, in closed form. It is desirable to simplify computations by transforming the system of evolution equations into terms of the original state variables, x_i, and the control variables, y_i.

The $n \times n$ matrix $H = [\partial^2\varphi/\partial x_i \partial x_j]$, the Hessian of φ, is symmetric and, by assumption, is invertible at all points of the process. Also, because of the form of the generalized energy, $H = [\partial X_i/\partial x_j]$. Its inverse is $H^{-1} = [\partial x_i/\partial X_j]$ or by symmetry $H^{-1} = [\partial x_j/\partial X_i]$.

Since X_i is a function of x_i for fixed y_i, taking the derivatives of $X_i = X_i(x_1,\ldots,x_n)$ with respect to time, t, produces

$$\begin{bmatrix} \frac{dX_1}{dt} \\ \frac{dX_2}{dt} \\ \cdots \\ \frac{dX_n}{dt} \end{bmatrix} = \begin{bmatrix} \frac{\partial X_1}{\partial x_1} & \frac{\partial X_1}{\partial x_2} & \cdots & \frac{\partial X_1}{\partial x_n} \\ \frac{\partial X_2}{\partial x_1} & \frac{\partial X_2}{\partial x_2} & \cdots & \frac{\partial X_2}{\partial x_n} \\ \cdots & \cdots & \cdots & \cdots \\ \frac{\partial X_n}{\partial x_1} & \frac{\partial X_n}{\partial x_2} & \cdots & \frac{\partial X_n}{\partial x_n} \end{bmatrix} \begin{bmatrix} \frac{dx_1}{dt} \\ \frac{dx_2}{dt} \\ \cdots \\ \frac{dx_n}{dt} \end{bmatrix}. \tag{22}$$

By the chain rule applied to φ'^* and the definition, Eq. (12) of the affinity X_i,

$$\begin{bmatrix} \frac{\partial \varphi'^*}{\partial X_1} \\ \frac{\partial \varphi'^*}{\partial X_2} \\ \cdots \\ \frac{\partial \varphi'^*}{\partial X_n} \end{bmatrix} = \begin{bmatrix} \frac{\partial x_1}{\partial X_1} & \frac{\partial x_2}{\partial X_1} & \cdots & \frac{\partial x_n}{\partial X_1} \\ \frac{\partial x_1}{\partial X_2} & \frac{\partial x_2}{\partial X_2} & \cdots & \frac{\partial x_n}{\partial X_2} \\ & \cdots\cdots\cdots & & \\ \frac{\partial x_1}{\partial X_n} & \frac{\partial x_2}{\partial X_n} & \cdots & \frac{\partial x_n}{\partial X_n} \end{bmatrix} \begin{bmatrix} X_1 \\ X_2 \\ \cdots \\ X_n \end{bmatrix}. \tag{23}$$

Since the matrix H is symmetric, its inverse is symmetric also. This means that the matrix on the right is also the inverse of H.

The relaxation process, Eq. (17), as a system of first-order nonlinear ordinary differential equations in terms of the original state variables, x_i, and the control variables, y_i, is equivalently

$$\begin{bmatrix} \frac{dx_1}{dt} \\ \frac{dx_2}{dt} \\ \cdots \\ \frac{dx_n}{dt} \end{bmatrix} = -k \begin{bmatrix} \frac{\partial^2 \varphi}{\partial x_1^2} & \frac{\partial^2 \varphi}{\partial x_1 \partial x_2} & \cdots & \frac{\partial^2 \varphi}{\partial x_1 \partial x_n} \\ \frac{\partial^2 \varphi}{\partial x_2 \partial x_1} & \frac{\partial^2 \varphi}{\partial x_2^2} & \cdots & \frac{\partial^2 \varphi}{\partial x_2 \partial x_n} \\ & \cdots\cdots\cdots\cdots & & \\ \frac{\partial^2 \varphi}{\partial x_n \partial x_1} & \frac{\partial^2 \varphi}{\partial x_n \partial x_2} & \cdots & \frac{\partial^2 \varphi}{\partial x_n \partial x_n} \end{bmatrix}^{-2} \begin{bmatrix} y_1 + \frac{\partial \varphi}{\partial x_1} \\ y_2 + \frac{\partial \varphi}{\partial x_2} \\ \cdots \\ y_n + \frac{\partial \varphi}{\partial x_n} \end{bmatrix}. \tag{24}$$

This form of the evolution equations does not explicitly involve the generalized energy function.

When the derivatives of the state variables with respect to time are zero, the system is in equilibrium. Eq. (24) reduces to the set of equilibrium equations given in Eq. (2),

$$y_i = -\frac{\partial \varphi}{\partial x_i} \quad \text{for} \quad i = 1, \ldots, n \tag{25}$$

since the fact that H is invertible implies that each $X_i = 0$.

Note that the right hand column consists of the affinities, again emphasizing that the affinities drive the non-equilibrium process. When the affinities are zero, the evolution ceases, and the process has reached M_e. The relation (24) is valid under small deformations in which the spatial and reference configurations are assumed to coincide. If the deformations are large, a gradient relaxation process in terms of spatial thermodynamic variables is pulled back to the reference state and (24) is applied.

Example The development of the evolution directly from the definition gives the same evolution equation as (24). For example, if the long-term behavior is linear elastic, the uniaxial linear elastic generalized energy density function, φ^*, for elastic modulus, E_1, is

$$\varphi^*(\epsilon, \sigma) = \frac{1}{2} E_1 \epsilon^2 - \sigma \epsilon.$$

The affinity is $X = \partial\varphi^*/\partial\epsilon = E_1\epsilon - \sigma$ and the generalized energy is rewritten as

$$\varphi'^*(X, \sigma) = \frac{1}{2E_1}(X + \sigma)^2 - \frac{1}{E_1}\sigma(X + \sigma).$$

The relaxation process has flux $dX/dt = -k(\partial\varphi'^*/\partial X) = -kX/E_1$ by the maximum dissipation criterion. Since $dX/dt = E_1(d\epsilon/dt)$, the flux equation becomes the evolution equation (24)

$$\frac{d\epsilon}{dt} = -\frac{k}{E_1^2}(E_1\epsilon - \sigma), \tag{26}$$

or

$$\sigma = E_1\epsilon + \frac{E_1^2}{k}\frac{d\epsilon}{dt}, \tag{27}$$

which is in the form required to solve using an integrating factor.

4.4 Relaxation Convergence to Equilibrium

Nonlinear dynamics guarantees that each gradient relaxation process converges to a point in the long-term behavior manifold, the equilibrium point associated with the fixed control variables (Haslach, 1996).

Theorem 12 *Suppose that for a choice of local control variables, $y_1, \ldots, y_n$, φ^* has a unique stable critical state, z. Then any solution of the associated gradient flow with fixed controls, $y_1, \ldots, y_n$ converges to the critical point, z.*

Proof This result is a consequence of the following three lemmas.

On a compact phase space, an orbit of a gradient flow with isolated critical points converges to a critical point as $t \to \infty$ (Katok and Hasselblatt, 1995, p. 38). To prove this result for relaxation processes, for which the orbit lies in n-dimensional space, $\Re^n$, the compact assumption is replaced by the fact that the gradient flow for $-k\varphi^{*\prime}$ is dissipative since the trace of the Hessian of $\varphi^{*\prime}$ is positive.

Definition 13 A system $dx/dt = f(t, x)$ is dissipative in a region $[a, b] \times R$ of $\Re \times \Re^n$ if the Euclidean norm $\langle f(t, x) - f(t, x'), x - x' \rangle \leq 0$ for all $x, x' \in R$ and all $t \in [a, b]$.

Any dissipative system is contractive, *i.e.* satisfies $|x(t_2) - x'(t_2)| \leq |x(t_1) - x'(t_1)|$ for two solutions $x(t)$ and $x'(t)$ with different initial conditions for all $a \leq t_1 \leq t_2 \leq b$, because $|x(t) - x'(t)|^2$ is a non-increasing function of time if $\langle f(x) - f(x'), x - x' \rangle \leq 0$.

Put $F(X_1, \ldots, X_n) = \varphi^{*\prime}(X_1, \ldots, X_n, y_1, \ldots, y_n)$. Let ϕ_F^t be the gradient flow for $F : \Re^n \to \Re$ with respect to the affinities. This F is not the deformation gradient tensor. The limit set, $\omega_F(x)$, of $x \in \Re^n$ is the set of points z such that there exists a sequence $\{t_n\}$ and $\phi_F^{t_n}(x)$ has limit z.

Lemma 14 *The limit set, $\omega_F(x)$, for the gradient flow, ϕ_F^t of $F : \Re^n \rightarrow \Re$ consists of critical points of F.*

Proof This follows because $F(\phi_F^t(x))$ is an increasing function of time (Katok and Hasselblatt, 1995, p. 37).

Lemma 15 *If a dissipative vectorfield in $\Re^n$ has critical point, x_0, every orbit has a limit point.*

Proof Let $x(t)$ be an arbitrary orbit. $x'(t) = x_0$ is also an orbit. Define $\mathcal{D}$ as the ball in $\Re^n$ centered at x_0 with radius $|x(0) - x_0|$. Pick a time sequence $\{t_n\}$. Since the vectorfield is dissipative, $|x(t_{n+1}) - x_0| < |x(t_n) - x_0|$ and the sequence $\{x(t_n)\}$ is contained in $\mathcal{D}$. The closure, $\bar{\mathcal{D}}$, in $\Re^n$ of $\mathcal{D}$ is compact since it is a closed subspace of $\Re^n$. Therefore the sequence $\{x(t_n)\}$ has a limit point in $\bar{\mathcal{D}}$.

Lemma 16 *If the generalized energy function φ^* has a unique equilibrium for fixed local controls, $y_1, \ldots, y_n$, then the origin, $X_i = 0$, is the unique critical state for the gradient relaxation process, Eq. (17).*

Proof The affinities $X_i = \partial\varphi^*/\partial x_i$ are zero exactly at equilibrium. By Eq. (24), when the Hessian of φ is invertible, the affinities are zero if and only if the states are stationary. The chain rule and the invertibility of the matrix $(\partial X_i/\partial x_j)$ imply that the affinities are stationary iff the states are stationary. The unique critical point for a relaxation process, $X_i(t)$, occurs when the affinities are zero, at equilibrium.

This result also follows from the Liapunov theorem in the case that the manifold M_e is a set of minimums for φ^*.

Proposition 17 *If the thermostatic state $\boldsymbol{X} = \boldsymbol{O}$ is a stable stationary state of φ'^* for fixed controls $\boldsymbol{y}$, then it is also an asymptotically stable state for the system of equations $dX_i/dt = -k_i(\partial\varphi'^*/\partial X_i)$, $i = 1, \ldots, n$, defining a homogeneous non-equilibrium gradient process.*

Proof Case 1: The thermodynamic variables are defined in the reference configuration. $\varphi'^*(\boldsymbol{y}, \boldsymbol{O})$ is a minimum since stable equilibria occur at minimums for φ^*. Define a Liapunov function $V(\boldsymbol{X}) = \varphi'^*(\boldsymbol{y}, \boldsymbol{X}) - \varphi'^*(\boldsymbol{y}, \boldsymbol{O}) \geq 0$ in a neighborhood of $X = \boldsymbol{O}$). Then by Eq. (17), since each $k_i > 0$,

$$\frac{dV}{dt} = \sum_{i=1}^{n} \frac{\partial V}{\partial X_i}\frac{dX_i}{dt} = -\sum_{i=1}^{n} k_i \left(\frac{\partial\varphi'^*}{\partial X_i}\right)^2 < 0.$$

Case 2: The thermodynamic variables are defined in the current configuration. Let Z_i be the pull-back of X_i under the deformation diffeomorphism χ. Then the gradient relaxation evolution equation pulls back to $DZ_i/Dt = -k_i(\partial\varphi'^*/\partial Z_i)$, i = $1, \ldots, n$, where the left-side is the material derivative. Repeat the calculation of Case 1. The asymptotic behavior is preserved under the push-forward by χ, since χ is a diffeomorphism. Convergence in the reference space is equivalent to convergence in the current space. In both cases, the stable point is asymptotically stable by a Liapunov theorem. A gradient relaxation process with fixed control variables evolves towards M_e, where each $X_i = 0$.

4.5 The Gibbs One-Form

The path of a process in thermodynamic space is given in local coordinates by $\gamma(t) = \big(\boldsymbol{x}(t), \boldsymbol{y}(t), \varphi^*[\boldsymbol{x}(t), \boldsymbol{y}(t)]\big)$. Here t is viewed as an arbitrary parameter; there is no question of objectivity in computing the action of a Gibbs form on a tangent vector.

The tangent to a process path at a given point on the graph of the generalized energy function describes its speed and direction. The graph is a $2n$-dimensional surface lying in $\Re^{2n+1}$. There are arbitrarily many possible directions and speeds. The set of these forms a $2n$-dimensional vector space and this vector space changes from point to point. Since the tangent is a vector and since it is not desired to write it in terms of the thermodynamic space $\Re^{2n+1}$, a new vector space at each point on the graph corresponding to $(x_1, \ldots, x_n, y_1, \ldots, y_n)$ is defined with basis denoted by $\{\partial/\partial x_i; \partial/\partial y_i\}$ in which the abstract tangent vectors are placed, the tangent space at the point (Chapter 7). The basis for the dual vector space is $\{dx_i; dy_i\}$. The elements of the dual space are called one-forms. Recall that a one-form is a scalar-valued function acting on tangents to paths. The one-form acts by $dx_i(\partial/\partial x_j) = \delta_{ij}$ etc, where δ_{ij} is the Kronecker delta. The tangent vector, $\boldsymbol{t}_p$, at each point p on a non-equilibrium path is in local coordinates

$$\boldsymbol{t}_p = \frac{d\varphi^*}{dt}\frac{\partial}{\partial z} + \sum_{i=1}^{n} \frac{dx_i}{dt}\frac{\partial}{\partial x_i} + \sum_{i=1}^{n} \frac{dy_i}{dt}\frac{\partial}{\partial y_i}. \tag{28}$$

Interpret the Gibbs classical expression $d\mathcal{U} = \theta d\mathcal{S} - \mathcal{P} d\mathcal{V}$ as the one-form $d\mathcal{U} - \theta d\mathcal{S} + \mathcal{P} d\mathcal{V}$ which is zero on the equilibrium surface. This idea is generalized to a Gibbs one-form associated with the graph of φ^* in $\Re^{2n+1}$, which has coordinates $(x_1, \ldots, x_n; y_1, \ldots, y_n, z)$.

Definition 18 The Gibbs one-form, ω, associated with this structure is

$$\omega = dz - \sum_{i=1}^{n} x_i dy_i. \tag{29}$$

The pull-back of a spatial Gibbs form is $F^{-t}\omega$; the push-forward of a reference Gibbs form is $F^t\omega$. The requirement that this one-form be non-positive when applied to the tangent to a process path lying on the graph of the generalized function will be shown to be equivalent to the Clausius-Duhem inequality.

In Chapter 7 a differential topology interpretation is adopted. The Gibbs one-form is viewed as an element of the cotangent space and the tangent to a process is viewed as an element of the tangent space to thermodynamic space. While a differential topology viewpoint is not required for this construction, a geometric model for Gibbs' view of the thermostatic surface takes place in a contact bundle (Haslach 1997).

4.6 Maximum Dissipation in Gradient Processes

The physical hypothesis defining the class of materials considered here depends on the dissipation measured by the time rate of change of the associated generalized energy function. The gradient relaxation process is characterized by the instantaneous maximum decrease of the generalized thermodynamic function written as a function of the affinities.

Definition 19 In a process governed by the generalized energy function φ^*, the dissipation at time t is given by $d\varphi^*/dt$.

There is no dissipation during an equilibrium process because

$$\frac{d\varphi^*}{dt} = \sum_i \frac{\partial \varphi^*}{\partial x_i}\dot{x}_i = 0 \tag{30}$$

when the equilibrium conditions, $\partial\varphi^*/\partial x_i = 0$, hold for all i.

Dissipation only occurs during non-equilibrium processes. The expression given by the chain rule

$$\frac{d\varphi^*}{dt} = \sum_i \frac{\partial \varphi'^*}{\partial X_i}\dot{X}_i \tag{31}$$

is the dot product of the gradient of φ'^* with respect to the affinities and the vector $\dot{\vec{X}}$. It can be viewed as the directional derivative in the direction $\dot{\vec{X}}$, which is maximal in the direction of the gradient.

By a calculation in Haslach (1997), the action of the Gibbs form on the tangent vector (28) is

$$\omega(t_p) = \frac{d\varphi^*}{dt} - \sum_{i=1}^{n} x_i \frac{dy_i}{dt}. \tag{32}$$

Since $d\varphi^*/dt = d\varphi/dt + \sum_{i=1}^{n}\left[(dx_i/dt)y_i + (dy_i/dt)x_i\right]$,

$$\omega(t_p) = \frac{d\varphi}{dt} + \sum_{i=1}^{n} y_i \frac{dx_i}{dt}. \tag{33}$$

Furthermore, since $d\varphi/dt = \sum_{i=1}^{n}(\partial\varphi/\partial x_i)(dx_i/dt)$,

$$\omega(\boldsymbol{t}_p) = \sum_{i=1}^{n} \frac{dx_i}{dt}\left(\frac{\partial\varphi}{\partial x_i} + y_i\right) = \sum_{i=1}^{n} \frac{dx_i}{dt}\frac{\partial\varphi^*}{\partial x_i} = \nabla_x(\varphi^*)\cdot\frac{d\boldsymbol{x}}{dt}, \tag{34}$$

where $\nabla_x(\varphi^*)$ is the gradient of φ^* with respect to the state variables.

At the long-term states, $X_i = \partial\varphi/\partial x_i + y_i = 0$ so that $\nabla_x(\varphi^*) = \boldsymbol{O}$ and $\omega(\boldsymbol{t}_p) = 0$. The process is on M_e at a particular state iff $\omega(\boldsymbol{t}_p) = 0$. Otherwise, it is a non-equilibrium process.

The scalar $\omega(\boldsymbol{t}_p)$ measures the instantaneous dissipation when φ is a specific energy. The dissipation in this construction by the second equality in (34) is

$$\omega(\boldsymbol{t}_p) = \sum_{i=1}^{n} \frac{dx_i}{dt}\frac{\partial\varphi^*}{\partial x_i}, \tag{35}$$

which has a form similar to that for theories using a separate dissipation potential. Again, the affinity X_i corresponding to the state x_i serves as the thermodynamic force. Geometrically, the magnitude of $\omega(\boldsymbol{t}_p) < 0$ defines the obtuse angle between the vectors $\nabla_x(\varphi^*)$ and the tangent to the non-equilibrium path, $\boldsymbol{t}_p$. Maximum dissipation of a generalized function is achieved if the tangent vector to an admissible path is parallel to $\nabla_x(\varphi^*)$. The gradient relaxation process is therefore a process inducing maximum dissipation of the generalized thermodynamic function. For example, if the control variables are held fixed during a non-equilibrium process, then $\omega(\boldsymbol{t}_p) = d\varphi^*/dt$ by Eq. (32).

4.7 The Gibbs Form and the Clausius-Duhem Inequality

The internal energy density, u, is a function of specific entropy, η, the strains, ϵ_{ij}, and internal state variables, a_i. For a homogeneous material, the second law in the form of the Clausius-Duhem inequality is

$$0 \geq du/dt - \theta(d\eta/dt) - \sigma_{ij}(d\epsilon_{ij}/dt) - A_i(da_i/dt), \tag{36}$$

where θ is the temperature and A_i are the affinities associated to the a_i.

Theorem 20 *For any process, the Clausius-Duhem inequality is equivalent to* $\omega(\boldsymbol{t}_x) \leq 0$.

Proof The free, or complementary, energy required in the construction of the generalized internal energy, U^*, is $\psi(\theta, \sigma_{ij}, A_i) = u - \theta\eta - \sigma_{ij}\epsilon_{ij} - A_i a_i$. Therefore by substituting for u in Eq. (36),

$$0 \geq d\psi/dt + \eta(d\theta/dt) + \epsilon_{ij}(d\sigma_{ij}/dt) + a_i(dA_i/dt).$$

The generalized internal energy is, by Eq. (4),

$$U^*(\theta, \sigma_{ij}, a_i; \eta, \epsilon_{ij}, A_i) = \psi(\theta, \sigma_{ij}, A_i) + \theta\eta + \sigma_{ij}\epsilon_{ij} + A_i a_i.$$

The associated Gibbs one-form is $\omega = dz - \theta d\eta - \sigma_{ij} d\epsilon_{ij} - A_i da_i$. For any process on the graph of U^* with tangent to the path given by Eq. (28), the action of the Gibbs form is given by Eq. (33)

$$\omega(\boldsymbol{t}_p) = d\psi/dt + \eta(d\theta/dt) + \epsilon_{ij}(d\sigma_{ij}/dt) + a_i(dA_i/dt).$$

Therefore $\omega(\boldsymbol{t}_p) \leq 0$ iff the Clausius-Duhem inequality holds.

4.8 Admissible Processes

The second law of thermodynamics determines which non-equilibrium processes are thermodynamically admissible. In this construction, admissibility is described in terms of the Gibbs one-form, ω, Eq. (29). Let $\boldsymbol{t}_p$ be the tangent to the process path at point $\boldsymbol{p} = (p_1, p_2, p_3)$.

Definition 21 An admissible thermodynamically homogeneous non-equilibrium process in the thermodynamic system defined by $\varphi^* : \Re^{2n} \to \Re$ is a curve $\gamma : \Re \to \Re^{2n+1}$ whose image lies on the graph of φ^* and for which $\omega(\boldsymbol{t}_p) < 0$ at each point along the path.

The admissibility definition (Definition 21) provides a geometric interpretation of the second law of thermodynamics for homogeneous non-equilibrium processes. Admissibility is independent of the magnitude of the tangent to the path $\boldsymbol{t}_p$, the speed at which the process takes place, since ω is linear. The path actually taken is defined by constitutive equations, such as the gradient relaxation process, giving the evolution of the state variables.

Proposition 22 *A gradient process is an admissible non-equilibrium process.*

Proof It is to be shown that $\omega(\boldsymbol{t}) \leq 0$ whether in spatial or reference coordinates. The Gibbs forms ω_r and ω_s in the reference and spatial configurations respectively are related by $\omega_r = F^t \omega_s$ and the tangent vectors by $\boldsymbol{t}_r = F^{-1}\boldsymbol{t}_s$. Then since the action of the Gibbs one-form on a tangent vector is a scalar product,

$$\omega_r(\boldsymbol{t}_r) = \langle \omega_r, \boldsymbol{t}_r \rangle = \langle F^t \omega_s, F^{-1}\boldsymbol{t}_s \rangle = \langle \omega_s, FF^{-1}\boldsymbol{t}_s \rangle = \langle \omega_s, \boldsymbol{t}_s \rangle = \omega_s(\boldsymbol{t}_s).$$

Therefore the proof may be caried out in the reference configuration. Since $\boldsymbol{y}$ is fixed, Eq. (32) implies that $\omega(\boldsymbol{t}_p) = d\varphi^*/dt$. But, in turn,

$$\frac{d\varphi^*}{dt} = \frac{d(\varphi'^* \cdot h)}{dt} = \sum_{i=1}^{n} \frac{\partial \varphi'^*}{\partial X_i}\dot{X}_i = -\sum_{i=1}^{n} \frac{1}{k_i}\dot{X}_i^2 \leq 0.$$

Therefore, $\omega(\boldsymbol{t}_p) \leq 0$ if the control variables are held fixed.

Example Consider again the uniaxial linearly elastic generalized energy density function, φ^*, with elastic modulus, E_1,

$$\varphi^*(\epsilon, \sigma) = \frac{1}{2}E_1\epsilon^2 - \sigma\epsilon.$$

From Eq. (32), $\omega(\boldsymbol{t}_p) = X\dot{\epsilon}$. Therefore

$$\omega(\boldsymbol{t}_p) = -(\sigma - E\epsilon)\dot{\epsilon}.$$

Since by Eq. (26), the terms $(\sigma - E\epsilon)$ and $\dot{\epsilon}$ have the same sign, $\omega(\boldsymbol{t}_p) < 0$. Therefore, this process satisfies the second law.

Example For the example of Newton's Law of Cooling a similar pattern occurs. The contact form acting on the tangent,

$$\omega(\boldsymbol{t}_p) = X\dot{\theta} = \frac{c_0}{\theta_0}(\theta - \theta_B)\dot{\theta} < 0,$$

since $(\theta - \theta_B)$ and $\dot{\theta}$ have opposite signs by Eq. (21). Therefore this process is admissible.

5 Forced Non-equilibrium Processes

A forced process is a non-equilibrium process in which the control variables vary with time. The second law of thermodynamics must be augmented by at least one additional assumption. A gradient relaxation process was assumed when the control variables are fixed over time. This physical hypothesis is generalized for arbitrary equilibrium processes. The gradient relaxation process in the case that the control variables are fixed over time is generalized for arbitrary non-equilibrium processes so that any admissible process is locally a maximum dissipation process. This condition is the constitutive restriction defining the class of materials considered.

Definition 23 Each non-equilibrium process is locally a maximum dissipation process, in the sense that it is a gradient relaxation process over each very short time interval.

The evolution equations have the same form as Eq. (24) except that each control variable is a function of the current time.

$$\begin{bmatrix} \frac{dx_1}{dt} \\ \frac{dx_2}{dt} \\ \cdots \\ \frac{dx_n}{dt} \end{bmatrix} = -k \begin{bmatrix} \frac{\partial^2\varphi}{\partial x_1^2} & \frac{\partial^2\varphi}{\partial x_1 \partial x_2} & \cdots & \frac{\partial^2\varphi}{\partial x_1 \partial x_n} \\ \frac{\partial^2\varphi}{\partial x_2 \partial x_1} & \frac{\partial^2\varphi}{\partial x_2^2} & \cdots & \frac{\partial^2\varphi}{\partial x_2 \partial x_n} \\ \cdots & \cdots & \cdots & \cdots \\ \frac{\partial^2\varphi}{\partial x_n \partial x_1} & \frac{\partial^2\varphi}{\partial x_n \partial x_2} & \cdots & \frac{\partial^2\varphi}{\partial x_n \partial x_n} \end{bmatrix}^{-2} \begin{bmatrix} y_1(t) + \frac{\partial\varphi}{\partial x_1} \\ y_2(t) + \frac{\partial\varphi}{\partial x_2} \\ \cdots \\ y_n(t) + \frac{\partial\varphi}{\partial x_n} \end{bmatrix}. \tag{37}$$

One important test of the local maximum dissipation assumption is the type of response it predicts under forced loadings, such as sinusoidal variation of the control variables, y_i, over time. In some polymeric materials and in rubber, experiment shows there is a steady state sinusoidal response to such sinusoidal loading.

Example A uniaxial linear viscoelastic model is obtained from the generalized energy, $\varphi^*(\epsilon, \sigma) = 0.5E_1\epsilon^2 - \epsilon\sigma$, where E_1 is a constant. The response to the sinusoidal forcing function $\sigma(t) = A\sin(\omega t)$ is defined by the evolution equation, Eq. (37), for small displacements,

$$\frac{d\epsilon}{dt} = -\frac{k}{E_1^2}[E_1\epsilon - \sigma(t)], \tag{38}$$

where $\sigma(t) = A\sin(\omega t)$. The response, obtained using an integrating factor, is

$$\epsilon(t) = \frac{Ak}{E_1\sqrt{k^2 + \omega^2 E_1^2}}\sin(\omega t + \phi) + \exp\left(\frac{k}{E_1}t\right)\left[\epsilon(0) + \frac{Ak\omega}{k^2 + \omega^2 E_1^2}\right], \tag{39}$$

where the phase shift is given by the tangent modulus, $\tan(\phi) = -E_1\omega/k$. In this case, the relaxation modulus, k, can be experimentally determined from the tangent modulus.

This non-equilibrium dissipative process tends to the limit cycle in $\epsilon - \sigma$-space

$$\left(\frac{Ak}{E_1\sqrt{k^2 + \omega^2 E_1^2}}\sin(\omega t + \phi),\ A\sin(\omega t)\right),$$

which does not intersect the equilibrium manifold, the set $(\epsilon, E_1\epsilon)$. It is not a gradient process because it does not tend to a critical point (Lemma 14). The strain response is steady under the time varying control stress rather than evolving to the long-term state as it would under a constant control stress. The choices of k give different linear viscoelastic models, as discussed in Chapter 4.

In terms of the affinities, a forced gradient relaxation process is one determined by direct substitution of time varying control variables in the gradient evolution equations $dX_i/dt = -k\partial\varphi'^*/\partial X_i$. The initial state for such a process must be a non-equilibrium state so that $X_i(0) \neq 0$ for some i. If all $X_i(0) = 0$, the process starts in equilibrium and remains there.

5.1 Numerical Methods

The response of a forced process, Eq. (37), can be approximated by a numerical method based on a sequence of gradient relaxation processes. The idea is that the controls have moved to new values before the system can come to equilibrium with respect to the original values. Divide time into intervals delineated by the sequence $\{t_m\}$. At time t_m, as initial condition for the state take $x_i(t_m)$ to be the state value at the end of the previous interval. Let the initial condition of the control, $y_i' \neq y(t_m)$, depend on the values of y_i in the interval $[t_m, t_{m+1}]$. Apply the gradient relaxation process to this non-equilibrium state. The value $x_i(t_{m+1})$ is the value of $x_i(t)$ in the relaxation process after time $t_{m+1} - t_m$. For example, the numerical method for a viscoelastic process with linear long-term behavior is, over the time interval $h = t_{m+1} - t_m$,

$$\epsilon(t_{m+1}) = \epsilon(t_m)\exp\left(-\frac{k}{E}h\right) + \frac{\sigma'}{E}\left[1 - \exp\left(-\frac{k}{E}h\right)\right]. \tag{40}$$

The choice of σ' determines the type of integrator. The method is a backwards algorithm if $\sigma' = \sigma(t_{m+1})$; it is a midpoint algorithm if $\sigma' = [\sigma(t_{m+1}) + \sigma(t_m)]/2$. These are exactly the asymptotic algorithms studied by Haslach, Freed and Walker (1994, 1995) for systems of functions of the form $\dot{\vec{x}} = c(\vec{x})[a(\vec{x}, t) - \vec{x}]$. Here, $c(\vec{x}) = k/E$ and $a(\vec{x}, t) = \sigma(t)/E$.

Some numerical methods may be easier to apply if the evolution can be described by the system of integral equations. By the fundamental law of calculus,

$$x_i(t) = \int_0^t dx_i = \int_0^t \frac{dx_i}{dt}dt = \int_0^t -k\left[\sum_j (H^{-2})_{ij}\left(y_j(s) + \frac{d\varphi(s)}{dx_j}\right)\right]ds, \tag{41}$$

where H is the Hessian of φ^* with respect to the state variables.

6 Generalized Nonlinear Onsager-Type Relations

Continuum thermodynamicists attacked those who used the linear Onsager relations because the relations were not believed to be well-justified (see Chapter 1). But the present construction whose admissibility condition is equivalent to the Clausius-Duhem inequality, and so encompasses continuum thermodynamics, also includes a nonlinear version of the Onsager relations if the variables are correctly interpreted. Present day Onsagerists have redefined the argument so that the issue in non-equilibrium thermodynamics is the range of validity of the classical Onsager relations (Garcìa-Colìn and Uribe, 1991). A generalized non-linear form of this relation can be written for a non-equilibrium relaxation process. This form reduces to the classical linear form in a small enough neighborhood of the limit equilibrium point on the non-equilibrium path.

Classically, non-equilibrium fluxes, denoted by J_i, are time rates of change of thermodynamic variables and are driven by generalized forces, or affinities, denoted by X_i. Each J_i is assumed to depend on all affinities, X_i. The first term of the Taylor series of J_i around the point $X_i = 0$ yields the linearization,

$$J_i = \sum_j L_{ij} X_j,$$

where the constants L_{ij} are called phenomenological coefficients and measure the coupling of the ith and j^{th} processes. The constant L_{ij} is zero if the two processes are uncoupled. The Onsager reciprocal relations state that these constants are symmetric, $L_{ij} = L_{ji}$.

This idea can be generalized for a relaxation process. Since $H_{ij} = \partial X_i/\partial x_j = \partial^2 \varphi^*/\partial x_i \partial x_j$,

$$J_i = -k\frac{\partial \varphi'^*}{\partial X_i} = -k\sum_j \frac{\partial \varphi^*}{\partial x_j}\frac{\partial x_j}{\partial X_i} = -k\sum_j X_j H_{ij}^{-1}, \tag{42}$$

where H_{ij}^{-1} is the i-j entry in the inverse of the Hessian. Denote $-H_{ij}^{-1}$ by L_{ij}, called the generalized phenomenological coefficients, for $1 \le i, j \le n$. Then L_{ij} is a function of the X_i and the y_i, and $J_i = \sum_j L_{ij} X_j$.

This relationship extends the Onsager relations for non-equilibrium states near the equilibrium manifold. In those states, L_{ij} can be approximated by H_{ij}^{-1} evaluated at the limit equilibrium point, so that L_{ij} is a constant. Furthermore since $H_{ij}^{-1} = H_{ji}^{-1}$, $L_{ij} = L_{ji}$. The construction is consistent with the fact that the linear Onsager relations are assumed valid only near equilibrium.

Example A relaxation process for an ideal gas with temperature and pressure as controls is modeled by the generalized Gibbs energy function. The coefficients L_{ij} can be determined either by linearizing the gradient relaxation path equations or by approximating L_{ij} by $(G^{*-1})_{ij}(\boldsymbol{x}, \boldsymbol{y})$. In either case, the resulting linear differential equations determine the time dependence of the relaxation path near the equilibrium surface. The generalized Gibbs energy function for an ideal gas is $G^*(P, T; V, S) = N^{5/3} V^{-2/3} \exp(2S/3NR) + PV - TS$. Then put

$$X_1 = \frac{\partial G^*}{\partial V} = -\frac{2}{3} N^{5/3} V^{-5/3} \exp\left(\frac{2S}{3NR}\right) + P.$$
$$X_2 = \frac{\partial G^*}{\partial S} = \frac{2}{3R} N^{2/3} V^{-2/3} \exp\left(\frac{2S}{3NR}\right) - T.$$

In terms of the X_i and the control variables T and P,

$$G'^* = \frac{3}{2} NR(X_2 + T) - PNR\frac{X_2 + T}{X_1 - P}$$
$$-\frac{3}{2} NRT \ln\left[3(X_1 - P)^{-2/3}\frac{X_2 + T}{2R^{-5/3}}\right].$$

The first derivatives are

$$\frac{\partial G'^*}{\partial X_1} = NR\frac{PX_2 + TX_1}{(X_1 - P)^2},$$
$$\frac{\partial G'^*}{\partial X_2} = \frac{3}{2}NR - NR\frac{P}{X_1 - P} - \frac{5}{2}NR\frac{T}{X_2 + T};$$

and these derivatives are zero on the equilibrium manifold defined by $X_1 = X_2 = 0$. A gradient relaxation process is modeled by $\partial X_1/\partial t = -\partial G'^*/\partial X_1$ and $\partial X_2/\partial t = -\partial G'^*/\partial X_2$. If this pair of coupled differential equations is linearized near the equilibrium surface, i.e. near $X_1 = X_2 = 0$, the following system is obtained.

$$\frac{\partial X_1}{\partial t} = -\frac{NRT}{P^2}X_1 - \frac{NR}{P}X_2$$
$$\frac{\partial X_2}{\partial t} = -\frac{NR}{P}X_1 - \frac{5NR}{2T}X_2,$$

where the coefficients of the X_i are the second derivatives of G^* with respect to the X_i evaluated at the origin. The same set of equations is obtained from

$$\frac{\partial X_i}{\partial t} = -\frac{\partial E'^*}{\partial X_i} = -\sum_j L_{ij}X_j = -\sum_j (G^{*-1})_{ij}X_j,$$

when approximated by evaluating L_{ij} at the origin since the matrix $(\partial x_i/\partial X_j) = (G^{*-1}_{ij})$ evaluated at equilibrium is

$$G^{*-1} = \begin{pmatrix} \frac{NRT}{P^2} & \frac{NR}{P} \\ \frac{NR}{P} & \frac{5NR}{2T} \end{pmatrix}.$$

The relaxation path asymptotically approaches the origin, i.e. equilibrium. The solution to the system is

$$X_1 = C_1 \exp(-\lambda_1 t)$$
$$X_2 = C_2 \exp(-\lambda_2 t),$$

where λ_1 and λ_2 are the eigenvalues defined by

$$\lambda = \frac{NR}{2}\left[\left(\frac{T}{P^2} + \frac{5}{2T}\right) \pm \sqrt{D}\right],$$

and $D = (T/P^2 - 5/2T)^2 + 4/P^2 > 0$ so that the eigenvalues are real and are positive. The coefficients C_1 and C_2 are the coordinates to which the system is perturbed before it is allowed to relax towards equilibrium.

References

M. Fabrizio and A. Morro (2003). *Electromagnetism of Continuous Media*, Oxford University Press, Oxford.

J. W. Gibbs (1873). A method of geometrical representation of the thermodynamic properties of substances by means of surfaces. *Transactions of the Connecticut Academy* II, 382–404. Also in *The Collected Works*, Vol. 1, Yale University Press, New Haven 1948, 33–54.

L. S. Garcìa-Colìn and F. J. Uribe (1991). Extended irreversible thermodynamics, beyond the linear regime: a critical overview. *Journal of Non-Equilibrium Thermodynamics* **16**, 89–128.

J. Hale and H. Koçak (1991). *Dynamics and Bifurcations*, Springer-Verlag, New York.

H. W. Haslach, Jr. (1996). Nonlinear thermoviscoelastic relaxation. In *Contemporary Research in the Mechanics and Mathematics of Materials*, eds. R. C. Batra and M. F. Beatty, pp. 384–396. International Center for Numerical Methods in Engineering, Barcelona, Spain.

H. W. Haslach, Jr. (1997). Geometrical structure of the non-equilibrium thermodynamics of homogeneous systems. *Reports on Mathematical Physics* **39**, 147–62.

H. W. Haslach, Jr. (2000). Constitutive models and singularity types for an elastic biaxially loaded rubber sheet. *Mathematics and Mechanics of Solids* **5**, 41–73.

H. W. Haslach, Jr. (2002). A non-equilibrium thermodynamic geometric structure for thermoviscoplasticity with maximum dissipation. *International Journal of Plasticity* **18**(2), 127–153.

H. W. Haslach, Jr. (2005). Nonlinear viscoelastic, thermodynamically consistent, models for biological soft tissue. *Biomechanics and Modeling in Mechanobiology* **3**(3), 172–189.

H. W. Haslach, Jr. (2009). Thermodynamically consistent, maximum dissipation, time-Dependent models for non-equilibrium behavior. International Journal of Solids and Structures 46, 3964–3976. DOI: 10.1016/j.ijsolstr.2009.07.017

H. W. Haslach, Jr. (2010). A non-equilibrium thermodynamic model for the crack propagation rate. *Mechanics of Time-Dependent Materials* **14**, 91–110. DOI: 10.1007/s11043-009-9094-9

H. W. Haslach, Jr. and N-N Zeng (1999). Maximum dissipation evolution equations for nonlinear thermoviscoelasticity. *International Journal of Non-linear Mechanics* **34**(2), 361–385.

H. W. Haslach, Jr., A. D. Freed, and K. P. Walker (1994). Nonlinear asymptotic integration algorithms for one-dimensional autonomous dissipative first-order ODEs. *NASA Technical Memorandum*, NASA TM 106780, 1–26.

H. W. Haslach, Jr., A. D. Freed, and K. P. Walker (1995). Nonlinear asymptotic integration algorithms for two-dimensional autonomous dissipative first-order ODEs. *NASA Technical Memorandum*, NASA TM 106837, 1–21.

P. Haupt (2002). *Continuum Mechanics and Theory of Materials*, 2nd ed., Springer, Berlin.

G. A. Holzapfel (2000). *Nonlinear Solid Mechanics*, 2005 reprinting. Wiley, Chichester.

A. Katok and B. Hasselblatt (1995). *Introduction to the Modern Theory of Dynamical Systems*, Cambridge University Press, New York.

J. R. Magnus (2010). On the concept of matrix derivative (Feb 21, 2010) http://center.uvt.nl/staff/magnus/wip12.pdf Accessed April 24, 2010.

L. E. Malvern (1969). *Introduction to the Mechanics of a Continuous Medium*, Prentice-Hall, NY.

J. E. Marsden and T. J. R. Hughes (1994). *Mathematical Foundations of Elasticity*, Dover, New York.

R. W. Ogden (1984). *Non-linear Elastic Deformations*, Dover, Mineola, NY.

J. G. Oldroyd (1950). On the formulation of rheological equations of state. *Proceedings of the Royal Society A* **200**, 523–541.

J. R. Rice (1971). Inelastic constitutive relations for solids: an internal variable theory and its application to metal plasticity. *Journal of the Mechanics and Physics of Solids* **19**, 433–453.

M. Šilhavý (1997). *The Mechanics and Thermodynamics of Continuous Media*, Springer-Verlag, Berlin.

J. C. Simo and K. S. Phister (1984). Remarks on rate constitutive equations for finite deformation problems: computational implications. *Computer Methods in Applied Mechanics and Engineering* **46**, 201–215.

Chapter 4
Viscoelasticity

1 Introduction

Viscoelastic processes are history dependent processes in which some energy is stored and some energy is dissipated. A simple viscoelastic relaxation process is one in which the state variables respond over time to constant control variables. For example, in creep, the stress is held constant, and in stress relaxation, the strain is held constant. Each viscoelastic material has a long-term behavior towards which a relaxation process tends. The structure of a viscoelastic material is perturbed only slightly during time-dependent responses. This is in contrast to a viscoplastic material (see Chapter 5) which involves mechanisms, such as dislocation multiplication, that permanently change the structure of the material. The creep of a non-aging polymer is a qualitatively different phenomenon than that of a metal creep, which may involve mechanisms such as dislocation climb. Other than relaxation processes such as creep or stress relaxation, viscoelastic process may be forced, such as in rate-dependent loading or in sinusoidal loading. The goal is to describe the application of the maximum dissipation evolution construction to produce a nonlinear viscoelastic model that can represent both relaxation and forced loadings. The class of models developed here assumes that dissipation occurs in a viscoelastic material without affecting the equilibrium, or long-term state.

The construction is shown to apply to many different types of viscoelastic materials. Some classical viscoelastic models may be interpreted as, or slightly modified to be, maximum dissipation models. The construction includes temperature effects in a simple manner in comparison to classical continuum models. In the rubber examples below, the Gough-Joule effects are easily captured in the nonlinear maximum dissipation model. The construction generates one-, two- or three-dimensional models as desired. As a second major category of examples, the construction produces nonlinear viscoelastic models for soft biological tissue.

2 Brief History of Viscoelastic Models

Early viscoelastic models were ad hoc, and were followed in the nineteenth century by linear viscoelastic models. In the middle of the twentieth century, restrictions on the form of possible constitutive models were obtained from continuum

H.W. Haslach Jr., *Maximum Dissipation Non-Equilibrium Thermodynamics and its Geometric Structure*, DOI 10.1007/978-1-4419-7765-6_4,

thermodynamics principles, and nonlinear viscoelastic models were generated. Many experimentalists today who work with nonlinear soft biological materials such as brain tissue have reverted to proposing ad hoc models because the continuum mechanics viscoelastic models are so complex.

The viscoelastic response of solids has been known since at least the early eighteen hundreds. For example, Wilhelm Weber studied the *elastiche nachwirkung* of silk threads in 1835 to show experimentally that its mechanical behavior was not elastic but rather time-dependent. While early workers proposed several empirical constitutive models to capture the history dependence, Boltzmann in 1874 used a superposition principle assuming small deformations to produce an integral model for shear stress, which in the uniaxial case is

$$\tau(t) = \int_{-\infty}^{t} G(t-s)\frac{d\gamma(s)}{ds}ds = G(0)\gamma(t) + \int_{0}^{\infty} \gamma(t-s)\frac{G(s)}{ds}ds, \tag{1}$$

where τ and γ are the shear stress and small shear strain respectively. The second equality is obtained by integration by parts. The function, $G(s)$, is material dependent and is to be determined by experiment. A linear history-dependent viscoelastic model for normal strain has the Boltzmann integral form in which the history is assumed to begin at $t = 0$ and depends on the material function $J(s)$,

$$\epsilon(t) = \int_{0}^{t} J(t-s)\dot{\sigma}(s)ds. \tag{2}$$

A very nice discussion of Boltzmann's work on viscoelasticity is given by Markovitz (1977).

Prior to this in 1867 Maxwell had proposed his differential equation, which can also be written in the Boltzmann form. The spring and damper analogy was introduced by Thomson and Poynting in 1903 and led to spring and dashpot based differential equation generalizations of the models proposed by Maxwell and by Voigt. Thomson and Poynting showed that the Kelvin-Voigt model is related to a spring and dashpot in parallel. Such models are linear viscoelastic. But the physical interpretation of the individual relaxation times is often controversial, and the times may only be an artifact of the model. Some models use nonlinear springs and dashpots; but these are only ad hoc models with no obvious basis in the physical structure of the material. In fact, the Kelvin-Voigt model was first given as a differential equation.

2.1 Contemporary Linear Viscoelasticity

Commercial FEM packages typically provide a capability to model viscoelastic materials, often either linear or nonlinear viscoelastic. The linear viscoelastic model is based on a Boltzmann-type hereditary integral

$$\sigma(t) = \int_0^t 2G(s-t)\frac{de}{ds}ds + I\int_0^t 2K(s-t)\frac{d\psi}{ds}ds, \tag{3}$$

where $G(s)$ and $K(s)$ are the shear and bulk moduli respectively. The strains are the deviatoric, e, and volumetric, ψ. Prony in 1795 developed a method that requires only a finite number of samples to analyze a signal that is the current basis of the viscoelastic representation of the time-dependent relaxation moduli, $G(s)$ and $K(s)$. Since the functional forms of the moduli are not known a priori, the Prony series expansion is assumed and the coefficients are obtained by curve fitting. The relaxation functions for the bulk and shear moduli respectively, $K(s)$ and $G(s)$, are of the form $G(s) = G_\infty + \sum_{i=1}^{nG} g_i \exp(-s/s_i)$, where the s_i are relaxation times.

Empirical viscoelastic evolution equations derived from a Prony series are commonly used in industry today to represent polymer viscoelastic behavior (e.g. Park and Kim, 2001). One does not know how many terms of the Prony series to take. In any case, almost any data can be fit if one allows a sufficient number of empirical parameters in the model.

The Laplace operator establishes a correspondence between linear viscoelastic theories and linear elasticity by which viscoelastic boundary value problems can be more easily solved (Christensen, 1971). Lee (1955) gives a nice discussion with examples. However, such a correspondence principle is not available for nonlinear thermoviscoelasticity, except for small deformations. Schapery (1984) proposed a construction to extend the correspondence principle to nonlinear viscoelasticity, but it does not satisfy frame-indifference. The construction was shown later to also fail to satisfy angular momentum balance for large deformations (Ragagopal and Srinivasa, 2005).

Linear viscoelastic models include the standard linear solid. Many combinations of three or four springs and dashpots are called the standard linear solid. One version of the standard linear solid is a three-element spring and dashpot model formed by a Maxwell model in parallel with a linear spring. Another three-element model often called a standard linear solid is formed from a Kelvin-Voigt model in series with a linear spring. The four-element model formed by combining the Maxwell and the Voigt models has also been called a standard linear solid. The shear relaxation modulus for the standard linear solid is $G(t) = G_\infty + (G_o - G_\infty)\exp(-\beta t)$, where G_o is the short term shear modulus ($t = 0$), G_∞ is the long term shear modulus ($t \to \infty$) and β is the decay parameter. This is a one-term Prony series. Therefore if the experimental stress relaxation data is not exponential, the standard linear solid model is not acceptable.

When accounting for temperature, a goal is to reduce the number of experiments needed to characterize the material as a function of temperature. In some materials, the idea of reduced time may simplify the experimental work needed. In general, the reduced time τ_s is given by

$$\tau_s = \int_0^t \frac{dt'}{A\big(\theta(t')\big)}, \tag{4}$$

where θ is temperature and A is the shift function. A commonly used shift function, near the glass transition temperature, is the Williams-Landell-Ferry (WLF) equation,

$$A(\theta) = -\exp\left(\frac{c_1(\theta - \theta_g)}{c_2 + (\theta - \theta_g)}\right), \tag{5}$$

where c_1 and c_2 are constants and θ_g is the glass transition temperature, the temperature at which the polymer changes from glassy to rubbery. Often the WLF shift function is modified by replacing the glass transition with a different reference temperature θ_o as which the relaxation data is known. Then the result is valid near θ_o.

If the reduced time stress or strain curves when plotted versus temperature (or moisture content, etc) form a single curve, that curve is called a master curve. Such a curve does not exist for all viscoelastic materials. If it does, then the temperature dependent behavior is known for a wide range of temperatures without doing a large number experiments.

A rheologically simple material is defined by Christensen as one for which a master curve exists for temperature. The constitutive equation defining a thermorheologically simple linear viscoelastic material satisfies a shift relationship in which the stress is

$$\sigma(t) = \int_{-\infty}^{t} G\big(s - A(\theta)\big)\dot{\epsilon}(s)ds, \tag{6}$$

where the shift function $A(\theta)$ is an increasing function of temperature θ. Essentially, the stress relaxation function shifts to the right or left when plotted against $\log(t)$. Therefore tests need only be conducted at one temperature (Christensen, 1971, p. 95). A master curve can then be constructed, portions of which give the stress response at various temperatures.

2.2 Ad Hoc Non-integral Creep Models Explicit in Time

In parallel with the development of viscoelasticity, simple history dependent creep models were devised by combining functions whose graphs have the same shape as experimental data. In low homologous temperatures, primary creep has been represented by

$$\epsilon = A\ln(1 + t) + C, \tag{7}$$

where A and C are material constants that depend on stress or temperature.

The 1921 Nutting uniaxial equation has been applied to the primary creep of soft materials like rubber (Findley et al., 1989, p. 15),

$$\epsilon(t) = k\sigma^p t^n, \tag{8}$$

where k, p, and n are material constants. However, the predictions are only good for small times.

Often these creep models include a time-independent summand, which is sometimes interpreted as the instantaneous strain as the creep load is initially applied. However, many models like

$$\epsilon(t) = \epsilon_o + a\log(t) + bt \tag{9}$$

for material constants ϵ_o, a and b, only apply after long times since the equation is not defined at $t = 0$. Those which are intended for all times include, for example, the hyperbolic sine model proposed by Findley, Adams and Worley (Findley et al., 1989, p. 14)

$$\epsilon(t) = a\sinh(\sigma/\sigma_o) + bt^n \sinh(\sigma/\sigma_1), \tag{10}$$

where a, b, n, σ_o, and σ_1 are material constants at constant temperature. The models, to be convincing, should be derived from physical principles because many models, including non-hereditary ones, can fit the same data.

2.3 Viscoelasticity in Classical Continuum Thermodynamics

Many constitutive models for viscoelastic behavior in continuum thermodynamics are in integral form in an apparent attempt to make the history dependence explicit. These models are consistent with continuum thermodynamics because they are required to satisfy the principle of material frame indifference, mass and momentum balances, as well as the Clausius-Duhem inequality. Energy balance is not satisfied because some energy is stored and some is dissipated. However, the constraints available are not sufficient to precisely define the model. An excellent reference for the continuum thermodynamics viscoelastic theory is the book by Christensen (1971).

A material is simple if its response to all homogeneous processes determines its response to all arbitrary processes, those in which gradients of the thermodynamic variable exist (Truesdell, 1984, p. 144). A simple material is one for which the current values of the thermodynamic variables depend on the past history of the deformation gradient tensor, F, as well as the current value. While in principle the evolution of a particular state variable may depend on the past history of all variables, a simple material is one in which the stress, for example, depends only on the past history of the strain measure and the current values of other variables such as temperature. Coleman (1964) specialized the theory of simple materials to those with fading memory. A material is said to have fading memory if its current behavior is most strongly influenced by its behavior in the recent past. These assumptions seem to have been made to reduce the complexity of the fully general viscoelastic integral models produced in continuum mechanics.

One class of models is a series expansion generalizing linear viscoelasticity. An example is the isothermal Green-Rivlin polynomial integral model that assumes that the strain and temperature may be approximated as linear functionals of time and expands in a series based on these functionals,

$$\epsilon(t) = \int_0^t G_1(t-s_1)\dot{\sigma}(s_1)ds_1 + \int_0^t \int_0^t \int_0^t G_3(t-s_1, t-s_2, t-s_3)\dot{\sigma}(s_1)\dot{\sigma}(s_2)\dot{\sigma}(s_3)ds_1ds_2ds_3 + \ldots. \quad (11)$$

These models require experimentally fitting the large number of empirical time-dependent functions, G_i.

Models of this type that include thermal effects have proved more difficult to generate. Christensen mentions that a "coupled thermoviscoelastic theory which includes the temperature dependence of mechanical properties is necessarily nonlinear" (1971, p. 99). Crochet and Naghdi (1969), in their work on solids with fading memory, show that the strain at zero stress, denoted by $\beta(\theta(t))$, as a function of temperature, θ, only depends on the current, but not on the past, time. Christensen (1971, p. 101) makes the hypothesis for a thermoviscoelastic material in non-constant temperatures that the stress functional, $\mathcal{F}$, is of the form

$$\sigma(t) = \mathcal{F}^t_{\tau_s=0}\big(E(t-\tau_s) - \beta(t-\tau_s), E(t) - \beta(t)\big), \quad (12)$$

where τ_s is a temperature dependent modified time scale. The Green strain, $E = \frac{1}{2}(C - I)$, where $C = F^t F$, for deformation gradient, F, and I is the identity matrix, in an isothermal model is replaced by strain minus the strain at zero stress as a function of temperature. If it is assumed that the temperature variation is known, the problem is uncoupled. The linearization of the model is a generalization of the thermorheologically simple model. The temperature is assumed to be constant and homogeneous, i.e. the same at all points of the body.

To reduce the number of material functions that must be determined from experiment in nonlinear viscoelastic models such as the Green-Rivlin (11), some researchers have proposed single integral models. Lockett (1972, p. 115) indicates how the material functions in the Bernstein, Kearsley and Zapas three-dimensional isothermal model may be chosen so that the long-term behavior model for an incompressible material is the Mooney-Rivlin model. The functional satisfies material frame indifference since it is written in terms of the Green strain E. Assuming that the constitutive functional has an integral polynomial approximation,

$$\mathcal{F}^t_{-\infty} = mI + I\int_{-\infty}^t a(t-s)Tr(E(s))ds + 2\int_{-\infty}^t b(t-s)E(s)ds. \quad (13)$$

The constitutive model becomes for stress relaxation, where $E(t) = E_o H(t)$ and $H(t)$ is the Heaviside function so that $E(t)$ is held constant,

$$\sigma(t) = -pI + F(t)[mI + I\bar{a}(t)Tr(E_o) + 2\bar{b}(t)E_o]F^t(t), \tag{14}$$

where m is a scalar constant and

$$\bar{a}(t) = \int_0^t a(s)ds \quad \text{and} \quad \bar{b}(t) = \int_0^t b(s)ds.$$

Let $F(\infty)$ be the long-term deformation gradient. Put $a = \lim_{t\to\infty} \bar{a}(t)$ and $b = \lim_{t\to\infty} \bar{b}(t)$. The long-term, equilibrium, stress is

$$\sigma(t) = -pI + F(\infty)\frac{\partial W}{\partial E_o}F^t(\infty), \tag{15}$$

where

$$W = mTr(E_o) + \frac{1}{2}a(Tr(E_o))^2 + bTr(E_o^2) = \left(\frac{1}{2}m + b\right)(I_1 - 3)$$
$$+\frac{1}{4}\left(\frac{1}{2}a + b\right) - \frac{1}{2}b(I_2 - 3),$$

and $I_1 = Tr(C)$ and $I_2 = \frac{1}{2}\left(I_1^2 - Tr(C^2)\right)$ are the strain invariants of C. Therefore the long-term behavior is described by the Mooney-Rivlin model if $\frac{1}{2}a + b = 0$ and by the neo-Hookean if $a = b = 0$. Many functions $a(t)$ and $b(t)$ may have the required limits. This analysis gives no insight into what form the functions may take; $a(t)$ and $b(t)$ are to be determined by experiment.

The integral models are clearly quite difficult to work with. However, they do make explicit the dependence of the current behavior on the past loading history. These models are not used very often today because of their complexity.

2.4 Recent Ad Hoc Nonlinear Viscoelastic Models

Because of the complexities of applying the integral models, contemporary workers in viscoelasticity continue to seek simpler models. Several strategies have been employed to generate nonlinear viscoelastic models, but none of these is explicitly based on thermodynamics. One strategy is to write a term as a product of a time-dependent function and a function of other variables. This strategy is illustrated by the work of Brüller. For example to model PMMA, Brüller (1985, 1991) has proposed an empirical nonlinear creep compliance, generalizing the linear Voigt compliance, of the form

$$J(t, \sigma) = g_o(\sigma)J_o + g(\sigma)\sum_{i=1}^{m} J_i\left[1 - \exp\left(-\frac{t}{\tau_i}\right)\right], \tag{16}$$

in which only one function, $g(\sigma)$, a nonlinear function of stress, must be determined for all times, i.e. independent of time. The J_i are constants so that the time-dependent function has the form of a Prony series. Brüller takes the function, $g_o = 1$ and thinks of J_o as the instantaneous elastic creep compliance. The long-term behavior is given by the nonlinear relation $\epsilon(\sigma) = g(\sigma) \sum_{i=1}^{m} J_i \sigma$. Brüller (1987) successfully fits torsional creep curves of PVC and tensile creep curves of PMMA by taking the function $g(\sigma)$ to be an increasing, concave up, second order polynomial.

A second strategy previously employed to construct nonlinear viscoelastic models is to add a dissipation function that tends to zero over time to a classical hyperelastic strain energy density for the long-term behavior. A constitutive model devised by Tanaka and Yamada (1990) specifically to represent the stress-strain hysteresis in the loading-unloading of blood vessels avoids the elaborate spring and dashpot constructions of others and is based on evolution equations for the internal variables such as those used in models for viscoplasticity. But like those models, their model assumes that the total strain splits into the sum of an elastic and a viscoelastic part. Most of the material coefficients were found by numerically fitting the hysteresis curve, and may depend on the chosen strain range, so that the model as presented may not be predictive.

To explicitly capture finite viscoelastic long-term behavior, Holzapfel et al. (2000, p. 282ff) split the free energy into a sum of the volumetric elastic energy, isochoric elastic energy, and a viscoelastic contribution as described in Section 4 in Chapter 1. The idea is a development of the Bonet (2001) large strain viscoelastic model based on splitting the deformation gradient into elastic and viscous components and on a generalized Maxwell model for the viscous component. This model was extended to artery tissue by Holzapfel et al. (2002). This three-dimensional viscoelastic model for artery tissue, viewed as a two layer (media and adventitia) cylinder with collagen reinforcing fibers, that explicitly shows the long-term behavior, uses the Spencer hyperelastic composite construction presented in Holzapfel (2000). The isochoric elastic function is taken as the sum of a neo-Hookean model and an exponential model for the collagen from Holzapfel et al. (2000). Their goal was to achieve a weakly frequency independent hysteresis response, which they take to mean that the area between the loading and unloading uniaxial curves is nearly independent of frequency. Their example dissipative energy uses five Maxwell models with different relaxation times for each to cover a four decade time domain of 0.01 – 100 Hz. Another five factors are computed to make the dissipation energy nearly constant over the design frequency range. Therefore this model requires a large number of empirical dissipation constants in addition to those for the elastic terms.

A third strategy, especially common in models for soft biological tissue, is to make time-dependent the empirical coefficients in a hyperelastic model. Linear viscoelasticity may be viewed as starting with the linear elastic equilibrium relation $\tau = G\gamma$ and making the coefficient time-dependent to represent shear stress relaxation by $\tau(t) = G(t)\gamma$. Here τ is the shear stress, γ is the shear strain and $G(t)$ is the linear viscoelastic relaxation modulus. Other modelers seem to be inspired by this viewpoint to let the coefficients in a nonlinear hyperelastic, large displacement, equilibrium model be time-dependent in order obtain a candidate model for nonlinear viscoelasticity. Because the tissue is assumed incompressible, a common strategy is to take a standard hyperelastic model for rubber like the neo-Hookean, Mooney-Rivlin, or Ogden models and then allow the material parameters to be time-dependent. The Ogden model is generally agreed to be the most useful model for rubber. The rubber models are hyperelastic and therefore only describe equilibrium states. Such nonlinear viscoelastic models are ad hoc modifications of a hyperelastic strain energy density function developed for static behavior. Therefore the modified models really assume that a dynamic process can be viewed as a sequence of quasi-static states. Whether or not such models nearly fit the data, they are not based on thermodynamic principles.

For example, Prange and Margulies (2002) make the coefficients in the Ogden rubber model be time-dependent to represent their stress relaxation data for porcine brain tissue. They start with the simplest Ogden hyperelastic strain energy density.

$$\varphi(\lambda_1, \lambda_2, \lambda_3) = \frac{2\mu}{\alpha^2}[\lambda_1^\alpha + \lambda_2^\alpha + \lambda_3^\alpha - 3]. \tag{17}$$

To account for the time-dependent behavior, the shear modulus is assumed to be a function of time, in fact a two-term Prony series,

$$\mu(t) = \mu_o \left\{ 1 - \sum_{i=1}^{2} C_i[1 - \exp(-t/\tau_i)] \right\}. \tag{18}$$

For brain tissue, Miller and Chinzei (2002) also define a strain energy density based on making the coefficients in the Ogden model time-dependent. They obtain the stress at each time by differentiating this modified energy density function with respect to the stretch to obtain a uniaxial stress-strain relation both in tension and compression, a procedure that is thermodynamically valid only in equilibrium. No matter how well these models fit the particular test data used to obtain the form of the time-dependent coefficients, these models have not always been applied to predict different test data as a validation.

Another strategy is to split the stress into a sum of viscous and deviatoric terms, or more specifically to break the Cauchy stress into a volumetric stress, a long-term elastic stress and a viscoelastic stress parts (e.g. for brain tissue, Bilston et al., 2001; Shen et al., 2006). The latter two are deviatoric. The long-term elastic stress is obtained from the Mooney-Rivlin strain energy density. The viscoelastic stress is a form of the Maxwell model modified so that the relaxation coefficients are strain-dependent, and a damping function taken as the coefficient of the viscous stress accounts for the nonlinear response and apparent yielding as exhibited experimentally by a stress maximum as a function of strain. This viscoelastic model requires some 32 empirical coefficients. There is a question of whether the Mooney-Rivlin model applies because the brain tissue has different properties in tension and compression. Bilston et al. argue that their model is acceptable since the only difference arises in the volumetric stress because the bulk modulus term is affected, while the deviatoric stresses are not affected. Such a splitting has not been directly supported by experiment.

3 Nonlinear, Maximum Dissipation, Viscoelastic Model

The maximum dissipation thermodynamically consistent construction of nonlinear viscoelastic models avoids the complexity of the integral models and avoids the pitfalls of the ad hoc models described in the previous subsection. Design work and analysis involving thermoviscoelastic materials is considerably simpler with a model that can represent the material hereditary effects without requiring prohibitively time consuming experiments to determine time-dependent empirical functions. The construction of Chapter 3 requires only a thermostatic constitutive model, φ, that represents the long-term, equilibrium behavior of a viscoelastic material and the thermodynamic relaxation modulus. The long-term behavior is assumed to be non-aging. Two criteria, a physical hypothesis based on maximum dissipation during relaxation and the organizing of the viscoelastic response around the long-term equilibrium states, reduce the number of experiments required to determine a model.

Recall from Section 4 in Chapter 3 that the nonlinear maximum dissipation evolution equation as a system of first-order nonlinear ordinary differential equations in terms of the original state variables, x_i, and the control variables, y_i, is equivalently in terms of the thermostatic function for the long-term behavior, φ, and the thermodynamic relaxation modulus,

$$\begin{bmatrix} \frac{dx_1}{dt} \\ \frac{dx_2}{dt} \\ \cdots \\ \frac{dx_n}{dt} \end{bmatrix} = -k \begin{bmatrix} \frac{\partial^2 \varphi}{\partial x_1^2} & \frac{\partial^2 \varphi}{\partial x_1 \partial x_2} & \cdots & \frac{\partial^2 \varphi}{\partial x_1 \partial x_n} \\ \frac{\partial^2 \varphi}{\partial x_2 \partial x_1} & \frac{\partial^2 \varphi}{\partial x_2^2} & \cdots & \frac{\partial^2 \varphi}{\partial x_2 \partial x_n} \\ & \cdots\cdots\cdots\cdots & & \\ \frac{\partial^2 \varphi}{\partial x_n \partial x_1} & \frac{\partial^2 \varphi}{\partial x_n \partial x_2} & \cdots & \frac{\partial^2 \varphi}{\partial x_n \partial x_n} \end{bmatrix}^{-2} \begin{bmatrix} y_1(t) + \frac{\partial \varphi}{\partial x_1} \\ y_2(t) + \frac{\partial \varphi}{\partial x_2} \\ \cdots \\ y_n(t) + \frac{\partial \varphi}{\partial x_n} \end{bmatrix}. \tag{19}$$

The class of maximum dissipation viscoelastic models assumes that dissipation occurs in a viscoelastic material without affecting the equilibrium, or long-term state. Therefore, in viscoelastic behavior, the thermodynamic relaxation modulus, k, is assumed constant as the control and state variables change.

4 Classical Models That May Be Interpreted as a Maximum Dissipation Models

As examples, the construction is applied to simple linear and nonlinear models for which closed form calculations can be made and then to nonlinear models of rubber and of biological soft tissue for which numerical solutions are required. The viscoelastic evolution equations can be illustrated by two uniaxial processes; in one the long-term behavior is linear and in the other nonlinear. As examples in closed form, several uniaxial models including the linear Kelvin-Voigt and the standard linear solid as well as a nonlinear model in a form similar to that of Brüller are derived from the maximum dissipation hypothesis.

4.1 Linear Uniaxial Long-Term Behavior

Several classical spring and dashpot, uniaxial, linear viscoelastic models are recovered from this construction when the long-term manifold is linear elastic. Perhaps surprisingly, some of the classical linear viscoelastic differential equations can be interpreted as maximum dissipation relaxation process models. The choice of the relaxation modulus, k, determines the particular model. The general forms of creep and stress relaxation are first derived from a linear long-term constitutive relation.

In uniaxial stress relaxation, where the state variable is the stress, σ, and the conjugate control variable is the strain, ϵ, the energy density function $\varphi(\sigma) = 0.5\sigma^2/E_1$ for a linear elastic long-term manifold so that the generalized energy density is $\varphi^*(\sigma;\epsilon) = 0.5\sigma^2/E_1 - \epsilon\sigma$. The evolution equation (19) for the state variable is then

$$\dot{\sigma} = -k_\sigma E_1^2\left[-\epsilon(t) + \frac{\sigma}{E_1}\right]. \tag{20}$$

In the dual case of uniaxial creep when the state variable is the strain, ϵ, and the conjugate control variable is the stress, σ, the linear elastic energy function for the long-term manifold is $\varphi(\epsilon) = 0.5E_1\epsilon^2$ so that the generalized energy density is $\varphi^*(\epsilon;\sigma) = 0.5E_1\epsilon^2 - \epsilon\sigma$. The evolution of ϵ is predicted by (19) to be

$$\dot{\epsilon} = -k_\epsilon E_1^{-2}[-\sigma(t) + E_1\epsilon]. \tag{21}$$

The solution is

$$\sigma = E_1\epsilon + \frac{E_1^2}{k}\frac{d\epsilon}{dt}. \tag{22}$$

In the long-term, rate independent case for which $d\epsilon/dt = 0$, Eq. (22) reduces to $\sigma = E_1\epsilon$, which is also the constitutive model for the long-term behavior.

Since $E_1\epsilon$ is the elastic portion of the stress, $\sigma - E_1\epsilon$ is the anelastic portion of the stress. The rate of change of the generalized energy function for this relaxation process is exactly the intrinsic dissipation, i. e. the mechanical dissipation, which is defined to be the product of the anelastic stress and the rate of change of strain (Lemaitre and Chaboche, 1990, p. 145),

$$\frac{d\varphi^*}{dt} = E_1\epsilon\dot{\epsilon} - \sigma\dot{\epsilon} = (E_1\epsilon - \sigma)\dot{\epsilon}.$$

The process would stop if the anelastic stress would reach zero, so that $\sigma = E_1\epsilon$.

The Kelvin-Voigt model of a spring and dashpot in parallel is obtained when $k_\epsilon = E_1^2/c$, where c is the dashpot viscous constant and E_1 is the spring constant. The Kelvin-Voigt model is locally maximum dissipative when strain is the state variable because (21) agrees with the Kelvin-Voigt differential equation, $c\dot{\epsilon}+E\epsilon = \sigma$, where the dot indicates the time derivative. The Kelvin-Voigt model does not admit stress relaxation because there is no $\dot{\sigma}$-term. If the strain is held fixed so that $\dot{\epsilon} = 0$, the stress remains unchanged, $\sigma = E\epsilon$. Therefore to represent this fact, take $k_\sigma = 0$ in (20). The Kelvin-Voigt equation is maximum dissipation and is locally maximum dissipation for time-dependent control variables.

The standard linear solid defined as a spring with constant, E_1, in parallel with a Maxwell model, which is itself a dashpot in series with a second spring is a maximum dissipation model but is not locally maximum dissipation when the control variables are time-dependent. The spring constant and dashpot viscosity in the Maxwell model are E_2 and c respectively. The differential equation that the applied stress, σ, and the total strain, ϵ, must satisfy is

$$E_2\sigma + c\dot{\sigma} = E_1E_2\epsilon + c(E_1 + E_2)\dot{\epsilon}. \tag{23}$$

In stress relaxation for fixed ϵ, after putting $\dot{\epsilon} = 0$ in (23) and integrating, the stress is

$$\sigma(t) = \sigma(0)\exp(-kt) + \epsilon E_1\big(1 - \exp(-E_2t/c)\big). \tag{24}$$

Over the long term, the stress-strain relation becomes linear, $\sigma = \epsilon E_1$.

Creep is predicted by (23) with $\dot{\sigma} = 0$. The solution is

$$\epsilon(t) = \frac{\sigma}{E_1} + \exp\left(\frac{-E_1E_2t}{c(E_1 + E_2)}\right)\left(\epsilon(0) - \frac{\sigma}{E_1}\right). \tag{25}$$

Creep again has a linear elastic long-term manifold with $\sigma = E_1\epsilon$. The long-term manifold predicted by (23) is therefore the same in both creep and stress relaxation.

The evolution form of stress relaxation of the standard linear solid is given by (19) with the state variable, σ, and fixed ϵ with long-term manifold defined by $\varphi(\sigma) = 0.5\sigma^2/E_1$. The result is given by (20) with the relaxation modulus $k_\sigma = E_2/(E_1c)$.

$$\dot{\sigma} = -\frac{E_1E_2}{c}\left(-\epsilon + \frac{\sigma}{E_1}\right). \tag{26}$$

This equation agrees with setting $\dot{\epsilon} = 0$ in (23).

Creep of the standard linear solid is obtained as well from (21) when strain is the state variable by taking $k_\epsilon = E_1^2E_2/c(E_1 + E_2)$. The long-term behavior of the standard linear solid is $\sigma = E_1\epsilon$ so that $\varphi(\epsilon) = 0.5E_1\epsilon^2$.

$$\dot{\epsilon} = \frac{E_2}{c(E_1 + E_2)} \left(-\sigma + E_1 \epsilon\right). \tag{27}$$

This agrees with the equation obtained by setting $\dot{\sigma} = 0$ in (24).

The standard linear solid is not, for a general time-dependent control variable, locally maximum dissipative because (23) does not agree with the maximum dissipation evolution equation from (19) under stress control,

$$\dot{\epsilon} = \frac{E_2}{c(E_1 + E_2)} \left(-\sigma(t) + E_1 \epsilon\right), \tag{28}$$

when the control variable $\sigma(t)$ is not constant; the evolution equation (28) does not include the $\dot{\sigma}$-term which appears in (23).

The viscoelastic models (20) and (21) in general, have no springs and dashpots interpretation, an idea that should be abandoned in constitutive modeling of viscoelastic materials.

The relaxation time in these parallel element linear models is defined as $t_R = E_1/k$. Increasing the material constant, k, as expected, reduces the relaxation time. The strategy of assuming k is a constant implies that there is a single relaxation time. Other researchers have defined models, especially multiple spring and dashpot combinations in series, with several relaxation times. This is done when it appears that different phenomena each cause relaxation. Such linear series models can be dealt with in this context by superposing several relaxation processes, each with its individual relaxation time. But in many cases the various relaxation times have no physical interpretation.

An example of a forced process under sinusoidal loading for a uniaxial linearly elastic long-term behavior is given in Section 5 in Chapter 3.

4.1.1 Relation to Classical Linear Viscoelasticity

The thermodynamic relaxation modulus, k, is not explicitly time-dependent as is the linear viscoelastic relaxation modulus, $G(t)$, but they are related. For example, the viscoelastic model obtained from a linear elastic long-term behavior is linear viscoelastic. The generalized energy density for a uniaxial loading in strain control is $\varphi^* = \sigma^2/2E_1 - \sigma\epsilon$ which induces the evolution equation $\dot{\sigma} = -kE^2(\sigma/E_1 - \epsilon)$. For stress relaxation with a fixed strain, ϵ_o, this equation integrates to

$$\sigma(t) = [\sigma(0) - E_1\epsilon_o]\exp(-kE_1 t) + E_1\epsilon_o.$$

The linear viscoelastic relaxation modulus is obtained as $G(t) = \sigma(t)/\epsilon_o$. Therefore,

$$G(t) = E_1[1 - \exp(-kE_1 t)] + \frac{\sigma(0)}{\epsilon_o}\exp(-kE_1 t).$$

Linear viscoelasticity assumes that the ratio $\sigma(0)/\epsilon_o = G(0)$ is the same for all initial strains. The value of $G(t)$ as $t \to \infty$ tends to the equilibrium value E_1 as expected. The thermodynamic modulus, k, plays a role akin to that of the decay parameter β in the Prony series.

4.2 Nonlinear Uniaxial Examples Solvable in Closed Form

A form very close to the nonlinear Brüller empirical uniaxial creep compliance model for PMMA (Eq. 16) restricted to a single relaxation time can be obtained from the maximum dissipation assumption. Brüller was able to fit the creep of some polymers with a compliance that was a quadratic function of stress. As a simple case which can be computed in closed form, consider the long-term constitutive model, $\epsilon = A\sigma^3$, where A is a constant. The corresponding generalized energy with the strain as the state variable is

$$\varphi^*(\epsilon;\sigma) = \frac{3}{4A^{1/3}}\epsilon^{4/3} - \epsilon\sigma.$$

The affinity is $X = \epsilon^{1/3}/A^{1/3} - \sigma$, and the generalized energy after transformation of the state variable to the affinity is

$$\varphi'^*(X,\sigma) = \frac{3A}{4}(X+\sigma)^4 - A\sigma(X+\sigma)^3.$$

The gradient relaxation evolution equation in terms of the affinity is $dX/dt = -3kAX(X+\sigma)^2$ so that since $X < 0$,

$$\ln\left(\frac{-X}{X+\sigma}\right) + \frac{\sigma}{X+\sigma} = -3kA\sigma^2 t + \sigma^2 c.$$

or

$$\frac{X}{X+\sigma}\exp\left(\frac{\sigma}{X+\sigma}\right) = -\exp(\sigma^2 c)\exp(-3kA\sigma^2 t), \tag{29}$$

where $\exp(\sigma^2 c) = [(A^{1/3}\sigma - \epsilon(0)^{1/3})/\epsilon(0)^{1/3}]\exp(A^{1/3}\sigma/\epsilon(0)^{1/3})$ depends on the initial condition. Make the approximation

$$\exp\left(\frac{\sigma}{X+\sigma}\right) \sim 1 + \frac{\sigma}{X+\sigma}$$

in Eq. (29) so that

$$(X+\sigma)^2 - \sigma^2 = -(X+\sigma)^2\exp(\sigma^2 c)\exp(-3kA\sigma^2 t).$$

When the state variable is substituted for the affinity,

$$\epsilon(t) = A\sigma^3 \left(1 + \exp(\sigma^2 c)\exp(-3kA\sigma^2 t)\right)^{-3/2}$$
$$\sim A\sigma^3 \left(1 - \frac{3}{2}\exp(\sigma^2 c)\exp(-3kA\sigma^2 t)\right), \qquad (30)$$

using a Taylor series. This form is similar to the Brüller creep model (Eq. 16), but the relaxation time, $\tau = 1/3kA\sigma^2$, is inversely proportional to a function of the stress, rather than independent of stress as in Brüller's model. The stress dependence is given by the derivative of the long term strain, $\epsilon(\sigma)$, with respect to the stress. This modification of Brüller's model is a maximum dissipation model, but the original Brüller model is not.

5 Nonlinear Maximum Dissipation Viscoelastic Model for Rubber

The maximum dissipation construction is applied to generate new viscoelastic models for rubber. An example involving temperature dependence and applicable to rubber assumes that the long-term quasi-static behavior is described by the Zeng and Haslach (1996) nonlinear thermoelastic generalization of the Mooney-Rivlin model for rubber. The resulting thermoviscoelastic model constructed from the thermoelastic version of the Mooney-Rivlin model is applied to two degree of freedom problems including the biaxial stretch behavior of a rubber sheet, including biaxial creep, and to the temperature-stretch behavior of a rod. The model reproduces the unusual temperature-stretch interaction exhibited by rubber, called the adiabatic Gough-Joule effect. The following results first appeared in Haslach and Zeng (1999).

5.1 Uniaxial Dynamic Response of Isothermal Rubber

The standard isothermal experimental uniaxial tests of rubber under harmonic loading show that, after a transient region, the response, which is out of phase with the load, approaches a steady state, producing a limit cycle in stress-strain space. The tangent modulus is the tangent of the phase shift angle. If the model is linear viscoelastic, loss and storage moduli can be calculated from the measured response amplitude. The tangent modulus and the various loss and elastic moduli have been used to characterize rubber, but rubber is not linear viscoelastic.

A test for any proposed nonlinear viscoelastic model is that it reproduces steady state solutions of the correct type under sinusoidal loading. The uniaxial evolution equation (Eq. (19)) for the stretch, λ, is periodic in time with period, $\tau = 2\pi/\omega$.

$$\frac{d\lambda}{dt} = -k\left(\frac{\partial^2\varphi}{\partial\lambda^2}\right)^{-2}\left(-A\sin(\omega t) + \frac{d\varphi}{d\lambda}\right). \tag{31}$$

Any periodic model produces a steady state solution for a bounded response. Let $\varphi(t, 0, \lambda_o)$ be a solution with initial conditions $t_o = 0$ and $\lambda(0) = \lambda_o$. Define $\phi_n(t) = \varphi(t + n\tau, 0, \lambda_o)$ for $0 \le t \le \tau$. Then because of the uniqueness of the solution and the periodicity of the differential equation, if $\phi_1(0) \ge \phi_0(0)$, then $\phi_{n+1}(t) \ge \phi_n(t)$ for all n and for $0 \le t \le \tau$. Alternatively, if $\phi_1(0) \le \phi_0(0)$, then $\phi_{n+1}(t) \le \phi_n(t)$ for all n and for $0 \le t \le \tau$. This observation shows that if a solution is bounded, then there exists a periodic limit solution, $\Phi(t)$, to which $\varphi(t + n\tau, 0, \lambda_o)$ converges monotonically and uniformly (Hale and Koçak, 1991, p. 115). Its initial condition is $\Phi(0) = \lim_{n\to\infty}\varphi(\tau, 0, \phi_n(0))$. The solution, $\Phi(t)$, is the steady state response.

This behavior can be observed in the response to sinusoidal loading for the evolution equation with long-term behavior given by the isothermal neo-Hookean or the Mooney-Rivlin models discussed in Section 5.2 in Chapter 2. The isothermal Mooney-Rivlin model for an incompressible material in terms of the principal stretches, λ_i, is

$$\varphi(\lambda_1, \lambda_2) = C_1\left(\lambda_1^2 + \lambda_2^2 + \lambda_1^{-2}\lambda_2^{-2} - 3\right) + C_2\left(\lambda_1^2\lambda_2^2 + \lambda_1^{-2} + \lambda_2^{-2} - 3\right) \tag{32}$$

where C_1 and C_2 are material constants. If $C_2 = 0$, the material is called neo-Hookean. The incompressibility condition is $\lambda_1\lambda_2\lambda_3 = 1$ so that $\lambda_3 = \lambda_1^{-1}\lambda_2^{-1}$.

In the uniaxial case, the incompressibility condition requires that $\lambda_2 = \lambda_3 = \lambda_1^{-1/2}$. Dropping the subscripts, the uniaxial Mooney-Rivlin model is

$$\varphi(\lambda) = C_1\left(\lambda^2 + 2\lambda^{-1} - 3\right) + C_2\left(\lambda^{-2} + 2\lambda - 3\right). \tag{33}$$

The uniaxial stretch evolution equation corresponding to the maximum dissipation assumption for the Mooney-Rivlin model under sinusoidal stress is

$$\dot{\lambda} = \frac{k}{[2C_1(1 + 2\lambda^{-3}) + 6C_2\lambda^{-4}]^2}\left[A\sin(\omega t) - 2C_1(\lambda - \lambda^{-2}) - 2C_2(-\lambda^{-3} + 1)\right], \tag{34}$$

a non-autonomous periodic first order equation.

Experiment shows that the steady state stretch response for rubber to a stress $A\sin(\omega t)$ should be sinusoidal with a mean stretch of one, the unstretched state, and a phase shift with respect to the load. A numerical solution of Eq. (34) verifies that the model constructed from maximum dissipation processes with long-term behavior given by the isothermal Mooney-Rivlin model does generate such a steady state stretch response. The Mooney-Rivlin stretch response to a stress of $(750 \times 10^3)\sin(16t)$ Pa when $C_1 = 2.240 \times 10^5$ Pa and $C_2 = 8.512 \times 10^3$, for $k = 600 \times 10^3$ Pa/s, tends to a steady state (Fig. 4.1). When

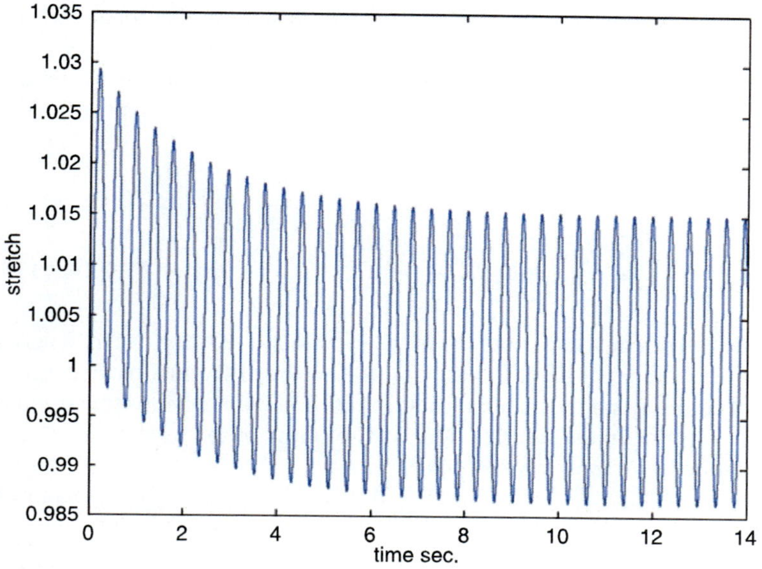

Fig. 4.1 Uniaxial stretch response to a sinusoidal stress for the gradient Mooney-Rivlin model, λ versus time

the stress is cycled about the zero stress, the strain response is asymptotically stable, and the limit solution is periodic of the same period as the forcing function, has mean stretch of one, but with a phase shift. In both cases, $\phi_{n+1}(t) \leq \phi_n(t)$ for all n. Numerical experiments indicate that increasing C_2 in the Mooney-Rivlin model reduces the amplitude of the stretch response, as does increasing k. The transient is reduced by increasing k. The phase shift depends primarily on C_1.

5.2 A Thermostatic Constitutive Model for Rubber

Few constitutive models have been proposed for rubber which include its thermal behavior; most are isothermal models. To include the influence of temperature, this class of thermoviscoelastic evolution equations requires a temperature dependent long-term behavior model. A material, such as rubber, even though it is incompressible under forces, exhibits thermal expansion and other thermal responses. A mechanically incompressible material has volume expansion, under variations in temperature, of the form

$$J(\theta_c) = \lambda_1\lambda_2\lambda_3 = 1 + H(\theta_c), \tag{35}$$

where the λ_i are the principal stretches and θ_c is the change in temperature from a reference temperature, θ_0.

A model can be constructed from any isothermal thermostatic Helmholtz free energy, φ, using the experimentally measured dependence on temperature of the shear modulus, of the volume expansion, and of the specific heat. As an example, the Helmholtz free energy is obtained for rubberlike materials by generalizing the isothermal Mooney-Rivlin model (32). The generalized thermoelastic Mooney-Rivlin equation is of the form (Zeng and Haslach, 1996),

$$\begin{aligned}\varphi(\lambda_1, \lambda_2, J, \theta_c) = C_1(\theta_c)\left[\lambda_1^2 + \lambda_2^2 + \lambda_1^{-2}\lambda_2^{-2}J^2(\theta_c) - 3\right] \\ + C_2\left[\lambda_1^2\lambda_2^2 + (\lambda_1^{-2} + \lambda_2^{-2})J^2(\theta_c) - 3\right] + G(\theta_c). \quad (36)\end{aligned}$$

Consistency with the isothermal model requires that $C_1(0) = C_1$, the material constant in the isothermal model, and that $G(0) = 0$. All thermal dependence in this model occurs in either the material coefficients of the stretch terms (the strain invariants), in $J(\theta_c)$ or in the function $G(\theta_c)$. The forms of the moduli are determined from their behavior in purely thermal expansion.

In this thermoelastic extension of the Mooney-Rivlin model, the function, $J(\theta_c)$, has the form,

$$J(\theta_c) = \bar{\lambda}^3(\theta_c) \sim (1 + \alpha\theta_c)^3, \quad (37)$$

where $\bar{\lambda}(\theta_c)$ is the stretch in thermal expansion, and α is the axial thermal expansion coefficient, which here is approximated as temperature independent and constant. The stress relations obtained from Eq. (36) in the case of unconstrained thermal expansion are equal to zero and have solutions consistent with $J(\theta_c) = \bar{\lambda}^3(\theta_c)$ since the material is isotropic. The experimental volume thermal expansion coefficient is approximately constant (Scott, 1935) and therefore so is α. Integration of $\alpha = d \ln L/d\theta_c$ yields $\bar{\lambda} \sim 1 + \alpha\theta_c$.

The shear moduli are by definition

$$\mu_{ij} = \frac{1}{2}\frac{\partial^2\varphi}{\partial\lambda_i\partial\lambda_j}.$$

The form of Eq. (36) and the definition of the shear modulus in the case of thermal expansion requires that the function $C_1(\theta_c)$ have the form

$$C_1(\theta_c) = \frac{1}{2}\mu(\theta_c) - C_2\bar{\lambda}^2(\theta_c). \quad (38)$$

Experiment (Shen, 1969; Wood, 1973) shows that the isotropic shear modulus is linear in temperature, $\mu(\theta_c) = A + B\theta_c$, where A and B are material constants. Therefore, $C_1(\theta_c)$ is quadratic in temperature. At the reference state, the function $C_1(\theta_c)$ reduces to C_1 as required since the isothermal shear modulus is twice the sum of the isothermal material constants, $\mu(0) = 2(C_1 + C_2)$, and $\bar{\lambda}(0) = 1$.

The function $G(\theta_c)$, which ensures the proper entropy expression, can be determined by the relation

$$\begin{aligned}\frac{d^2G(\theta_c)}{d\theta_c^2} &= -\frac{c(\theta_c)}{\theta_0+\theta_c} - \left[\bar{\lambda}^2(\theta_c) + 2\bar{\lambda}^{-1}(\theta_c)J(\theta_c) - 3\right]\frac{d^2C_1(\theta_c)}{d\theta_c^2} \\ &- 4\bar{\lambda}^{-1}(\theta_c)\frac{dC_1(\theta_c)}{d\theta_c}\frac{dJ(\theta_c)}{d\theta_c} - 2C_1(\theta_c)\bar{\lambda}^{-1}(\theta_c)\frac{d^2J(\theta_c)}{d\theta_c^2} \\ &- 2C_2\bar{\lambda}^{-2}(\theta_c)\left[\left(\bar{\lambda}^3(\theta_c)+J(\theta_c)\right)\frac{d^2J(\theta_c)}{d\theta_c^2} + \left(\frac{dJ(\theta_c)}{d\theta_c}\right)^2\right], \quad (39)\end{aligned}$$

where the specific heat under constant deformation is by definition

$$c(\theta) = -\theta\frac{\partial^2\varphi}{\partial\theta^2}.$$

Experiment (Wunderlich and Baur, 1970) shows that for rubber $c(\theta_c) = C + D\theta_c$, where C and D are material constants.

This model predicts two experimental thermal responses, called the Gough-Joule effects, that distinguish rubber from other polymers. Rubber emits heat when stretched and recovers its original temperature when unstretched. A rubber specimen held at a constant stress contracts if heated, the opposite behavior of most other materials. Both natural and vulcanized rubber exhibit such behavior; it is not affected by sulfur crosslinking of the rubber molecules.

5.3 A Nonlinear Thermoviscoelastic Model for Rubber

The viscoelastic evolution equations constructed from the classical Mooney-Rivlin isothermal model are tested on a rubber sheet subject to biaxial, in-plane stress perturbations. This is a biaxial creep process in which the evolution of the two in-plane stretches is viewed as a non-equilibrium relaxation process. The same behavior is modeled at various uniform sheet temperatures to examine how the Gough-Joule effects influence the predicted viscoelastic response. A uniaxial rod which is deformable and which is subjected to temperature changes is viewed as a two-dimensional system with state variables the stretch and the uniform temperature of the rod. Its response to perturbations of stress and entropy is viewed as a relaxation process. Finally the uniaxial rod subjected to a stress perturbation at constant entropy is investigated to test the ability of the thermoviscoelastic model to predict the adiabatic Gough-Joule phenomenon. The relaxation modulus, k, has not been measured experimentally; values are chosen for the purpose of illustration.

In each case, the system of evolution equations is solved by the Runge-Kutta method, with the initial conditions given by the original state variables and the perturbed control variables. Convergence is verified by repeating the calculation with smaller time steps.

5.4 Sudden Stress Perturbations in an Isothermal Rubber Sheet

Most linear viscoelastic models, like the Kelvin-Voigt, produce a relaxation path which is an exponential function of time. The nonlinear model based on the isothermal Mooney-Rivlin strain energy density function does not predict such an exponential relaxation. To apply the viscoelastic model, Eq. (19), to an isothermal rubber sheet, choose the principal stretches, λ_1 and λ_2, as the state variables, the negative engineering stresses, s_1 and s_2, as the control variables, and the isothermal Mooney-Rivlin equation as the function φ. Then Eq. (19) becomes,

$$\begin{bmatrix} \frac{d\lambda_1}{dt} \\ \frac{d\lambda_2}{dt} \end{bmatrix} = k \begin{bmatrix} \frac{\partial^2\varphi}{\partial\lambda_1^2} & \frac{\partial^2\varphi}{\partial\lambda_1\partial\lambda_2} \\ \frac{\partial^2\varphi}{\partial\lambda_2\partial\lambda_1} & \frac{\partial^2\varphi}{\partial\lambda_2^2} \end{bmatrix}^{-2} \begin{bmatrix} s_1 - \frac{\partial\varphi}{\partial\lambda_1} \\ s_2 - \frac{\partial\varphi}{\partial\lambda_2} \end{bmatrix}. \tag{40}$$

The values of the Mooney-Rivlin constants, $C_1 = 2.240 \times 10^5$ Pa and $C_2 = 8.512 \times 10^3$, are those used in Zeng and Haslach (1996).

If the sudden stress perturbation, s_1 and s_2, is given, Eq. (40) is solved for the evolution of λ_1 and λ_2. As a simple example of a biaxial creep test with unequal loads, an initially unloaded rubber sheet is stressed by $s_1 = 9.456 \times 10^5$ Pa and $s_2 = 7.359 \times 10^5$ Pa with various relaxation moduli, $k = 5.0 \times 10^7, 2.0 \times 10^7$, and 1.0×10^7 Pa/s respectively, for the initial conditions $\lambda_1 = \lambda_2 = 1.0$ and $s_1 = s_2 = 0$ at $t = 0$. Figs. 4.2 and 4.3 show that the principal stretches relax asymptotically to the equilibrium values of $\lambda_1 = 2.0$ and $\lambda_2 = 1.5$, corresponding to the perturbed stresses as calculated from the Mooney-Rivlin equation. Furthermore, the larger the relaxation modulus, k, the faster the relaxation to equilibrium. The relaxation rate is initially slow, then increases, and finally slows during the asymptotic approach. This generates a change of concavity in the stretch as a function of time. The smaller stretch, λ_2, in the direction of the lower stress, shows an initial contraction so that it does not relax monotonically to equilibrium.

The short initial compressive stretch response is not apparent in a biaxial creep test for which the initial stress perturbations are equal (Fig. 4.4). However, the relaxation stretch as a function of time still exhibits a change in concavity. Here $k = 3.0 \times 10^7$ Pa/s. In the midrange of the relaxation, the shape of stretch curve as a function of time is nearly linear. This region would be the best in which to experimentally measure k. Numerical calculations not presented in a figure show that, as expected, the relaxation time decreases with an increase in the magnitudes of the sudden stress perturbations for fixed k.

Conversely, a classical relaxation test is performed by stretching the sheet unequally to $\lambda_1 = 2.0$ and $\lambda_2 = 1.5$ at $t = t_0$ by the corresponding biaxial equilibrium stresses. If these biaxial stresses are suddenly released, i.e. perturbed

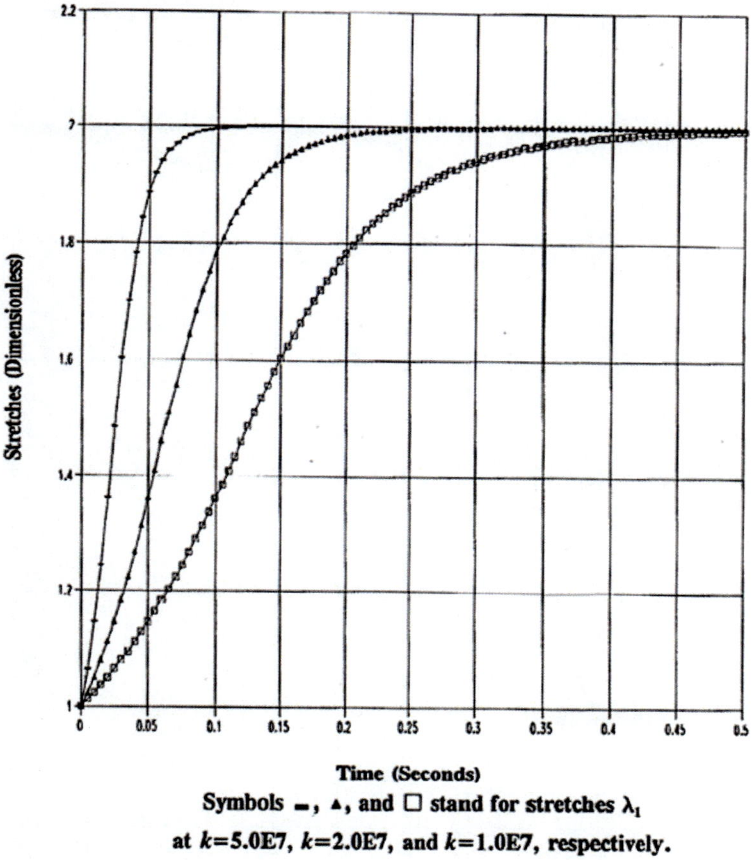

Fig. 4.2 Rubber sheet loaded by unequal biaxial tensile stress perturbations at various k- biaxial creep: λ_1 versus time (Zeng, 1995)

to zero, the stretches relax to zero along a curve which has the qualitative shape of an exponential curve (Fig. 4.5). The relaxation times of the two principal stretches are nearly equal. Again, $k = 3.0 \times 10^7$ Pa/s.

5.5 The Sheet Response at Different Constant Temperatures

The influence of the Gough-Joule effects on the thermoviscoelastic relaxation of rubber can be seen by repeating some of the previous calculations at various constant temperatures. The response of biaxial creep in a sheet to a change in the equilibrium temperature is consistent with the fact that a rubber specimen held at a constant stress contracts if the uniform temperature of the material is increased.

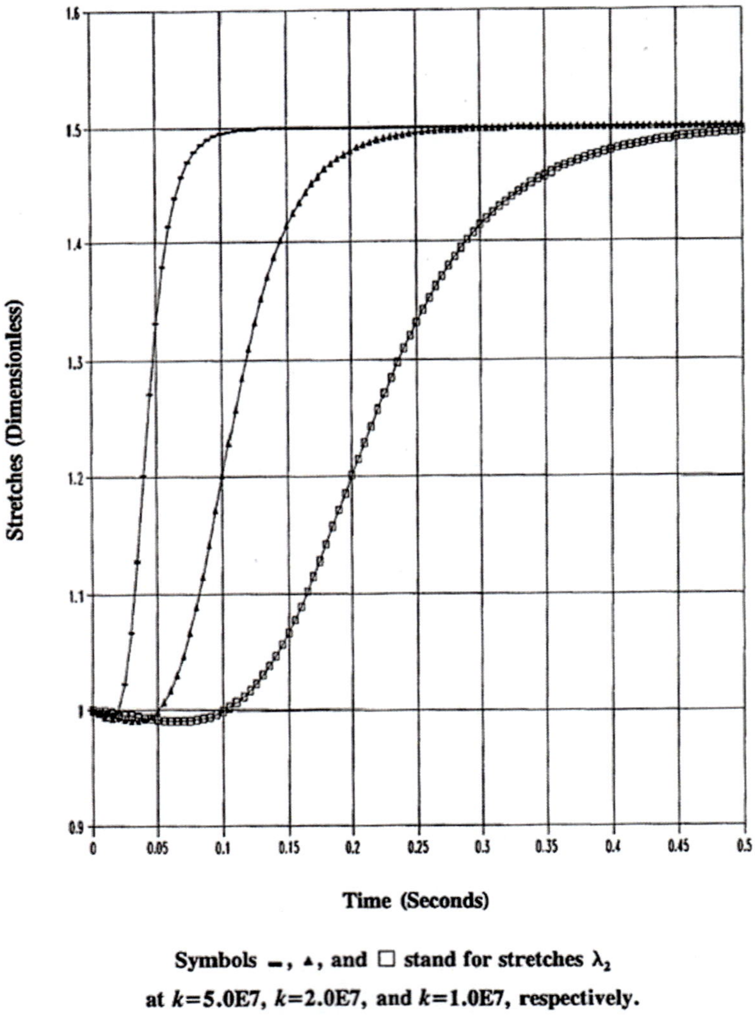

Fig. 4.3 Rubber sheet loaded by unequal biaxial tensile stress perturbations at various k- biaxial creep: λ_2 versus time (Zeng, 1995)

Varying the uniform temperatures in the transient viscoelastic response of a rubber sheet requires using Eq. (40) with the Mooney-Rivlin equation replaced by the thermoelastic constitutive model of Eq. (36). The unequal stress perturbations in a biaxially loaded sheet chosen are those corresponding at equilibrium at four different temperatures, 20°C, 50°C, 80°C, and 120°C and the stretches, $\lambda_1 = 2.0$ and $\lambda_2 = 1.5$. The initial conditions and the Mooney-Rivlin constants are the same as those above and the relaxation modulus is, $k = 3.0 \times 10^7$ Pa/s.

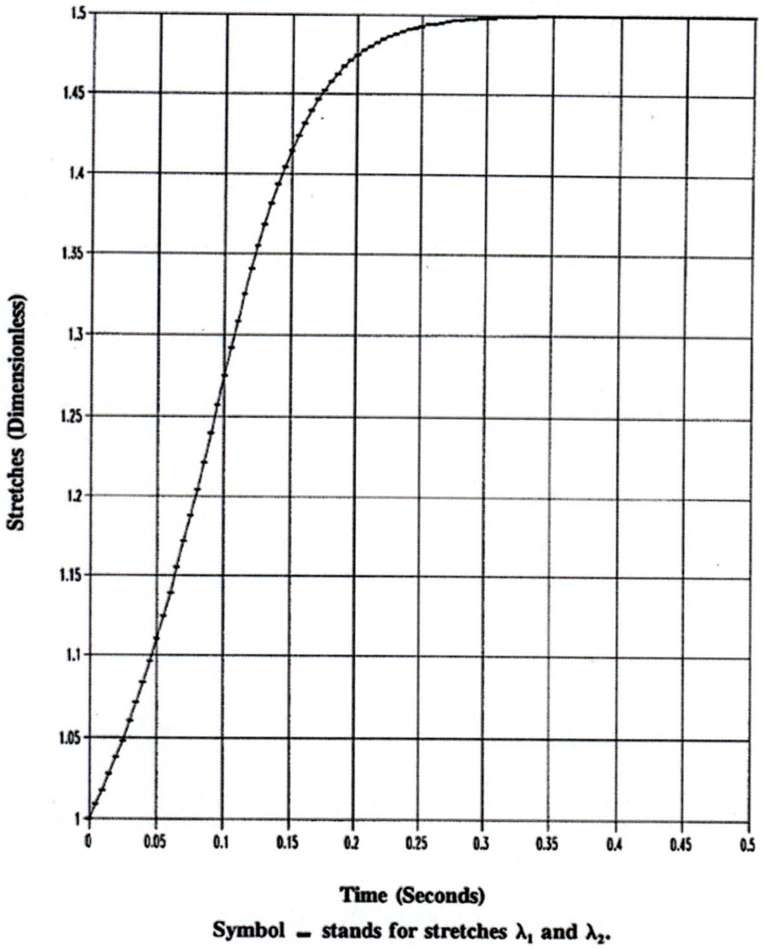

Fig. 4.4 Rubber sheet loaded by equal biaxial tensile stress perturbations- relaxation: λ_1 and λ_2 versus time (Zeng, 1995)

The shear modulus as a function of temperature θ measured on the Kelvin scale is $\mu(\theta) = 4.07 \times 10^{-1} + 8.9 \times 10^{-4}\theta$ MPa, and the linear thermal expansion coefficient is $\alpha = 6.0 \times 10^{-4}/K^{\circ}$. The relaxation time for both stretches increases slightly as the temperature is increased (Figs. 4.6 and 4.7). On the other hand, in polymers described by the Kelvin-Voigt model, for example, the relaxation time decreases at higher constant temperatures. The opposite response is predicted here because of the exceptional thermal response of rubber, the Gough-Joule effects. The rubber, which contracts more at higher temperatures under a constant tensile stress, resists the relaxation to equilibrium.

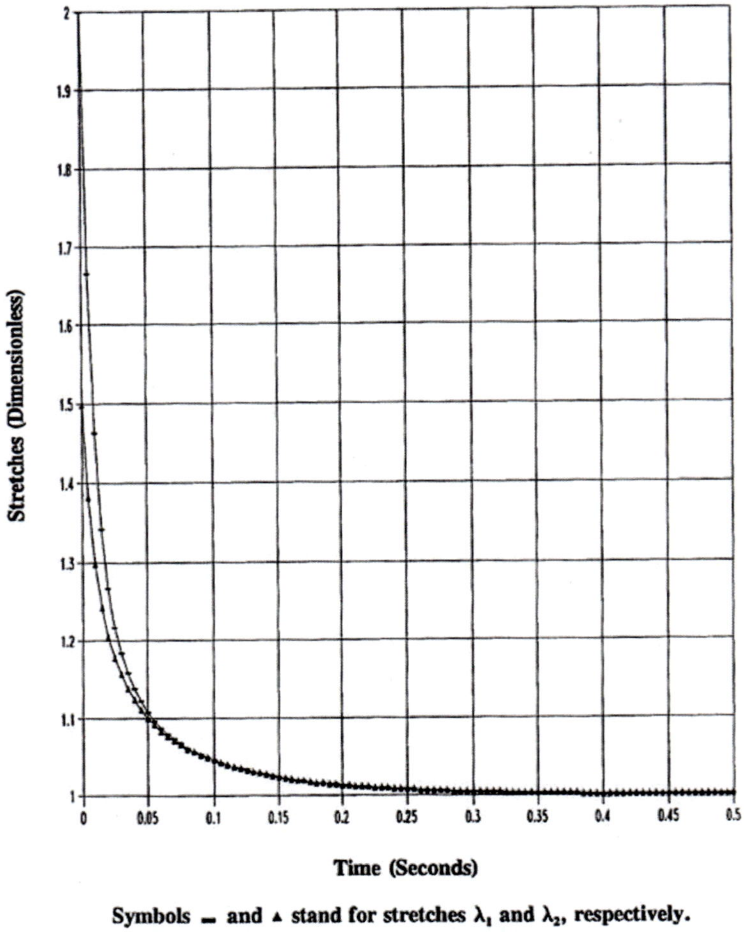

Fig. 4.5 Rubber sheet unloaded from unequal biaxial stress states- relaxation: λ_1 and λ_2 versus time (Zeng, 1995)

5.6 The Nonlinear Thermoviscoelastic Behavior of a Rubber Rod

The system of equations, Eq. (19), is a nonlinear thermoviscoelastic model when temperature is one of the state or control variables and when the generalized energy function is built from a thermoelastic equilibrium energy function. A two-dimensional example is given by the temperature and stretch response of a rubber rod subjected to a sudden perturbation from an equilibrium state. The uniaxial stretch, λ, and the temperature change, θ, are the state variables. Thus, in the generalized energy function, the corresponding control variables are the negative engineering stress, $-s$, and the positive entropy, η, respectively. Then, Eq. (19) becomes,

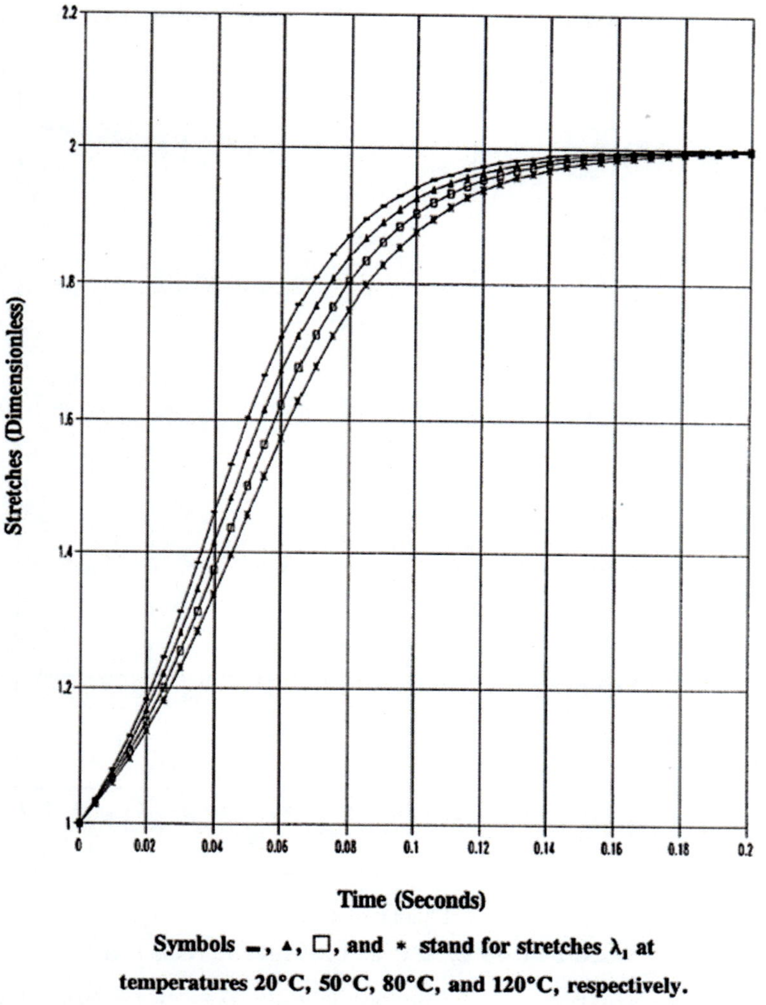

Fig. 4.6 Rubber sheet loaded by unequal biaxial tensile stress perturbations at various temperatures- biaxial creep: λ_1 versus time (Zeng, 1995)

$$\begin{bmatrix} \frac{d\lambda}{dt} \\ \frac{d\theta}{dt} \end{bmatrix} = k \begin{bmatrix} \frac{\partial^2\varphi}{\partial\lambda^2} & \frac{\partial^2\varphi}{\partial\lambda\partial\theta} \\ \frac{\partial^2\varphi}{\partial\theta\partial\lambda} & \frac{\partial^2\varphi}{\partial\theta^2} \end{bmatrix}^{-2} \begin{bmatrix} s - \frac{\partial\varphi}{\partial\lambda} \\ \eta + \frac{\partial\varphi}{\partial\theta} \end{bmatrix}. \tag{41}$$

where s and η are the sudden stress perturbation and the sudden entropy perturbation, respectively. This presumes that some physical means exists to perturb the entropy. The Helmholtz free energy φ is given by Eq. (36). Equilibrium material constants were chosen to be consistent with Joule's data (1859).

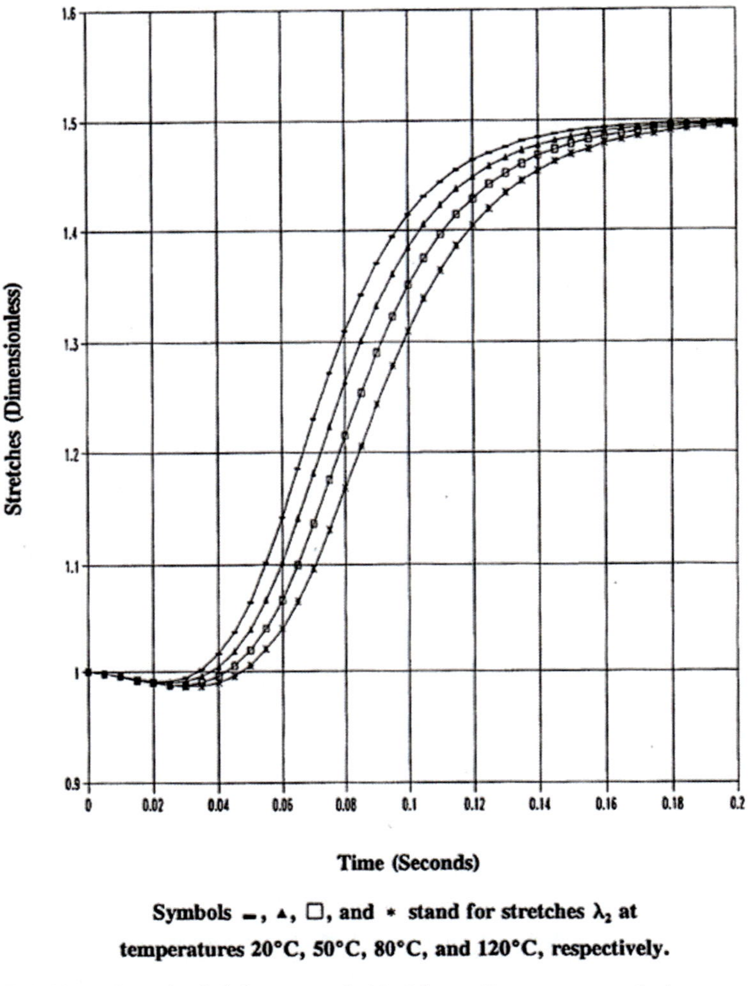

Fig. 4.7 Rubber sheet loaded by unequal biaxial tensile stress perturbations at various temperatures- biaxial creep: λ_2 versus time (Zeng, 1995)

The shear modulus is $\mu(\theta_c) = 4.36 \times 10^{-1} + 7.35 \times 10^{-4}\theta_c$ MPa, and the volume thermal expansion coefficient is $\beta = 5.26 \times 10^{-4}/\mathrm{K}^\circ$. The specific heat is $c(\theta_c) = 964.37 + 1.59\theta_c$ J/kg-K°.

The stress and entropy perturbations were chosen to correspond to an equilibrium at $\lambda = 1.5$ and temperature change of $\theta_c = 10\mathrm{C}^\circ$ using Eq. (36). Since the temperature change relaxes to equilibrium more rapidly than the stretch, a smaller value of the relaxation modulus $k = 6.0 \times 10^4$ is used to make the transient response of the system more visible by increasing the relaxation time of the temperature change. The stretch relaxes to $\lambda = 1.5$ with a slight change in concavity (Fig. 4.8). The temperature change relaxes to $\theta_c = 10^o$ along a curve with the qualitative shape

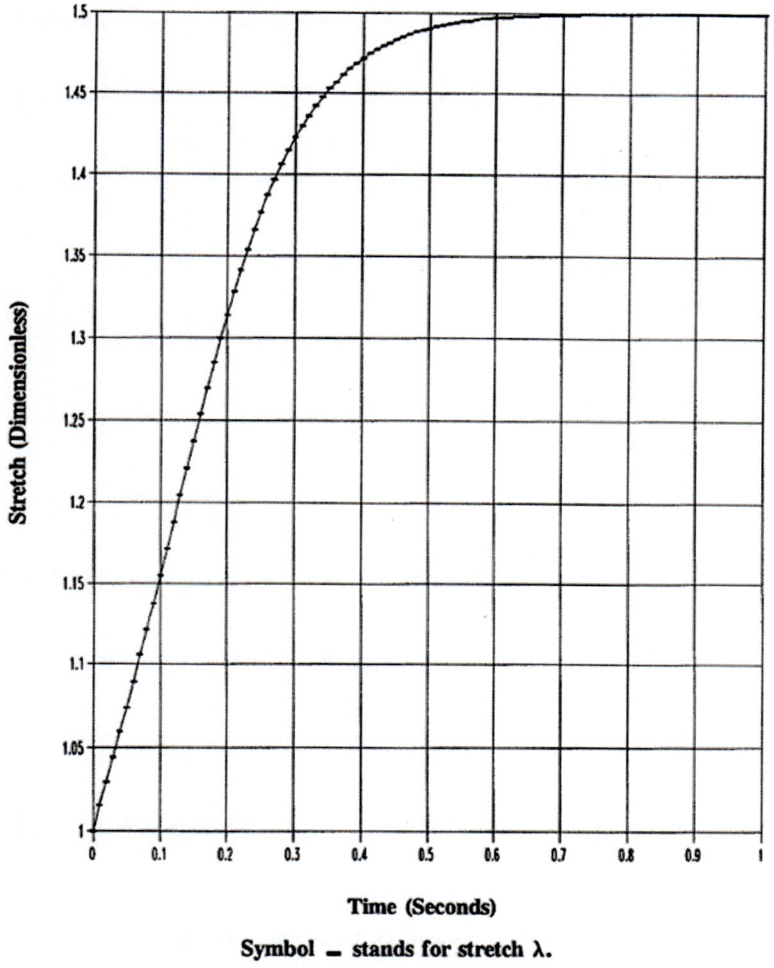

Fig. 4.8 Rubber rod under stress and entropy perturbations: λ versus time (Zeng, 1995)

of an exponential (Fig. 4.9). The relaxation time of the temperature change, about 3.0×10^{-3} sec, is much shorter than that of the stretch, about 0.67 s.

5.7 The Adiabatic Gough-Joule Effect as a Non-equilibrium Relaxation Process

If a rubber uniaxial specimen, which is initially in equilibrium, is quickly stretched under adiabatic conditions, the temperature of the specimen rises measurably. This is called the adiabatic Gough-Joule effect. Joule (1859) produced a graph of the rise

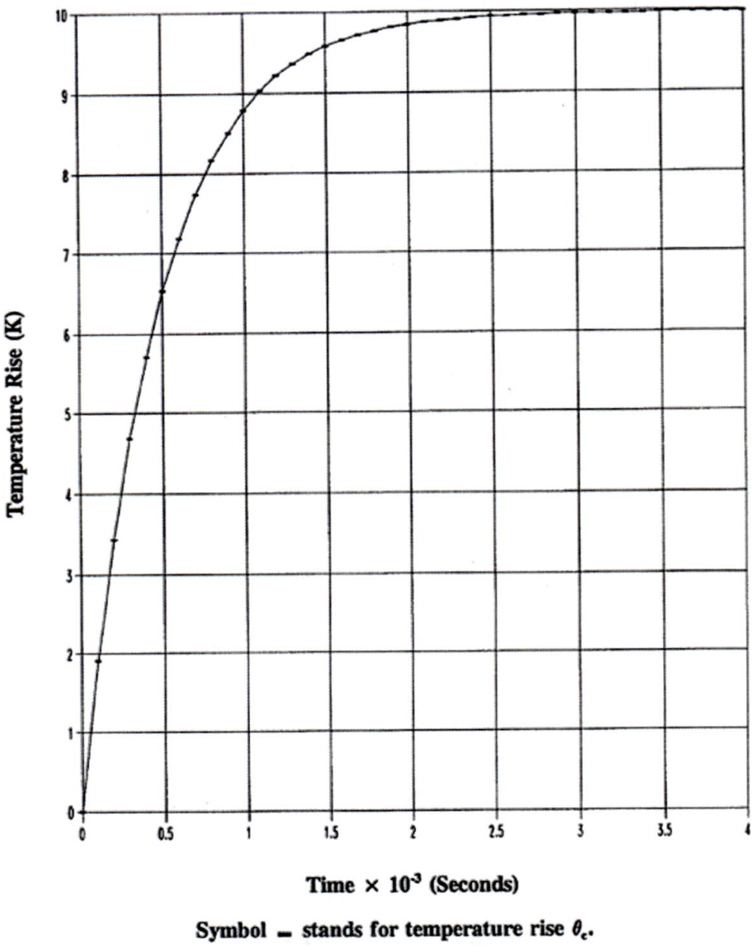

Fig. 4.9 Rubber rod under stress and entropy perturbations: θ versus time (Zeng, 1995)

in temperature for a range of stretches. Presumably the measurements of the temperature by a thermocouple were made after the specimen reached equilibrium. Only for small stretches does the temperature decrease. At about $\lambda = 1.09$, a minimum, called the adiabatic inversion point, in the equilibrium stretch-temperature curve is reached. This process is usually assumed to be isentropic. Under the assumption of constant entropy, Chadwick (1974) was able to recover from his equilibrium constitutive model the qualitative shape of Joule's test data but did not obtain close quantitative agreement. Kelvin (1857) proposed a differential equation for adiabatic processes which can also be used to reproduce Joule's test data under the assumption of constant entropy.

$$\theta_c = \theta_0 \int_{L_0}^{L} \frac{1}{c_L} \frac{\partial s}{\partial \theta} dL. \tag{42}$$

Joule's experiment can be alternatively viewed as a relaxation process in which the control variable stress is suddenly perturbed, while the control variable entropy is held constant to create a constant entropy non-equilibrium process. Subsequently, the state variables, stretch and temperature, relax to new equilibria.

The thermoviscoelastic model based on the thermoelastic generalization of the Mooney-Rivlin model, Eq. (36), is used to recover Joule's test data. All the material constants are the same as in the previous section. Each of the points is obtained by solving Eq. (41) with the constant entropy calculated by Eq. (36) from $\lambda = 1.0$ and $\theta_c = 0^o$ C and the sudden stress perturbation at $\theta_c = 0^o$ C and a chosen stretch. The comparison of the computed final equilibrium values of the temperature change and the stretch for the adiabatic Gough-Joule effect with Joule's and the James (1975) data in Fig. 4.10 shows that the predictions from the relaxation model, calculated at stretch increments of 0.01, are nearly identical to the experimental data. The system of nonlinear thermoviscoelastic equations together with the thermoelastic model Eq. (36) can reproduce the adiabatic Gough-Joule effect.

6 Nonlinear Maximum Dissipation Viscoelastic Models for Soft Biological Tissue

The mechanical behavior of biological soft tissue is time-dependent and nonlinear. A full description of the mechanical response of soft biological tissue to loads requires a nonlinear viscoelastic model. The Holzapfel et al. (2000) three-dimensional two-layer elastic model for healthy artery tissue, when used as the long-term constitutive model generates evolution equations by the maximum dissipation construction for biaxial loading of a flat specimen. A simplified version of the Shah and Humphrey (1999) model for the elastodynamical behavior of a saccular aneurysm is extended to viscoelastic behavior. These evolution models first appeared in Haslach (2005).

The soft tissue is viscoelastic because, according to the small number of experiments available in the literature, the uniaxial stress-strain response of soft tissue exhibits hysteresis in a loading-unloading test. The hysteresis is believed to be only weakly dependent on the loading rate and therefore is only weakly dependent on frequency in cyclic load testing. This belief seems to be founded on a set of experimental results performed by Fung on rabbit papillary muscle. Fung repeated in several papers that the loading and unloading uniaxial curves are almost strain-rate independent, however his data at three different rates, as given in (fig. 2, 1972; fig. 1a, 1973), only approximately supports this statement. A moderate sensitivity for tests under strain control was observed for articular cartilage by Woo et al. (1980) because in a loading and unloading test "the peak stresses increased slightly with

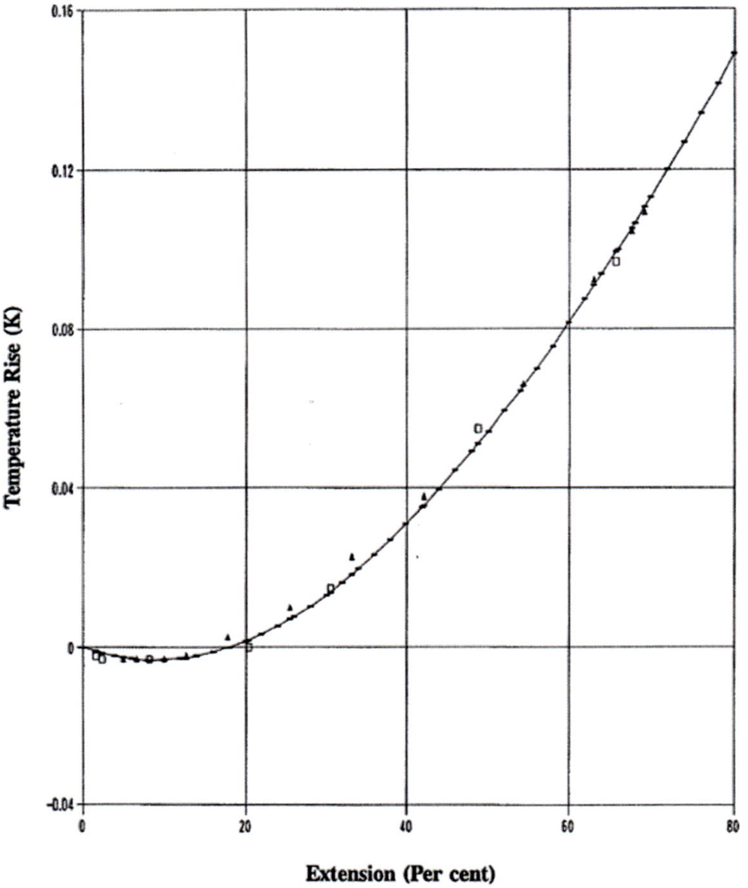

Fig. 4.10 Comparison of experimental data of Joule (1859) and James (1975) and the relaxation model data for the adiabatic Gough-Joule effect (Zeng, 1995)

increased strain rates". The response may not be the same if load rather than strain is controlled. A study by Bennett et al. (1986) showed that the energy dissipated per cycle in a dolphin tendon was nearly independent of frequency; however a close examination of their data for cyclic tests at 2.2 Hz but at different load amplitudes shows that the strain response is not independent of load rate. Even so, to simplify the modeling problem, some have assumed that the stress-strain behavior is time-independent, or at least strain rate insensitive, and have proposed nonlinear elastic models for soft tissue.

Hyperelastic (i.e. time-independent) models have commonly been used to represent the stress-strain relation of biological soft tissue in the special case of monotonically increasing loads (e.g. Skalak et al., 1975). Fung and his coworkers generated hyperelastic models for skin and for arterial tissue (Fung, 1993, pp. 302–306) that are of the form of an exponential. The constants in this model depend on whether the material is being loaded or unloaded. Humphrey (1992) suggested that a two-dimensional model of the form of the Fung exponential model can approximately represent the stress-stretch relation for canine epicardal tissue under monotonic loading, and Kyriacou and Humphrey (1996) fit the model to the stress-stretch data of Scott et al. (1972) for intracranial aneurysmal tissue. Humphrey and co-workers also developed such models for pleura and pericardium (e.g. Humphrey, 1987). After a clear and detailed description of the physical structure of an artery wall, Holzapfel et al. (2000) presented a two-layer hyperelastic model for the composite arterial tissue which is orthotropic in each layer. The model accounts for residual stresses and requires three material parameters for each layer.

However, hyperelastic models cannot represent time-dependent processes such as creep or stress relaxation. Viscoelastic models proposed for soft tissue have included isothermal linear viscoelastic models in the form of spring and dashpot configurations, such as the modified standard linear solid used by Apter (1964) for the viscoelastic properties of strain controlled loadings, either stress relaxation or harmonic loading, in aortic and pulmonary artery tissue of dogs. Unfortunately, the linear long-term relation between stress and strain predicted by most spring and dashpot models is not bourn out by experiment.

Soft tissue viscoelastic models using an infinite number of spring and dashpots have been of two major types, those which produce a continuous relaxation spectrum and those intended to represent the tissue as a network of fibers. A nonlinear Kelvin-Voigt type model was constructed by Viidik (1968) using a sequence of springs of different lengths so that more springs are stretched as the material deforms, to account for the increasing stiffness of artery tissue with stretch. Lanir (1983) built a model from the properties of the structural components of the tissue, the elastin and collagen. A probability distribution function is required to describe the density of each type of fiber. The behavior of a fiber is determined by a fading memory function. The complexity of the Lanir type models and the difficulty of determining the fading memory functions limit their usefulness. Further, the constituents may behave differently in isolation than they do in the composite tissue. Other investigators in the middle of the twentieth century examined the dynamic moduli under harmonic loads of different frequencies. Such analyses assume linear viscoelastic behavior.

To produce large deformation, three-dimensional evolution models using the maximum dissipation construction, a choice must be made in the long-term constitutive model of a strain measure and of a stress that are conjugate. One possible pair is the Green strain, $E = 0.5(C - I)$, and the second Piola-Kirchhoff stress, S, since $S = \partial\varphi/\partial E$. Recall that by objectivity and the polar decomposition $\varphi(F) = \varphi(C) = \varphi(E)$. Another possibility is the deformation tensor, F, and the first Piola-Kirchhoff stress, P.

6.1 Uniaxial Nonlinear Viscoelastic Models for Biological Tissue

To verify that the maximum dissipation construction of a nonlinear viscoelastic model can in fact apply to biological tissue, the Fung and the Skalak et al. hyperelastic models are chosen as long-term constitutive equations. None of the models taken here as examples for the long-term behavior was originally proposed for that purpose.

6.1.1 Long-Term Behavior Represented by the Fung Exponential Model

The Fung exponential model for biological membranes has been applied to both arterial and lung membrane tissue by Fung and coworkers. The strain energy density for uniaxial loading is of the form, $\varphi(E) = c[\exp(c_1 E^2) - 1]$, where the uniaxial Green strain $E = (\lambda^2 - 1)/2$, for the stretch λ. For this strain energy, the second Piola-Kirchhoff stress, S, is given by

$$S = \frac{d\varphi}{dE} = 2cc_1 E \exp(c_1 E^2). \tag{43}$$

Tong and Fung (1976) emphasize that the exponential model, applied to rabbit abdominal skin, requires different empirical constants in loading and unloading. This lack of universality is a consequence of their use of a hyperelastic rather than a viscoelastic model. Most applications of this model have been to fitting monotonic stress-strain tests. Here, the construction is instead based on $\varphi(E) = c\exp(c_1 E^2)$ taken as the model for the long-term behavior, even though no data exists applying it to the long-term behavior. In stress control, at a given stress S, the instantaneous strain E is less than its value on the long-term manifold. In strain control at a given E, the instantaneous stress, S, is greater than the value on the long-term manifold. An estimate is needed to relate available stress-strain data to the long-term behavior.

The maximum dissipation evolution equation (19) for Green strain as a state variable and the second Piola-Kirchhoff stress, S, as a control variable is

$$\dot{E} = -k_E[c(2c_1 + 4c_1^2 E^2)\exp(c_1 E^2)]^{-2}[2cc_1 E \exp(c_1 E^2) - S(t)]. \tag{44}$$

The creep strain versus time graph has a shape which appears similar to a logarithmic curve. That it is not logarithmic is verified by plotting the strain against $\log(t)$. Then an S-shaped curve is produced as in the Tanaka and Fung (1974) data. The S-shape of the curve is not affected by the magnitude of the applied stress. As k_E is increased, the curve approaches its long-term value faster so that the initial portion is steeper.

The model (44), when $S(t) = st$ for constant stress rate s, predicts the stress-stretch curve of Fig. 4.11 under stress control for $s = 0.05$. The construction here captures the loading-unloading hysteresis that, by its very nature, the hyperelastic model cannot. In this representation, the same value of k_E works for both loading and unloading in contrast to the Tanaka-Fung model which required two sets

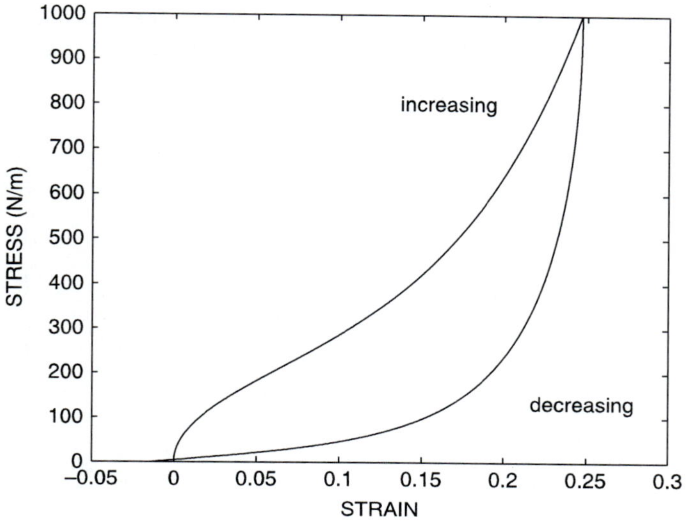

Fig. 4.11 Prediction of the uniaxial stress-strain curve controlled by a monotonically increasing stress. The long-term Fung exponential model has constants $c = 0.88$ and $c_1 = 13$. The upper curve is the loading at a stress rate of 0.05 Pa/sec for 20,000 seconds, and the lower curve is the unloading at the same rate. The relaxation modulus is $k_\epsilon = 0.00001$. The solution is by MATLAB ode23

of material constants to predict loading-unloading hysteresis. In stress control, this model predicts a short initial concave down portion unless k_E is taken larger than in Fig. 4.11, where $k_E = 0.00001$. Smaller k_E produce the S-shaped stress-stretch curve which is typical of rubber. Therefore, the numerical experimentation suggests that perhaps the magnitude of the relaxation modulus is one factor that distinguishes rubber from soft tissue.

The inverse function for $S = d\varphi/dE = 2cc_1 E \exp(c_1 E^2)$ is needed to consider the case that the stress is the state variable. The strain as a function of stress is written in terms of the product log function, which is the inverse of the function $z = x\exp(x)$. The product log of z, which is denoted $h(z) = x$, exists because $x\exp(x)$ is monotonically increasing, but $h(z)$ must be computed numerically because it has no closed form. Then the strain is written in terms of the stress,

$$E = \frac{1}{\sqrt{2c_1}}\sqrt{h(S^2/a)}, \tag{45}$$

where $a = 2c^2 c_1$.

The evolution (19) for stress as a state variable and strain as a control variable is

$$\dot{S} = -k_S\left(\frac{2c_1 S^2\left(1 + h(S^2/a)\right)^2}{h(S^2/a)}\right)\left[\frac{1}{\sqrt{2c_1}}\sqrt{h(S^2/a)} - E(t)\right]. \tag{46}$$

The stress-strain curve (Fig. 4.12) obtained from the evolution equation (46) with $k_S = 0.0001$ in strain control for $E(t) = 0.002t$ captures concave up behavior as well as loading-unloading hysteresis.

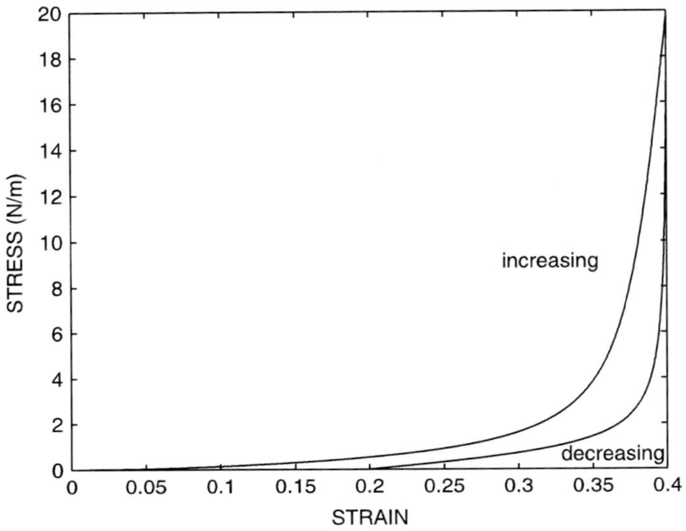

Fig. 4.12 Prediction of the uniaxial stress-strain curve controlled by a monotonically increasing strain. The long-term Fung exponential model has constants $c = 0.88$ and $c_1 = 13$. The upper curve is the loading at a strain rate of 0.0002/sec for 2,000 s, and the lower curve is the unloading at the same rate. The relaxation modulus is $k_\sigma = 0.0001$. The solution is by MATLAB ode23

6.1.2 Skalak Model

The Skalak et al. (1975) model has strain energy, which is here taken as the long-term manifold relation,

$$\varphi(\lambda) = c\left[2(\lambda^4 - 2\lambda^2 + 1) + \Gamma(\lambda^4 - 1)^4\right], \tag{47}$$

where $c > 0$ has units of force per length and $\Gamma > 0$ is dimensionless.

The first Piola-Kirchhoff stress, P, versus stretch relation is given by

$$P = d\varphi/d\lambda = c[2(-4\lambda + 4\lambda^3) + 16\Gamma\lambda^3(-1 + \lambda^4)^3],$$

whose graph has the same qualitative shape as that of the Fung models. Increasing Γ moves the curve slightly to the left and increases the magnitude for a given stretch. The evolution (19) for λ the state and P the control variables is

$$\begin{aligned}\dot{\lambda} &= k\left(\frac{d^2\varphi}{d\lambda^2}\right)^{-2}\left[\frac{d\varphi}{d\lambda} - P\right] \\ &= (k/c^2)[2(-4+12\lambda^2) + 192\Gamma\lambda^6(-1+\lambda^4)^2 + 48\Gamma\lambda^2(-1+\lambda^4)^3]^{-2} \times \\ &\quad c[2(-4\lambda+4\lambda^3) + 16\Gamma\lambda^3(-1+\lambda^4)^3 - P/c]\end{aligned} \tag{48}$$

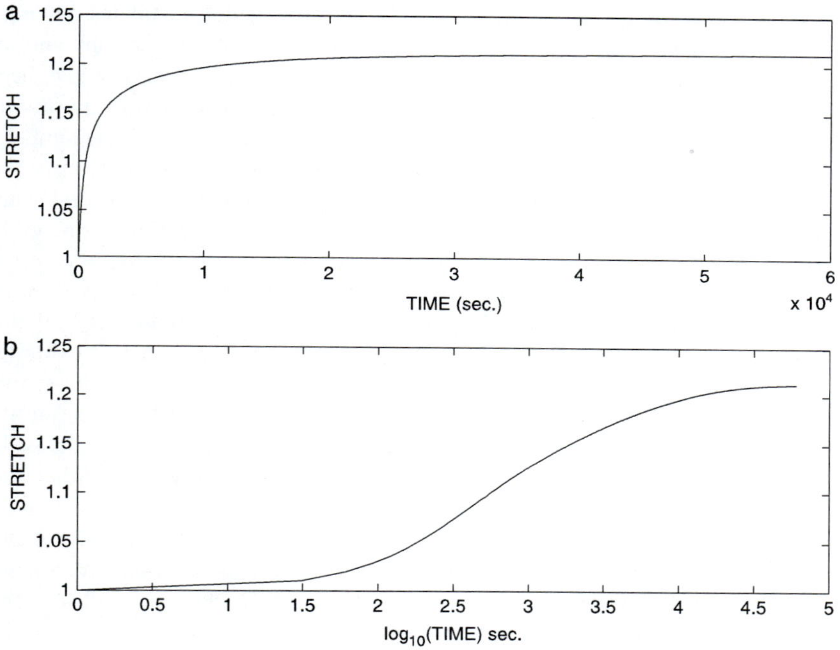

Fig. 4.13 Creep response if the long-term manifold is represented by a Skalak model, under a stress of $\sigma = 9$, with $\lambda(0) = 1$, $c = 1$, $\Gamma = 0.1$, and $k = 0.01$. **a** Creep stretch versus time; **b** Creep stretch versus $\log_{10}$(time). The solution is by MATLAB ode23

The creep behavior predicted by (48) is shown in Fig. 4.13 for $\lambda(0) = 1$, $\Gamma = 0.1$, $k = 0.01$, $c = 1$, and $P = 9$. This model reproduces the qualitative shape of the experimental creep data given by Tanaka and Fung (1974) because the stretch versus $\log_{10}$(time) curve (Fig. 4.13b) is S-shaped. If k is made larger, say $k = 1$, then the stretch versus $\log_{10}$(time) curve is nearly a straight line until the long-term manifold is reached; at that point the stretch oscillates about its long-term value. Therefore in this case the transient response predicted by the evolution equation with Skalak long-term model is nearly logarithmic.

A stress relaxation model is mathematically harder to obtain because the long-term stretch must be written as a function of the stretch. A closed form seems impossible.

6.2 Temperature Dependence in Uniaxial Loading

Biological soft tissue normally operates in a narrow temperature range. The viscoelastic evolution equation construction can represent the evolution of temperature. Soft biological tissues contract when heated (Roy 1880–1881), just as rubber does, in contrast to metals. Also soft tissue temperature increases slightly when tissue is stretched, as does rubber. These facts suggested to Fung (1972) the possibility of applying the thermodynamics of entropic elasticity taken from classical rubber theory (see also Fung, chapter 7.4, 1993). The entropic theory postulates that elastic loading may be resolved into two components, one due to the entropy and the other to the internal energy (Anthony et al., 1942). This idea seems to have been based on an analogy with the van der Waals gas model. The component due to internal energy was claimed to be insignificant. Experimental data on the temperature dependence of stresses is collected for this theory by measuring the stress on a specimen held at constant length and volume as the temperature is varied. Such a test on rubber produces a nearly linear stress-temperature relation. Apparently, little data describing the temperature dependence of stress in soft tissue exists in the literature. Rigby (1964) reported that, for a rat-tail tendon in a 0.9% NaCl solution, the force decreased only slightly as temperature was increased at 0.5°/min from 20–60°C. At about 60°, the shrinkage temperature of collagen, it exhibited a sharp increase. Unfortunately, the entropic theory is controversial even for rubber elasticity (Rosa et al., 1992). Fung (p. 269, 1993) points out that the entropic theory does not work if the material structure changes with temperature, as is the case for biological materials such as elastin which adsorbs moisture as temperature changes.

For the purpose of demonstrating the construction of an evolution model which represents the evolution of the two state variables, stretch and temperature, assume that soft biological tissue has a long-term energy function which depends on temperature, θ_c, measured from a reference state and the Green strain, $E = 0.5(\lambda^2 - 1)$, where λ is the uniaxial stretch.

$$\varphi(E, \theta_c) = c\theta_c \left\{ \exp[c_2(\theta_c)E^2] - 1 \right\} + f(\theta_c). \tag{49}$$

The function $f(\theta_c)$, which ensures the proper entropy expression, must be determined from additional experimental data. Then $d^2 f/d\theta_c^2$ is determined from the relation defining the specific heat at constant deformation, $c_L(\theta) = -\theta(\partial^2\varphi/\partial\theta^2)$. This information for soft tissue is apparently not yet available in the literature. Therefore a linear expression will be taken that is similar to rubber for this illustration; $c_L(\theta) = 964.4 + 1.59\theta$. Then $f(\theta_c) = -964.4[\theta\ln(\theta) - \theta] - 0.5(1.59)\theta^2 + C\theta + D$. The constants $c = 0.88/25$ and $c_2 = 26$ are based on those of Shah and Humphrey (1999) where c is adjusted for the reference temperature of 25°C, and c_2 is assumed independent of temperature.

This Helmholtz function implies that the generalized energy, φ^*, is

$$\varphi^*(E,\theta;S,\eta) = c\theta\exp(c_2E^2) - 964.4[\theta\ln(\theta)-\theta] - 0.5(1.59)\theta^2 + C\theta + D - SE - \theta\eta. \quad (50)$$

If the second Piola-Kirchhoff stress S and entropy are control variables, then the isentropic temperature evolution is coupled with the creep strain response by Eq. (19) which becomes,

$$\begin{bmatrix} \frac{dE}{dt} \\ \frac{d\theta}{dt} \end{bmatrix} = k \begin{bmatrix} \frac{\partial^2\varphi}{\partial E^2} & \frac{\partial^2\varphi}{\partial E\partial\theta} \\ \frac{\partial^2\varphi}{\partial\theta\partial E} & \frac{\partial^2\varphi}{\partial\theta^2} \end{bmatrix}^{-2} \begin{bmatrix} S - \frac{\partial\varphi}{\partial E} \\ -\bar{\eta} - \frac{\partial\varphi}{\partial\theta} \end{bmatrix}, \quad (51)$$

where S and η are the sudden stress perturbation and the sudden entropy perturbation, respectively, which are held fixed during the process, and where $\bar{\eta} = \eta - C$. The evolution of the stretch is obtained by setting $dE/dt = \lambda(d\lambda/dt)$ in (51). The evolution of stretch and temperature is shown in Fig. 4.14 for $S = 22,000\,\text{Pa}$ and $\bar{\eta} = 3,300$. The stretch asymptotically approaches the equilibrium value and the temperature increases about 0.3° for $k = 0.01$ in 500 s. Larger values of k make the stretch approach its asymptotic value more quickly and increase the temperature rise. There is a slight initial decrease temperature after the load is applied; this dip

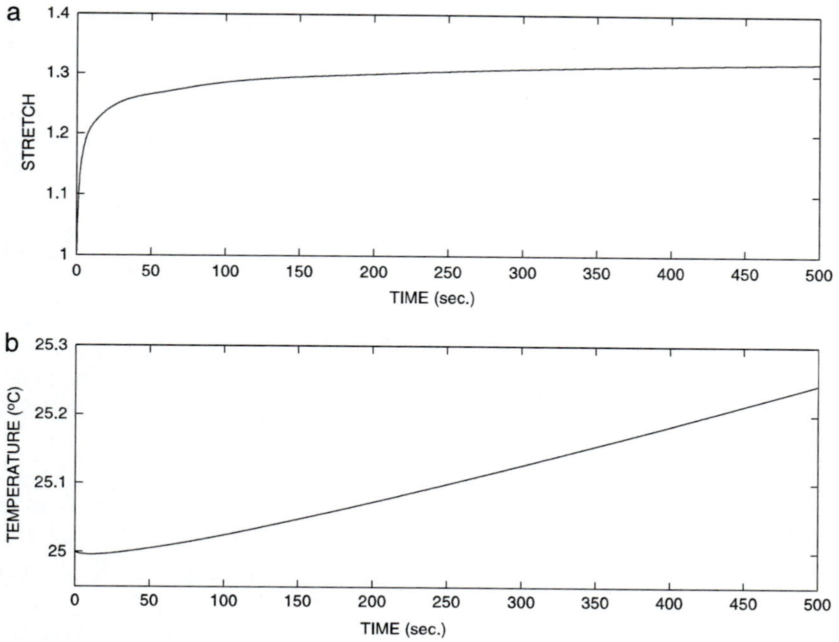

Fig. 4.14 Temperature change during uniaxial loading of a material with a Fung type long-term model with $c = 0.88/25$ and $c_2 = 26$. **a** The evolution of the stretch under a constant stress of 22,000 Pa. **b** The time dependent increase in temperature. The initial conditions are $\lambda(0) = 1$ and $\theta(0) = 25^o$C. The solution is by MATLAB ode15s

is exaggerated at smaller values of k. The temperature increase predicted by this model during an axial loading is qualitatively the same as that found experimentally for biological soft tissue.

6.3 Evolution Equations Based on the Holzapfel et al. Long-Term Three-Dimensional Model for Healthy Artery Tissue

The construction of the evolution equations can be applied to represent the time-dependent behavior when the elastic Holzapfel et al. (2000) composite three-dimensional, energy density function for healthy artery tissue is the model for the long-term behavior. This model assumes a two-layer tissue with reinforcing collagen fibrils. Tensors $\mathbf{A}_1$ and $\mathbf{A}_2$ define the orientation of the two families of parallel collagen fibers in each layer. In cylindrical coordinates (r, θ, z), the tensors describing the collagen fibers are defined by $\mathbf{A}_i = \mathbf{a}_i \bigotimes \mathbf{a}_i$, where

$$\mathbf{a}_1 = (0, \cos\beta, \sin\beta) \text{ and } \mathbf{a}_2 = (0, \cos\beta, -\sin\beta).$$

This tensor implies that any small components of the fibers in the radial direction are neglected. The angle β is on the order of 29° for the media; it is about 62° in the adventitia.

Holzapfel et al. (2000) work in terms of isochoric stresses and strains. If the deformation gradient tensor $\mathbf{F} = J\bar{\mathbf{F}}$, where $J = \det(\mathbf{F})$, then by definition $\bar{\mathbf{C}} = \bar{\mathbf{F}}'\bar{\mathbf{F}}$. Their long-term anisotropic elastic energy for a two-layer tissue is $\Psi = \psi_M + \psi_A$, where the functions ψ_M and ψ_A represent the strain energy in the media and adventitia layers, respectively. These functions have the form

$$\psi = \frac{c}{2}(\bar{I}_1 - 3) + \frac{k_1}{2k_2} \sum_{i=4,6} \{\exp[k_2(\bar{I}_i - 1)^2] - 1\}, \tag{52}$$

where $\bar{I}_1 = Tr(\bar{\mathbf{C}})$, $\bar{I}_4 = \bar{\mathbf{C}} : \mathbf{A}_1$, $\bar{I}_6 = \bar{\mathbf{C}} : \mathbf{A}_2$, and k_1, k_2, and c are material constants. The first term, a neo-Hookean model, accounts for the isotropic matrix, which is mostly elastin and muscle cells, and the exponential terms account for the anisotropic collagen reinforcing fibers.

Assume $\det(\mathbf{F}) = 1$. If $\mathbf{F}$ is diagonal, then $\bar{I}_4 = \bar{\mathbf{C}} : \mathbf{A}_1 = a_{11}\lambda_1^2 + a_{22}\lambda_1^2 + a_{33}\lambda_1^2$, where $a_{11} = 0$, $a_{22} = \cos^2\beta$ and $a_{33} = \sin^2\beta$. Also $\bar{I}_6 = \bar{\mathbf{C}} : \mathbf{A}_2 = \bar{I}_4$ in this case. Then, for example, $\bar{I}_4 - 1 = \cos^2(\beta)\lambda_2^2 + \sin^2(\beta)\lambda_3^2 - (\sin^2\beta + \cos^2\beta) = 2\cos^2(\beta)E_2 + 2\sin^2(\beta)E_3$. The exponential term in the Holzapfel et al. model is

$$\exp[k_2(\bar{I}_4 - 1)^2] = \exp[k_2(4\cos^4(\beta)E_2^2 + 4\sin^4(\beta)E_3^2 + 8\cos^2(\beta)\sin^2(\beta)E_2E_3)]. \tag{53}$$

The term $\exp[k_2(\bar{I}_6 - 1)^2]$ is identical. Therefore in the case that $\mathbf{F}$ is diagonal, the Holzapfel et al. model is of the same form as, and is essentially a three-dimensional version of, the Fung model, except that the material constants are given a physical interpretation.

The evolution equations (19) do not maintain incompressibility during a deformation. This behavior is similar to the plastic deformation of metals, for which experiments show that the volume is not preserved during plastic deformation, although it is preserved from before the load is applied to after the load is removed. These evolution equations define a material which is incompressible in the long-term.

Definition 1 A material is said to be long-term incompressible under a load if its long-term behavior is described by an incompressible energy function in the sense that $\lambda_1\lambda_2\lambda_3 = 1$ in the long-term.

The calculations here are based on the principal stretches and presented in terms of the conjugate pair composed of the Green tensor $\mathbf{E}$ and the second Piola-Kirchhoff stress, $\mathbf{S}$. The evolution equations (19) are, for stress control,

$$\dot{\mathbf{E}} = -k\left(\frac{\partial^2\Psi}{\partial\mathbf{E}^2}\right)^{-2}\left[-\mathbf{S}(t) + \frac{\partial\Psi}{\partial\mathbf{E}}\right]. \tag{54}$$

By the discussion in the Appendix, Eq. (54) becomes, in the case that $\mathbf{F}$ is diagonal,

$$\begin{bmatrix} \frac{dE_2}{dt} \\ \frac{dE_3}{dt} \end{bmatrix} = k\begin{bmatrix} \frac{\partial^2\Psi}{\partial E_2^2} & \frac{\partial^2\Psi}{\partial E_2\partial E_3} \\ \frac{\partial^2\Psi}{\partial E_3\partial E_2} & \frac{\partial^2\Psi}{\partial E_3^2} \end{bmatrix}^{-2}\begin{bmatrix} S_2 - \frac{\partial\Psi}{\partial E_2} \\ S_3 - \frac{\partial\Psi}{\partial E_3} \end{bmatrix}, \tag{55}$$

where $\dot{E}_i = \lambda_i\dot{\lambda}_i$ and $\Psi = \psi_M + \psi_A$ for which ψ_M and ψ_A have the form

$$\begin{aligned}\psi(E_2, E_3) &= \frac{c}{2}(\bar{I}_1 - 3) + \frac{k_1}{2k_2}\sum_{i=4,6}\{\exp[k_2(\bar{I}_i - 1)^2] - 1\} \\ &= c(E_2 + E_3) + \frac{k_1}{k_2}\{\exp[c_1E_2^2 + c_2E_3^2 + c_3E_2E_3] - 1\},\end{aligned} \tag{56}$$

and $c_1 = 4k_2\cos^4\beta$, $c_2 = 4k_2\sin^4\beta$, and $c_3 = 8k_2\cos^2\beta\sin^2\beta$. The 2 in front of the exponential in the second line accounts for both the terms involving $(\bar{I}_4 - 1)$ and $(\bar{I}_6 - 1)$. The derivatives are

$$\frac{\partial \psi}{\partial E_2} = c + \frac{k_1}{k_2}[2c_1E_2 + c_3E_3]\exp[c_1E_2^2 + c_2E_3^2 + c_3E_2E_3];$$
$$\frac{\partial \psi}{\partial E_3} = c + \frac{k_1}{k_2}[2c_2E_3 + c_3E_2]\exp[c_1E_2^2 + c_2E_3^2 + c_3E_2E_3];$$
$$\frac{\partial^2 \psi}{\partial E_2^2} = \frac{k_1}{k_2}[2c_1 + (2c_1E_2 + c_3E_3)^2]\exp[c_1E_2^2 + c_2E_3^2 + c_3E_2E_3];$$
$$\frac{\partial^2 \psi}{\partial E_3^2} = \frac{k_1}{k_2}[2c_2 + (2c_2E_3 + c_3E_2)^2]\exp[c_1E_2^2 + c_2E_3^2 + c_3E_2E_3];$$
$$\frac{\partial^2 \psi}{\partial E_2 \partial E_3} = \frac{k_1}{k_2}[c_3 + (2c_1E_2 + c_3E_3)(2c_2E_3 + c_3E_2)] \times \exp[c_1E_2^2 + c_2E_3^2 + c_3E_2E_3].$$

Experimenters sometimes excise tissue and apply biaxial loads to a flat specimen. Creep of a two-layer specimen cut along the θ and z directions under biaxial dead loads causing fixed stresses, S_θ and S_z, in the respective principal directions is described by the evolution model, Eq. (19). Choose a coordinate system in which the stretches are principal, and assume the isothermal Holzapfel et al. two-layer model for the long-term behavior. As an example, the constants given for rabbit carotid tissue by Holzapfel et al. (2000, p. 36) are applied for an equibiaxial stress of

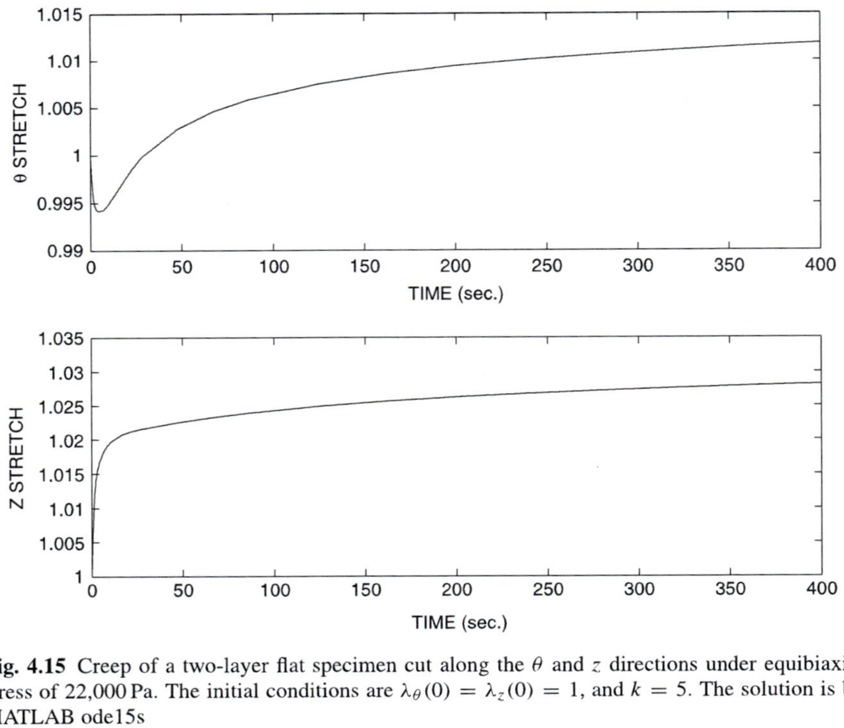

Fig. 4.15 Creep of a two-layer flat specimen cut along the θ and z directions under equibiaxial stress of 22,000 Pa. The initial conditions are $\lambda_\theta(0) = \lambda_z(0) = 1$, and $k = 5$. The solution is by MATLAB ode15s

22,000 Pa. The transient response of the stretches, λ_θ and λ_z, is shown in Fig. 4.15. The specimen is stiffer in the θ-direction. The calculations were made in MATLAB using the stiff integrator, ode15s.

The evolution equations (55) differ from those presented in Holzapfel et al. (2002) where the Holzapfel et al. (2000) model was used for the long-term behavior term in the Bonet construction that adopts Maxwell models to represent dissipation. The Bonet-Holzapfel viscoelastic model assumes a splitting of the stress into a sum of the long-term stress and a transient (viscoelastic) stress which tends to zero in the long term. Apparently, a generalized Maxwell evolution model was assumed because in that model, as opposed to other simple spring and dashpot models like the Kelvin-Voigt, the stress relaxes to zero under a constant strain. It, therefore satisfies the splitting of the stress into a sum of the long-term stress and a transient (viscoelastic) stress which tends to zero in the long term.

The Holzapfel et al. (2002) model assumes an energy function of the form

$$\Psi = U(\mathbf{X}, J) + \bar{\Psi}(\mathbf{X}, \bar{\mathbf{C}}, \mathbf{A}_1, \mathbf{A}_2) + \sum_\alpha \Psi_\alpha(\mathbf{X}, \bar{\mathbf{C}}, \mathbf{A}_1, \mathbf{A}_2, \Gamma_\alpha), \tag{57}$$

where Γ_α are viscous variables and $\mathbf{X}$ denotes the point in the body. For computational reasons, they allow the tissue to be slightly compressible, as accounted for in the term, $U(\mathbf{X}, J)$.

The assumed dissipation is given by the set of linear differential equations

$$\dot{\mathbf{Q}} + \frac{\mathbf{Q}}{\tau} = \beta^\infty \dot{\mathbf{S}}_{iso},$$

for constants τ and β^∞, that are valid for a time interval $0 < t \leq T$ and for small perturbations from equilibrium. Here, $\mathbf{Q}$ is the non-equilibrium viscoelastic contribution to the stress and is defined from the energy function as

$$\mathbf{Q}_\alpha = 2\frac{\Psi_\alpha(\mathbf{X}, \bar{\mathbf{C}}, \mathbf{A}_1, \mathbf{A}_2, \Gamma_\alpha)}{\partial \mathbf{C}}.$$

The evolution equations in Holzapfel et al. (2002) are linear involving two new coefficients, τ and β^∞ for each $\mathbf{Q}_\alpha$, in contrast to the single value, k, in Eq. (19). The nonlinear evolution equations (19) proposed here are not restricted to small perturbations from equilibrium.

6.4 Viscoelastic Saccular Aneurysm Model

This viscoelastic construction is applied to the Shah and Humphrey (1999) elastodynamical model for a saccular aneurysm as an illustration. Shah and Humphrey (1999) represented a thin-necked aneurysm as a thin-walled, isotropic, nonlinear elastic sphere without a hole which is surrounded by the viscous, Newtonian, cranial spinal fluid and forced by a known intra-saccular pressure. The cranial spinal fluid is neglected in this illustration.

The sphere is in load control subjected to a transmural radial pressure, $p(t)$. The sphere is assumed to deform uniformly as measured by a single state variable, the in-plane membrane stretch $\lambda(t) = 2\pi a(t)/2\pi A = a(t)/A$, where $a(t)$ is the current deformed radius and A is the original undeformed radius. Here ρ denotes the constant mass density of the membrane, and H the undeformed thickness of the membrane. The incompressibility constraint on the membrane volume requires that the deformed thickness of the membrane is $h = H/\lambda^2$ because the volume is preserved, $4\pi A^2 H = 4\pi a^2 h$. The problem is assumed by Shah and Humphrey to be two-dimensional with deformation gradient $\mathbf{F}$ having diagonal entries equal to λ.

The control variable, the second Piola-Kirchhoff stress $\mathbf{S}$, is related to the transmural pressure and the stretch acceleration by the equation of motion in the r-direction. The Shah and Humphrey (1999, equation (4)) is

$$p(t) - \frac{2T}{A\lambda} = \rho A h \ddot{\lambda}, \tag{58}$$

where T is the Cauchy stress. The second Piola-Kirchhoff stress is related to the Cauchy stress by $\mathbf{S} = J\mathbf{F}^{-1}\mathbf{T}\mathbf{F}^{-t}$. In this two-dimensional model, $J = \lambda^2$ so that $T = S$. Therefore

$$S = \frac{A\lambda}{2}\left[p(t) - \rho A h \ddot{\lambda}\right] = \frac{A\lambda}{2}\left[p(t) - \frac{\rho A H}{\lambda^2}\ddot{\lambda}\right]. \tag{59}$$

The energy function representing the long-term behavior is

$$\varphi(E_2, E_3) = c\{\exp[c_1 E_2^2 + c_2 E_3^2 + c_3 E_2 E_3] - 1\}. \tag{60}$$

But $E_2 = E_3 = 0.5(\lambda^2 - 1)$. So

$$\varphi = c[\exp(c_t E_2^2) - 1]. \tag{61}$$

where $c_t = c_1 + c_2 + c_3$.

$$\frac{\partial \varphi}{\partial E_2} = 2cc_t E_2[\exp(c_t E_2^2)]$$
$$\frac{\partial^2 \varphi}{\partial E_2^2} = 2cc_t(1 + 2cc_t E_2^2)[\exp(c_t E_2^2)]$$

The evolution equation (19) becomes

$$\dot{E}_2 = k\left(\frac{\partial^2 \varphi}{\partial E_2^2}\right)^{-2}\left(S - \frac{\partial \varphi}{\partial E_2}\right). \tag{62}$$

Using $\dot{E}_2 = \lambda\dot{\lambda}$, writing the equation in terms of λ and rearranging yields the governing differential equation,

$$\ddot{\lambda} = \frac{2\lambda}{\rho A^2 H}\left[-\frac{\lambda\dot{\lambda}}{k}\left(\frac{\partial^2\varphi}{\partial E_2^2}\right)^2 + \frac{A\lambda}{2}p(t) - \frac{\partial\varphi}{\partial E_2}\right]$$
$$= \frac{2\lambda}{\rho A^2 H}\left[-\frac{\lambda\dot{\lambda}}{k}\left[2cc_t(1+0.5cc_t(\lambda^2-1)^2)\exp[0.25c_t(\lambda^2-1)^2]\right]^2\right.$$
$$\left. + \frac{A\lambda}{2}p(t) - cc_t(\lambda^2-1)\exp[0.25c_t(\lambda^2-1)^2]\right]. \quad (63)$$

Shah and Humphrey (1999) assume the values $A = 0.003$ m, $H = 0.000278$ m, $\rho = 1050$ kg/m^3, $c = 0.88$ and $c_t = 13$ obtained from the Kyriacou and Humphrey (1996) numerical fit to the experimental data of Scott et al. (1972), and these values are used in the following two computations for creep and sinusoidal loading.

If the internal pressure in an undeformed saccular aneurysm is suddenly increased from 0 to 22,000 Pa and if $\dot{\lambda}(0) = 0$ and $k = 15$, then the stretch increases asymptotically from 1 to nearly $\lambda = 1.31$, the long-term stretch at this load (Fig. 4.16). The MATLAB stiff integrator ode15s was used to solve the evolution equation.

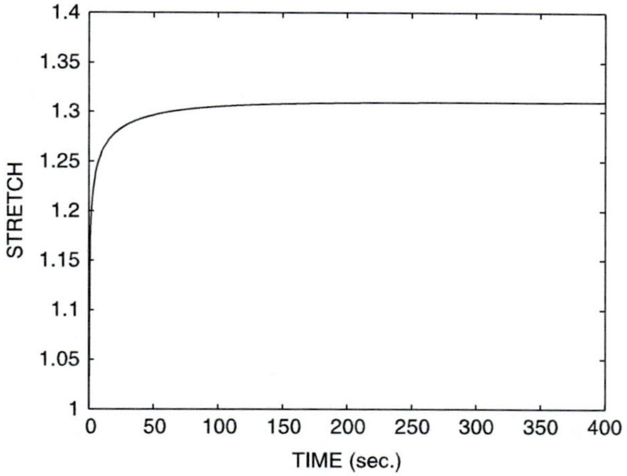

Fig. 4.16 The viscoelastic creep behavior of the modified Shah-Humphrey saccular aneurysm model as the internal pressure is suddenly increased from zero to 22,000 Pa with initial conditions $\lambda = 1$ and $\dot{\lambda}(0) = 0$ and $k = 15$. The MATLAB stiff integrator ode15s was used to solve the evolution equation

The maximum dissipation construction is applied to a sinusoidal pressure loading $22,000 + 7,000\sin(\omega t)$, where $\omega = 2\pi/1.5$ to represent the blood pressure. The initial conditions in Fig. 4.17 were $\lambda(0) = 1.3095$ and $\dot{\lambda}(0) = 0$, with $k = 15$. Fig. 4.17 for a short portion of the evolution time shows that the stretch lags the applied sinusoidal pressure load, as expected. The lag is initially about $\pi/4$ but increases over time because of the nonlinear viscoelastic behavior. In this example, the rate

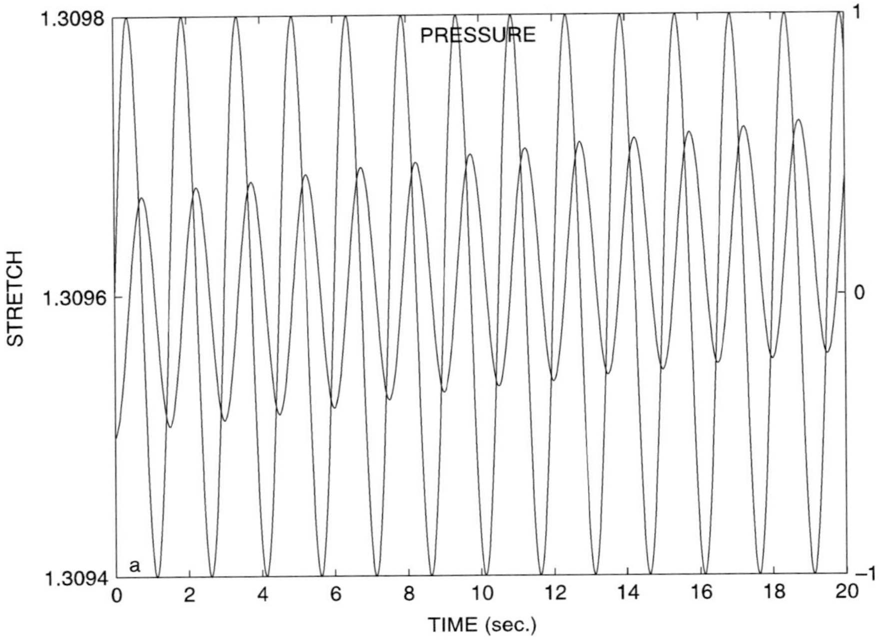

Fig. 4.17 Stretch response of the modified Shah-Humphrey saccular aneurysm model under a sinusoidal pressure loading $22,000 + 7,000\sin(\omega t)$, where $\omega = 2\pi/1.5$. The initial conditions are $\lambda(0) = 1.3095$ and $\dot{\lambda}(0) = 0$, with $k = 15$. The MATLAB stiff integrator ode15s was used to solve the evolution equation. The large amplitude curve is the scaled pressure to exhibit the stretch lag behind the applied pressure in a short range of times.

of change of the stretch, $\dot{\lambda}$, is very small. The MATLAB stiff integrator ode15s was used to solve the evolution equation. The evolution equation was also solved for values of k up to 500. The larger that k is, the larger the amplitude of the limit cycle. The stretch rate increases somewhat with an increase in k. At smaller k, the mean continues to grow approaching the asymptote for some time. The response to cyclic loading is expected to approach a stable cycle because several experimenters have conditioned soft tissue specimens by cycling the control load until a stable response is observed before taking measurements.

Appendix: Evolution Equation When the Strain Energy Is a Function of a Tensor

Let $\mathbf{E}$ be a second order tensor of state variables, and let $\mathbf{H}$ be the second order tensor of conjugate control variables. Assume that the energy function has the form

$$\psi(\mathbf{E}; \mathbf{H}) = \phi(\mathbf{E}) + \mathbf{E} : \mathbf{H}.$$

The affinities are the second order tensor, $\mathbf{X} = \partial\psi/\partial\mathbf{E}$. The affinities are a function of $\mathbf{E}$, so that $\mathbf{X} = h(\mathbf{E})$. Assume that h is invertible so that $\psi(\mathbf{E};\mathbf{H})$ can be written in terms of the affinities for fixed control variables, $\bar{\psi}(\mathbf{X};\mathbf{H})$.

The gradient dissipation condition is that the evolution of the affinities obeys $\dot{\mathbf{X}} = -k\partial\bar{\psi}/\partial\mathbf{X}$ for fixed controls, where $\dot{\mathbf{X}}$ is a second order tensor. By the chain rule,

$$\dot{\mathbf{X}} = \frac{\partial\mathbf{X}}{\partial\mathbf{E}}\dot{\mathbf{E}}. \tag{64}$$

Here $\partial\mathbf{X}/\partial\mathbf{E}$ is a fourth order tensor. Again by the chain rule,

$$\frac{\partial\bar{\psi}}{\partial\mathbf{X}} = \frac{\partial\psi}{\partial\mathbf{E}}\frac{\partial\mathbf{E}}{\partial\mathbf{X}}. \tag{65}$$

Substitution of (64) and (65) into $\dot{\mathbf{X}} = -k\partial\bar{\psi}/\partial\mathbf{X}$ yields the evolution equation in terms of the state variables,

$$\dot{\mathbf{E}} = -k\left(\frac{\partial\mathbf{X}}{\partial\mathbf{E}}\right)^{-2}\frac{\partial\psi}{\partial\mathbf{E}} = -k\left(\frac{\partial\mathbf{X}}{\partial\mathbf{E}}\right)^{-2}\left(\frac{\partial\phi}{\partial\mathbf{E}} + \mathbf{H}\right). \tag{66}$$

This form reduces to that of Eq. (19) in the case that $\mathbf{E}$ has only diagonal non-zero entries. Then $\mathbf{H}$ also has only diagonal non-zero entries. The affinity is a second order tensor defined by

$$(\mathbf{X})_{ij} = \left(\frac{\partial\psi}{\partial\mathbf{E}}\right)_{ij} = \frac{\partial\psi}{\partial\mathbf{E}_{ij}}. \tag{67}$$

If ψ only depends on E_{11}, E_{22}, and E_{33}, then the only non-zero terms of $\mathbf{X}$ are the three terms $\mathbf{X}_{ii}$.

The fourth order tensor $\partial\mathbf{X}/\partial\mathbf{E}$ is defined by

$$\left(\frac{\partial\mathbf{X}}{\partial\mathbf{E}}\right)_{ijkm} = \frac{\partial\mathbf{X}_{ij}}{\partial\mathbf{E}_{km}}. \tag{68}$$

The only non-zero term of this tensor, if only the diagonal entries of $\mathbf{E}$ are non-zero, have the form

$$\left(\frac{\partial\mathbf{X}}{\partial\mathbf{E}}\right)_{iikk} = \frac{\partial\mathbf{X}_{ii}}{\partial\mathbf{E}_{kk}} = \frac{\partial^2\psi}{\partial\mathbf{E}_{ii}\partial\mathbf{E}_{kk}}. \tag{69}$$

The evolution equation therefore has the form of Eq. (19).

References

R. L. Anthony, R. H. Caston, and E. Guth (1942). Equations of state for natural and synthetic rubber-like materials. I. *Journal of Physical Chemistry* **46**, 826–840.

J. T. Apter (1964). Mathematical development of a physical model of some visco-elastic properties of the aorta. *Bulletin of Mathematical Biophysics* **26**, 267–288.

J. T. Apter, M. Rabinowitz, and D. H. Cummings (1966). Correlation of visco-elastic properties of large arteries with microscopic structure. I, II, III. *Circulation Research* **19**, 104–121.

J. T. Apter and E. Marquez (1968). Correlation of visco-elastic properties of large arteries with microscopic structure. IV. *Circulation Research* **22**, 393–404.

M. B. Bennett, R. F. Ker, N. J. Dimery, and R. McN. Alexander (1986). Mechanical properties of various mammalian tendons. *Journal of Zoology, London A* **209**, 537–548.

L. E. Bilston, Z. Liu, and N. Phan-Thien (2001). Large strain behavior of brain tissue in shear: Some experimental data and differential constitutive model. *Biorheology* **38**, 335–345.

J. Bonet (2001). Large strain viscoelastic constitutive models. *International Journal of Solids and Structures* **38**, 2953–2968.

O. S. Brüller (1985). On the nonlinear response of polymers to periodical sudden loading and unloading. *Polymer Engineering and Science* **25**, 604–607.

O. S. Brüller (1987). On the nonlinear characterization of the long term behavior of polymeric materials. *Polymer Engineering and Science* **27**, 144–148.

O. S. Brüller (1991). Some New Results Concerning the Nonlinear Viscoelastic Characterization of Polymeric Materials. In *Constitutive Laws for Engineering Materials*, eds. C. S. Desai, E. Krempl, G. Frantziskonis, and H. Saadatmanesh, 241–246. ASME Press, New York, NY.

P. Chadwick (1974). Thermo-mechanics of rubberlike materials. *Philosophical Transactions of the Royal Society London* **A 276**, 371–403.

R. M. Christensen (1971). *Theory of Viscoelasticity: An Introduction*, Academic Press, New York, London.

B. D. Coleman (1964). Thermodynamics of materials with memory. *Archive for Rational Mechanics and Analysis* **17**, 1–46.

M. J. Crochet and P. M. Naghdi (1969). A class of simple solids with fading memory. *International Journal of Engineering Science* **7**, 1173.

W. N. Findley, J. S. Lai, and K. Onaron (1989). *Creep and Relaxation of Nonlinear Viscoelastic Materials*, Dover, New York, NY.

Y. C. Fung (1972). Stress-strain history relations of soft tissues in simple elongation. In *Biomechanics: Its Foundations and Objectives*, eds. Y. C. Fung, N. Perrone, and M. Anliker, Prentice-Hall, Englewood Cliffs, NJ, 181–208.

Y. C. Fung (1973). Biorheology of soft tissues. *Biorheology* **10**, 139–155.

Y. C. Fung (1993). *Biomechanics. Mechanical Properties of Living Tissues*, 2nd ed., Springer, New York, NY.

J. Hale and H. Koçak (1991). *Dynamics and Bifurcations*, Springer, New York, NY.

H. W. Haslach, Jr. (2005). Nonlinear viscoelastic, thermodynamically consistent, models for biological soft tissue. *Biomechanics and Modeling in Mechanobiology* **3**(3), 172–189.

H. W. Haslach, Jr. and N-N. Zeng (1999). Maximum dissipation evolution equations for nonlinear thermoviscoelasticity. *International Journal of Non-linear Mechanics* **34**, 361–385.

G. A. Holzapfel (2000). *Nonlinear Solid Mechanics*, John Wiley and Sons, Chichester, UK.

G. A. Holzapfel, T, C. Gasser, and R. W. Ogden (2000). A new constitutive framework for arterial wall mechanics and a comparative study of material models. *Journal of Elasticity* **61**, 1–48.

G. A. Holzapfel, T, C. Gasser, and M. Stadler (2002). A structural model for the viscoelastic behavior of arterial walls: continuum formulation and finite element analysis. *European Journal of Mechanics A/Solids* **21**, 441–463.

J. D. Humphrey, D. L. Vawter, and R. P. Vito (1987). Pseudoelasticity of excised visceral pleura. *Journal of Biomechanics* **109**, 115–120.

J. D. Humphrey, R. K. Strumpf, and F. C. P. Yin (1992). A constitutive theory for biomembranes: Application to epicardal mechanics. *Journal of Biomechanical Engineering* **114**, 461–466.

A. G. James, A. Green, and G. M. Simpson (1975). Strain energy functions of rubber. I. Characterization of gum vulcanizates. *Journal of Applied Polymer Science* **19**, 2033–2058.

J. P. Joule (1859). Some thermo-dynamic properties of solids. *Philosophical Transactions of the Royal Society London* **A 149**, 91–131.

L. Kelvin (W. Thomson) (1857). On the thermo-elastic and thermo-magnetic properties of matter. *Quarterly Journal of Pure and Applied Mathematics* **1**, 57.

S. K. Kyriacou and J. D. Humphrey (1996). Influence of size, shape and properties on the mechanics of axisymmetric saccular aneurysms. *Journal of Biomechanics* **29**(8), 1015–1022.

Y. Lanir (1983). Constitutive equations for fibrous connective tissues. *Journal of Biomechanics* **16**, 1–12.

E. H. Lee (1955). Stress analysis in visco-elastic bodies. *Quarterly of Applied Mathematics* **13**, 183–190.

J. Lemaitre and J.-L. Chaboche (1990). *Mechanics of Solid Materials*, Cambridge University Press, Cambridge.

F. J. Lockett (1972). *Nonlinear Viscoelastic Solids*, Academic Press, New York, NY.

H. Markovitz (1977). Boltzmann and the beginnings of linear viscoelasticity. *Transactions of the Society of Rheology* **21**, 381–398.

K. Miller and K. Chinzei (2002). Mechanical properties of brain tissue in tension. *Journal of Biomechanics* **35**, 483–490.

R. W. Ogden (1972). Large deformation isotropic elasticity - on the correlation of theory and experiment for incompressible rubberlike solids. *Proceedings of the Royal Society London. A* **326**, 565–584.

S. W. Park and Y. Kim (2001). Fitting Prony-series viscoelastic models with power-law presmoothing. *Journal of Materials in Civil Engineering* **13**, 26–32.

M. T. Prange and S. S. Margulies (2002). Regional, directional, and age-dependent properties of the brain undergoing large deformation. *Journal of Biomechanical Engineering* **124**, 244–252.

K. R. Rajagopal and A. R. Srinivasa (2005). A note on a correspondence principle in nonlinear viscoelastic materials. *International Journal of Fracture* **131**, 147–152.

B. J. Rigby (1964). The effect of mechanical extension upon the thermal stability of collagen. *Biochimica et Biophysica Acta* **79**, 634–636.

E. Rosa, Jr., A. R. Altenberger, and J. S. Dahler (1992). Model calculations based on a new theory of rubber elasticity. *Acta Physica Polonica B* **23**, 337–356.

C. S. Roy (1880–1881). The elastic properties of the arterial wall. *Journal of Physiology* **3**, 125–159.

S. Scott, G. G. Ferguson, and M. R. Roach (1972). Comparison of the elastic properties of human intracranial arteries and aneurysms. *Canadian Journal of Physiology and Pharmacology* **50**, 328–332.

A. H. Scott (1935). Specific volume, compressibility, and volume thermal expansivity of rubber - sulpher compounds. *Journal of Research of the National Bureau of Standards Washington* **14**, 99–20.

A. D. Shah and J. D. Humphrey (1999). Finite strain elastodynamics of intracranial saccular aneurysms. *Journal of Biomechanics* **32**, 593–599.

R. A. Schapery (1984). Correspondence principles and generalized J integral for large deformation and fracture analysis of viscoelastic media. *International Journal of Fracture* **25**, 195–223.

M. Shen (1969). Internal energy contribution to the elasticity of natural rubber. *Macromolecules* **2**, 358–364.

F. Shen, T. E. Tay, J. Z. Li, S. Nigen, P. V. S. Lee, and H. K. Chan (2006). Modified Bilston Nonlinear Viscoelastic model for Finite Element head injury studies. *Journal of Biomechanical Engineering* **128**, 797–801.

R. Skalak, A. Rachev, and A. Tozeren (1975). Stress-strain relations for membranes under finite deformations. *Bulgarian Academy of Sciences: Biomechanics* **2**, 24–29.

T. T. Tanaka and Y. C. Fung (1974). Elastic and inelastic properties of the canine aorta and their variations along the aortic tree. *Journal of Biomechanics*, **7**, 357–370.

P. Tong and Y. C. Fung (1976). The stress-strain relationship for the skin. *Journal of Biomechanics* **9**, 649–657.

E. Tanaka and H. Yamada (1990). An inelastic constitutive model of blood vessels. *Acta Mechanica* **82**, 21–30.

C. Truesdell (1984). *Rational Thermodynamics*, Springer, New York, NY.

A. Viidik (1968). A rheological model for uncalcified parallel-fibered collogeneous tissue. *Journal of Biomechanics* **1**, 3–11.

L. A. Wood (1973). Modulus of natural rubber crosslinked by dicumyl peroxide. II. Comparison with Theory. *Rubber Chemistry and Technology* **46**, 1287–1298.

S. L.-Y. Woo, B. R. Simon, S.C. Kuel, and W. H. Akeson (1980). Quasi-linear viscoelastic properties of normal articular cartilage. *Journal of Biomechanical Engineering* **102**, 85–90.

B. Wunderlich and H. Baur (1970). Heat capacities of linear high polymers. III. Results of experimental heat capacity measurements. *Advances in Polymer Science* **7**, 260.

N. Zeng (1995). *A Thermoelastic Constitutive Equation for Rubberlike Materials and a System of Non-equilibrium Evolution Equations*, Ph.D. Thesis, University of Maryland, College Park, MD.

N-N. Zeng and H. W. Haslach, Jr. (1996). Thermoelastic generalization of isothermal elastic constitutive models for rubberlike materials. *Rubber Chemistry and Technology* **69** (2), 313–324.

Chapter 5
Viscoplasticity

1 Introduction

Viscoplastic behavior is a phenomenon of metals, while viscoelasticity is a phenomenon of polymers. The goal of this chapter is to apply the maximum dissipation evolution equation construction to represent thermoviscoplasticity. The construction then provides a means for distinguishing between thermoviscoplasticity and thermoviscoelasticity in a unified theory. Behavior is assumed homogeneous, in the sense that there are no spatial gradients of the variables; the thermodynamic variables do not vary from point to point in the body.

Two conditions distinguish viscoelastic from viscoplastic behavior in the maximum dissipation construction. First, the viscoelastic relaxation modulus does not vary with time, while the viscoplastic relaxation modulus is a function a time through the instantaneous values of some internal variables. In the viscoplastic case, the norm of the plastic strain rate, its magnitude, is a measure of the instantaneous dissipation of the material, and is used as the thermodynamic relaxation modulus. Second, the long-term states are the set of equilibria for viscoelasticity and of steady states in viscoplasticity. In the latter case, the internal variables come to equilibrium on the long-term states so that the internal dissipative structure ceases to evolve, but the plastic strain continues to evolve at a constant rate of change.

Most time-dependent models in the literature represent either viscoelasticity or viscoplasticity; rarely are both included in a unified model. One exception is that of Valanis (1971) for whom viscoelastic evolution equations depend on time but the viscoplastic equations depend on an "intrinsic time scale" postulated as a material property to account for the rate of change of deformation states. There is no yield surface, with the advantage that no loading function is needed to distinguish loading and unloading, thereby avoiding the discontinuities existing in theories with a yield surface. Valanis argues that one cannot identify the yield surface in any case because the transition is gradual.

Rate independent plasticity theory is characterized by the assumption of a yield surface. The construction of Chapter 3 avoids the long-standing controversy over the existence of a yield surface by treating plastic deformation as a time-dependent non-equilibrium process. No yield surface is postulated for the maximum

H.W. Haslach Jr., *Maximum Dissipation Non-Equilibrium Thermodynamics and its Geometric Structure*, DOI 10.1007/978-1-4419-7765-6_5,

dissipation evolution equation construction because dislocation motion begins at stresses much smaller than the classical yield stress. The critical resolved shear stress is the minimum shear stress required to initiate dislocation motion by slip in a slip system, defined by a direction in a plane of the metal lattice. The value depends on both the particular slip system and on the material. For example experiment shows that in copper (FCC) the slip system $(111)[1\bar{1}0]$ has a critical resolved shear stress of 0.64 MPa, in α-iron (BCC) the slip system $(110)[\bar{1}11]$ has 27.5 MPa, and in zinc (HCP) the slip system $(0001)[\bar{1}1\bar{2}0]$ has 0.18 MPa (Haslach and Armstrong, 2004, p. 327). All of these values are considerably smaller than the corresponding generally accepted yield stresses, suggesting that slip is possible at very low loads. Such evidence argues for not including the idea of a yield stress in any mathematical model involving the viscoplastic response of metals. If the plastic strain is small, say less than 10 μ during the initial loading, the observer probably cannot distinguish this plastic strain from the accompanying elastic strain and the initial portion of the stress-strain curve appears to be linear. Such a situation occurs in one of the examples given below.

In those models that assume a plastic yield surface, a viscoplastic process lies outside the convex yield surface (e.g. Maugin, 1992). The rate of plastic strain is related to the distance between the stress state and the yield surface in stress space. The Perzyna (1971) model is also based on this idea in the sense that the rate of change of the plastic deformation tensor is a function of the excess of stress over the static yield condition. Some relate such models to plasticity by suggesting that viscoplasticity smooths out a stress jump at the yield surface (e.g. Chaboche, 1991; Maugin, 1992).

Various researchers have postulated potential functions to control the evolution of the internal variables in analogy with thermodynamic energies. A model in which the evolution of the viscoplastic strain and of the internal variables are both obtained from the same potential function is often called a model of Perzyna type after Perzyna (1966). It is a generalization of associative plasticity. Chaboche (1991) produced a similar thermodynamically based system of viscoplastic constitutive models in which the potential for the kinematic hardening internal variables was quadratic with no coupling.

Early research on viscoplasticity, and especially creep, was concerned with the existence of a potential function for the long-term behavior. Rice (1970) proves the existence of a flow potential for the plastic strain rate, but points out that a steady state creep potential is not implied by the flow potential. However, this fact does not preclude the existence of a long-term potential. Ponter and Leckie (1976) base their steady state creep model on the idea that the long term stationary creep, that with a constant strain rate, occurs when the processes of strain hardening and thermal softening are in balance. Although their model does not account for the transient creep approaching the stationary state of creep, they show that a long-term potential exists for their model. Mandel (1973) makes the hypothesis of normal dissipativity that the evolution of the internal variables is given by a potential function, Ω, of their conjugates. If A is the thermodynamic conjugate to a with respect to an energy function, then $\dot{a} = \partial\Omega/\partial A$.

Time-dependent models for solid behavior, including thermoviscoplasticity, are distinguished by the manner in which they deal with dissipation as a non-equilibrium thermodynamic process. The evolution of internal thermodynamic variables can describe dissipation. Mandel (1973) relates internal variables to history dependent functional constitutive equations. Most recent modelers adopt internal variables (e.g. Kratochvil and Dillon, 1969; Perzyna, 1971; Lubliner, 1972). However, in early and in many subsequent models, not every thermodynamic variable had a conjugate, and the variables were not identified as either state or control variables.

The plastic potential is a function g such that in rate independent plasticity, $\epsilon_{ij}^{p} = \dot{\lambda}(\partial g/\partial \sigma_{ij})$. If g is the function defining the yield surface, the plastic model is called associated. Some investigators have generalized the classical rate independent plastic multiplier construction to viscoplasticity (e.g. Maugin and Muschik, 1994). The multiplier construction has only a superficial resemblance to the gradient relaxation process; they are not the same.

Simo and coworkers (Simo and Honein, 1990; Simo and Hughes, 1998) have developed a theory of viscoplasticity around a maximum plastic dissipation principle, in part, because useful computational techniques arise from the variational formulation of the theory. Simo defines the dissipation function for a given plastic strain rate, $\dot{\boldsymbol{\epsilon}}^{p}$, and conjugate internal state variables, $\boldsymbol{q}, \boldsymbol{\alpha}$, as

$$\mathcal{D}^{p}(\boldsymbol{\tau}, \boldsymbol{q}; \dot{\boldsymbol{\epsilon}}^{p}, \dot{\boldsymbol{\alpha}}) = \boldsymbol{\tau} : \dot{\boldsymbol{\epsilon}}^{p} + \boldsymbol{q} : \dot{\boldsymbol{\alpha}}. \tag{1}$$

Their maximum dissipation principle says that the actual state of stress, $\boldsymbol{\sigma}$, is that which maximizes the dissipation function over all admissible states of stress, $\boldsymbol{\tau}$.

Simo and Hughes (1998, p. 105 ff) obtain a Perzyna type model for viscoplasticity from a penalty formulation of their maximum dissipation principle. The yield function, $f(\boldsymbol{\sigma}, \boldsymbol{q})$, is allowed to be positive so that the viscoplastic model accepts states outside the closure of the elastic region, $f(\boldsymbol{\sigma}, \boldsymbol{q}) \leq 0$. Their model requires a material dependent, monotonically increasing real function, g, defined on the values of $f(\boldsymbol{\sigma}, \boldsymbol{q})$, and a fluidity parameter, η.

$$\dot{\boldsymbol{\epsilon}}^{vp} = \gamma \frac{\partial f}{\partial \boldsymbol{\sigma}}; \qquad \dot{\boldsymbol{q}} = -\gamma \boldsymbol{D} \frac{\partial f}{\partial \boldsymbol{q}}; \qquad \text{and} \qquad \gamma = \frac{\langle g(f(\boldsymbol{\sigma}, \boldsymbol{q}))\rangle}{\eta}, \tag{2}$$

where $\boldsymbol{D}$ is the matrix of hardening moduli, and where

$$\langle x \rangle = \begin{cases} x & \text{if} \quad x \geq 0 \\ 0 & \text{if} \quad x < 0 \end{cases}. \tag{3}$$

In these models, the viscoplastic strain is thought of as an internal state variable dual to the stress.

Most efforts to create a thermodynamic foundation for viscoplasticity have involved ensuring that the evolution equations are consistent with the Clausius-Duhem inequality and the continuum mechanics balance laws. Unfortunately, this technique of limiting constitutive models does not produce enough information to determine the precise evolution of a non-equilibrium process given a set of initial conditions. The Clausius-Duhem inequality only provides a weak constraint for admissible processes. For example, Kratochvil and Dillon (1969) define the evolution of their dislocation rearrangement variables as a product of the plastic strain rate and a function which must satisfy a dissipation inequality related to the Clausius-Duhem inequality. They also took the plastic strain as a state variable. Freed et al. (1991) explicitly construct some constraints on the plastic potential, as distinguished from a dissipation potential defining the plastic strain rate, from a dissipation inequality analogous to the Clausius-Duhem inequality.

Some modelers assume that the total strain is the sum of three components, elastic, plastic and viscoplastic, $\epsilon = \epsilon^e + \epsilon^p + \epsilon^{vp}$. No experimental evidence supports such a splitting. The only evidence for splitting the total strain arises from a uniaxial loading into the plastic region followed by unloading, and followed by reloading. The uniaxial unloading curve is nearly parallel to the initial straight line curve of the first loading, and the reloading curve nearly is superimposed on the unloading curve. Such an experiment approximately justifies the common assumption of rate independent plasticity that the total strain is the sum of a plastic and an elastic component. Apparently, some models simply assume that this idea generalizes to viscoplasticity. In recent finite deformation theories, such a splitting allows the definition of intermediate configurations. No such splittings are necessary in the construction of the maximum dissipation evolution equations for thermoviscoplasticity.

In the class of thermoviscoplastic models described here, the elastic yield stress is the minimum stress at which creep initiates; this stress may be zero. No traditional plastic yield surface need be assumed. After developing this thermoviscoplastic model within a thermodynamic context, it is shown that some models previously proposed in the literature fit into this class. This class of thermoviscoplastic models includes the three-dimensional Freed-Chaboche-Walker model in which the internal state variables are the back stress, the drag strength and the elastic yield strength for creep initiation. These results first appeared in Haslach (2002).

2 Maximum Dissipation Models for Viscoplasticity

Internal state variables represent the material microstructure variation during a non-equilibrium process in many contemporary thermodynamically based viscoplasticity models. The evolution equations for the response in the maximum dissipation non-equilibrium thermodynamics model give the rate of change of the state variables in terms of the internal state variables, a_i, and their conjugates, A_i, as well as the conjugate pairs stress and strain, and the entropy, η, and temperature, θ. The internal stress variables assumed are the back stress, B_{ij}, the elastic yield stress, Y, and the drag stress, D. The back stress tensor, $\boldsymbol{B}$, accounts for kinematic hardening. It measures stiffness change as the material plastically deforms and is caused by the non-homogeneous nature of the dislocation structures. Krempl (1987) states that this stress, introduced by Nowick and Machlin (1947), represents "the inner stress field of the lattice of stuck dislocations". The scalars drag stress, D, and the elastic yield stress, Y, account for hardening due to the accumulation of plastic deformation. The drag stress, a scalar strength parameter, is proportional to the square root of the dislocation density (Freed and Walker, 1993). These internal stresses, $\boldsymbol{B}$, Y and D, have been taken to be conjugate to the plastic strain tensor, $\boldsymbol{\epsilon}^p$, the accumulated plastic strain, p, and the accumulated plastic work, respectively (Chaboche, 1991). However, these are not conjugate pairs in this model. In fact, the accumulated plastic work plays no role here as a thermodynamic variable. The overstress, $\boldsymbol{\sigma} - \boldsymbol{B}$, has been given the role of an independent internal variable in theories such as those of Krempl and coworkers (e.g. Yao and Krempl, 1985; Krempl, 1987). Another possibility is to represent the dislocation structure with a finite or infinite set of dislocation arrangement tensors, $\Gamma^{(k)}$, as proposed by Kröner (1963) (e.g. Kratochvil and Dillon, 1969; Perzyna, 1971). The plastic strain does not appear as a state variable in the energy functions. The plastic strain rate may be derived from a dissipation potential, but need not be.

In the linear elastic case, the affinity associated with the pair $\epsilon, -\sigma$, is $X = \partial\varphi^*/\partial\epsilon = \partial\varphi/\partial\epsilon - \sigma$, the difference between the equilibrium stress associated with ϵ and the current stress. A similar affinity has been used as an internal state variable in a viscoelastic model by Papoulia and Kelly (1997).

2.1 Thermoviscoplastic Generalized Energy

To represent a thermoviscoplastic material, the form of the generalized energy density must be modified to include the plastic energy. Let σ_{ij} be the stress. The strain is assumed to decompose into the sum of the elastic and plastic strain, $\epsilon_{ij} = \epsilon^e_{ij} + \epsilon^p_{ij}$. In this model, the plastic strain, $\boldsymbol{\epsilon}^p$, is *not* a state variable.

As an example, let the state variables be $\epsilon_{ij}, a_i, \theta$ and the controls be their conjugates, σ_{ij}, A_i, η. The Helmholtz energy which defines the long-term states is assumed to have been Legendre transformed to a function, φ, of the state variables. The energy density φ is also assumed to be a sum of elastic and plastic portions,

$$\varphi(\theta, \epsilon^e_{ij}, a_i) = \varphi^e(\theta, \epsilon^e_{ij}) + \varphi^p(\theta, a_i),$$

where $\partial\varphi/\partial\theta = -\eta$, $\partial\varphi/\partial\epsilon^e_{ij} = \sigma_{ij}$, and $\partial\varphi/\partial a_i = A_i$. No temperature gradient appears because the states are assumed to be homogeneous. The function $\varphi^p(\theta, a_i)$ is not the plastic potential considered in other models. It represents energy stored due to material hardening.

Assume that the material has linear thermoelastic behavior. Then

$$\varphi^e(\epsilon^e_{ij}, \theta) = \frac{1}{2}\epsilon^e_{ij} D_{ijkl}\epsilon^e_{kl} - \alpha_{ij}\epsilon^e_{ij}(\theta - \theta_o) - \eta_o(\theta - \theta_o) - c\left\{\theta - \theta_o\left[1 + \ln\left(\frac{\theta}{\theta_o}\right)\right]\right\}, \tag{4}$$

where D_{ijkl} is the tensor of elastic moduli, θ_o and η_o are the reference temperature and entropy, α_{ij} is the coefficient of thermal expansion and c is the heat capacity. The index summation rules are followed here and below.

The associated generalized energy function is then

$$\varphi^*(\epsilon^e_{ij}, a_i, \theta; \sigma_{ij}, A_i, \eta) = \varphi(\theta, \epsilon^e_{ij}, a_i) - \epsilon^e_{ij}\sigma_{ij} - A_i a_i + \eta\theta - \epsilon^p_{ij}\sigma_{ij}. \quad (5)$$

Put $\bar{\varphi}^* = \varphi(\theta, \epsilon^e_{ij}, a_i) - \epsilon^e_{ij}\sigma_{ij} - A_i a_i + \eta\theta$ so that $\varphi^* = \bar{\varphi}^* - \epsilon^p_{ij}\sigma_{ij}$.

The splitting of the total strain into elastic and plastic components in the generalized energy density function allows the computation of the dissipation on the long-term steady states as well as for the non-equilibrium process.

The long-term states are defined by the zero gradient condition for φ^*.

$$\frac{\partial\varphi^*}{\partial\epsilon^e_{ij}} = 0, \quad \frac{\partial\varphi^*}{\partial a_i} = 0, \quad \text{and} \quad \frac{\partial\varphi^*}{\partial\theta} = 0. \quad (6)$$

These relations are equivalent to the conjugate relations for φ.

Dissipation is measured by the Gibbs one-form (Section 4.5 in Chapter 3). The Gibbs form, ω, associated with φ is the differential one-form, with energy coordinate z,

$$\omega = dz - \theta d\eta + \epsilon_{ij} d\sigma_{ij} + a_i dA_i. \quad (7)$$

In elastic behavior, the internal variables are fixed so that the Gibbs form reduces to that for elasticity, $\omega = dz - \theta d\eta + \epsilon^e_{ij} d\sigma_{ij}$.

2.2 Admissible Thermodynamic Processes and Dissipation

The analysis presented in Chapter 3 (Section 4.6) is modified slightly to account for the plastic strain. A thermodynamic process, with controls, σ_{ij}, η, A_i, and states, $\epsilon_{ij}, \theta, a_i$, is a time dependent path $\gamma : \Re \to \Re^{2n+1}$ given in local coordinates on the energy surface at time t by

$$\gamma(t) = \left(\sigma_{ij}(t), \eta(t), A_i(t), \varphi^*(t), \epsilon_{ij}(t), \theta(t), a_i(t)\right). \quad (8)$$

The tangent vector, $\boldsymbol{t}_p$, at each point p on a non-equilibrium path is

$$\begin{aligned} \boldsymbol{t}_p = {} & \tfrac{d\varphi^*}{dt}\tfrac{\partial}{\partial z} + \textstyle\sum_{i,j=1}^{3} \tfrac{d\sigma_{ij}}{dt}\tfrac{\partial}{\partial\sigma_{ij}} + \tfrac{d\eta}{dt}\tfrac{\partial}{\partial\eta} + \textstyle\sum_{i=1}^{n} \tfrac{dA_i}{dt}\tfrac{\partial}{\partial A_i} \\ & + \textstyle\sum_{i,j=1}^{3} \tfrac{d\epsilon_{ij}}{dt}\tfrac{\partial}{\partial\epsilon_{ij}} + \tfrac{d\theta}{dt}\tfrac{\partial}{\partial\theta} + \textstyle\sum_{i=1}^{n} \tfrac{da_i}{dt}\tfrac{\partial}{\partial a_i}, \end{aligned} \quad (9)$$

where the vectors, $\partial/\partial z$, etc., are the basis for the tangent space at the point p in local coordinates. For any set of variables x_i, the differential form acts on the tangent space by $dx_i(\partial/\partial x_j) = \delta_{ij}$, the Kronecker delta. Therefore, the Gibbs form, Eq. (7), acting on $\boldsymbol{t}_p$ is

$$\omega(\boldsymbol{t}_p) = \frac{d\varphi^*}{dt} - \theta\frac{d\eta}{dt} + \sum_{i,j=1}^{3} \epsilon_{ij}\frac{d\sigma_{ij}}{dt} + \sum_{i=1}^{n} a_i\frac{dA_i}{dt}. \tag{10}$$

After substituting for $d\varphi^*/dt$ using Eq. (5),

$$\omega(\boldsymbol{t}_p) = \frac{d\varphi}{dt} + \eta\frac{d\theta}{dt} - \sum_{i=1}^{n} A_i\frac{da_i}{dt} - \sum_{i,j=1}^{3} \sigma_{ij}\frac{d\epsilon_{ij}^e}{dt} - \sum_{i,j=1}^{3} \sigma_{ij}\frac{d\epsilon_{ij}^p}{dt}. \tag{11}$$

Furthermore, applying the chain rule to $d\varphi/dt$ and rearranging yields

$$\omega(\boldsymbol{t}_p) = \frac{d\theta}{dt}\left(\frac{\partial\varphi}{\partial\theta} + \eta\right) + \sum_{i,j=1}^{3} \frac{d\epsilon_{ij}^e}{dt}\left(\frac{\partial\varphi}{\partial\epsilon_{ij}^e} - \sigma_{ij}\right) + \sum_{i=1}^{n} \frac{da_i}{dt}\left(\frac{\partial\varphi}{\partial a_i} - A_i\right)$$
$$- \sum_{i,j=1}^{3} \sigma_{ij}\frac{d\epsilon_{ij}^p}{dt} = \sum_{i=1}^{n} \frac{dx_i}{dt}\frac{\partial\bar{\varphi}^*}{\partial x_i} - \sum_{i,j=1}^{3} \sigma_{ij}\frac{d\epsilon_{ij}^p}{dt} = \frac{d\bar{\varphi}^*}{dt} - \sum_{i,j=1}^{3} \sigma_{ij}\frac{d\epsilon_{ij}^p}{dt},$$

where x_i are the state variables. In terms of the dot product, at each point of the process,

$$\omega(\boldsymbol{t}_p) = \nabla_x(\bar{\varphi}^*) \cdot \frac{d\boldsymbol{x}}{dt} - \sum_{i,j=1}^{3} \sigma_{ij}\frac{d\epsilon_{ij}^p}{dt}, \tag{12}$$

where $\nabla_x(\bar{\varphi}^*)$ is the gradient of $\bar{\varphi}^*$ with respect to the state variables. The expression $\omega(\boldsymbol{t}_p)$ measures the dissipation. For example, on the long-term steady states, the dissipation is $\omega(\boldsymbol{t}_p) = -\sum_{i,j=1}^{3} \sigma_{ij}\dot{\epsilon}_{ij}^p$ because $d\boldsymbol{x}/dt = 0$. A similar construction is valid for other choices of states and controls.

Definition 1 A thermodynamically admissible non-equilibrium process in the thermodynamic system defined by $\varphi^* : \Re^{2n} \to \Re$ is a curve $\gamma : \Re \to \Re^{2n+1}$ of the form of Eq. (8) for which $\omega(\boldsymbol{t}_p) < 0$ at each point along the path.

2.3 Maximum Dissipation and Gradient Relaxation Processes

For thermoviscoplasticity, the gradient relaxation processes of the maximum dissipation evolution equation construction must be defined in terms of the function, $\bar{\varphi}^*$. Again, a coordinate transformation expesses $\bar{\varphi}^*$ as a function of the affinities, $\bar{\varphi}'^*$.

The affinities are the state variables for $\bar{\varphi}'^{*}$. Alternatively, they may be thought of as internal state variables. Lubliner (1973, Eq. (5)) briefly mentions, as an example of an internal variable, an analogous transformation of rate dependent models into terms of an internal variable which is the difference between the current and steady state values of the original variable.

Definition 2 A gradient relaxation process in thermoviscoplasticity is one for which the evolution of the affinities, X_i, is given by

$$\frac{dX_i}{dt} = -k\frac{\partial \bar{\varphi}'^{*}}{\partial X_i}, \qquad \text{for} \qquad i = 1, \ldots, n, \tag{13}$$

for a thermodynamic relaxation modulus, k.

For such a process,

$$\omega(t_p) = \frac{d\bar{\varphi}^{*}}{dt} - \sum_{i,j=1}^{3} \sigma_{ij}\frac{d\epsilon_{ij}^{p}}{dt} \leq 0.$$

Perzyna (1971) defines a relaxation process as one in which the deformation tensor and the temperature are held constant, but his model is not based on a maximum dissipation principle. He proves that the asymptotically stable equilibrium state is reached only by an isothermal relaxation process. While Perzyna requires that the evolution equations be zero at equilibrium, he does not develop the evolution equations from the equilibrium, but rather the equilibrium from the assumed evolution equations.

2.4 The Thermodynamic Relaxation Modulus

In thermoviscoplasticity, even though the stress and total strain are conjugates for the generalized energy function, there is no relation between stress and strain on the long-term states for stress controlled relaxation, as is assumed for the internal variables. In contrast, in the generalized energy model for the relaxation of thermoviscoelastic materials under either stress or strain control, there is a long-term strain state. Physically, this is because the dissipative structure of nonlinear elastic materials and viscoelastic materials remains fixed over time. However, in thermoviscoplastic behavior, the material dislocation structure changes with time due to mechanisms such as diffusion or dislocation multiplication. This behavior is accounted for in the non-equilibrium thermodynamic model for thermoviscoplasticity by a form for the positive thermodynamic relaxation modulus, k, which varies with the change in dislocation structure. The function, k, is the norm of the plastic strain rate,

$$k = ||\dot{\epsilon}^{p}||. \tag{14}$$

The evolution of $||\dot{\epsilon}^p||$, which depends on the dislocation structure, cannot arise from the generalized energy function which is defined from the long-term state. The form of the evolution equation for the plastic strain may either be postulated or derived from an assumed dissipation potential. Any such dissipation potential plays no role in the thermodynamic admissibility of a gradient relaxation process, which requires only that k be positive.

A dissipation potential, $\Omega(\theta, \sigma_{ij}, B_{ij}, D, Y)$, may be assumed to give the rate of plastic strain as

$$\dot{\epsilon}^p_{ij} = \frac{\partial \Omega}{\partial \sigma_{ij}}. \tag{15}$$

Rice (1970 or 1971, section 2.2) proves that such a potential must exist under the assumption that the rate of structural rearrangement, as measured by an internal state variable, is fully determined by the corresponding thermodynamic force.

When any relaxation process ceases evolving, the long-term states are fixed. Consequently, the thermodynamic relaxation modulus $k = ||\dot{\epsilon}^p||$, which is a function of the states, also ceases evolving there. The dissipation becomes constant so that the long-term states are steady states. A dissipation potential is valid both on and off the set of long-term states. Its form is determined experimentally from creep, a relaxation process.

Zener and Holloman (1944) suggest that the thermal dependence in $||\dot{\epsilon}^p||$ can be multiplicatively decomposed into a product of a thermal (static) and a stress (dynamic) term. In the one-dimensional case,

$$\dot{\epsilon}^p = \exp\left(-\frac{Q}{R\theta}\right)\sigma^n,$$

where Q is the activation energy. They verify this fact for the stress-strain curve of moderate carbon steel at various constant temperatures. Also see Freed and Walker (1990).

The dissipation potential which generalizes this decomposition is

$$\Omega(\sigma, B, Y, D) = A\exp\left(\frac{-Q}{R\theta}\right)\left(\left\langle\frac{||\sigma - B|| - Y}{D}\right\rangle\right)^n. \tag{16}$$

Equation (16) is a variation of the classical creep power law when B, D, and Y are constant, as in the long term. Other models such as the logarithmic and hyperbolic sine can be constructed in a similar manner by making a different assumption about the form of the dissipation function. For example,

$$\Omega(\sigma, B, Y, D) = A\exp\left(\frac{-Q}{R\theta}\right)\sinh^n\left(\left\langle\frac{||\sigma - B|| - Y}{D}\right\rangle\right). \tag{17}$$

Freed and Walker (1993) suggest that one must account for the dependence of the creep activation energy on the mechanism. For low stress and high temperature, there is diffusion controlled dislocation climb. At high stress or lower temperature, the mechanism is obstacle controlled dislocation glide. So the plastic strain rate is modified by an appropriate Arrhenius function of temperature, which has different forms for low or high temperature.

2.5 Relaxation Examples

A model requires the choice of the state and control variables, of the elastic and plastic potential for the long-term behavior and of a dissipation potential. Thermoviscoplastic creep is a relaxation process occurring after the stress magnitude is suddenly changed to a new constant value. The strain evolves under the constant control, the stress, and the internal states evolve to their long-term state. The complexity of a model is directly determined by the form of the potential for the long-term states and by the number of conjugate pairs of variables. As an easily computed illustration, the energy for a single scalar internal variable is chosen as quadratic so that the equations defining the long-term states are linear; however, the evolution is nonlinear.

Example A simple uniaxial creep model is one in which the state variables are the scalars, a back stress, B, and the total strain, ϵ. The required thermodynamic dual of B is a back strain, b. The back strain is a residual local strain associated with the back stress at a point. The back strain and the applied stress are the control variables. The model assumes that the stress and the back strain b are constant, but the back stress and plastic strain evolve. The drag strength, D, is assumed to be fixed. The creep yield strength is assumed to be $Y = 0$. The elastic and plastic energies on the long-term states are assumed to be uncoupled quadratics,

$$\psi^e = \frac{1}{2}\bar{E}(\epsilon^e)^2 \qquad \text{and} \qquad \psi^p = \frac{1}{2}HB^2, \tag{18}$$

where $\bar{E}$ and H are positive constants. On the long-term states, the back stress and back strain are related by

$$b \equiv \frac{\partial \psi^p}{\partial B} = HL,$$

where L is the value of the back stress in the long term. This is the value to which the back stress tends over time.

The generalized energy is

$$P^* = \frac{1}{2}\bar{E}(\epsilon^e)^2 + \frac{1}{2}HB^2 - \sigma\epsilon^e - Bb - \epsilon^p\sigma.$$

Because the stress is constant, the elastic strain is also constant; its value is an initial condition and equal to that on the long-term states. The affinity associated with the back stress is

$$X \equiv \frac{\partial P^*}{\partial B} = HB - b. \tag{19}$$

Then, $B = (X + b)/H$ and the generalized energy in terms of the affinity is

$$P^{*\prime} = \frac{1}{2H}(X + b)^2 - \frac{b(X + b)}{H} - \epsilon^p \sigma,$$

neglecting the elastic term. Because the process is maximum dissipation, the evolution of the affinity is

$$\dot{X} = -k\frac{\partial P^{*\prime}}{\partial X} = -k\left[\frac{1}{H}(X + b) - \frac{b}{H}\right]. \tag{20}$$

But $\dot{X} = H\dot{B}$, so that the maximum dissipation process in response to a constant stress σ is given by the evolution equation

$$\dot{B} = -k\left(\frac{B}{H} - \frac{b}{H^2}\right). \tag{21}$$

The positive second term accounts for strain hardening and the negative first term for strain-induced dynamic recovery. The relaxation modulus is $k = ||\dot{\epsilon}^p||$ and the isothermal dissipation potential,

$$\Omega(\sigma, B, D) = A\left(\frac{||\sigma - B||}{D}\right)^n, \tag{22}$$

gives

$$\dot{\epsilon}^p = \frac{\partial \Omega}{\partial \sigma} = \frac{nA}{D}\left(\frac{||\sigma - B||}{D}\right)^{n-1}. \tag{23}$$

The system to be solved is then

$$\begin{aligned} \dot{B} &= -\frac{nA}{D}\left(\frac{||\sigma - B||}{D}\right)^{n-1}\left(\frac{B}{H} - \frac{b}{H^2}\right) \\ \dot{\epsilon}^p &= \frac{nA}{D}\left(\frac{||\sigma - B||}{D}\right)^{n-1} \end{aligned} \tag{24}$$

Figure 5.1 shows the creep curve for copper at 700° K with $A = 2.3785 \times 10^{-8}$; $n = 4.5$; $H = 1/32{,}000$ MPa^{-1}; $\sigma = 30$ MPa; $L = 20$ MPa; $D = 13$ MPa.

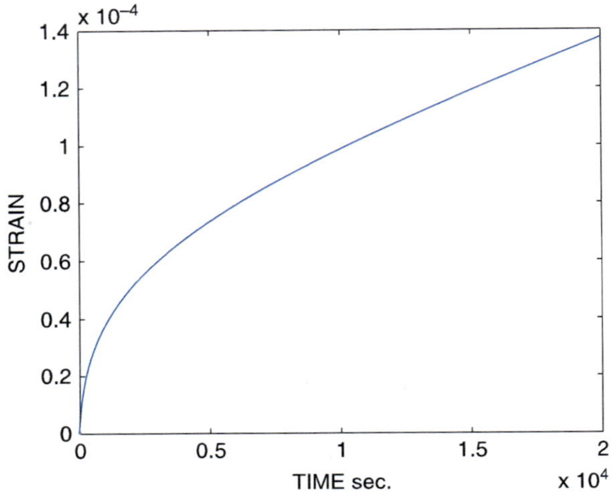

Fig. 5.1 The time dependence of the strain in uniaxial creep of copper at 700° K under $\sigma = 30\,\text{MPa}$

Notice that B only asymptotically approaches the steady state. But once it is close, the rate of change of the plastic strain is nearly constant. It would appear constant in experiments. This model can be related to classical uniaxial creep. The primary region corresponds to the relaxation to the long-term states; this region is viewed as the transient portion of creep. In the long term, the strain rate, $\dot{\epsilon}^p$, is constant because all the internal state variables are constant. Therefore this corresponds to the secondary creep region. The tertiary region involves necking and fracture and so is not described by this homogeneous model.

This simple model could be viewed as isothermal. However, a complete isothermal model with the temperature as a control variable would include an evolution equation for its dual, the entropy.

Example The relaxation process dual to creep is stress relaxation. In the simplest one-dimensional strain model, the state variables are the scalars, a back stress, B, and the total stress, σ. Again, the thermodynamic dual of B is a back strain, b. Because $Y = 0$, if the uniaxial specimen is loaded to (ϵ_0, σ_0), creep is initiated immediately so that $\epsilon_0 > \sigma_0/\bar{E}$, where $\bar{E}$ is the elastic modulus.

In this case, the elastic thermostatic energy must be written as a function of the state variable, the stress σ. The generalized energy is taken, for illustration, as the simplest quadratic,

$$P^* = \frac{\sigma^2}{2\bar{E}} + \frac{1}{2}H_B B^2 - \sigma\epsilon - bB. \tag{25}$$

Assume that stress relaxation is the process of the fixed strain, ϵ_0, becoming fully plastic as the stress relaxes to zero from σ_0. In order for the stress relaxation to end, it must be true that $\dot{\epsilon}^p = 0$. At such a terminal point then, $\sigma = B$ at the conclusion of relaxation due to the expression (23) for $\dot{\epsilon}^p$. Both σ and B relax to zero. The evolution of the back stress is obtained as before from P^* where the control is set as $b = 0$ (See Eq. (24)).

On the other hand, in this model because there is no long-term relation between the stress and strain, the evolution of the stress cannot be obtained as a gradient relaxation process in the sense of Eq. (13). However, the evolution of the stress is related to that of the plastic strain. The assumption $\epsilon = \epsilon^e + \epsilon^p$ implies that, when the total strain is fixed, $\dot{\sigma}/\bar{E} = -\dot{\epsilon}^p$. Therefore, in stress relaxation

$$\dot{\sigma} = -\frac{nA\bar{E}}{D}\left(\frac{||\sigma - B||}{D}\right)^{n-1}. \tag{26}$$

The initial condition for the plastic strain is $\epsilon_0 - \sigma_0/\bar{E}$ and for the stress is σ_0. The initial condition for the back stress is its value after the material is loaded to (ϵ_0, σ_0). As time tends to infinity, the stress approaches zero and the plastic strain increases to ϵ_0 as required.

Coupling is possible in the long-term potential, ψ^p, if it depends on more than one variable. This coupling makes the evolution equations more complex, as is illustrated by a simple coupled deformation and temperature evolution model. Chaboche (1991, p. 212) says that an energy, ψ^p, with non-quadratic kinematic hardening variables leads to abnormal results, but coupling with temperature was not discussed there.

Example An isentropic, back stress model for uniaxial creep assumes that the long-term elastic behavior is linear elastic and that for fixed uniaxial stress, σ, the state variables are the back stress, B, and temperature, θ. The conjugate controls are the back strain, b, and the entropy, η, which are constant. Neglect the elastic strain since it is fixed. The plastic potential is assumed to be quadratic,

$$\psi^p(\theta, B) = \frac{1}{2}H_\theta\theta^2 + \frac{1}{2}HB^2 + H_{\theta B}\theta B, \tag{27}$$

for H_θ, H and $H_{\theta B}$ material constants. The fixed entropy corresponds to a perturbation of the entropy corresponding to the initial temperature, and the back strain is fixed at its value in the long term, $b = HL$, where L is the assumed long-term limit of the back stress.

The induced generalized energy function is, ignoring the constant terms,

$$P^*(\theta, B, \eta, b) = \frac{1}{2}H_\theta\theta^2 + \frac{1}{2}HB^2 + H_{\theta B}\theta B - Bb + \theta\eta - \epsilon^p\sigma. \tag{28}$$

The affinity associated with the back stress is

$$X_B \equiv \frac{\partial P^*}{\partial B} = HB + H_{\theta B}\theta - b. \tag{29}$$

The affinity associated with the temperature is

$$X_\theta \equiv \frac{\partial P^*}{\partial \theta} = H_\theta \theta + H_{\theta B} B + \eta. \tag{30}$$

Solve these two equations simultaneously to obtain

$$B = \frac{1}{HH_\theta - H_{\theta B}^2}[H_\theta(X_B + b) - H_{\theta B}(X_\theta - \eta)]; \tag{31}$$

$$\theta = \frac{1}{HH_\theta - H_{\theta B}^2}[-H_{\theta B}(X_B + b) + H_\theta(X_\theta - \eta)]. \tag{32}$$

Substitute into $P^*(\theta, B, \eta, b)$ to obtain $P^{*\prime}(X_\theta, X_B, \eta, b)$.

The generalized energy in terms of the two affinities is

$$\begin{aligned} P^{*\prime} = {} & \tfrac{1}{2H}(X_B + b - H_{\theta B}\theta)^2 + \tfrac{1}{2H_\theta}(X_\theta - \eta - H_{\theta B}B)^2 \\ & + \tfrac{1}{HH_\theta}(X_\theta - \eta - H_{\theta B}B)(X_B + b - H_{\theta B}\theta) - b\tfrac{1}{H}(X_B + b - H_{\theta B}\theta) \\ & + \eta\tfrac{1}{H_\theta}(X_\theta - \eta - H_{\theta B}B) - \epsilon^p\sigma. \end{aligned} \tag{33}$$

Because the process is maximum dissipation, a gradient relaxation process, the evolution of the affinities is

$$\begin{aligned} \dot{X}_B &= -k\frac{\partial P'}{\partial X_B} \\ &= -k\left[\frac{1}{H}(X_B + b - H_{\theta B}\theta) + \frac{1}{HH_\theta}(X_\theta - \eta - H_{\theta B}B) - \frac{b}{H}\right]; \end{aligned} \tag{34}$$

$$\begin{aligned} \dot{X}_\theta &= -k\frac{\partial P'}{\partial X_\theta} \\ &= -k\left[\frac{1}{H_\theta}(X_\theta - \eta - H_{\theta B}B) + \frac{1}{HH_\theta}(X_B + b - H_{\theta B}\theta) + \frac{\eta}{H_\theta}\right]. \end{aligned} \tag{35}$$

The evolution equations can be obtained either by substituting the expression for X_B and X_θ and solving for $\dot{B}$ and $\dot{\theta}$, assuming b and η are constant, or directly by the Eq. (23) of Chapter 3.

$$\begin{bmatrix} \dot{B} \\ \dot{\theta} \end{bmatrix} = -k\frac{1}{(HH_\theta - H_{\theta B}^2)^2}\begin{bmatrix} H_\theta^2 + H_{\theta B}^2 & -H_{\theta B}(H + H_\theta) \\ -H_{\theta B}(H + H_\theta) & H^2 + H_{\theta B}^2 \end{bmatrix}$$

$$\begin{bmatrix} HB + H_{\theta B}\theta - b \\ H_\theta\theta + H_{\theta B}B + \eta \end{bmatrix}. \tag{36}$$

Therefore,

$$\dot{B} = -k\frac{1}{(HH_\theta - H_{\theta B}^2)^2}[(H_\theta(HH_\theta - H_{\theta B}^2)B - H_{\theta B}(HH_\theta - H_{\theta B}^2)\theta$$
$$-(H_\theta^2 + H_{\theta B}^2)b + H_{\theta B}(H + H_\theta)\eta]; \tag{37}$$
$$\dot{\theta} = -k\frac{1}{(HH_\theta - H_{\theta B}^2)^2}[-H_{\theta B}(HH_\theta - H_{\theta B}^2)B + H(HH_\theta - H_{\theta B}^2)\theta$$
$$+H_{\theta B}(H + H_\theta)b + (H^2 + H_{\theta B}^2)\eta. \tag{38}$$

The assumed dissipation potential, Ω, is given by Eq. (16), where Ω also depends on the temperature, θ, through B, D, and Y. Therefore, because $\dot{\epsilon}^p = \partial\Omega/\partial\sigma$,

$$k = ||\dot{\epsilon}^p|| = \frac{nA}{D}\exp\left(\frac{-Q}{R\theta}\right)\left(\left\langle\frac{||\sigma - B|| - Y}{D}\right\rangle\right)^{n-1}.$$

Note that the evolution equation for the back stress, in contrast to that for the previous isothermal example, now has thermally driven terms. Also the temperature evolution equation has a back stress contribution. Such cross-terms do not occur in most models proposed in the literature. If there is no interaction of the back stress and temperature in the long-term states, then $H_{\theta B} = 0$. In this case, the system reduces to uncoupled equations. If B increases from an initial zero state to a positive limit L, then the temperature decreases to maintain constant entropy.

It appears that for the terms involving θ and η in the evolution equation for $\dot{B}$ or the terms involving B and b in the evolution equation for $\dot{\theta}$ to be recovery terms, $H_{\theta B}$ must be negative. In this case, these two terms can be thought of as thermal softening terms.

If the elastic long term energy Eq. (4) is not neglected, then equilibrium thermal terms such as the heat capacity, c, appear in the expression for $\dot{\theta}$.

3 Forced Non-equilibrium Processes

A forced process is a non-equilibrium process in which the control variables vary with time. Recall from Chapter 3, Section 5.

Definition 3 Each non-equilibrium process is locally a maximum dissipation process, in the sense that it is a gradient relaxation process over each very short time interval.

This additional condition is the constitutive restriction defining a class of thermo-viscoplastic materials.

3.1 Simple Monotonic Loading

If the isothermal, monotonic hardening of a material depends on the load rate, then the process can be viewed as viscoplastic. The local maximum dissipation assumption is used to construct the response by viewing the stress as applied in increments. The elastic yield surface, Y, is fixed in monotonic loading. In general, the back and drag stresses require solving a coupled system of differential equations. Finally, the plastic strain rate is integrated to obtain the stress-strain curve in which the total strain is $\epsilon = \epsilon^e + \epsilon^p$. In stress control, the elastic strain is known to be $\epsilon(t) = \sigma(t)/E$, but once the load exceeds the creep yield, the plastic strain evolves with the time-varying stress.

Example The simplest possible monotonic loading model is one in which the state variables are the back stress, B, and the total strain, ϵ. The applied stress and the back strain, b, are the control variables. The stress is assumed to increase linearly in time, $\sigma(t) = \dot{\sigma}t$, where $\dot{\sigma}$ is a constant. The drag strength, D, is assumed to be fixed. Also $Y = 0$. The energies defining the long-term states are assumed to be the uncoupled quadratics in the simple creep example, and again the back strain $b = HL$ is held fixed. By the same development which leads to Eqns. (24), the non-autonomous system to be solved is then

$$\dot{B} = -\frac{nA}{D}\left(\frac{||\dot{\sigma}t - B||}{D}\right)^{n-1}\left(B - \frac{b}{H}\right)$$

$$\dot{\epsilon}^p = \frac{nA}{D}\left(\frac{||\dot{\sigma}t - B||}{D}\right)^{n-1} \tag{39}$$

Figure 5.2 shows the response of copper at 43°C obtained from this model. The coefficients are those based on the data in Freed and Walker (1993), $A\exp(-200,000/(316R)) = 1.345 \times 10^{-20}$, $n = 4.5$ and $D = 0.57\,\text{MPa}$. $H = 1/38000\,\text{MPa}^{-1}$ and $L = 1\,\text{MPa}$. H is the inverse of the shear modulus. The slow load rate is $\dot{\sigma} = 0.0061$ MPa/s. The elastic modulus is $E = 125$ GPa. The initial condition is $B(0) = 0$. The response closely reproduces their Fig. 6(b).

According to this model, not only is there no well defined plastic yield stress, but the initial portion of the curve is not linear. However, the deviation from linear, due to the instantaneous creep, is very small. The plastic strain is about 10 microstrain at a stress of 20 MPa. Therefore, the model may be most appropriate for FCC metals. This is not a unique idea. The Valanis (1971) theory, based on stress as a functional of strain history and a concept of intrinsic time, has no yield surface and no purely elastic deformation.

4 A Three-Dimensional Model

The three-dimensional thermoviscoplastic model of Freed and Walker (1993) can be viewed as a member of the general class of the maximum dissipation viscoplastic models. Their model for polycrystalline metals was developed with the goal of

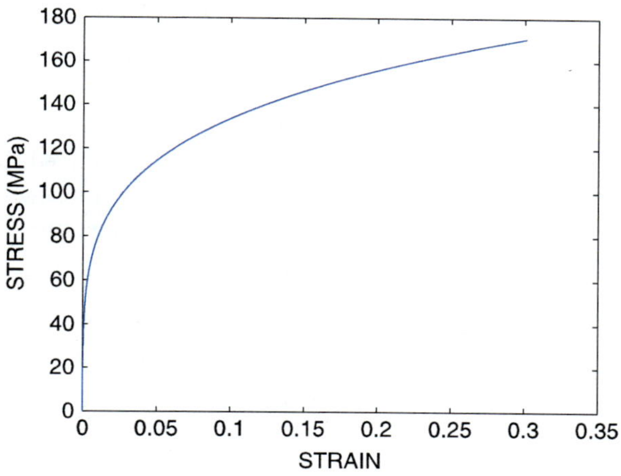

Fig. 5.2 The stress-strain response of copper at 43°C under a stress $\sigma = 0.0061t$ MPa

reducing to the Odqvist creep theory at high temperatures and a Prager type plasticity theory at low temperatures. This model is a maximum dissipation model, even though it was not developed from that viewpoint. The evolution of their internal variables is based on the Chaboche type evolution equations. To reproduce their evolution equations, which are similar to Freed, Chaboche (1991), as a maximum dissipation model, the state and control variables must be distinguished. The state variables are the strain, ϵ, the back stress tensor, $\boldsymbol{B}$, the drag strength, D, the yield strength, Y, and the entropy, η. The control variables are the stress tensor, $\boldsymbol{\sigma}$, the back strain, $\boldsymbol{b}$, the drag strain, δ, and the yield strain, y, and the temperature, θ. The deviatoric stress is $\boldsymbol{S}$.

Assume that the long term behavior is described by the elastic Helmholtz energy of Eq. (4) and that the plastic Helmholtz energy defining the long term states is quadratic,

$$\psi^p(B_{ij}, D, Y, \theta) = \frac{1}{2} H Tr(\boldsymbol{B} \cdot \boldsymbol{B}) + \frac{1}{2} H_t \theta^2 + \frac{1}{2} H_Y Y^2 + \frac{1}{2} H_D D^2, \tag{40}$$

for H, H_t, H_D, and H_Y material constants. The quadratic energy in which the various internal variables are not coupled is sufficiently complex to recover this model.

The long-term internal strain variables are defined by

$$b_{ij} = \frac{\partial \psi^p}{\partial B_{ij}}, \qquad \delta = \frac{\partial \psi^p}{\partial D}, \qquad \text{and} \qquad y = \frac{\partial \psi^p}{\partial Y}. \tag{41}$$

Off the long-term states, these strains are allowed to take arbitrary values.

The control variables stress and temperature are measurable. However, a form obtainable from a measurable quantity must be assumed for the internal back strain, drag strain, and yield strain. The back strain is assumed to be the tensor with components, in which no summation is intended,

$$b_{ij} = HL_{ij}\frac{\dot{\epsilon}^p_{ij}}{||\dot{\boldsymbol{\epsilon}}^p||}. \tag{42}$$

This agrees with Eq. (41) because in the long term $\dot{\boldsymbol{\epsilon}}^p$ is constant.

In the simplest model, the drag and yield strain remain fixed throughout the process. These control variables are not forced. They therefore must have values corresponding to the values of the internal stresses in the long term. The drag strain is $\delta = l_d H_D$, and the yield strain is $y = l_y H_Y$.

The norm is

$$||\dot{\boldsymbol{\epsilon}}^p|| = \sqrt{\frac{2}{3}\dot{\epsilon}^p_{ij}\dot{\epsilon}^p_{ij}}. \tag{43}$$

No dissipation potential was proposed in Freed and Walker (1993), but one which is consistent with a plastic strain rate assumption is easily appended by integrating the plastic strain rate expression with respect to the stress. The evolution of plastic strain is

$$\dot{\epsilon}^p_{ij} = ||\dot{\boldsymbol{\epsilon}}^p||\frac{1}{2}\frac{S_{ij} - B_{ij}}{||\boldsymbol{S} - \boldsymbol{B}||}, \tag{44}$$

where the norm of the plastic strain rate has the form,

$$||\dot{\boldsymbol{\epsilon}}^p|| = f(\theta)Z\left(\left\langle\frac{||\boldsymbol{S} - \boldsymbol{B}|| - Y}{D}\right\rangle\right). \tag{45}$$

Note that in the $||\cdot||$ - norm,

$$\frac{S_{ij} - B_{ij}}{||\boldsymbol{S} - \boldsymbol{B}||}$$

is a unit vector.

Typically $f(\theta) = \exp(-Q/R\theta)$, the Arrhenius form, and the function $Z(||S||/C) = A(||S||/C)^n$, for $||S|| \leq C$ and $n > 0$, $A > 0$ are material constants. For example, the dissipation potential may be given by Eq. (16).

The generalized energy is

$$P^* = \psi^e(\epsilon^e_{ij}, \theta) + \psi^p(B_{ij}, D, Y, \theta) - \sigma_{ij}\epsilon_{ij} - \boldsymbol{B}\cdot\boldsymbol{b} - D\delta - Yy + \theta\eta. \tag{46}$$

Put

$$\Psi = \psi^e(\epsilon^e_{ij}, \theta) + \psi^p(B_{ij}, D, Y, \theta).$$

The evolution equations, when the temperature evolution is neglected, are

$$\begin{bmatrix} \dot{B}_{11} \\ \cdots \\ \dot{B}_{ij} \\ \cdots \\ \dot{B}_{33} \\ \dot{D} \\ \dot{Y} \end{bmatrix} = -k \begin{bmatrix} \frac{\partial^2 \Psi}{\partial B_{11}^2} & \cdots & \frac{\partial^2 \Psi}{\partial B_{11} \partial B_{ij}} & \cdots & \frac{\partial^2 \Psi}{\partial B_{11} \partial B_{33}} & \frac{\partial^2 \Psi}{\partial B_{11} \partial D} & \frac{\partial^2 \Psi}{\partial B_{11} \partial Y} \\ & & \cdots\cdots\cdots & & & & \\ \frac{\partial^2 \Psi}{\partial B_{11} \partial B_{ij}} & \cdots & \frac{\partial^2 \Psi}{\partial B_{ij}^2} & \cdots & \frac{\partial^2 \Psi}{\partial B_{ij} \partial B_{33}} & \frac{\partial^2 \Psi}{\partial B_{ij} \partial D} & \frac{\partial^2 \Psi}{\partial B_{ij} \partial Y} \\ & & \cdots\cdots\cdots & & & & \\ \frac{\partial^2 \Psi}{\partial B_{11} \partial B_{33}} & \cdots & \frac{\partial^2 \Psi}{\partial B_{33} \partial B_{ij}} & \cdots & \frac{\partial^2 \Psi}{\partial B_{33}^3} & \frac{\partial^2 \Psi}{\partial B_{33} \partial D} & \frac{\partial^2 \Psi}{\partial B_{33} \partial Y} \\ \frac{\partial^2 \Psi}{\partial B_{11} \partial D} & \cdots & \frac{\partial^2 \Psi}{\partial B_{ij} \partial D} & \cdots & \frac{\partial^2 \Psi}{\partial D \partial B_{33}} & \frac{\partial^2 \Psi}{\partial D^2} & \frac{\partial^2 \Psi}{\partial D \partial Y} \\ \frac{\partial^2 \Psi}{\partial B_{11} \partial Y} & \cdots & \frac{\partial^2 \Psi}{\partial Y \partial B_{ij}} & \cdots & \frac{\partial^2 \Psi}{\partial Y \partial B_{33}} & \frac{\partial^2 \Psi}{\partial Y \partial D} & \frac{\partial^2 \Psi}{\partial Y^2} \end{bmatrix}^{-2} \begin{bmatrix} -b_{11} + \frac{\partial \Psi}{\partial B_{11}} \\ \cdots \\ -b_{ij} + \frac{\partial \Psi}{\partial B_{ij}} \\ \cdots \\ -b_{33} + \frac{\partial \Psi}{\partial B_{33}} \\ -\delta + \frac{\partial \Psi}{\partial D} \\ -y + \frac{\partial \Psi}{\partial Y} \end{bmatrix}. \tag{47}$$

Because, the square matrix is diagonal for the assumed energy on the long-term states, the inverse square is easily computed.

In the Freed-Walker model, the evolution of the back stress, $\boldsymbol{B}$, is given by

$$\dot{B}_{ij} = 2H_B \left(\dot{\epsilon}_{ij}^p - \frac{B_{ij}}{2L} ||\dot{\boldsymbol{\epsilon}}^p|| - f(\theta) R_b \frac{B_{ij}}{2||\boldsymbol{B}||} \right), \tag{48}$$

where $H_B > 0$, L is the limiting back stress, and $R_b > 0$ are moduli and $f(\theta)$ accounts for thermal recovery. The drag strength is given by

$$\dot{D} = h \left(||\dot{\boldsymbol{\epsilon}}^p|| - \frac{D}{l_d} ||\dot{\boldsymbol{\epsilon}}^p|| + f(\theta) r \right), \tag{49}$$

where $h > 0$, $l_d > 0$ and $r \geq 0$. The evolution of the yield strength is

$$\dot{Y} = h_y \left[||\dot{\boldsymbol{\epsilon}}^p|| - \frac{(Y - Y_o)}{l_y} ||\dot{\boldsymbol{\epsilon}}^p|| - f(\theta) r_y \right], \tag{50}$$

where Y_o is the annealed yield strength and $h_y > 0$, $l_y > 0$, and $r_y > 0$. Here Y_0, the annealed yield, could be written in terms of the Hall-Petch empirical relation of the yield stress to the grain size, $\sigma_Y = \sigma_0 + k_o l^{-1/2}$, where σ_0 and k_o are material constants. In these three evolution equations, the first term is strain hardening, the second is strain softening or dynamic recovery, and the third is thermal recovery, which is independent of the mechanical dissipation. The thermal softening (recovery) moduli, R_b, r, and r_y, are functions of state.

If the Freed-Walker thermal softening terms are neglected, the maximum dissipation model with an uncoupled, quadratic potential for the long-term states recovers the three-dimensional Freed-Walker model, assuming that the internal control variables, δ and y are held constant. The back stress terms predicted by the model here are

$$\dot{B}_{ij} = -||\dot{\boldsymbol{\epsilon}}^p|| \left[\frac{1}{H^2} (H B_{ij} - b_{ij}) \right]. \tag{51}$$

The model agrees with that of Freed-Walker if each $L_{ij} = L$ and if

$$\frac{1}{H} = \frac{H_B}{L} \quad \text{and} \quad \frac{||\dot{\boldsymbol{\epsilon}}^p|| b_{ij}}{H^2} = H_B \dot{\epsilon}_{ij}^p.$$

For the drag, D, an analogous result is obtained.

$$\dot{D} = -||\dot{\boldsymbol{\epsilon}}^p|| \left[\frac{1}{H_D^2}(H_D D - \delta) \right]. \tag{52}$$

The model agrees with that of Freed-Walker if

$$\frac{1}{H_D} = \frac{h_d}{l_d} \quad \text{and} \quad \frac{\delta}{H_D^2} = h_d.$$

For Y, the model predicts

$$\dot{Y} = -||\dot{\boldsymbol{\epsilon}}^p|| \left[\frac{1}{H_Y^2}(H_Y Y - y) \right]. \tag{53}$$

The model agrees with that of Freed-Walker if

$$\frac{1}{H_Y} = \frac{h_y}{l_y} \quad \text{and} \quad \frac{y}{H_Y^2} = h_y.$$

The steady state, slow rate long term limit of the viscoplastic processes described above is $dB_{ij}/dt = dD/dt = dY/dt = 0$ for $||d\epsilon_{ij}^p/dt|| > 0$ and $\dot{\epsilon}_{ij}^p$ is constant.

The relaxation term is not a sum of strain softening and thermal softening. Both effects are included in a single softening term. The simple additive form of the Freed-Chaboche-Walker expressions for thermal softening cannot be recovered by this construction. In Freed and Walker (1993), the recovery function R_b is taken to be zero so that recovery of B occurs implicitly through D. r is assumed to be a function of only D, and r_y is assumed to be a function of only Y. Thermal recovery cannot be obtained in the construction here unless each internal stress is coupled to the temperature. In that case, each evolution equation involves all variables.

In Freed and Walker (1993), the long-term limit, L, of the back stress is assumed to depend on the drag strength, D. Such a relationship cannot be recovered from a simple uncoupled, quadratic long term energy function.

A primary advantage of these maximum dissipation models is that, given the energy function for the long-term states and the relaxation modulus, they predict a specific and well-defined non-equilibrium process described by evolution equations proceeding from given initial conditions and by the control variables. In this sense, these models go beyond continuum thermodynamics. The presence of cross terms in the long-term energy, ψ^p, produces much more complicated models than those in the literature. These terms destroy the Chaboche form of the evolution equations. Within the resulting evolution equations, the recovery terms are no longer uncoupled as in the Freed-Chaboche-Walker model. The need for these complications can only be assessed by experimentation. Similarly, the need for non-quadratic energy terms involving the internal variables, which this construction allows, remains to be investigated.

References

J.-L. Chaboche (1991). Thermodynamically based viscoplastic constitutive equations: theory versus experiment. In *High Temperature Constitutive Modeling - Theory and Application*, MD-Vol. 26/AMD-Vol. 121, ASME, pp. 207–226.

A. D. Freed and K. P. Walker (1990). Steady-state and transient Zener parameters in viscoplasticity: drag strength versus yield strength. *Applied Mechanics Reviews* **43**, S328–S337.

A. D. Freed and K. P. Walker (1993). Viscoplasticity with creep and plasticity bounds. *International Journal of Plasticity* **9**, 213–242.

A. D. Freed, J.-L. Chaboche and K. P. Walker (1991). A viscoplastic theory with thermodynamic considerations. *Acta Mechanica* **90**, 155–174.

H. W. Haslach, Jr. (2002). A non-equilibrium thermodynamic geometric structure for thermoviscoplasticity with maximum dissipation. *International Journal of Plasticity* **18**(2), 127–153.

H. W. Haslach, Jr. and R. W. Armstrong (2004). *Deformable Bodies and the Material Behavior*, Wiley, New York.

J. Kratochvil and O. W. Dillon, Jr. (1969). Thermodynamics of elastic-plastic materials as a theory with internal state variables. *Journal of Applied Physics* **40**, 3207–3218.

E. Krempl (1987). Models of viscoplasticity, Some comments on equilibrium (back) stress and drag stress. *Acta Mechanica* **69**, 25–42.

E. Kröner (1963). Dislocation: A new concept in the continuum theory of plasticity. *Journal of Mathematics and Physics* **42**, 27–37.

J. Lubliner (1972). On the thermodynamic foundations of non-linear solid mechanics. *International Journal of Non-linear Mechanics* **7**, 237–254.

J. Lubliner (1973). On the structure of the rate equations of materials with internal variables. *Acta Mechanica* **17**, 109–119.

J. Mandel (1973). Thermodynamics and Plasticity. In *Foundations of Continuum Thermodynamics*, ed. J. J. Delgado Domingos, M. N. R. Nina, J. H. Whitelaw, John Wiley and Sons, New York, pp. 283–311.

G. A. Maugin (1992). *The Thermodynamics of Plasticity and Fracture*, Cambridge University Press, Cambridge.

G. A. Maugin, and W. Muschik (1994). Thermodynamics with internal variables Part 1. General concepts. *Journal of Non-Equilibrium Thermodynamics* **19**, 217–249.

A. S. Nowick, and E. S. Machlin (1947). Dislocation theory as applied by NACA to the creep of metals. *Journal of Applied Physics* **18**, 79–87.

K.-D. Papoulia and J. M. Kelly (1997). Visco-hyperelastic model for filled rubbers used in vibration isolation. *Journal of Engineering Materials and Technology* **119**, 292–297.

P. Perzyna (1963). The constitutive equations for rate sensitive plastic materials. *Quarterly of Applied Mathematics* **20**, 321–332.

P. Perzyna (1966). Fundamental problems in viscoplasticity. In *Advances in Applied Mechanics*, Vol. 9, ed. C.-S. Yih, Academic Press, New York, pp. 243–377.

P. Perzyna (1971). Thermodynamic theory of viscoplasticity. *Advances in Applied Mechanics*, Vol. 11, ed. C.-S. Yih, Academic Press, New York, pp. 313–354.

A. R. S. Ponter and F. A. Leckie (1976). Constitutive relationships for the time-dependent deformation of metals. *Journal of Engineering Materials and Technology* **98**, 47–51.

J. R. Rice (1970). On the structure of stress-strain relations for time-dependent plastic deformation in metals. *Journal of Applied Mechanics* **37**, 728–737.

J. R. Rice (1971). Inelastic constitutive relations for solids: an internal variable theory and its application to metal plasticity. *Journal of the Mechanics and Physics of Solids* **19**, 433–455.

J. C. Simo and T. Honein (1990). Variational formulation, discrete conservation laws, and path-domain independent integrals for elasto-viscoplasticity. *Journal of Applied Mechanics* **57**, 488–497.

J. C. Simo and T. J. R. Hughes (1998). *Computational Inelasticity*. Springer-Verlag, New York.

K. C. Valanis (1971). A theory of viscoplasticity without a yield surface. Parts I and II. *Archives of Mechanics* **2–3**, 517–535.

D. Yao and E. Krempl (1985). Viscoplasticity theory based on overstress. The prediction of monotonic and cyclic proportional and non-proportional loading paths of an aluminium alloy. *International Journal of Plasticity* **1**, 259–274.

C. Zener and J. H. Holloman (1944). Effect of strain rate on plastic flow of steel. *Journal of Applied Physics* **15**, 22–32.

Chapter 6
The Thermodynamic Relaxation Modulus as a Multi-Scale Bridge from the Atomic Level to the Bulk Material

1 Introduction

The construction of the maximum dissipation evolution equation requires only the thermodynamic relaxation modulus, k, in addition to the long-term quasi-static energy density function. As indicated in the discussion of viscoelastic and viscoplastic materials (Chapters 4 and 5), the thermodynamic relaxation modulus depends on the material microstructure and controls the relaxation speed. Therefore the thermodynamic relaxation modulus is the primary tool for multi-scale modeling of spatial scales in the material response. Such scales include the molecular and substructures in the body. For example, such substructures within the normally well-hydrated artery wall include elastin lamellae, collagen fibers, smooth muscle bundles, and the interface between the blood and artery wall. The mechanical response of soft biological tissue is influenced by intermolecular forces between water and biopolymers. The maximum dissipation construction is applied to represent the viscoelastic response of the arterial elastin-water system.

The relaxation modulus is assumed constant for viscoelastic materials (Chapter 4) since the microstructure of the long-chain molecules changes only slightly and for viscoplastic materials depends indirectly on the dislocation structure which changes during loading (Chapter 5). The thermodynamic relaxation modulus is not explicitly time-dependent, but it may depend on parameters that do change with time. A dynamic model in the particular case of the arterial elastin-water system depends on the thermodynamic relaxation modulus for elastin based on the moisture content through a postulated non-equilibrium intermolecular bonding energy function. The relaxation modulus provides a bridge from the molecular to the elastin substructure scale in the artery. In this model, the fibered structure of an elastin lamella has not been modeled so that only two spatial scales are considered. Others, such as Porter (1995) have related the molecular scale and the static response of the system.

H.W. Haslach Jr., *Maximum Dissipation Non-Equilibrium Thermodynamics and its Geometric Structure*, DOI 10.1007/978-1-4419-7765-6_6,

1.1 Multi-Scale and Dynamic Modeling

Traditionally, different models are used at each of the different scales. Information may be passed from one level to the next highest level through moduli in the higher level model or by taking the sum of energy functions for each level. Modeling of multi-scale phenomena requires bridging from molecular to continuum models for the large-scale system. The importance of such bridging has been emphasized by many observers. "Analysis of multi-scale phenomena, while apparently beyond the horizon of contemporary capabilities, is one of the most fundamental challenges of research in the next decade and beyond. So-called scale bridging, in which the careful characterization of mechanical phenomena require that the model bridge the representations of events that occur at two or more scales, require the development of a variety of new techniques and methods" (Oden et al., 2003). One can also consider different time scales. For example, in an artery the fast scale is the periodic variation in blood pressure; the slow scale is the lifetime of the organism during which the development of disease such as hypertension or the formation of aneurysms may influence the dynamic response.

1.2 The Viscoelastic Response of the Elastin-Water System

Almost all dynamic mechanical responses of soft tissues are non-equilibrium processes. The mechanical response of the elastin-water system in an artery wall is viscoelastic. A dynamic multi-scale time-dependent evolution equation for the mechanical response of the elastin-water system captures the effect of moisture content on the glass transition of the elastin. The thermodynamic relaxation modulus is defined to relate spatial scales, the molecular scale including moisture bonding and the bulk material scale. The maximum dissipation model reproduces published experimental data on the elastin glass transition behavior with respect to load frequency and to ambient relative humidity but is not merely empirical in the sense of being a fit to such data because it predicts dynamic responses such as non-physiological creep and physiological rate-dependent stress-stretch relations. The maximum dissipation viscoelastic model predicts the influence of moisture content and the glass transition of the elastin on the time-dependent response of the circumferential stretch and the change in radius of a hydrated arterial cylindrical elastin lamella under cyclic radial pressure loads in the hemodynamic range. Such an elastin cylinder approximates the behavior of the elastin substructure in an elastic artery wall.

Water is known to play a key role in modulating dynamic biomolecular interactions, for example by lowering the glass transition temperature of elastin in the artery below body temperature so that the elastin is rubbery. Elastin in vivo must operate in the rubbery region, i.e. above the glass transition, which depends on moisture content. Dry elastin exhibits little elasticity, but hydrated elastin exhibits significant elasticity. Some seem to attribute most of the observed viscoelastic response of an artery wall to the hydrated proteoglycan gel (e.g. Raines, 2000). But the elastin in the artery wall is also a contributor to the viscoelastic response (e.g. Lillie and Gosline, 1990, 2002). Yet, mechanical analyses of soft tissues continue to employ static methods. No model for the dynamic mechanical response of the water-elastin system has been presented in the literature.

Poroelastic and mixture theories usually postulate that the properties of the multiphase material are determined by the properties of the constituents. Not only is a good understanding of the structure of dry elastin unavailable, but elastin properties change drastically in the presence of water. Most importantly, these models do not capture the influence of water on the glass transition of elastin. Even in poroelastic or poroviscoelastic models for the transmural fluid flow in an artery wall, elastin is rarely distinguished and little consideration has been given to whether the elastin is in a glassy or in a rubbery state.

Biphasic models to account for moisture have been extensively applied to cartilage, ligaments, spinal disks (e.g. Iatrides et al., 2003) but only rarely to artery walls. Such models are combinations of Darcy's law, linear or nonlinear elasticity and the Navier-Stokes equation. Poroelastic models, which are related to biphasic and triphasic models, for moisture transport in arterial walls assume a steady state Darcy law because steady flow mechanics is most familiar, and the material is assumed isotropic (Simon, 1992; Simon et al., 1996, 1998). Typically, the wall is viewed as a thick wall cylinder. The stress is obtained from the linear elastic Lamé solution for a thick-walled cylinder even though the artery wall is certainly not linear elastic. The pressure drop is obtained from Pouseuille's law, and the force driven diffusion of solutes through the wall is obtained from Darcy's law (or the slightly more general Brinkman equation for porous materials), both of which assume steady behavior. The steady state assumption may underestimate the actual non-steady situation (Ritman and Lerman, 2007). See Chapter 9 for a non-steady version of Darcy's law based on the maximum dissipation evolution equation construction.

As preparation for the maximum dissipation evolution equation, the background section summarizes necessary published experimental work on some mechanical properties of the elastin-water system. The construction of the viscoelastic evolution equation requires modifying the long-term quasi-static constitutive model for the presence of moisture at equilibrium and accounting for swelling. Experimental data from the literature is used to generate numerical values of the proposed multiscale thermodynamic relaxation modulus at different constant moisture contents. No direct measurements of some elastin-water bonding properties are available and so these must be deduced from bulk experimental data available in the literature. The maximum dissipation dynamic model for the elastin-water system relates the elastin molecular scale to the elastin lamella substructure scale in an artery and neglects the fiber scale making up the elastin lamella. The model can recover experimental data of Lillie and Gosline for the frequency relation to the glass transition of hydrated elastin. Finally, the proposed model is applied in numerical computations for non-physiological creep and for a pressure-loaded elastin cylinder to indicate the predicted behavior of elastin in the artery wall.

2 Background

To put the known properties of the elastin-water system in context, arterial elastin composing part of the media is first located in its physical setting. Further, because the construction of the evolution equation requires the long-term quasi-static response, the somewhat contradictory stress-strain relations reported in the literature are summarized. This lack of agreement for the quasi-static stress-strain behavior is accounted for in the numerical applications of the proposed model by considering different constitutive models for the long-term quasi-static behavior. Further, a summary is given of some published experimental facts about the glass transition of the elastin-water system needed to make computations with the proposed model.

2.1 *The Structure of Arterial Elastin*

Arterial elastin operates at many spatial scales including its role as a component in the artery wall, its fiber structure, and its molecular structure. The media of an elastic artery is a multi-layered structure with each layer made of elastin sheets, smooth muscle cells (SMC), ground material, and collagen fibers. Each cylindrical elastin sheet, which alternates with smooth muscle cell layers and glycoproteins, rarely forms a complete cylinder around the wall. Elastin, guided by smooth muscle cells, is the main mechanism to control radial dilations of the artery. Collagen fibers are coiled at low stress and are nearly inextensible when fully recruited, mechanically serving primarily as a reinforcement mechanism when the load is too high. The wall contains significant moisture that hydrates the components. Somewhat contradictory physical models of the elastin-SMC interaction within the structure of the wall have been proposed in the musculo-elastic fascicle of Clark and Glagov (1985) and in the contractile-elastic unit (CEU) of Davis (1993). Both physical models include the observed alternating layers of SMC and elastin, as many as 40 to 70 such elements in the radial direction. A primary difference in the physical models is whether the contractile SMC lie parallel or obliquely to the elastin sheets. Surrounding this structure between the parallel elastin lamellae in either physical model is hydrophilic ground material, which is sometimes thought of as a hydrated gel.

The elastin membrane in the artery has several spatial scales ranging from the molecular to the fibril and to the fiber. The elastin molecule is created in the extracellular matrix by crosslinking tropoelastin molecules (Brown-Augsburger et al., 1995). The crosslinks make the elastin molecule insoluble in water, in contrast to the elastin precursor tropoelastin. Most reports on elastin view the elastin molecule to be built as an alternating chain of hydrophobic and crosslinking segments. The fibrils are composed of 4 nm diameter microfibrils built from aligned and crosslinked elastin molecules. A bundle of 100–200 nm diameter fibrils forms the elastin fiber, with a 1.5–15 μm diameter (Winlove and Parker, 1987). The fibers are assembled into the elastin lamella.

The elasticity of hydrated elastin is called entropic because a deformation lowers the entropy. When the load is removed, the system relaxes to its original configuration of higher entropy as required by the second law of thermodynamics. Although elastin behaves like rubber in that both exhibit entropic elasticity, rubber does not need to be hydrated to become highly elastic as elastin does. Disagreements persist over the elastin molecule structure and the source of the entropy change upon loading and unloading (e.g. Li and Daggett, 2002). The crosslinks are rigid and so the elastic behavior must occur in the hydrophobic segments between crosslinks. Most physical models proposed to explain the source of elasticity at the molecular scale involve a mechanism by which an applied force moves hydrophobic residues to the outer part of the structure. The molecule recovers its configuration when the force is removed. These physical models attribute the elastic recovery of the elastin molecule to segmental entropy changes that involve hydrophobic hydration, in which a non-polar molecule is surrounded by a net of water molecules. Gosline (1978) argues that the hydrophobic interaction is only significant for very small strains and that the behavior of elastin in the physiological range may be best described by the random network model developed for rubber. Even though disputes remain about the molecular structure of elastin and its behavior in water and about the causes of elastin elasticity, the general consensus appears to be that the kinetic theory of rubber can accurately model the bulk behavior of hydrated elastin in the physiological range (e.g. Li and Daggett, 2002). Competing proposals for the structure of the hydrated elastin molecule may be found in the following articles: Aaron and Gosline (1981); Debelle and Alix (1999); Gray et al. (1973); Grut and McCrum (1974); Weinberg et al. (1995); Weiss-Fogh and Anderson (1970); Hoeve and Flory (1974); Urry (1983); Wasserman and Salemme (1990).

2.2 Experimental Stress-Strain Relations in Elastin

Stress-strain curves for elastin in the older literature were obtained from the very elastic neck ligament (ligamentum nuchae), usually from an ox. The tests Wöhlisch (1939) conducted in a water bath show a concave up J-shaped curve. Elastin does not crystallize as rubber and some other polymers do at high extensions, and so crystallization cannot contribute to the J-shape, as suggested by Wöhlisch. Later, staining showed that collagen fibers are present in the ligamentum nuchae fiber bundles (Hoeve and Flory, 1958). The fiber bundle structure of the ligament may affect the results; certainly any collagen present does. Hoeve and Flory (1958) obtain linear engineering stress-engineering strain results for extensions up to 70%, but discount this result as fortuitous.

All material except the elastin is commonly removed by autoclaving a sample and then placing it in 0.1M NaOH. Gotte et al. (1968) compared the effect of such treatment on the stress-strain curves. Large untreated ligamentum nuchae specimens produce a concave up stress-strain curve. The treated specimens hydrated to 40% moisture content and to 60% moisture content both yield a linear stress-strain curve up to a 0.7 strain. In tests of water-swollen elastin fiber bundles obtained from the ligamentum nuchae that had been treated, Gosline (1978) obtains linear force-extension curves up to an extension of 7 mm for a 10 mm specimen. He argues that the concave up curves obtained by others result from the recruitment of fine, slack elastin fibers in their samples.

The elastin in an artery is structured into a membrane of fibers rather than a fiber bundle. To examine if the arterial elastin membrane structure affects the stress-strain curve, an elastin sample in membrane form is usually obtained by digesting all material except the elastin from an artery specimen. In one such series of tests on dog and sheep thoracic aorta, after treatment, the circumferential Cauchy stress - engineering strain curve in a distilled water bath is a concave up curve in the strain range of 0–0.3 (Sherebin et al., 1983). Longitudinal stress-strain curves are also concave up, at least to a strain of 1.2. In more recent tests, the circumferential engineering stress-engineering strain curves obtained from treated artery rings and immersed in water have a concave up J-shape (Lillie et al., 1998; Lillie and Gosline, 2002).

The elastin sheet is mechanically anisotropic, according to the uniaxial stress-strain tests of Sherebin et al. (1983). The elastic moduli and ultimate stress are significantly greater in the circumferential than in the longitudinal direction. Further, their micrographs show that, under a strain, the lamella separates into a fibrous net with many fibers oriented in the strain direction. Elastin membrane anisotropy has also been observed in biaxial tests on bovine aorta, with the circumferential direction stiffer than the axial (Zou and Zhang, 2009).

The experimental data does not definitively eliminate a linear quasi-static stress-strain curve for elastin. But the elastin fibers in ligamentum nuchae samples may be more axially oriented than those in arterial elastin membranes, resulting in a more linear response. Some have simply concluded that elastin from the ligamentum nuchae is linearly elastic while arterial elastin membranes are nonlinear (e.g. Dorrington and McCrum, 1977).

2.3 The Glass Transition Temperature of the Moisture-Elastin System

The glass transition is the temperature at which the polymer changes from rubbery to glassy; it therefore indicates a change in the physical properties of the material. Arterial elastin must operate in the rubbery region to ensure the necessary large deformation elastic behavior in response to blood pressure. The glass transition temperature of dry elastin is about 200°C, but water interaction with elastin lowers the glass transition in vivo to well below the 37°C body temperature. The glass transition value of many polymers is changed by a diluent such as water (Ferry, 1961, p. 223). The total amount of water in an elastin specimen is described by the moisture content, $m = m_w/m_e$, the ratio of the mass of the water to the mass of the elastin in the specimen, measured in grams of water per grams of dry elastin. The elastin glass transition temperature data of Kakivaya and Hoeve (1975) as a function of moisture content, m, (in percent) have been fit by the equation

$$T_g = 201 - 126 \log m \tag{1}$$

over the range of 1 to 31% moisture content, where the glass transition temperature T_g is in C (Lillie and Gosline, 1990). These data were obtained from treated ligamentum nuchae specimens by locating the peak in heat capacity measurements as a function of temperature with various constant moisture contents. At higher moisture contents, the glass transition is below 0°C and so the method does not work because of ice formation. But the result has the advantage that it is independent of a load frequency. Since the specimens were tested under a dead load, the frequency could be taken as zero. The relation (1) implies that a moisture content of about 26% is needed to have the glass transition at room temperature (23°C). Lillie and Gosline (1990) argue that in order to function in the rubbery region at 37° C between 1 and 10 Hz, the water content of the elastin must be about 50%. The relation (1) predicts that at 50% moisture content, the glass transition is –13C if the range is extended to higher moisture contents.

The loss tangent, $\tan\delta$, is obtained from the phase shift δ between a sinusoidal loading and the sinusoidal response component of the same frequency. It has been known for a long time that the loss tangent takes a maximum in a neighborhood of the glass transition where the dissipation is maximal, whether the polymer is crosslinked or not (e.g. Ferry, 1961, p. 36). The loss tangent is sometimes viewed as the fraction of mechanical energy converted to heat (Porter et al., 2005). In linear viscoelasticity, the storage modulus, G', and the loss modulus, G'', are related to the loss tangent by $\tan\delta = G''/G'$.

As is well known, the observed glass transition temperature depends on load frequency, because the frequency of an applied load affects the time scale. To compare the behavior as a function of load frequency, using the Kakivaya and Hoeve relation (1) for $\omega = 0$ at 37°C gives a moisture content $m = 10^{164/126} = 20.02\%$. The corresponding relative humidity is 95.29% from the sorption isotherm (10) discussed below, but at 100 Hz and 37°C, Lillie et al. (1996) measured the glass transition to occur at 97.6% relative humidity. So at the higher cyclic loading frequency, higher moisture content is required to produce the same glass transition temperature. Lillie and Gosline (1993, figure 1) experimentally verify that, at a fixed temperature, the load frequency to produce a glass transition at that temperature increases with an increase in moisture content. The loss tangent has its maximum at the glass transition temperature when plotted against temperature (Porter et al., 2005), against frequency (Gosline and French, 1979; Birge and Nagel, 1985; Lillie and Gosline, 1990) and against moisture content (Lillie et al., 1996). In summary, the glass transition temperature of elastin depends on load frequency and moisture content. Porter (1995, p. 274) suggests that the glass transition results from thermal energy overcoming the cohesive energy to permit mer unit motion on a large scale. Such an idea implies that the loss tangent (tangent modulus) is a maximum at the glass transition because, with the release of cohesive energy, a maximum amount of dissipation occurs.

3 The Maximum Dissipation Multi-Scale Viscoelastic Model for the Elastin-Water System

Recall the construction of the maximum dissipation evolution equations from Chapter 3. The version of the non-equilibrium system applied here is homogeneous so that the thermodynamic variables are the same at all points of the body. The construction therefore gives a constitutive model for the bulk behavior of the body.

$$\begin{bmatrix} \frac{dx_1}{dt} \\ \frac{dx_2}{dt} \\ \cdots \\ \frac{dx_n}{dt} \end{bmatrix} = -k \begin{bmatrix} \frac{\partial^2\varphi}{\partial x_1^2} & \frac{\partial^2\varphi}{\partial x_1 \partial x_2} & \cdots & \frac{\partial^2\varphi}{\partial x_1 \partial x_n} \\ \frac{\partial^2\varphi}{\partial x_2 \partial x_1} & \frac{\partial^2\varphi}{\partial x_2^2} & \cdots & \frac{\partial^2\varphi}{\partial x_2 x_n} \\ & \cdots\cdots\cdots\cdots & & \\ \frac{\partial^2\varphi}{\partial x_n \partial x_1} & \frac{\partial^2\varphi}{\partial x_n \partial x_2} & \cdots & \frac{\partial^2\varphi}{\partial x_n^2} \end{bmatrix}^{-2} \begin{bmatrix} y_1(t) + \frac{\partial\varphi}{\partial x_1} \\ y_2(t) + \frac{\partial\varphi}{\partial x_2} \\ \cdots \\ y_n(t) + \frac{\partial\varphi}{\partial x_n} \end{bmatrix}. \tag{2}$$

The right-hand column is the set of affinities that drive the process; when they are zero, the process ceases evolving. The only experimental data needed to apply the model are the time-independent long-term model, φ, and the thermodynamic relaxation modulus tensor, k. The form of the relaxation modulus is chosen below to bridge from the molecular to the bulk behavior, making this construction a multi-scale model.

The magnitude of the relaxation modulus determines whether the response is instantaneous (elasticity), or takes time (viscoelastic). As discussed in Chapter 3, the smaller the value of $k > 0$, the more viscous and the slower is the uniaxial response. Therefore k is a minimum at the glass transition.

For an example of (2), the uniaxial first Piola stress, P, and stretch, λ, are conjugate thermodynamic variables. The uniaxial non-equilibrium maximum dissipation thermodynamic evolution equation has the nonlinear form,

$$\frac{d\lambda}{dt} = -k \left(\frac{\partial^2\varphi}{\partial\lambda^2}\right)^{-2} \left(\frac{\partial\varphi}{\partial\lambda} - P(t)\right). \tag{3}$$

3.1 The Multi-Scale Thermodynamic Relaxation Modulus

The thermodynamic relaxation modulus measures the speed of relaxation and has stress rate units when the stretch is the control variable. For the elastin-water system, this speed is assumed to describe the rate of change of the instantaneous steady state bond force. To derive an expression for this non-equilibrium speed, the idea is to start with a molecular bond potential that gives equilibrium behavior and modify it to account for the non-equilibrium state in time-dependent behavior. This gives a non-equilibrium bond function that generalizes that for equilibrium behavior.

When modeling the influence of the molecular scale on the bulk behavior, the discrete molecular chain behavior is typically represented by a continuum average bond energy function. In equilibrium, the Lennard-Jones potential is commonly assumed to account for the attractive and repulsive forces in hydrophobic hydration (e.g. Paschek, 2009). This bond energy forms the starting point for the spatial multi-scale thermodynamic relaxation modulus for the viscoelastic relaxation of elastin. The Lennard-Jones potential is

$$\psi = 4C_e\left[\left(\frac{r_o}{r}\right)^{12} - \left(\frac{r_o}{r}\right)^{6}\right], \tag{4}$$

where r is average separation distance between the elastin chain molecules, r_o is the van der Waals separation for the Lennard-Jones potential (where the energy is zero). The first term is the repulsive force term and the second term is the attractive force term. The stretch is defined to be $\lambda \equiv r/r_o$. The net bond force is $d\psi/d\lambda$, which is zero at equilibrium where the stretch is $\lambda_e = r_e/r_o = 2^{1/6}$. The coefficient $C_e > 0$ is the magnitude of the equilibrium energy for the Lennard-Jones potential, as is verified by substituting λ_e in (4). The elastic modulus is $d^2\psi/d\lambda^2$ evaluated at λ_e; or more generally, the bond stiffness at a given separation is $d^2\psi/d\lambda^2$. The binding force between chain molecules is a maximum if the second derivative $d^2\psi/d\lambda^2 = 0$; *i.e.* the bond stiffness is zero. This maximum occurs at a separation $\lambda_T \equiv r_T/r_o = 1.244$.

Modifying the Lennard-Jones potential to account for hydrogen bonded moisture is known to be a difficult problem even for equilibrium behavior (Porter and Vollrath, 2008), due to the complexity of the electron energy distribution in quantum chemistry. Here, the bond energy is modified to account for temperature, θ, by taking the coefficient $C_e(\theta)$ to be a function of temperature and for moisture content by adjusting the attractive force by a unitless function of ambient relative humidity $g(rh)$ to obtain $\widehat{\psi} = 4C_e(\theta)[\lambda^{-12} - g(rh)\lambda^6]$. The presence of water molecules and hydrophobic hydration must affect the attractive energy and thus the equilibrium separation. The equilibrium stretch becomes $\lambda_e = 2^{1/6}g^{-1/6}$. The predicted van der Waals radius including the moisture function is $r_{vw}^6 = r_o^6/g(rh)$. The stretch $\lambda_T = 1.244g^{-1/6}$ now gives the glass transition separation where $d^2\widehat{\psi}/d\lambda^2 = 0$. The separation should be maximal with respect to moisture content near the glass transition and therefore $g(rh)$ should have a minimum near the glass transition moisture content when the temperature and load frequency are fixed.

The observation frequency, ω, affects the measured glass transition temperature. The faster the test rate, the larger the value of the glass transition temperature (Porter, 1995, p. 310). The skeletal modes are the vibration modes of longer chain segments normal to the chain in contrast to vibrations of small atomic groups on the chain. The glass transition is due to the development of new skeletal modes of vibration. But a higher observation frequency reduces the likelihood that new skeletal modes are occupied. An observation at $\omega = 0$ produces more occupied modes. Because the bond energy function gives the energy of interaction between chains, this effect is accounted for in the present model adjusting the magnitude of $\widehat{\psi}$ by a unitless function $f(\omega)$ that has a minimum near the glass transition frequency. The non-equilibrium bond energy, ϕ, resulting from a dynamic load is approximated by

$$\phi = 4C_e(\theta) f(\omega)[\lambda^{-12} - g(rh)\lambda^{-6}]. \tag{5}$$

Steady state occurs at the separation, λ_s, making $d\phi/d\lambda = 0$, $\lambda_s^6 = 2/g(rh)$.

The Born stability criterion (Born, 1939) states that the glass transition occurs at the separation for which the binding force between chain molecules is a maximum because at that point the force can do nothing but decrease. Since the binding force is the first derivative of the bond energy with respect to the separation between chain molecules r, in equilibrium the Born criterion means that the second derivative with respect to r is zero; i.e. the molecular elastic modulus is zero. The Born stability criterion has been applied to biopolymers to obtain a relation of amide-amide hydrogen bonding and amide-water bonding to the glass transition temperature at the molecular level (e.g. Porter et al., 2005; Porter and Vollrath, 2008). By the Born criterion, the steady state separation $\lambda = \lambda_s$ is not the glass transition separation.

The thermodynamic relaxation modulus, k, a multi-scale bridge from the molecular to the membrane, is taken to be a function of the temperature θ, the load frequency ω, and the ambient relative humidity, rh. The thermodynamic relaxation modulus is defined to capture the behavior on the molecular level.

Definition 1 The thermodynamic relaxation modulus, k, is the time derivative of the binding force evaluated at the steady state separation $\lambda = \lambda_s$.

$$k = \frac{d}{dt}\left(\frac{d\phi}{d\lambda}\right)\bigg|_{\lambda=\lambda_s} = \frac{d^2\phi}{d\lambda^2}\bigg|_{\lambda=\lambda_s}\frac{d\lambda}{dt}. \tag{6}$$

The form of the thermodynamic relaxation modulus is

$$k = \frac{d^2\phi}{d\lambda^2}\bigg|_{\lambda=\lambda_s}\frac{d\lambda}{dt} = \frac{72}{2^{1/3}}C_e(\theta) f(\omega) g(rh)^{7/3}\frac{d\lambda}{dt}. \tag{7}$$

Collecting terms into k_o, write

$$k = f(\omega) g(rh)^{7/3} k_o. \tag{8}$$

4 Modification of the Long-Term Energy Density Function to Account for Moisture Content

The viscoelastic model requires a long-term quasi-static energy density function. In the elastin-water system, the quasi-static response is influenced by the water content. This section first summarizes some published data describing the relation between swelling of elastin, its moisture content and the ambient relative humidity. A published general expression for the effect of swelling on a quasi-static energy density function and its associated shear modulus is reviewed. The results are applied to the neo-Hookean and Zulliger hyperelastic models, which are to be studied as alternatives for the long-term quasi-static energy density in the proposed model for the elastin-water system.

Moisture may be held in different portions of the specimen, and the location has physical consequences. Moisture may be held in the extrafibrillar (interfibrillar) region between and around the elastin fibers forming the membrane or may be held in the intrafibrillar region formed by pores of diameter $\sim$ 3 nm in the elastin fiber (Weinberg et al., 1995). Large voids created in the tissue by removing all non-elastin material increase the interfibrillar volume in treated specimens compared to the in vivo interfibrillar volume. The interfibrillar moisture in treated specimens includes that held in the spaces formerly occupied by the digested material (Lillie and Gosline, 1993).

To estimate the amount of intrafibrillar and interfibrillar moisture, specimens were tested in both high humidity (95–99.5%) and in a bath by Lillie et al. (1996). They conclude that in an ambient high humidity at 37° C, the 50.5% water content at the highest humidity is taken up in the intrafibrillar space while specimens in a bath took up 200% water, the difference being held in the interfibrillar regions.

4.1 Water-Induced Swelling of Elastin

The swelling of an elastin specimen due to the uptake of water strongly influences the mechanical response. The swelling ratio, v, is the ratio of the specimen volume at its current state to the volume at a reference state. The stretch λ is the ratio of the current length to the reference length. The current volume in terms of the stretch λ_i, $i = 1, 2, 3$ in three principal directions, where each λ_i depends on the moisture content, is $V = \lambda_1\lambda_2\lambda_3 V_o$, where V_o is the reference volume. Therefore $v = \lambda_1\lambda_2\lambda_3$ is the volumetric swelling ratio, V/V_o. The swelling is isotropic if $\lambda_1 = \lambda_2 = \lambda_3 = v^{1/3}$.

Gosline and French (1979) attribute elastin swelling to the hydrophobic segments of the elastin molecule. The involvement of hydrophobic segments suggests that the swelling volume is not the sum of the volume of solid elastin and the volume of water adsorbed; some free volume may be generated.

In contrast, the uniaxial swelling ratio, s_u, is defined to be the ratio of a dimension at the current moisture content divided by the same dimension at the reference moisture content, both measured at the same fixed temperature. This ratio is much easier than the volume ratio to obtain experimentally. Experimental uniaxial swelling data for arterial elastin in circumferential specimens, at 37°C, given by Lillie et al. (1996, figure 8) is reasonably fit in the moisture content range 25–300% by

$$s_{bath} = 1 - [0.306 \exp(-0.0114m)], \tag{9}$$

where s_{bath} is the uniaxial swelling ratio referred to the length in a bath and m is the moisture content in percent. The uniaxial swelling ratio referred to the length at 25% moisture content, s_{bath}^{-1}, for moisture content 50% is about 0.8/0.77=1.04, from their figure 8, and the corresponding isotropic volume swelling ratio is about

1.12, about half that in a bath. Presumably, the uniaxial swelling of a dry specimen raised to 25% moisture content is quite small, but no direct measurement is available.

The swelling of elastin from ligamentum nuchae is anisotropic (Gotte et al., 1968). The transverse swelling is approximately twice that in the axial direction, presumably because the elastin fibers are parallel to the ligament axis. However, this observation does not necessarily imply that the swelling of arterial elastin membranes is anisotropic.

These data relate moisture content and swelling volume. Because some of the published data give the tangent modulus, tan δ, versus ambient relative humidity, a relationship between ambient relative humidity and moisture content, called a sorption isotherm, is required. An experimental sorption isotherm for elastin at 37° C in the relative humidity range from 95 to 99.5% obtained by measuring the equilibrium moisture content at different values of ambient relative humidity was produced by Lillie et al. (1996, figure 5). A parabolic refit of the data by the author produces the sorption isotherm,

$$m(rh) = 9872.72 - 206.667rh + 1.08444rh^2, \tag{10}$$

where m and rh are in percents. Note that $m(100) = 50.5\%$, $m(97) = 29.6\%$, and $m(95) = 26.5\%$. In the case of vapor hydration, other hydrophilic materials exhibit hysteresis in the sorption isotherm when adsorption is compared to desorption, but such hysteresis is ignored here.

4.2 Model to Account for Swelling in the Strain Energy Density Function

The effect of swelling on the dynamic response is partially captured by modifying the long-term energy density function, φ, to account for the swelling. Because the long-term energy density represents the equilibrium behavior, the equilibrium swelling is sufficient. The effect of a dynamic load on swelling is extremely difficult to measure experimentally. The strain energy density to be modified for swelling is defined in terms of strain invariants. Let F be the deformation gradient tensor. The three tensor invariants of the right Cauchy strain tensor $C = F^t F$, a strain defined with respect to the reference state, in principal stretch coordinates are $I_1 = \mathrm{Tr}(C) = \lambda_1^2 + \lambda_2^2 + \lambda_3^2$, the second $I_2 = \lambda_1^2\lambda_2^2 + \lambda_1^2\lambda_3^2 + \lambda_2^2\lambda_3^2$, and the third $I_3 = \det(C) = (\lambda_1\lambda_2\lambda_3)^2$. If the material is mechanically incompressible, $I_3 = 1$.

Swelling is accounted for by including in the long-term energy density function a scalar swelling field, $\hat{V}$, that represents the swelling volume at each point of the elastin, with respect to a choice of reference state. Uniform swelling means that the swelling field, $\hat{V} = v$, is independent of position in the elastin lamella. In the unloaded isotropic swelling case, the first invariant $I_1 = 3v^{2/3}$ at equilibrium and the second invariant is $I_2 = 3v^{4/3}$, with respect to the unswollen state.

Treloar (1972) proposed accounting for swelling in the neo-Hookean model, $\varphi(I_1) = 0.5\mu(I_1 - 3)$, by assuming $\varphi(I_1, \hat{V}) = 0.5\mu(I_1 - 3v^{2/3})$. This relation of Treloar is generalized (Pence and Tsai, 2006) by modifying the quasi-static strain energy density without swelling, $\widetilde{\varphi}(I_1, I_2)$, to account for isotropic swelling

$$\varphi(I_1, I_2, \hat{V}) = m(\hat{V})\widetilde{\varphi}(\hat{V}^{-2/3}I_1, \hat{V}^{-4/3}I_2), \tag{11}$$

where $m(\hat{V}) > 0$ satisfies $m(1) = 1$. The unswollen ($\hat{V} = 1$) and unloaded state is the reference state. Putting $m(\hat{V}) = v^{2/3}$ and $\hat{V} = v$, recovers the Treloar model when $\widetilde{\varphi} = 0.5\mu(I_1 - 3)$. Assume that arterial elastin swells isotropically even though such swelling behavior remains to be convincingly determined by experiment.

4.3 Shear Modulus as a Function of Swelling Ratio

The shear modulus, μ, is expected to decrease with moisture content. The shear modulus for a polymer isotropically swollen by a liquid, when the strain energy density is a function of the invariants, $\varphi(I_1, I_2)$, is by definition (Pence and Tsai, 2006), away from the glass transition,

$$\mu = 2\left(v^{-1/3}\frac{\partial\varphi}{\partial I_1} + v^{1/3}\frac{\partial\varphi}{\partial I_2}\right)\Bigg|_{I_1=3v^{2/3},\, I_2=3v^{4/3}}. \tag{12}$$

In the case of the neo-Hookean strain energy density, this definition implies that $\mu(m) = \mu v^{-1/3}$ is the shear modulus at moisture content m.

The shear modulus is $\mu = \rho R\theta/M_c$ in the Gaussian network theory, where ρ is the polymer density, R is the gas constant, θ is the temperature in Kelvin, and M_c is the molecular weight of the polymer, usually taken as that between crosslinks. This relation has been used to obtain a value for M_c from stress-strain data giving μ. For example, one experimentally obtained value is $\mu = 410$ kPa for a single water-swollen elastin fiber at 24°C (Aaron and Gosline, 1980). Then, using the kinetic theory of rubber relation, $\mu = \rho R\theta/M_c$, they computed $M_c = 7.1$ kDa assuming $\rho = 1.33$. Treloar (1972, p. 68) argues that the shear modulus for the swollen material is $\mu = (\rho R\theta/M_c)v_2^{1/3}$, where v_2 is the volume fraction of elastin in the system. Therefore, the swollen shear modulus is $v_2^{1/3} = v^{-1/3}$ times the unswollen shear modulus in agreement with (12).

A plot of the Young's modulus versus moisture content obtained from ligamentum nuchae specimens at 23° C shows a large drop at 33% moisture content, presumably because 23° C is the glass transition temperature at this moisture content (Gotte et al., 1968). (This result does not agree with the relation of Kakivaya and Hoeve (1) that predicts a glass transition temperature of 9.67° C for 33% moisture content). The modulus drops from 7×10^9 dyn/cm^2 (700 MPa) at 20% moisture content in the glassy region to 10^7 dyn/cm^2 (1 MPa) at 40% moisture content, which is in the rubbery region. After the drop, the elastic modulus continues to decrease slowly with increasing moisture content. Since for a neo-Hookean material, the elastic modulus $E = 3\mu$, an estimate of μ is 333 kPa at 40% moisture content. Fung (1993) has reported that the shear modulus of elastin ranges between 100-1000 kPa.

Clearly, the Gotte et al. experimental data does not agree with the Treloar analysis that $\mu(m) = \mu v^{-1/3}$, which does not account for the large variation in shear modulus at the glass transition. Therefore, in what follows, the modification of the long-term strain energy density includes the swelling as determined by (11) but replaces μ with the experimental value of the shear modulus at the moisture content in question.

4.4 Neo-Hookean Long-Term Strain Energy Density as a Function of Moisture Content

As an approximation, each principal stretch is assumed to be the product of the swelling stretch, λ_{s_i} and the mechanical stretch, λ_{m_i}, so that $\lambda_i = \lambda_{s_i}\lambda_{m_i}$. But for simplicity, the material is assumed mechanically incompressible so that $\lambda_{m_1}\lambda_{m_2}\lambda_{m_3} = 1$. Then for any mechanical load, when F is the deformation gradient with respect to the unswollen state, $J \equiv \det(F) = \lambda_{s_1}\lambda_{s_2}\lambda_{s_3} = v$, the volume swelling ratio.

The neo-Hookean model in principal stretch coordinates is $\varphi(\lambda_1, \lambda_2, \lambda_3) = 0.5\mu\left(\lambda_1^2 + \lambda_2^2 + \lambda_3^2 - 3\right)$, where μ is the shear modulus. The stretches are components of the deformation gradient so that the derivatives $\partial\varphi/\partial\lambda_i$ give the components of the first Piola-Kirchhoff stress tensor, P. Putting $\lambda_3 = \lambda$ the incompressibility condition $\lambda_1\lambda_2\lambda_3 = 1$ implies that $\lambda_1 = \lambda_2 = \lambda^{-1/2}$, and the uniaxial neo-Hookean model is $\varphi(\lambda) = 0.5\mu(\lambda^2 + 2\lambda^{-1} - 3)$. The long-term uniaxial Piola stress is $P = \partial\varphi/\partial\lambda = \mu(\lambda - \lambda^{-2})$. To get a concave up J-shaped stress-stretch curve, the Cauchy (not the Piola) stress, σ, where $P = J\sigma F^{-t}$, must be plotted against the stretch because the Piola versus stretch curve is concave down.

The isotropic neo-Hookean uniaxial strain energy including swelling is

$$\varphi(\lambda) = 0.5\mu(m)v^{2/3}(\lambda^2 + 2\lambda^{-1} - 3), \tag{13}$$

where λ is now the circumferential mechanical stretch with respect to the unloaded *swollen* state, and $\mu(m)$ is the shear modulus as a function of moisture content for the swollen elastin. An experimentalist is most likely to measure deformation with respect to the unloaded swollen state. The Eq. (13) is obtained directly by writing the total stretch as the product of the swelling and mechanical stretches, assuming isotropic swelling. This agrees with (11) with $m(\hat{V}) = v^{2/3}$ where the stretch is taken with respect to the unswollen state.

This strain energy density is called neo-Hookean because it is simply the incompressible Hookean strain energy density. If one assumes an isotropic linearly elastic strain energy density $\varphi = 0.5E(\epsilon_1^2 + \epsilon_2^2 + \epsilon_3^2)$, with elastic modulus E, where, by incompressibility, the sum of the engineering strains $\epsilon_1 + \epsilon_2 + \epsilon_3 = 0$, then

$\lambda_1 + \lambda_2 + \lambda_3 = 3$. Writing the linear elastic strain energy density in terms of stretch in the incompressible case yields $\varphi = 0.5E(\lambda_1^2 + \lambda_2^2 + \lambda_3^2 - 3)$. Therefore to account for swelling in isotropic incompressible linearly elastic materials, the factor $\upsilon^{2/3}$ applies, as in the neo-Hookean case, to give

$$\varphi = 0.5E(m)\upsilon^{2/3}(\epsilon_1^2 + \epsilon_2^2 + \epsilon_3^2). \tag{14}$$

4.5 *Zulliger Long-Term Strain Energy Density as a Function of Moisture Content*

The isotropic Zulliger et al. (2004) long-term strain energy density function for elastin adjusted for swelling is $\varphi = c(m)\upsilon(I_1 - 3)^{3/2}$, where the stretches are the mechanical stretches measured with respect to the unloaded swollen state. For uniaxial loading, the model becomes $\varphi = c(m)\upsilon(\lambda^2 + 2\lambda^{-1} - 3)^{3/2}$ so that

$$P = \frac{\partial \varphi}{\partial \lambda} = 3c(m)\upsilon(\lambda^2 + 2\lambda^{-1} - 3)^{1/2}(\lambda - \lambda^{-2}). \tag{15}$$

The Taylor series for P about $\lambda = 1$ is $3c(m)\upsilon[3\sqrt{3}(\lambda - 1)^2 - 4\sqrt{3}(\lambda - 1)^3 + \ldots]$, indicating that the tangent at $\lambda = 1$ is horizontal so that the apparent elastic modulus is $E = 0$ for the $P - \lambda$ curve. Apparently, the form of the strain energy was chosen to produce the initial non-load bearing unwrinkling of elastin lamellae. While the elastin lamellae are wrinkled in unloaded excised specimens, it is not clear they are ever wrinkled in vivo. The shear modulus with respect to the swollen state is $\mu = 3\upsilon^{-1/3}c(m)\upsilon^{-2/3}\sqrt{\upsilon^{-2/3}3\upsilon^{2/3} - 3} = 0$ by (12) independent of $c(m)$, in agreement with the zero elastic modulus. Zulliger et al. estimate $c = 53$ kPa for a normotensive artery. Assuming that the elastin normally operates at 50% moisture content in vivo and since $\mu = 333$ kPa is estimated at 50% moisture content from Gotte et al. (1968), the moisture dependent values will be taken as $c(m) = \mu(m)/6$, where $\mu = E/3$ from the Gotte et al. data.

5 Numerical Determination of the Multi-Scale Thermodynamic Relaxation Modulus

The second key component of the viscoelastic model for the elastin-water system is the thermodynamic relaxation modulus, (8). No direct experimental measurement of the binding force in hydrated elastin is available to determine the functions $f(\omega)$ and $g(rh)$. Therefore the relaxation modulus must be estimated from published data on the bulk behavior near the glass transition where elastin operates in vivo. In general, the glass transition temperature is $T_g = \mathcal{F}(\omega, m)$, a function of observation frequency and moisture content.

The experimental relations between the tangent modulus, $\tan \delta$, and either moisture content or load frequency for treated elastin specimens under small strain amplitude cyclic loads, discussed in Section 2.3, are used to estimate the thermodynamic relaxation modulus, k, near the glass transition. The experimental results of Lillie and Gosline (1990, figure 2) for elastin with a moisture content of 28.1% and $T_g = 33°$ C give the tangent modulus as a function of frequency, with a peak at $\omega_c = 1.25$ Hz. These conditions are not realistic for in vivo arterial elastin since greater moisture content would be needed to ensure that the elastin is significantly into the rubbery region under a hemodynamic loading near 1 Hz. The measures of $\tan\delta$ by Lillie et al. (1996) for arterial elastin as a function of ambient relative humidity show a peak at about 97.6% relative humidity in ambient 37° C and 100 Hz observational frequency.

5.1 Linear Elastic Long-Term Strain Energy Density

Since there is no consensus about the quasi-static uniaxial stress-strain behavior of elastin, the thermodynamic construction for the viscoelastic response (2) is first applied to the linear elastic hyperelastic model for the long-term behavior, as a guideline to the nonlinear model prediction and to the forms of $f(\omega)$ and $g(rh)$.

A quadratic thermostatic energy density function, φ, that yields a linear elastic thermostatic stress-strain curve always produces a linear viscoelastic model in this construction. For uniaxial loading in stress control, the long-term strain energy density is $\varphi(\epsilon) = 0.5E\epsilon^2$ so that the general evolution Eq. (2) for stress control is $\dot{\epsilon} = -k_\epsilon E^{-2}[E\epsilon - \sigma(t)]$. The elastic modulus, E, is a function of moisture content and temperature; the relaxation modulus k_ϵ must have units of stress/sec. For a dynamic loading, let the sinusoidal loading be $\sigma(t) = a\sin(\omega t)$. The evolution equation has solution

$$\epsilon(t) = \epsilon(0)\exp(-k_\epsilon t/E) + k_\epsilon a\omega\exp(-k_\epsilon t/E) + \frac{ak_\epsilon}{E}\left[\frac{k_\epsilon\sin(\omega t) - E\omega\cos(\omega t)}{k_\epsilon^2 + E^2\omega^2}\right]. \tag{16}$$

The third term is the steady state response after the transients die out as $t \to \infty$.

The phase shift between the input and output is obtained by comparing (16) and the input $\sigma(t) = a\sin(\omega t)$ as functions of time. The phase shift satisfies

$$\tan\delta = -E\omega/k_\epsilon, \tag{17}$$

so that the strain response lags the applied stress. This form includes strictly monotonic loadings since if $\omega = 0$ and if $k_\epsilon \neq 0$, then $\tan\delta = 0$ implies no phase shift for a monotonic load, as expected.

Since the phase shift δ varies inversely to k_ϵ by (17), to capture the tangent modulus having a maximum at the glass transition, the relaxation modulus, k, must have a minimum near the glass transition, where the dissipation is greatest. In the linear elastic long-term energy function case, the relation (17) gives k directly from experimental data for $\tan\delta$ at an observational frequency, $\omega > 0$, and for the dependence of E on moisture content.

If the relaxation modulus is assumed constant in ω, $\tan\delta$ does not have a peak between 0 and 8 Hz by (17), and so does not represent the response of the elastin-water system measured by Lillie and Gosline (1990) that has a peak value near 1 Hz. So the most common spring and dashpot models cannot reproduce the viscoelastic data around the glass transition because k is a constant in the construction of these maximum dissipation models (Chapter 4).

5.2 The Moisture Content Function, $g(rh)$

The tangent modulus versus ambient relative humidity data of Lillie et al. (1996, figure 6.6) is approximately fit by the relation $\tan\delta_{100} = 1.2\exp[-0.246(rh-97.6)^2]$, where the relative humidity $95 \leq rh \leq 99.5$ is measured in percent. The glass transition occurs at the moisture content of about 33%, which corresponds to about 97.6 % relative humidity. However, this value does not satisfy the Kakivaya and Hoeve relation (1) between the glass transition temperature of elastin and its moisture content because of the difference in load frequency. These data, for a treated artery specimen that therefore has many voids, may only approximate the response of a single elastin lamella.

Not enough data is available to seek a function $g(rh)$ for all relative humidity or moisture contents so a local fit near the glass transition relative humidity is sought. Using the linearly elastic expression (17) for $\tan\delta$ as a guide and using (8), the function $g(rh)$ is approximated, based on an exponential fit to the $\tan\delta$ data of Lillie et al. (1996, Figure 6), for positive constants b_1 and b_2, as

$$g(rh)^{7/3} = \frac{E(m)\omega}{f(\omega)k_o\tan(\delta)} = b_1\exp\{b_2[rh - h(\theta,\omega)]^2\}, \tag{18}$$

where $0.95 \leq rh \leq 0.995$, and $h(\theta,\omega)$ is the relative humidity at which the glass transition occurs for a tissue temperature θ and fixed frequency, ω, *i.e.* the relative humidity corresponding to the solution of $T_g = \mathcal{F}(\omega, m)$ for given T_g and ω. All terms except for the elastic modulus, $E(m)$, and $\tan\delta$ are independent of the moisture content. A second-order Taylor series expansion of this function about $h(\theta,\omega)$ gives a parabolic approximation $g(rh)^{7/3} \sim b_1\{1 + b_2[rh - h(\theta,\omega)]^2\}$. The function (18) has a local minimum at the relative humidity corresponding to the glass transition.

5.3 The Frequency Function, $f(\omega)$

The function $f(\omega)$ modifies the magnitude of the bond energy function to account for the non-equilibrium response under cyclic loads. Assume that the moisture content, $\widehat{m}$, and glass transition temperature, T_g, are fixed. Let ω_c be the solution of $T_g = \mathcal{F}(\omega, \widehat{m})$ for the critical observational frequency. Using the linearly elastic expression (17) for $\tan\delta$ as a guide and using (8), the function $f(\omega)$ is approximated near the glass transition for fixed $\widehat{m}$ and T_g as

$$f(\omega)/\omega = \frac{E(\widehat{m})}{g(rh)^{7/3}k_o \tan(\delta)} = a_1 \exp[a_2(\omega - \omega_c)^2], \tag{19}$$

where a_1 and a_2 are positive constants. All terms in the middle equality are independent of frequency except $\tan\delta$. A Taylor series about ω_c to the second order for (19) produces a concave up parabolic approximation $f(\omega)/\omega \sim a_1[1 + a_2(\omega - \omega_c)^2]$.

5.4 Estimated Values of the Multi-Scale Thermodynamic Relaxation Modulus and Other Parameters

In Table 6.1, the experimental swelling is with respect to a 25% moisture content reference state at 37° C (Lillie et al., 1996). Their data in terms of relative humidity is related to moisture content by (10). The glass transition temperatures, T_g, are obtained from the Kakivaya and Hoeve relation (1). The experimental shear modulus, $\mu(m)$, is taken to be one third of the corresponding elastic modulus given by Gotte et al. (1968). The thermodynamic relaxation modulus, k, is estimated from the linearly elastic Eq. (17), $k = -E\omega/\tan\delta$, using the Lillie et al., 1996) data for $\tan\delta$ at 100 Hz and 37C as a function of relative humidity and the elastic modulus data as a function of moisture content of Gotte et al. (1968). The smallest value of k, i.e. the slowest relaxation, is near the glass transition as required.

Table 6.1 Specimen properties

Moisture (%)	T_g (C)	Uniaxial swelling	v (isotropic)	Shear modulus (kPa/s)	k(kPa/s)
20	37	0.982	0.948	133,300	122, 474
30	14.88	1.016	1.050	66,700	17, 597
40	–0.8	1.047	1.147	367	133.6
50	–13	1.071	1.239	333	328.0
60	–23	1.098	1.325	300	1, 024

6 Recovering the Lillie-Gosline Data for the Frequency Dependence of the Glass Transition in the Elastin-Water System

The construction reproduces the $\tan\delta$ versus frequency data of Lillie and Gosline (1990, Figure 2) near the glass transition for elastin with a moisture content of 28.1% and $T_g = 33$°C. For a nonlinear long-term strain energy density, a sinusoidal load is applied in the model for a selection of frequencies between

0 and 6 Hz to determine if tan δ has a maximum in this range that corresponds to the maximum in their experiment.

The idea of a phase shift only makes sense if one compares harmonic signals of the same frequency. Therefore, the response of a nonlinear system is broken into its spectrum using the fast Fourier transform technique. The component with the same frequency as the excitation is compared to the excitation to determine the phase shift. To compute the phase shift of the ω component of the output, the chosen sinusoidal forcing is applied for 12 s to reduce the influence of transients. The result of (2), which is in terms of stretch, has a large zero-frequency component. To avoid having this "dc" component dominate the transform, stretch is converted to engineering strain by subtracting 1. The Matlab function *fft* is used to compute the discrete Fourier transform of this time-dependent engineering strain data to the frequency domain, $Y(\omega)$, and yields the power spectrum. The spectrum shows a large component at the input frequency and possibly other components. The phase shift, δ, is computed from the complex term, $Y(\omega) = a + bI$, corresponding to the input frequency as $\delta(\omega) = \arctan(b/a)$.

6.1 *Application for the Neo-Hookean and the Zulliger Long-Term Quasi-static Strain Energy Densities*

The neo-Hookean thermostatic uniaxial strain energy density, under the assumption that the applied Piola stress is sinusoidal, $P(t) = a\sin(\omega t)$, produces the uniaxial evolution equation,

$$\dot{\lambda} = -k_\lambda \left[\mu(m)v^{2/3}\left(1 + 2\lambda^{-3}\right)\right]^{-2}\left[\mu(m)v^{2/3}\left(\lambda - \lambda^{-2}\right) - a\sin(\omega t)\right]. \quad (20)$$

Assume $\mu(m)v^{2/3} = 150\,\text{kPa}$. A second order Taylor series expansion of k yields the parabola, $k_\lambda = 1.5\mu(m)v^{2/3}[20 + 50(\omega - 1)^2]$, where the coefficients were guessed by trial and error. The power spectrum (Fig. 6.1a) for a 2 Hz input shows a single frequency component with frequency equal to the input frequency. The computation of the phase shift over a range of frequencies (Fig. 6.1b) captures the qualitative form of the Lillie and Gosline (1990) data near the glass transition.

The analysis is repeated for the long-term Zulliger et al. (2004) uniaxial evolution equation for stress control (3),

$$\begin{aligned}\dot{\lambda} = &- k\left[\frac{3c(m)v}{\lambda^4}\left(5 - 6\lambda + 2\lambda^2 - 3\lambda^4 + 2\lambda^6\right)\right]^{-2}(\lambda^2 + 2\lambda^{-1} - 3)\\ &\times\left[3c(m)v(\lambda - \lambda^{-2})(\lambda^2 + 2\lambda^{-1} - 3)^{1/2} - a\sin(\omega t)\right]. \quad (21)\end{aligned}$$

The evolution Eq. (21) gives zero if the initial condition is $\lambda(0) = 1$. Therefore to obtain a numerical solution, the initial stretch is taken as slightly deformed, say $\lambda(0) = 1.001$. For this computation, assume $c(m)v = 25\,\text{kPa}$ and $k_\lambda =$

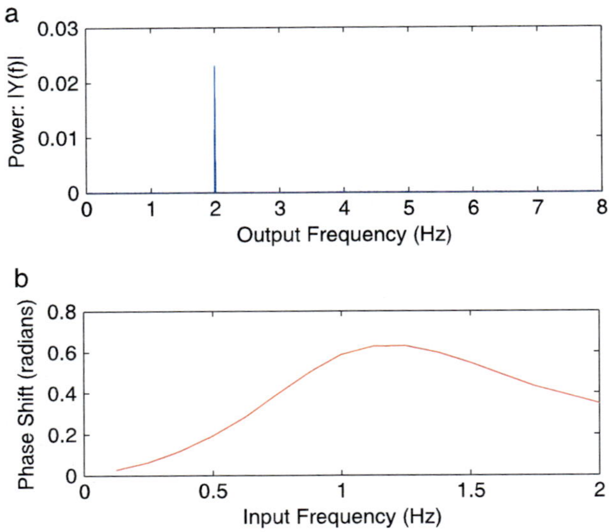

Fig. 6.1 Response of elastin to cyclic loading with the long-term behavior represented by the neo-Hookean model with $\mu(m)v^{2/3} = 150\,\text{kPa}$. **a** Single-sided amplitude spectrum showing a single frequency output for a 2 Hz input. **b** Phase shift versus observational frequency

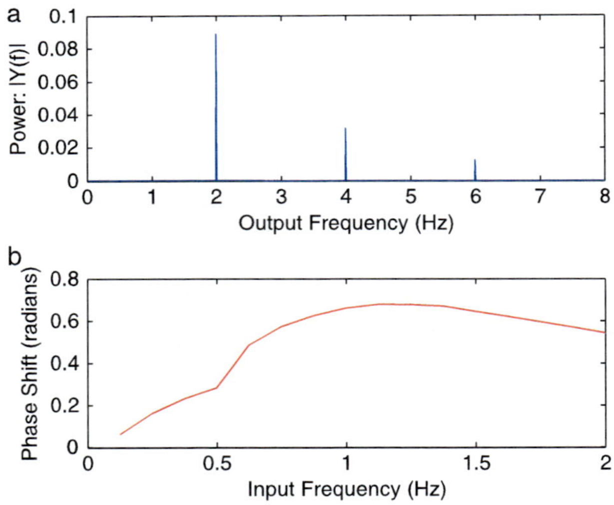

Fig. 6.2 Response of elastin to cyclic loading with the long-term behavior represented by the Zulliger model with $c(m)v = 25\,\text{kPa}$. **a** Single-sided amplitude spectrum showing higher harmonics for a 2 Hz input. **b** Phase shift versus observational frequency.

$0.15c(m)v[20 + 50(\omega - 1)^2]$. The power spectrum (Fig. 6.2a) shows higher order frequency components in the response to a 2 Hz input. In this sense, the Zulliger long-term strain energy density produces a more nonlinear behavior than the neo-Hookean. Again, a computation of the phase shift over a range of frequencies near the glass transition (Fig. 6.2b) captures the qualitative form of the Lillie and Gosline experimental results (1990, Fig. 6.2). Both computations reproduce the maximum of $\tan\delta$ near the glass transition. Therefore, in the case of a nonlinear long-term strain energy density function as well as in the linear case, the thermodynamic relaxation modulus provides a bridge from the molecular to the elastin lamella substructure scale.

7 Application to the Response of Arterial Elastin

Very few hyperelastic models for the mechanics of artery tissue have considered the influence of the moisture content on the mechanical response. To illustrate the predictive capabilities of the dynamic model, two loadings are examined: uniaxial creep and a sinusoidally pressurized hydrated elastin cylindrical membrane. The elastin is assumed isotropic.

7.1 Uniaxial Creep

The influence of moisture content on the viscoelastic behavior of elastin is exhibited in creep curves. Creep is not a physiological loading, but it is a simple test to explore the time-dependent mechanical properties of elastin, including the influence of the glass transition temperature. Although some creep data have been published for elastin-like synthetic polymers (Wu et al., 2008), no experimental creep stretch versus time curves appear to be available in the literature for natural elastin by itself. A brief mention of creep is made in Lillie and Gosline (2007), but their data are not in a useful form to validate the model presented here.

Creep curves are obtained using the data in Table 6.1 and an applied creep stress of 300 kPa. The creep stretch with respect to the swollen state is computed using Matlab ode23s at four fixed moisture contents (30–60%) for each of three incompressible long-term strain energy density functions: linear elastic, neo-Hookean, and the Zulliger. The linear elastic case is given in terms of engineering strain and the others are given in terms of the stretch. From (3) and the incompressibility condition, the evolution equations for fixed first Piola-Kirchhoff stress, P, applied to the swollen specimen are respectively,

$$\dot{\epsilon} = -k_\lambda (E(m)v^{2/3})^{-2}(E(m)v^{2/3}\epsilon - P); \tag{22}$$

$$\dot{\lambda} = -k_\lambda \left[\mu(m)v^{2/3}\left(1 + 2\lambda^{-3}\right)\right]^{-2}\left[\mu(m)v^{2/3}\left(\lambda - \lambda^{-2}\right) - P\right]; \tag{23}$$

$$\dot{\lambda} = -k_\lambda \left[\frac{3c(m)v}{\lambda^4} \left(5 - 6\lambda + 2\lambda^2 - 3\lambda^4 + 2\lambda^6\right) \right]^{-2} (\lambda^2 + 2\lambda^{-1} - 3)$$
$$\times \left[3c(m)v(\lambda - \lambda^{-2})(\lambda^2 + 2\lambda^{-1} - 3)^{1/2} - P \right]. \tag{24}$$

The same values of k_λ as a function of moisture content are used in all three cases to investigate the effect of the long-term strain energy density on the response. The moisture-dependent moduli in the long-term energy density functions are assumed to be related by $E = 3\mu$ and $c = \mu/6$ for purposes of illustration. In fact, k_λ may depend on the long-term strain energy density through the magnitude, C_e, in (5).

The linear elastic and the neo-Hookean long-term strain energy densities produce qualitatively similar results as shown in Figs. 6.3 and 6.4, respectively, since the only difference is the magnitude of the moduli $E(m)$ and $\mu(m)$ in the long-term energy densities. The equilibrium stretch increases with increasing moisture content, but the time for the stretch to reach equilibrium decreases with increasing moisture content. The neo-Hookean long-term strain energy model predicts an equilibrium stretch of only 1.0014 at 30% moisture content, but an equilibrium stretch of 1.37 at 60% moisture content. The creep slows as the moisture content gets closer to the glass transition moisture content at this temperature. The creep at 30% moisture content, for which the elastin is just in the glassy region but near the glass transition, is also slow.

The Zulliger long-term strain energy creep (Fig. 6.5) approaches equilibrium much more rapidly than the other two, perhaps because of the low initial slope of the stress-strain curve that represents unwrinkling. As in the linear and

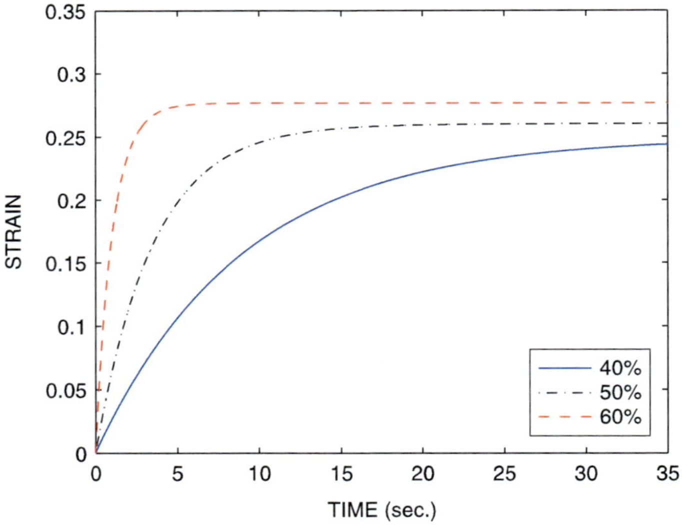

Fig. 6.3 The creep strain response as a function of moisture content for elastin with the long-term behavior represented by the incompressible, isotropic linear elastic model

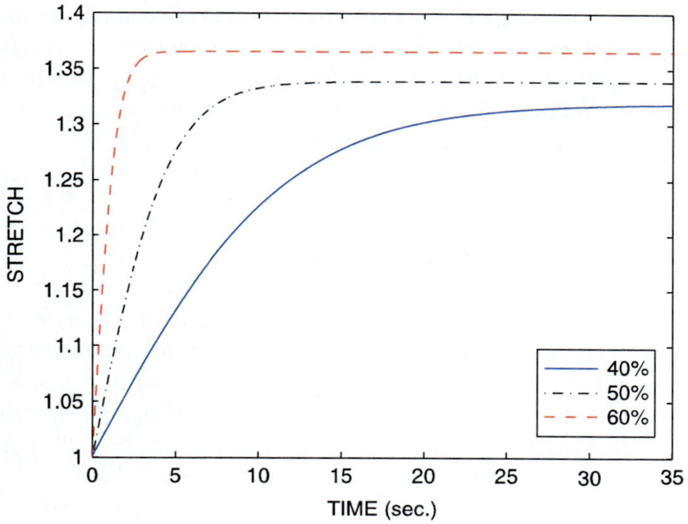

Fig. 6.4 The creep stretch response as a function of moisture content for elastin with the long-term behavior represented by the neo-Hookean model.

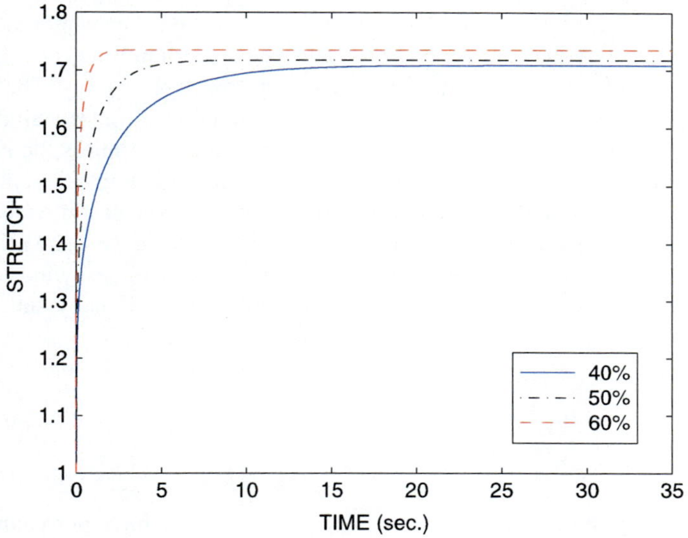

Fig. 6.5 The creep stretch response as a function of moisture content for elastin with the long-term behavior represented by the Zulliger model

neo-Hookean cases, the response is slower for moisture contents closer to the glass transition. Even tripling the value of c to make the long-term response stiffer does not change the qualitative behavior; it changes the equilibrium value to which the creep approaches but not the speed. The equilibrium stretch to which the creep is asymptotic is much larger for the Zulliger model.

7.2 Pressure Loaded Elastin Cylinder

A cylindrical membrane models a single elastin lamella in the artery wall. The symbol θ now represents the circumferential coordinate, not the temperature. The cylindrical elastin structure is in load control under the uniform time-dependent internal radial pressure, $p(t)$. During such a dynamic loading, the stresses such as $\sigma_{\theta\theta}$ cannot be obtained by substituting the instantaneous stretches in the hyperelastic function, φ, which is valid only for the long-term equilibrium states.

The elastin membrane in an artery does not form a complete cylinder, but for purposes of the numerical computation, the elastin membrane is assumed to be a full cylinder. The circumferential stretch is $\lambda_\theta = r/R$, where R is the original radius, and r is the current radius. The ends of the cylinder are modeled as open but anchored. Therefore, the axial stretch, λ_z, is assumed constant.

The two-dimensional stress state in the elastin membrane is described by the Cauchy stresses $\sigma_{\theta\theta}$, $\sigma_{\theta z}$, and σ_{zz}. The loading is applied to the membrane surface by the normal stress $\sigma_{rr} = p(t)$. A shear stress, $\sigma_{r\theta}$, on the membrane surface could be induced by the contraction of the attached smooth muscle cells. Further, a shear could result from σ_{rz} induced by the blood flow at the intima. These shears do not appear in the equation of motion in the r direction if $\sigma_{r\theta}$ is assumed to be constant in θ and if the shear, σ_{rz}, is assumed constant in z (unless the change in shear stress along the arterial tree is of interest). In order to examine the influence of the moisture content on the radial deformation of the swollen elastin lamella under a radial pressure, all shear stresses are neglected so that $\sigma_{\theta\theta}$ and σ_{zz} are principal.

Conservation of the radial direction linear momentum for a cylindrical body produces the following equation of motion in terms of the Cauchy stresses (e.g. Malvern, 1969, p. 668),

$$\frac{\partial \sigma_{rr}}{\partial r} + \frac{1}{r}\frac{\partial \sigma_{r\theta}}{\partial \theta} + \frac{\partial \sigma_{rz}}{\partial z} + \frac{1}{r}(\sigma_{rr} - \sigma_{\theta\theta}) + b_r = \rho a_r. \tag{25}$$

The acceleration in the radial direction is $a_r = \ddot{r}$. The body force per volume, b_r is neglected. For the two-dimensional membrane in which σ_{rr} does not vary with r, conservation of linear momentum in the r-direction is

$$\frac{1}{r}(\sigma_{rr} - \sigma_{\theta\theta}) + \frac{1}{r}\frac{\partial \sigma_{r\theta}}{\partial \theta} + \frac{\partial \sigma_{rz}}{\partial z} = \rho \ddot{r}. \tag{26}$$

In the full cylinder case, $\ddot{r} = R\ddot{\lambda}_\theta$. Then (26) is solved for $\sigma_{\theta\theta}$ after neglecting the shear terms.

$$\sigma_{\theta\theta}(t) = p(t) - \rho R^2 \lambda_\theta \ddot{\lambda}_\theta. \tag{27}$$

The evolution equation is, from (3),

$$\dot{\lambda}_\theta = -k \left(\frac{\partial^2 \varphi}{\partial \lambda_\theta^2}\right)^{-2} \left(\frac{\partial \varphi}{\partial \lambda_\theta} - P_{\theta\theta}\right), \tag{28}$$

where $P_{\theta\theta}$ is the component of the first Piola stress. Using $P = J\sigma F^{-t}$ and the fact that all the tensors involved are diagonal if the only load is the radial pressure, $P_{\theta\theta} = J\sigma_{\theta\theta}\lambda_\theta^{-1}$. The total volume is unchanged ($J = 1$) with respect to the swollen state since the elastin is assumed mechanically incompressible.

The resulting second-order ordinary differential equation for λ_θ is, from (27) and (28),

$$\ddot{\lambda}_\theta = \frac{1}{\rho R^2 J}\left[-\frac{\dot{\lambda}_\theta}{k}\left(\frac{\partial^2 \varphi}{\partial \lambda_\theta^2}\right)^2 - \frac{\partial \varphi}{\partial \lambda_\theta} + Jp(t)\lambda_\theta^{-1}\right] \tag{29}$$

If the long-term strain energy density is assumed to be isotropic linear elastic and the material is assumed incompressible so that $\epsilon_\theta + \epsilon_z + \epsilon_r = 0$, the strain energy density is $\varphi = 0.5E(m)v^{2/3}[\epsilon_\theta^2 + \epsilon_z^2 + (-\epsilon_\theta - \epsilon_z)^2]$ by (14), and (29) becomes

$$\ddot{\epsilon}_\theta = \frac{1}{\rho R^2 J}\left[-\frac{\dot{\epsilon}_\theta}{k}\left(E(m)v^{2/3}\right)^2 - E(m)v^{2/3}(2\epsilon_\theta + \epsilon_z) + Jp(t)\epsilon_\theta^{-1}\right]. \tag{30}$$

The neo-Hookean strain energy density for a cylindrical membrane is obtained by assuming that the axial stretch, λ_z, is given and that because of the mechanical incompressibility $\lambda_r = \lambda_\theta^{-1}\lambda_z^{-1}$. Therefore $\varphi = 0.5\mu(m)v^{2/3}(\lambda_\theta^2 + \lambda_z^2 + \lambda_\theta^{-2}\lambda_z^{-2})$, $d\varphi/d\lambda_\theta = \mu(m)v^{2/3}(\lambda_\theta - \lambda_\theta^{-3}\lambda_z^{-2})$, and $d^2\varphi/d\lambda_\theta^2 = \mu(m)v^{2/3}(1 + 3\lambda_\theta^{-4}\lambda_z^{-2})$. Equation (29) becomes

$$\ddot{\lambda}_\theta = \frac{1}{\rho R^2 J}\left\{-\frac{\dot{\lambda}_\theta}{k}\left[\mu(m)v^{2/3}(1 + 3\lambda_\theta^{-4}\lambda_z^{-2})\right]^2 - [\mu(m)v^{2/3}(\lambda_\theta - \lambda_\theta^{-3}\lambda_z^{-2})] + Jp(t)\lambda_\theta^{-1}\right\} \tag{31}$$

The stress in the z-direction is $d\varphi/d\lambda_z = \mu(m)v^{2/3}(\lambda_z - \lambda_\theta^{-2}\lambda_z^{-3})$, so that it depends on both stretches.

The Zulliger et al. model for elastin is $\varphi = c(I_1 - 3)^{3/2}$. So for a hydrated membrane and stretches with respect to the swollen state, $\varphi = c(m)v(\lambda_\theta^2 + \lambda_z^2 + \lambda_\theta^{-2}\lambda_z^{-2})^{3/2}$. Equation (29) becomes

$$\ddot{\lambda}_\theta = \frac{1}{\rho R^2 J}\left\{-\frac{\dot{\lambda}_\theta}{k}[c(m)v]^2\left[3(\lambda_\theta - \lambda_\theta^{-3}\lambda_z^{-2})(\lambda_\theta^2 + \lambda_z^2 + \lambda_\theta^{-2}\lambda_z^{-2})^{-1/2}\right.\right.$$
$$\left. +3(1 + 3\lambda_\theta^{-4}\lambda_z^{-2})(\lambda_\theta^2 + \lambda_z^2 + \lambda_\theta^{-2}\lambda_z^{-2})^{1/2}\right]^2$$
$$\left. - \left[3c(m)v(\lambda_\theta - \lambda_\theta^{-3}\lambda_z^{-2})(\lambda_\theta^2 + \lambda_z^2 + \lambda_\theta^{-2}\lambda_z^{-2})^{1/2}\right] + Jp(t)\lambda_\theta^{-1}\right\} \quad (32)$$

Once $\lambda_\theta(t)$ is determined, the time-dependent change in membrane radius is known since $r(t) = R\lambda_\theta(t)$.

Example The internal pressure within a complete single elastin cylindrical membrane is assumed sinusoidal, $p(t) = p_m + p_a[1 + \sin(2\pi t\omega - \pi/2)]$, where the minimum (diastolic) pressure is $p_m = 12000$ Pa (90 mm Hg) and the amplitude is $p_a = 4000$ Pa (30 mm Hg), using 1 mm Hg = 133.3 Pa. The circumferential stretch with respect to the swollen state is computed and compared to the applied hemodynamic pressure, $p(t)$, to examine its phase shift. Matlab ode23s is used to solve the viscoelastic model (29) applied to the linear, neo-Hookean, and Zulliger strain energy densities as given, respectively, in Eqs. (30), (31), and (32). The moisture-dependent thermodynamic relaxation modulus k at a particular moisture content from Table 6.1 is taken as the same for each long-term strain energy density. The long-term moduli are assumed related by $E = 3\mu$ and $c = \mu/6$. The reference radius of the lamella is taken as $R = 0.01$ m and $\lambda_z = 1$. No experimental data exist in the literature to compare to the predictions, but a parametric investigation shows that the results are reasonable.

The responses for the three long-term strain energy densities differ primarily in the amplitude of the response. In all cases, the stretch response lags the applied hemodynamic cyclic pressure because the model is viscoelastic. As would be expected, the phase shift in this model is primarily determined by the thermodynamic relaxation modulus, not by the long-term strain energy density. The phase shift in all cases is about 60° and varies slightly with moisture content. The phase shift for the different long-term energy functions is nearly the same at each moisture content since the values of k from Table 6.1 are used for each different strain energy density computation. This large phase shift for one elastin lamella acting by itself does not agree with Lillie and Gosline (1990) experiment for multiple elastin lamellae because the moisture contents are greater than their 28.1%. The phase shifts are greater than the Gow and Taylor (1968) measurements of the pressure versus diameter for the full artery wall of living dogs. Perhaps in vivo, the contractile forces from the smooth muscle cells cause a faster retraction that reduces the phase shift. As a numerical experiment, if k for the 50% moisture content is multiplied by 100, then the response for the neo-Hookean long-term energy is nearly in phase with the load, in agreement with the idea that the larger the thermodynamic relaxation modulus, the faster the response. The larger k also produces a larger amplitude response (about 0.0025 versus 0.0001).

The response amplitude varies between models based on the different long-term energy density functions and with the magnitude of the load. The maximum stretch for each particular strain energy density also depends on the moisture content. For example, if the minimum pressure is reduced to 8000 Pa, in the neo-Hookean based model, the respective stretch amplitudes at 40, 50, and 60% moisture content are approximately 0.00005, 0.0001, 0.0003 so that the deformation increases with moisture content as it moves away from the glass transition. This pattern is similar to that exhibited by the creep curves above. To illustrate the predicted response for the neo-Hookean based model (31), the curves for 50% moisture content are given in Fig. 6.6 for minimum pressure 12000 Pa. The model using the Zulliger long-term strain energy also gives approximately a 60° phase lag for 50% moisture content (Fig. 6.7) but a slightly larger stretch response amplitude than the neo-Hookean.

As shown by uniaxial creep calculations, the further the moisture content is from the value for the glass transition at the given temperature and observational frequency, the less viscoelastic the elastin is, in the sense that it relaxes more quickly to equilibrium. An inappropriate and naive viewpoint would be that the water merely acts as a lubricant. Moisture content is most important for the mechanical response of elastin because of its effect on the glass transition of elastin, not because water acts as a plasticizer. The stretch response lags the applied load for cylindrical elastin lamellae under a cyclic inflation pressure, representing the hemodynamic load, as reported in vivo measurements for dog arteries.

The model is empirical in the sense that the thermodynamic relaxation modulus is determined here by fitting uniaxial peak tangent modulus data reported by Lillie and Gosline. But the model is predictive because it is applied to predict the mechanical response to membrane loads unrelated to the uniaxial tangent modu-

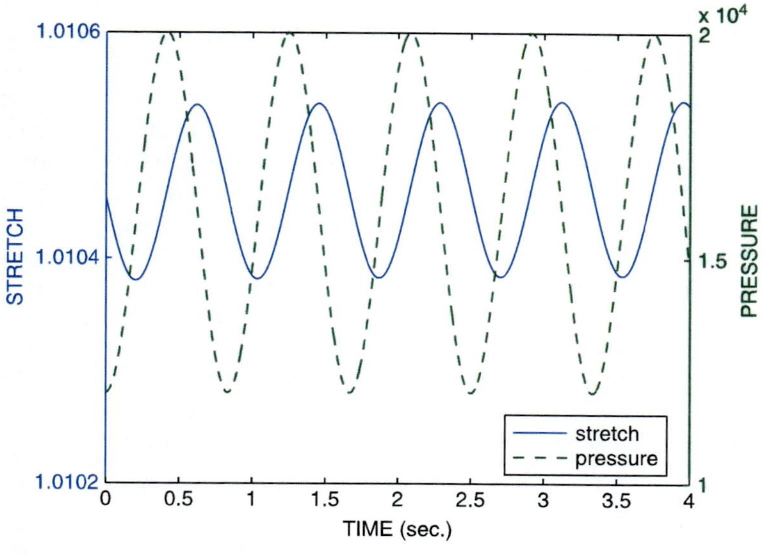

Fig. 6.6 The stretch response to a normal sinusoidal loading, $p(t) = 12{,}000 + 4{,}000[1 + \sin(2\pi t\omega - \pi/2)]$, for a cylindrical elastin membrane having 50% moisture content with the long-term behavior represented by the neo-Hookean model. The constants are obtained from Table 6.1, $R = 0.01$ m and the frequency is 1.2 Hz. The initial conditions are $\lambda(0) = 1.010455$ and $\dot{\lambda}(0) = 0$

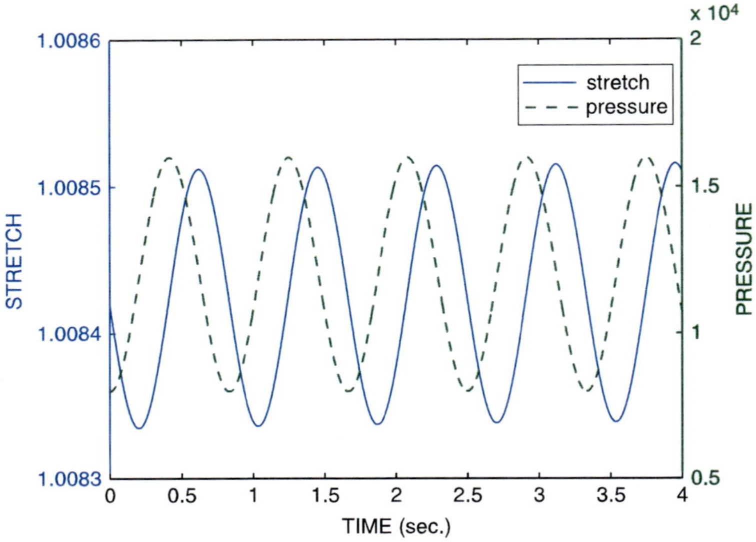

Fig. 6.7 The stretch response to a normal sinusoidal loading, $p(t) = 12000+4000[1+\sin(2\pi t\omega - \pi/2)]$, for a cylindrical elastin membrane having 50% moisture content with the long-term behavior represented by the Zulliger model. The constants are obtained from Table 6.1, $R = 0.01$ m and the frequency is 1.2 Hz. The initial conditions are $\lambda(0) = 1.011255$ and $\dot{\lambda}(0) = 0$

lus peak data. More detailed experimental data are required to precisely identify the empirical functions in the model. For example, the best strain energy density function to represent the quasi-static long-term behavior of arterial elastin remains undetermined even though uniaxial tensile tests have been performed by several research groups. The long-term quasi-static energy density has been assumed to be isotropic and incompressible since such functions have been used to describe elastin in the literature. Further, experimental work is required to determine whether an elastin lamella should be represented as anisotropic. Uniaxial creep or cyclic loadings of elastin membranes at different moisture contents could be compared to the predictions of this construction assuming different long-term strain energy functions to decide which one gives the best fit. This model can help guide those wishing to design and perform such experiments. For example, a direct measurement of $f(\omega)$ and $g(rh)$ would be very useful. The model should help predict the mechanical response of an elastic artery determined by the effects of the microstructures in the media since elastin is a major source of the viscoelastic response of the wall.

References

B. B. Aaron and J. M. Gosline (1981). Elastin as a random-network elastomer: a mechanical and optical analysis of single elastin fibers. *Biopolymers* **20**, 1247–1260.

B. B. Aaron and J. M. Gosline (1980). Optical properties of single elastin fibres indicate random protein conformation. *Nature* **287**, 865–867.

N. O. Birge and S. R. Nagel (1985). Specific-heat spectroscopy of the glass transition. *Physical Review Letters* **54**, 2674–2677.

M. Born (1939). Thermodynamics of crystals and melting. *Journal of Chemical Physics* **7**, 591–603.

P. Brown-Augsburger, C. Tisdale, T. Broekelmann, C. Sloan, and R. P. Mecham (1995). Identification of an elastin cross-linking domain that joins three peptide chains. *Journal of Biological Chemistry* **270**, 17778–17783.

J. M. Clark and S. Glagov (1985). Transmural organization of the arterial media. *Arteriosclerosis* **5**, 19–34.

E. C. Davis (1993). Smooth muscle cell to elastic lamella connections in developing mouse aorta. Role in aortic media organization. *Laboratory Investigation* **68**, 89–99.

L. Debelle and A. J. P. Alix (1999). The structures of elastins and their function. *Biochimie* **81**, 981–994.

K. L. Dorrington and N. G. McCrum (1977). Elastin as a rubber. *Biopolymers* **16**, 1201–1222.

J. D. Ferry (1961). *Viscoelastic Properties of Polymers*. Wiley, New York.

Y. C. Fung (1993). *Biomechanics. Mechanical Properties of Living Tissues*, 2nd Ed., Springer-Verlag, New York.

J. M. Gosline (1978). Hydrophobic interaction and a model for the elasticity of elastin. *Biopolymers* **17**. 677–695.

J. M. Gosline and C. J. French (1979). Dynamic mechanical properties of elastin. *Biopolymers* **18**, 2091–2103.

L. Gotte, M. Mammi and G. Pezzin (1968). Some Structural Aspects of Elastin Revealed by X-ray Diffraction and Other Physical Methods. In *Symposium on Fibrous Proteins, Australia 1967*, ed. W.G. Crewther, Butterworth and Co. Ltd., London.

B. S. Gow and M. G. Taylor (1968). Measurement of viscoelastic properties of arteries in the living dog. *Circulation Research* **23**, 111–122.

W. R. Gray, L. B. Sandberg and J. A. Foster (1973). Molecular model for elastin structure and function. *Nature* **246**, 461–466.

W. Grut and N. G. McCrum (1974). Liquid drop model of elastin. *Nature* **251**, 165.

C. A. J. Hoeve and P. J. Flory (1958). The elastic properties of elastin. *Journal of the American Chemical Society* **80**, 6523–6526.

C. A. J. Hoeve and P. J. Flory (1974). The elastic properties of elastin. *Biopolymers* **13**, 677–686.

J. C. Iatrides, J. P. Laible, and M. H. Krag (2003). Influence of fixed charge density magnitude and distribution on the intervertebral disc: application of a poroelastic and chemical electric (PEACE) model. *Journal of Biomechanical Engineering* **125**, 12–24.

S. R. Kakivaya and C. A. J. Hoeve (1975). The glass point of elastin. *Proceedings National Academy of Science* **72**, 3505–3507.

B. Li and V. Daggett (2002). Molecular basis for the extension of elastin. *Journal of Muscle Research and Cell Motility* **23**, 561–573.

M. A. Lillie and J. M. Gosline (1990). The effects of hydration on the dynamic mechanical properties of elastin. *Biopolymers* **29**, 1147–1160.

M. A. Lillie and J. M. Gosline (1993). The effects of polar solutes on the viscoelastic behavior of elastin. *Biorheology* **30**, 229–242.

M. A. Lillie and J. M. Gosline (2002). The viscoelastic basis for the tensile strength of elastin. *International Journal of Biological Macromolecules* **30**, 119–127.

M. A. Lillie and J. M. Gosline (2007). Limits to the durability of arterial elastic tissue. *Biomaterials* **28**, 2021–2031.

M. A. Lillie, G. W. G. Chalmers, and J. M. Gosline (1996). Elastin dehydration through the liquid and the vapor phase: a comparison of osmotic stress models. *Biopolymers* **39**, 627–639.

M. A. Lillie, G. L. David, and J. M. Gosline (1998). Mechanical role of elastin-asociated microfibrils in pig aortic elastic tissue. *Connective Tissue Research* **37**, 121–141.

L. E. Malvern (1969). *Introduction to the Mechanics of a Continuous Medium*, Prentice-Hall, Englewood Cliffs, NJ.

T. Oden, Belytschko, Babuska and Hughes (2003). Research directions in computational mechanics *Computer Methods in Applied Mechanics and Engineering* **192**, 913–922.

D. Paschek (2009). How does solute-polarization affect the hydrophobic hydration of methane. *Zeitschrift für Physikalische Chemie* **223**, 1091–1104. DOI 10.1524.zpch.2009.6060

T. J. Pence and H. Tsai (2006). Swelling-induced cavitation of elastic spheres. *Mathematics and Mechanics of Solids* **11**, 527–551.

D. Porter (1995). *Group Interaction Modelling of Polymer Properties*. Marcel Dekker, Inc., New York.

D. Porter, F. Vollrath and Z. Shao (2005). Predicting the mechanical properties of spider silk as a model nanostructured polymer. *European Physical Journal E* **16**, 199–206.

D. Porter and F. Vollrath (2008). The role of kinetics of water and amide bonding in protein stability. *Soft Matter* **4**, 328–336.

E. W. Raines (2000). The extracellular matrix can regulate vascular cell migration, proliferation, and survival: relationships to vascular disease. *International Journal of Experimental Pathology* **81**, 173–182.

E. L. Ritman and A. Lerman (2007). Review: The dynamic vasa vasorum. *Cardiovascular Research* **75**, 649–658.

M. H. Sherebin, S. H. Song, and M. R. Roach (1983). Mechanical anisotropy of purified elastin from the thoractic aorta of dog and sheep. *Canadian Journal of Physiology and Pharmacology* **61**, 539–545.

B. R. Simon (1992). Multiphase poroelastic finite element models for soft tissue structures. *Applied Mechanics Reviews* **45**, 191–218.

B. R. Simon, M. V. Kaufmann, M. A. McAfee, and A. L. Baldwin (1996). Porohyperelastic theory and finite element models for soft tissue with application to arterial mechanics. In *Mechanics of Poroelastic Media*, Ed. A.P.S. Selvadurai, Kluwer Academic Publishers. pp. 245–261.

B. R. Simon, M. V. Kaufman, J. Liu, and A. L. Baldwin (1998). Porohyperelastic-transport-swelling theory, material properties and finite element models for large arteries. *International Journal of Solids and Structures* **35**, 5021–5031.

L. R. G. Treloar (1972). *The Physics of Rubber Elasticity*, 3rd Ed, Clarendon, Oxford.

D. W. Urry (1983). What is elastin; what is not. *Ultrastructural Pathology* **4**, 227–251.

Z. R. Wasserman and F. R. Salemme (1990). A molecular dynamics investigation of the elastomeric restoring force in elastin. *Biopolmers* **29**, 1613–1631.

P. D. Weinberg, C. P. Winlove, and K. H. Parker (1995). The distribution of water in arterial elastin: Effects of mechanical stress, osmotic pressure, and temperature. *Biopolymers* **35**, 161–169.

T. Weiss-Fogh and S. O. Anderson (1970). New model for the long-range elasticity of elastin. *Nature* **227**, 718–721.

C. P. Winlove, and K. H. Parker (1987). The influence of the elastin lamellae on mass transport in the arterial wall. In *Advances in Microcirculation, Interstitial Lymphatic Liquid and Solute Movement*, eds., N. C. Staub, J. C. Hogg, A. R. Hargans, Karger, Basle, vol. **13**, 74–81.

E. Wöhlisch (1939). Statistisch-kinetisch theorie, thermodynamik und biologische bedeutung der kautschukartigen elastizität. *Kolloid-Zeitschrift* **89**, 239–270.

X. Wu, R. E. Sallach, J. M. Caves, V. P. Conticello and E. L. Chaikof (2008). Deformation responses of a physically cross-linked high molecular weight elastin-like protein polymer. *Biomacromolecules* **9**, 1787–1794.

Y. Zou and Y. Zhang (2009). An experimental and theoretical study on the anisotropy of elastin network. *Annals of Biomedical Engineering* **37**, 1572–1573.

M. A. Zulliger, P. Fridez, K. Hayashi, and N. Stergiopulos (2004). A strain energy function for arteries accounting for wall composition and structure. *Journal of Biomechanics* **37**, 989–1000.

Chapter 7
Contact Geometric Structure for Non-equilibrium Thermodynamics

1 Introduction

Thermodynamics is easier to understand if it is put in a geometric context. Arnold (1990) has stated, "Every mathematician knows that it is impossible to understand any elementary course in thermodynamics. The reason is that the thermodynamics is based – as Gibbs has explicitly proclaimed – on a rather complicted mathematical theory, on the contact geometry". This comment refers to thermostatics, but an extended contact structure applies to non-equlibrium thermodynamics as well. Homogeneous thermodynamics is geometrically represented in a contact manifold by a codimension one submanifold which is locally the graph of the generalized energy function, φ^*. The graph of the generalized function contains the thermostatic system as a Legendre submanifold. Equilibrium or non-equilibrium processes are paths on the corresponding portions of the graph. The relationship of the graph of the thermodynamic energy function and the graph of the thermostatic energy function is defined by a cross-section of a vector bundle. An alternative definition of the thermostatic manifold is obtained as a Lagrange submanifold defined by the symplectic two-form associated to the Gibbs contact one-form.

The geometric structure of thermodynamics builds on the geometry of continuum mechanics. Knowledge of the geometry of non-equilibrium thermodynamics allows visualization of the non-equilibrium processes. The geometry of continuum mechanics, expressed in terms of manifolds, helps keep track of the domain and range of the various tensor functions that are thermodynamic variables and their relation to the geometry of non-equilibrium thermodynamics. Another question that may be dealt with in this geometric context is a clarification of the spaces in which large and small deformations occur. In particular, the geometry clarifies the domain and range of the deformation gradient, F, and its transpose. The geometic structure permits a description of the mathematical structure underlying the form of the generalized thermodynamic energy function given in Chapter 3, Section 3.

H.W. Haslach Jr., *Maximum Dissipation Non-Equilibrium Thermodynamics and its Geometric Structure*, DOI 10.1007/978-1-4419-7765-6_7,

2 The Geometry of Continuum Mechanics

Classical continuum mechanics is represented in Euclidean three space, $\Re^3$, a space with an inner product. A spatial coordinate system is a one to one correspondence between points of a space and n-tuples $(p_1, \ldots, p_n)$. The point (p_1, p_2, p_3) in the space $\Re^3$ corresponds to the vector $p_1\mathbf{i} + p_2\mathbf{j} + p_3\mathbf{k}$ so that $\Re^3$ is viewed as a vector space. A description of the inner product requires the idea of the cross-product of vector spaces.

Definition 1 The cross-product, $X \times Y$, of two spaces X and Y is the set of ordered pairs of points, one from each space,

$$X \times Y = \{(x, y) \mid x \in X \quad \text{and} \quad y \in Y\}.$$

Here in this Section, x and y do not refer to thrmodynamic state and control variables as in previous Chapters.

Examples

1. $\Re^2 = \Re \times \Re$ since $\Re \times \Re = \{(x, y) \mid x \in \Re \quad \text{and} \quad y \in \Re\}$.
2. $\Re^n = \Re^{n-1} \times \Re$.
3. A two-dimensional torus is the cross product of two one-spheres, $S^1 \times S^1$.

An inner product is a bilinear function, $\langle \cdot, \cdot \rangle : V \times V \to \Re$, on the vector space, V. In contrast, the scalar product is a bilinear function $(\cdot, \cdot) : V^* \times V \to \Re$ defined by $(\mathbf{v}^*, \mathbf{u}) = \mathbf{v}^*(\mathbf{u})$, where V^* is the dual vector space to V. Note that if $\mathbf{v}^* = \langle \mathbf{v}, \cdot \rangle$, then $(\mathbf{v}^*, \mathbf{u}) = \langle \mathbf{v}, \mathbf{u} \rangle$.

For example, an inner product in $\Re^3$ is $\langle \mathbf{u}, \mathbf{v} \rangle = u_1 v_1 + u_2 v_2 + u_3 v_3$. Given an inner product, a distance is defined by $(\langle \mathbf{u} - \mathbf{v}, \mathbf{u} - \mathbf{v} \rangle)^{1/2}$. The vector space, $\Re^3$, therefore has such a distance. It is a metric space.

Other examples include the real line, $\Re$ and the real plane, $\Re^2$. A two sphere, S^2, in three-dimensional space, $\Re^3$, is not Euclidean. But it is locally Euclidean. The concept of locality is made precise in the idea of an open set.

Definition 2 In Euclidean space, a set U is open if for each point, $\mathbf{u}$, in U, there is a disk $N = \{\mathbf{v} \mid d(\mathbf{u}, \mathbf{v}) < \epsilon\}$, for some number ϵ such that N is contained in U. N is called a neighborhood of x.

Note An element $\mathbf{u} \in U$ is on the boundary if there is no such neighborhood N contained in U, but any N containing $\mathbf{u}$ intersects U.

2.1 Manifolds

A manifold is a more general space that is locally Euclidean. A manifold has no preferred coordinate system and so is an appropriate space in which to account for the fact that all physical laws are independent of the coordinate system in which they are expressed.

In classical continuum mechanics, the body is assumed to deform in a Euclidean space with a fixed reference frame (e.g. Malvern, 1969; Holzapfel, 2000, p. 58). The body inherits coordinates from the ambient Euclidean space in terms of this fixed frame. A point in the manifold is classically represented by a vector in the ambient space. In order to develop a coordinate-independent description and to carefully establish the domains of thermodynamic tensors, the physical body is viewed as a manifold with boundary, embedded in three-dimensional space. As the body deforms over time, the manifold moves in the ambient three-space. The definition of a manifold with boundary is reviewed to establish notation. A manifold with boundary has charts that are the closed half hyperplane, $\mathcal{H}^n = \{(p_1, \ldots, p_n) | p_1 \geq 0\}$ contained in $\Re^n$ (e.g. Conlon, 1993, p. 87). The chart coordinates are needed to apply results from differential topology.

Definition 3 A space M is an n-dimensional differentiable manifold with boundary if it is locally Euclidean in the following sense.

1. There is a collection of open sets U_i such that the union of all the U_i equals M so that the $\{U_i\}$ cover M. Each U_i is called a chart and the full set is called an atlas.
2. There is a one-to-one, infinitely differentiable function $\psi_i : U_i \to \mathcal{H}^n$, where $\mathcal{H}^n$ is the half space contained in $\Re^n$, the n-dimensional Euclidean space, such that the image of ψ_i is open in $\mathcal{H}^n$. For a point $\boldsymbol{p} \in U$, this one-to-one function $\psi_i(\boldsymbol{p}) = \big(p_1(\boldsymbol{p}), \ldots, p_n(\boldsymbol{p})\big)$ defines the coordinate system on U_i. A point $\boldsymbol{p}$ is a boundary point if $\psi(\boldsymbol{p})$ is contained in $\partial\mathcal{H}^n$.
3. On overlaps, the intersection $U_i \cap U_j$, there is a change of coordinate function, $\psi_j \psi_i^{-1}$ relating the coordinate system defined by U_i to that defined by U_j.

The body, $\mathcal{B}$, is assumed to deform in the same Euclidean ambient space, $\Re^3$, so that displacement of its points may be defined. There is an embedding map from the body to the ambient space $\Re^3$ that relates a chart of coordinates for the body to the coordinates of the ambient space (e.g. Conlon, 1993, p. 91).

Definition 4 A subset, $\mathcal{B}$, of $\Re^3$ is an embedded n-dimensional submanifold ($n \leq 3$) iff $\forall \boldsymbol{p} \in \mathcal{B}$, there is a $\Re^3$ chart (U, ψ) about $\boldsymbol{p} \in \Re^3$ such that $\psi(U \cap \mathcal{B}) = \psi(U) \cap \mathcal{H}^n$. In this case, $(U \cap \mathcal{B}, \psi|_{(U \cap \mathcal{B})})$ is a manifold chart for $\mathcal{B}$.

Open sets on the body $\mathcal{B}$ are intersections of open sets U in $\Re^3$ with B to produce $U \cap \mathcal{B}$. Around each point $\boldsymbol{p} \in \mathcal{B}$, there must be at least one such open set to which coordinates may be assigned by a chart, (U, ψ).

The embedding may be described in terms of the local coordinates on the body and the ambient coordinates, but the embedding is a coordinate-free one-to-one function, a diffeomorphism. A one-to-one function $f : M \to M'$ of manifolds is a diffeomorphism if it is smooth with a smooth inverse. The function f is smooth if the local function $\psi'_j f \psi_i^{-1} : \psi_i(U_i) \to \psi'_j(U'_j)$ is smooth. An embedding $\mathcal{B} \to \Re^3$ is a diffeomorphism that is homeomorphic onto its image. An immersion, by contrast, is only locally an embedding.

Two types of coordinate changes are possible, either of the ambient space or within the coordinate charts on the manifold representation of the body. Each observer chooses their own coordinates for the ambient space $\Re^3$. These coordinate choices are related by a Euclidean transformation, $Q \in SO(n)$, since all observers see the same distances and angles. Such transformations account for changes of observers. Alternatively, coordinate transformations may be made in the local chart. The chart gives local Euclidean space coordinates to U. The change of coordinates requirement for the thermodynamic variables is satisfied naturally in the manifold representation. Let the rotation $Q \in SO(n)$ be a coordinate transformation. Then the chart $\psi_i : U_i \rightarrow \Re^n$ in the old coordinates becomes $Q\psi_i : U_i \rightarrow \Re^n$ in the new coordinates. These local coordinate transformations are involved in the definition of the objectivity of a thermodynamic tensor.

A deformation is a map $\chi : \mathcal{B}_o \times \Re \rightarrow \mathcal{B} \subset \Re^3$ taking the reference state, $\mathcal{B}_o$, to the state, $\mathcal{B}$, at time t in ambient space. This map is thought of as a motion of the deforming body. A motion is regular if it has an inverse at all times t. The current point is denoted $\boldsymbol{p} = \chi(\boldsymbol{p}^0, t)$.

2.1.1 Examples of Manifolds

1. $\Re^n$ itself is the simplest example. The smallest atlas consists of one open set U and $U = \Re^n$. The map $f : U = \Re^n \rightarrow \Re^n$ is the identity function.
2. The sphere S^{n-1} contained in $\Re^n$ defined by

$$S^{n-1} = \{(p_1, \ldots, p_n) \mid \sum_i p_i^2 = 1\}$$

is an $n-1$-dimensional manifold with charts

$$U_i^+ = \{(p_1, \ldots, p_n) \mid (p_1, \ldots, p_n) \in S^{n-1} \quad \text{and} \quad p_i > 0\} \tag{1}$$

$$U_i^- = \{(p_1, \ldots, p_n) \mid (p_1, \ldots, p_n) \in S^{n-1} \quad \text{and} \quad p_i < 0\} \tag{2}$$

for all $i = 1, \ldots, n$. This is a covering by open hemispheres. In U_i^+, $p_i = (1 - \sum_{i \neq j} p_j^2)^{1/2}$ and in U_i^-, $p_i = -(1 - \sum_{i \neq j} p_j^2)^{1/2}$. The coordinate map in this case is $\psi_i : U_i^\pm \rightarrow \Re$ defined by $\psi_i(p_1, \ldots, p_n) = (p_1, \ldots, \hat{p}_i, \ldots, p_n)$ in which the ith-coordinate is simply dropped. The map, ψ_i, projects the hemisphere onto an $n-1$-dimensional disk, and is a one-to-one function. This atlas is not the smallest one possible.

Note that the $n-1$ sphere is not a vector subspace of $\Re^n$ since it does not contain zero and is not closed under addition. The only vector subspaces of $\Re^3$, for example, are the lines and planes through the origin.

A manifold of importance in the thermodynamics of solids is the real projective space.

3. The $n-1$-dimensional real projective space, RP^{n-1}, is the $n-1$-dimensional differentiable manifold defined by identifying antipodal points on the $n-1$ sphere.

Equivalently, the projective space, RP^{n-1}, can be thought of as the family of all lines through the origin in $\Re^n$. Each pair of antipodal points on the sphere determines a unique line through the origin, and conversely. For the case of RP^2, obtained from the 2-sphere in $\Re^3$, the lines through the origin can be distinguished by their direction cosines. Each line can be given three coordinates, $(\cos\theta_x, \cos\theta_y, \cos\theta_z)$, its directions cosines. But the set of lines is two-dimensional since $\cos^2\theta_x + \cos^2\theta_y + \cos^2\theta_z = 1$, so that only two coordinates are independent. This relation explicitly verifies that RP^2 is a 2-manifold.

The projective space can also be viewed as the family of all possible hyperplanes passing through the origin. A hyperplane in $\Re^n$ is the $n-1$ dimensional plane defined by $\sum_{i=1}^{n-1} a_i p_i = p_n$. Each hyperplane is determined by the line normal to the plane. This implies that there is a one-to-one correspondence between hyperplanes and lines passing through the origin. The family of hyperplanes passing through the origin in $\Re^n$ is another model for RP^{n-1}.

One may show, using a sketch in which the antipodal points are identified, that RP^1 is diffeomorphic to S^1. This relationship is not true for higher dimensional real projective spaces and spheres.

2.2 The Tangent Space of a Manifold

One goal of the concept of a tangent manifold for the manifold, M, is to create a coordinate free description of M and its tangent spaces. At each point p on an n-dimensional manifold, one can imagine an n-dimensional space of tangent vectors to curves through p. For example, if M is a surface in $\Re^3$, the space of such tangent vectors is the tangent plane to the surface at p. The tangent hyperplane to a manifold, M, at a point approximates the local structure of the manifold near the point. The total tangent space of M, denoted T_*M, combines M and the tangent space to each of its points into one mathematical object. The result is a vector fiber bundle $\pi : T_*M \to M$ whose fiber T_pM is the vector space of tangents to $p \in M$.

Definition 5 A fiber bundle is a map $\pi : \hat{E} \to \hat{B}$ where $\hat{E}$ and $\hat{B}$ are the total and base spaces respectively. The space $\pi^{-1}(p)$ for $p \in \hat{B}$ is called the fiber, one for each point of $\hat{B}$. The topology on $\hat{E}$ connects the fibers.

If the fiber is a vector space, the fiber bundle is called a vector bundle. The tangent vector bundle is the fiber bundle whose fiber at each point is the tangent space at the point. A vector bundle has an inner product if an inner product smoothly varying with respect to points in the base space is defined on each fiber (e. g. Conlon, 1993, p. 81).

To define the tangent bundle in a coordinate-free manner, let $\gamma : \Re \to M$, a function of time, be a path through the point $p \in M$. Its tangent at p is the vector $\mathbf{v}$ (e.g. Conlon, 1993). The tangent vector bundle, $\pi : T_*M \to M$, to a manifold may be thought of as locally a set of ordered pairs of points p of M and tangent vectors, $\mathbf{v}$, to M at p. The fiber at point p is the vector space, $T_p = \pi^{-1}(p)$. If the point $p \in M$ has local coordinates $(p_1, \ldots, p_n)$, then the corresponding n basis vectors for the tangent space T_p to M at p are denoted by $\partial/\partial p_1, \ldots, \partial/\partial p_n$. For example, the tangent space to the manifold $\Re^n$ is $\Re^n \times \Re^n = \Re^{2n}$. The tangent bundle map $\pi : \Re^n \times \Re^n \rightarrow \Re^n$ is defined by $\pi(p, \mathbf{v}) = p$. The tangent spaces to other manifolds are often hard to visualize. In such cases, it is best to think of them abstractly as sets of ordered pairs of points and vectors.

As a manifold, the tangent bundle of a manifold with boundary is defined from the atlas $\{U_i\}$ for the n-dimensional manifold M.

Definition 6 Let $\{U_i\}$ be an atlas for the n-dimensional manifold M. The tangent space of M is the $2n$-dimensional manifold T_*M defined by the atlas, $\{U_i \times \Re^n\}$, with maps $\psi_i \times D\psi_i : U_i \times \Re^n \rightarrow \mathcal{H}^n \times \Re^n$ subject to compatibility conditions on overlaps of the U_i which give a transformation of coordinates. The function $D\psi_i$, the derivative of ψ_i, is induced by the Jacobian of ψ_i, a linear map of vector spaces.

The concept of the tangent manifold for M creates a coordinate free description of M and its tangent spaces. The tangent space to the boundary, ∂M, of the manifold is a $2(n-1)$-dimensional submanifold of T_*M.

2.2.1 The Deformation Gradient as a Function on the Fiber of the Tangent Bundle

To represent the behavior of a body, the idea is to approximate the deformation $\chi : \mathcal{B}_o \times \Re \rightarrow \mathcal{B}$ locally by a linear map, F, the deformation gradient. In classical continuum mechanics, the domain of the deformation gradient is sometimes a source of confusion.

Definition 7 The deformation gradient is the differential of the deformation function at fixed time t, $D\chi(\boldsymbol{p}^0, t) \equiv F(\boldsymbol{p}^0, t) : T_{\boldsymbol{p}^0}B_o \rightarrow T_{\boldsymbol{p}}B$, where $D\chi(\boldsymbol{p}^0, t) = (\partial\chi_i/\partial \boldsymbol{p}_j^0)$ and $\chi(\boldsymbol{p}^0, t) = \boldsymbol{p}$.

The deformation gradient is therefore a linear map on the tangent space at a point. The deformation gradient F is always non-singular since physically no two point masses can combine into one point mass in a current state; the map, F, is an injection.

In the notation of classical continuum mechanics, the deformation gradient at the point X in the reference configuration is viewed as a map whose domain is a neighborhood of X in the manifold representing the reference state of the body (e.g. Malvern, 1969, p. 155) to a neighborhood of $\chi(X, t) = x$ in the manifold representing the deformed body, each embedded in the ambient space $\Re^3$. However, if F is to be a second order tensor, its domain and range must be vector spaces. The tangent space to the body at the point X plays the role of the vector space domain. To correlate such a viewpoint with the classical viewpoint requires an identification of a neighborhood of zero in the tangent bundle fiber at X, the tangent hyperplane, with a neighborhood of X in $\mathcal{B}_o$. The tangent space at a point, X, denoted $T_X(\mathcal{B}_o)$, is thought of as a hyperplane tangent to the body at that point. Therefore tangent vectors can be used to approximate the vector from X to a nearby point in the body. Because $\Re^n$ is a Lie group, the exponential map, $\exp : T_X(\mathcal{B}_o) \rightarrow \mathcal{B}_o$, makes such an identification that takes the zero vector in $T_X(\mathcal{B}_o)$ to X (e. g. Conlon, 1993, p. 133). A similar identification is made in the current configuration.

2.2.2 Vectorfields

The path of a process on a manifold M can be described by its tangents at each point. Equivalently, a system of first order differential equations defines the tangents to a path. The collection of tangent vectors to a path on M can be abstracted into one mathematical object by viewing the relationship between a point on the path and the tangent to the path at that point as a function.

Definition 8 A vectorfield on an n-dimensional manifold M is a function $\mathcal{X} : M \to T_*M$ such that for each $p \in M$, the vector $\mathcal{X}(p) \in T_pM$, the tangent space to M at p. Locally, in an open set, U of M, the vectorfield has the form $\mathcal{X} : U \to U \times \Re^n$ given by $\mathcal{X}(p) = (p, \xi(p))$, where $\xi(p)$ is thought of as a tangent vector at p to a path through p.

Example The second order differential equation, $\ddot{p} = -p^3 - p - \dot{p}$ can be written as a system of two first-order differential equations, by setting $p = p_1$ and $\dot{p} = p_2$,

$$\begin{aligned} \dot{p}_1 &= p_2 \\ \dot{p}_2 &= -p_1^3 - p_1 - p_2. \end{aligned} \tag{3}$$

The corresponding vectorfield on the plane (p_1, p_2) is

$$\mathcal{X}(p_1, p_2) = \left(p_1, p_2, \ p_2\frac{\partial}{\partial p_1} + (-p_1^3 - p_1 - p_2)\frac{\partial}{\partial p_1}\right) \in \Re^2 \times \Re^2 = T_*\Re^2.$$

Notice that the domain of $\mathcal{X}$ is $\Re^2$ and the range is four-dimensional consisting of the point and a two-dimensional vector at the point.

The solution to a differential equation can be viewed as a path on M whose tangents are given by the vectorfield representing the differential equation. The path is called an integral curve for the vectorfield.

2.3 The Cotangent Space of a Manifold

To define the cotangent bundle associated with an n-dimensional manifold, M, for p a point in M, denote by T_p^* the dual vector space to the n-dimensional vector space of tangent vectors T_p; this is the vector space of differential forms, ω, at p, i.e. the linear real-valued functions with domain T_p. The cotangent fiber bundle over M, $\pi : T^*M \to M$, has fiber, T_p^*. The space T^*M, is roughly the set of all pairs consisting of points $p \in M$ and differential forms to M at p. T^*M is a $2n$-dimensional manifold. Both T_p^* and T_p are isomorphic to $\Re^n$.

Definition 9 Let $\{U_i\}$ be an atlas for the n-dimensional manifold M with boundary. The cotangent manifold, T^*M, is the $2n$-dimensional manifold defined by the atlas, $\{U_i \times \Re^{n*}\}$, with maps $\psi_i \times D\psi_i^* : U_i \times \Re^{n*} \to \mathcal{H}^n \times \Re^n$ subject to compatibility conditions on overlaps which give a transformation of local coordinates.

Definition 10 A differential form of degree one (a one-form) on the manifold, M, is an infinitely differentiable real valued function on the tangent manifold to M, $\omega : T_*M \to \Re$, which is a linear real valued function on each tangent space T_pM in the sense that $\omega(p, \mathbf{v}) \equiv \omega_p(\mathbf{v})$ is real.

In local coordinates $p = (p_1, \ldots, p_n)$, given by a chart on M, such a form can be viewed as a linear combination of functions $dp_i : T_*M \to \Re$ defined on each fiber by $dp_i(\partial/\partial p_j) = \delta_{ij}$, the Kronecker delta. The basis for the dual algebra T_p^*, the cotangent space at a point, is often denoted by $\{dp_j\}$ in local coordinates $\{p_j\}$, $j = 1, \ldots, n$. In these terms, a one-form $\omega = \sum_{i=1}^n a_i dp_i$ for real functions, a_i, of the p_i.

If $f : M \to \Re$ is a differentiable real valued function on the manifold, M, the differential of f is the one-form, $df : T_*M \to \Re$, defined on each tangent space by $df_p : T_{pM} \to \Re$,

$$df_p = \sum_{i=1}^{n} \frac{\partial f}{\partial p_i} dp_i,$$

where the partial derivatives, $\partial f/\partial p_i$, are evaluated at the point $p \in M$.

Example As discussed in Chapter 3, Section 4.5, a Gibbs fluid has internal energy, $\mathcal{U}(\mathcal{S}, \mathcal{V})$, a function of the entropy $\mathcal{S}$ and the volume V, with equations of state for the temperature $\theta = \partial\mathcal{U}/\partial\mathcal{S}$ and pressure $\mathcal{P} = -\partial\mathcal{U}/\partial\mathcal{V}$. The Gibbs relation $d\mathcal{U} = \theta d\mathcal{S} - \mathcal{P}d\mathcal{V}$ can be thought of as the one-form, $\omega = d\mathcal{U} = \theta d\mathcal{S} - \mathcal{P}d\mathcal{V}$. This one-form is zero on the equilibrium states, viewed as the graph of $\mathcal{U}$ in $\Re^3$, because when the equations of state are satisfied

$$\omega = d\mathcal{U} - \theta d\mathcal{S} + \mathcal{P}d\mathcal{V} = \frac{\partial U}{\partial S} dS - \frac{\partial U}{\partial V} dV - \theta dS + Pd\mathcal{V} = 0.$$

Definition 11 The exterior product of differential forms, α and β, whose domain is T_pM is $\alpha \wedge \beta : T_pM \times T_pM \to \Re$ defined by

$$(\alpha \wedge \beta)(v_1, v_2) = 2[\alpha(v_1)\beta(v_2) - \alpha(v_2)\beta(v_1)],$$

for v_1 and v_2 in T_pM. The real valued bilinear function, $\alpha \wedge \beta$, is a two-form. This product is anti-commutative in the sense that $\alpha \wedge \beta = -\beta \wedge \alpha$. Consequently, $\alpha \wedge \alpha = 0$.

Definition 12 The exterior derivative of $\omega = \sum_{i=1}^n a_i dp_i$ is

$$d\omega = \sum_{i=1}^{n} da_i \wedge dp_i.$$

The two-form is needed below to relate a symplectic form to the Gibbs one-form.

Example In the case that the manifold is the plane, the exterior derivative of $\omega = a_1 dp_1 + a_2 dp_2$ is

$$\begin{aligned}
d\omega &= da_1 \wedge dp_1 + da_2 \wedge dp_2 \\
&= \left(\frac{\partial a_1}{\partial p_1} dp_1 + \frac{\partial a_1}{\partial p_2} dp_2\right) \wedge dp_1 + \left(\frac{\partial a_2}{\partial p_1} dp_1 + \frac{\partial a_2}{\partial p_2} dp_2\right) \wedge dp_2 \\
&= \frac{\partial a_1}{\partial p_2}(dp_2 \wedge dp_1) + \frac{\partial a_2}{\partial p_1}(dp_1 \wedge dp_2) \\
&= \left(\frac{\partial a_2}{\partial p_1} - \frac{\partial a_1}{\partial p_2}\right)(dp_1 \wedge dp_2).
\end{aligned}$$

The exterior derivative is used to express Stoke's theorem in terms of differential forms.

$$\begin{aligned}
\int_{\partial R} (a(p_1, p_2) dp_1 + b(p_1, p_2) dp_2) &= \int_{\partial R} \omega = \int_R d\omega \\
&= \int_R \left(\frac{\partial b}{\partial p_1} - \frac{\partial a}{\partial p_2}\right)(dp_1 \wedge dp_2).
\end{aligned}$$

The area $dp_1 dp_2$ is interpreted as the differential two-form, $dp_1 \wedge dp_2$. This allows Stoke's theorem to be extended from the plane to a two-dimensional manifold under a condition restricting the points on which ω is non-zero.

2.4 Tensors

The tensor algebra permits a description of a tensor by its spectral decomposition.

Definition 13 The tensor algebra on a vector space, V, is the sequence of tensor products, $\{T^n(V)\}$, where $T^0(V) = \Re$, the real numbers, and $T^n(V) = V \otimes \ldots \otimes V$, n times for $n > 0$. The space $T^n(V)$ is the vector space of nth order tensors. This is a graded algebra with multiplication $T^p \otimes T^q \to T^{p+q}$.

Definition 14 A $p + q$ order tensor is a linear function $T^p(V) \to T^q(V)$. Such a tensor is an element of $T^{p+q}(V)$.

In the tensor algebra, scalars are zero order tensors and vectors are first order tensors. A second order tensor is a linear map of vector spaces that satisfies a coordinate change relation. Several thermodynamic variables, such as the deformation gradient, the stress, the strain, or internal variables are second order tensors. The view of second order tensors as maps on the tangent or cotangent spaces adopted here was used by Marsden and Hughes (1994).

If V has basis, $\{\mathbf{e}_i\}$, the dyads $\{\mathbf{e}_i \otimes \mathbf{e}_j\}$ form a basis for $V \otimes V$. A dyad $(\mathbf{v} \otimes \mathbf{u}) \in T^2(V) = V \otimes V$ is viewed as a linear transformation $V^* \to V$, where V^* is the dual space of V. For $\mathbf{w}^* \in V^*$

$$(\mathbf{v} \otimes \mathbf{u})(\mathbf{w}^*) = \mathbf{u}[\mathbf{w}^*(\mathbf{v})]. \tag{4}$$

Likewise if $\mathbf{u}^*, \mathbf{v}^* \in V^*$ and $\mathbf{w} \in V$, then $(\mathbf{v}^* \otimes \mathbf{u}^*) : V \to V^*$ by

$$(\mathbf{v}^* \otimes \mathbf{u}^*)(\mathbf{w}) = \mathbf{u}^*[\mathbf{v}^*(\mathbf{w})]. \tag{5}$$

Traditionally $(\mathbf{v} \otimes \mathbf{u}) : V \to V$ is defined by using an inner product in place of the action of the dual space on the vector space. Recall that composition of tensors expressed as dyads follows the single-dot product of dyads (Malvern, 1969, p. 38). Not every second order tensor may be represented as a single dyad of two vectors, but is rather a sum of dyads.

2.4.1 Riemannian Metric and the Metric Tensor

Let a chart be $\psi : U \to \Re^3$ such that $\psi^{-1}(p_1, p_2, p_3) = \boldsymbol{p} \in U$ where $p_i = \psi^i(\boldsymbol{p})$. A coordinate surface is defined by holding one p_i fixed so that $\psi^i(\boldsymbol{p}) =$ constant. A coordinate curve is the intersection of two of the coordinate surfaces, defined by fixing two coordinates. The tangents, $\{\mathbf{g}_i\}$, to the coordinate curves form the natural basis of the tangent space at $\boldsymbol{p} \in U$. Then the dual basis is classically defined by $\{\mathbf{g}^i\} = \nabla\psi^i(\boldsymbol{p})$. Since the scalar product $(\mathbf{g}_i, \mathbf{g}^j) = \delta_{ij}$, the basis $\{\mathbf{g}^i\}$ can be thought of as a basis of the dual space, the cotangent space at $\boldsymbol{p} \in U$.

The metric tensor, G, is a linear function from the tangent to the cotangent space at $\boldsymbol{p} \in U$. Write the dual basis in terms of the natural basis and vice versa by putting $\mathbf{g}_i = g_{ij}\mathbf{g}^j$ and $\mathbf{g}^i = g^{ij}\mathbf{g}_j$. Then $g_{ij} = \langle \mathbf{g}_i, \mathbf{g}_j \rangle$ and $g^{ij} = \langle \mathbf{g}^i, \mathbf{g}^j \rangle$.

If $\mathbf{u} = u^i\mathbf{g}_i$ is a vector in the tangent space, $G(\mathbf{u}) = u^i g_{ij}\mathbf{g}^j$ in the cotangent space. The metric tensor can be expressed in terms of dyads (e.g. Holzapfel, 2000; Malvern, 1969) as $G = g^{ij}\mathbf{g}_i \otimes \mathbf{g}_j$. Then for $\mathbf{u} = u^k\mathbf{g}_k$, $G(\mathbf{u}) = u^k\langle \mathbf{g}_j, \mathbf{g}_k \rangle g^{ij}\mathbf{g}_i = u^k g_{kj}\mathbf{g}^j$ as above since $g_{jk} = g_{kj}$. Alternatively, the metric tensor is the unique symmetric tensor on V such that $\langle \mathbf{u}, \mathbf{v} \rangle = (\mathbf{u}, G\mathbf{v})$, where $(\cdot, \cdot)$ is the scalar product $V \times V^* \to \Re$.

When the metric, $G : T_*B \to T^*B$, is given by $G = I$, the identity tensor, the identification of the tangent and cotangent spaces is often implicitly assumed in classical continuum mechanics. A vector space isomorphism from the tangent space at a point for a finite dimensional manifold to the cotangent space at the same point in the manifold is given in local coordinates by $\partial/\partial p_i \to dp_i$.

Definition 15 The transpose T^t on V^* of a tensor, T on V, is that tensor satisfying $\langle T\mathbf{u}, \mathbf{v} \rangle = (\mathbf{u}, T^t G\mathbf{v})$, where the left side involves the inner product and the right side involves the scalar product.

The transpose of the deformation gradient, $F^t : T^*\mathcal{B} \to T^*\mathcal{B}_0$, is therefore a linear function from the cotangent space to the manifold at time t to the cotangent space to the reference body. Regularity of the deformation, χ, is needed for the transpose F^t to be well-defined.

Each chart $\psi : U \to \Re^3$, an embedding into Euclidean space, has an associated metric tensor. Let $I : T_*\Re^3 \to T^*\Re^3$ be the standard Euclidean space identification of vectors and one-forms. Denote the Jacobean of the embedding by $D\psi : T_*U \rightarrow T_*\Re^3$; and the transpose $(D\psi)^t : T^*\Re^3 \rightarrow T^*U$. The metric tensor is defined by $G = (D\psi)^t I(D\psi) : T_*U \rightarrow T^*U$. It is the pullback of I (Fig. 7.1). The length squared of a vector, $\mathbf{v} \in T_*U$ is obtained by taking the scalar product of $G\mathbf{v}$ and $\mathbf{v}$.

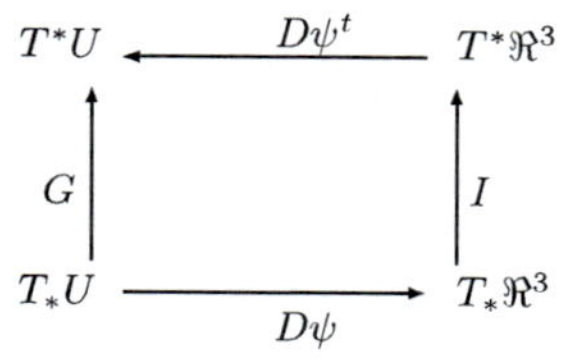

Fig. 7.1 Metric on charts for a manifold

2.4.2 Covariant and Contravariant Tensors

Let $p'_i = p'_i(p_1, \ldots, p_n),\ i = 1, \ldots, n$ be a change of coordinates. The change of coordinates gradient tensor is $\mathcal{F} = (\partial p'_i/\partial p_j)$, which is a function between tangent spaces. As defined traditionally in mathematical physics,

Definition 16 A vector v_i is covariant if it transforms by $v'_i = (\partial p_j/\partial p'_i)v_j$ so that $\mathbf{v}' = \mathcal{F}^{-t}\mathbf{v}$, and contravariant if it transforms by $v'_i = (\partial p'_i/\partial p_j)v_j$, so that $\mathbf{v}' = \mathcal{F}\mathbf{v}$.

A covariant vector is an element of the cotangent space and a contravariant vector is an element of the tangent space.

The definitions extend to second order tensors by viewing the second order tensors as sums of dyads. For covariant vectors, $\mathbf{u}$ and $\mathbf{v}$, the associated covariant second order tensor transforms by $(\mathcal{F}^{-t}\mathbf{u} \otimes \mathcal{F}^{-t}\mathbf{v}) = \mathcal{F}^{-t}(\mathbf{u} \otimes \mathbf{v})\mathcal{F}^{-1}$. For contravariant vectors, $\mathbf{u}$ and $\mathbf{v}$, the associated contravariant second order tensor transforms by $(\mathcal{F}\mathbf{u} \otimes \mathcal{F}\mathbf{v}) = \mathcal{F}(\mathbf{u} \otimes \mathbf{v})\mathcal{F}^t$. If the change of coordinates is a rigid body rotation so that $\mathcal{F} = Q$, then both covariant and contravariant tensors transform by $T' = QTQ^t$, since $Q^{-1} = Q^t$. For example using Eq. (4), the covariant tensor $\mathbf{u}\otimes\mathbf{v}$ satisfies for $(Q\mathbf{u} \otimes Q\mathbf{v})\mathbf{w}^* = Q\mathbf{u}\mathbf{w}^*(Q\mathbf{v}) = Q\mathbf{u}(Q^t\mathbf{w}^*)(\mathbf{v}) = Q(\mathbf{v} \otimes \mathbf{u})Q^t\mathbf{w}^*$.

A geometric definition of second order tensors is given by

Definition 17 A second order tensor is covariant if it is a fiberwise linear map from the tangent space to the cotangent space at a point. A second order tensor is contravariant if it is a fiberwise linear map from the cotangent space to the tangent space at a point. A second order tensor is a mixed tensor if it is a fiberwise linear map from the tangent space to the tangent space at a point or from the cotangent space to the cotangent space at a point.

The metric tensor is a covariant tensor from the vector space to its dual space, $G : V \rightarrow V^*$.

The covariant-contravariant nomenclature is not clearly related to that used in category theory. A category is a class of objects, C^*, and morphisms for each pair of objects. The morphisms may be composed and include the identity morphism,

1_{C^*}, on each object (MacLane, 1963, p. 25). The class of manifolds is a category as is the class of tangent bundles and the class of cotangent bundles. Let fg be the composition of morphisms in the domain category.

Definition 18 (MacLane, 1963, p. 28) A covariant functor is a map $\mathcal{F}^*$ of categories that satisfies $\mathcal{F}^*(1_{C^*}) = 1_{\mathcal{F}^*(C^*)}$ and $\mathcal{F}^*(fg) = \mathcal{F}^*(f)\mathcal{F}^*(g)$. A contravariant functor is a map $\mathcal{F}^*$ of categories that assigns to each morphism $f : C^* \to C'^*$ a morphism $\mathcal{F}^*(f) : T(C'^*) \to \mathcal{F}^*(C^*)$ that satisfies $\mathcal{F}^*(1_{C^*}) = 1_{\mathcal{F}^*(C^*)}$ and $\mathcal{F}^*(fg) = \mathcal{F}^*(g)\mathcal{F}^*(f)$.

A functor that relates the category of manifolds and diffeomorphisms to the category of tangent bundles and bundle maps is covariant, and the functor from the category of manifolds to their cotangent bundles is contravariant.

The deformation $\chi : \mathcal{B}_o \to \mathcal{B}$ induces a map $F : T_*\mathcal{B}_o \to T_*\mathcal{B}$ and a map $F^t : T^*B \to T^*\mathcal{B}_o$. In the category whose objects are the tangent bundles, the push-forward of $\mathbf{u} \in T_*\mathcal{B}_o$ is $\chi_*(\mathbf{u}) = F(\mathbf{u})$, and the pull-back of $\mathbf{u} \in T_*\mathcal{B}$ is $\chi_*^{-1}(\mathbf{u}) = F^{-1}(\mathbf{u})$. In the category whose objects are the cotangent bundles, the push-forward of $\omega \in T^*\mathcal{B}_o$ is $\chi_*(\omega) = F^{-t}(\omega)$, and the pull-back of $\omega \in T^*\mathcal{B}$ is $\chi_*^{-1}(\omega) = F^t(\omega)$.

The dyad $\mathbf{v} \otimes \mathbf{u}$, where $\mathbf{v}, \mathbf{u} \in T_*\mathcal{B}$, has push-forward $F\mathbf{v} \otimes F\mathbf{u} = F(\mathbf{v} \otimes \mathbf{u})F^t$ and its pull-back is $F^{-1}\mathbf{v} \otimes F^{-1}\mathbf{u} = F^{-1}(\mathbf{v} \otimes \mathbf{u})F^{-t}$. This defines the push-forward and pull-back of contravariant tensors since they are represented by sums of dyads in their range.

The dyad $\mathbf{v}^* \otimes \mathbf{u}^*$, where $\mathbf{v}^*, \mathbf{u}^* \in T^*B$, has push-forward $F^{-t}\mathbf{v}^* \otimes F^{-t}\mathbf{u}^* = F^{-t}(\mathbf{v}^* \otimes \mathbf{u}^*)F^{-1}$ and pull-back is $F^t\mathbf{v}^* \otimes F^t\mathbf{u}^* = F^t(\mathbf{v}^* \otimes \mathbf{u}^*)F$. This defines the push-forward and pull-back of covariant tensors since they are represented by sums of dyads in their range.

2.5 Strain Tensors

The covariant right Cauchy strain tensor is defined on the tangent bundle to the reference state to be $C = F^t gF$, where g is the metric tensor acting on $T_*\mathcal{B}$. Therefore C is the pull-back of g (Fig. 7.2).

The covariant Green strain tensor is defined on the reference state by $E = \frac{1}{2}(C - G)$, where G is the metric tensor acting on $T_*\mathcal{B}_o$. The contravariant left

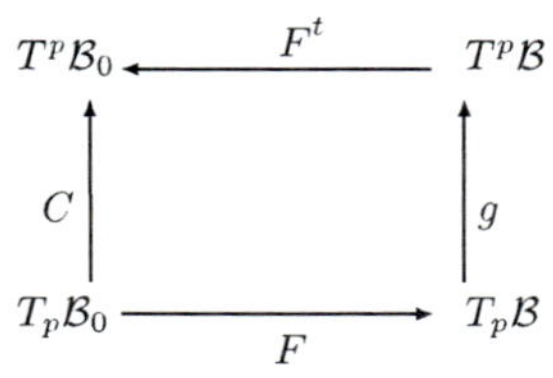

Fig. 7.2 Right Cauchy strain C

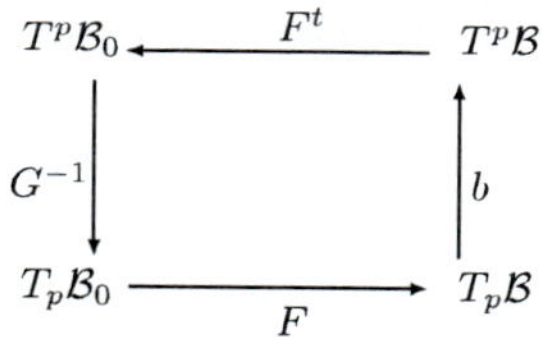

Fig. 7.3 Left Cauchy-green strain b

Cauchy-Green tensor is defined on $T^*\mathcal{B}$ by $b = FG^{-1}F^t$. Therefore b is the push-forward of G^{-1} (Fig. 7.3).

The reference rate of deformation covariant tensor D_r is defined by $2D_r = \frac{\partial}{\partial t}C$ and its push-forward is the spatial rate of deformation covariant tensor D. If $\boldsymbol{v}$ is the spatial velocity vectorfield of $\chi : B_o \to B$, then in terms of the Lie derivative $2D = \mathcal{L}_{\mathbf{v}}(\mathbf{g})$ (Marsden and Hughes, 1994, p. 98).

To write C in terms of the stretch tensor, consider the polar decomposition of $F : T_*\mathcal{B}_o \to T_*\mathcal{B}$ is $F = RU_s$. The rotation of the spatial state with respect to the reference state is given by $R : T_*\mathcal{B}_o \to T_*\mathcal{B}$ and the stretch is $U_s : T_*\mathcal{B}_o \to T_*\mathcal{B}_o$. Also $U_s^t : T^*\mathcal{B}_o \to T^*\mathcal{B}_o$. Then $C \equiv F^t gF = U_s^t R^t gRU_s$. But $R^t gR = G$ (e.g. Simo et al., 1984) so that $C = U_s^t GU_s : T_*\mathcal{B}_o \to T^*\mathcal{B}_o$ (Fig. 7.4).

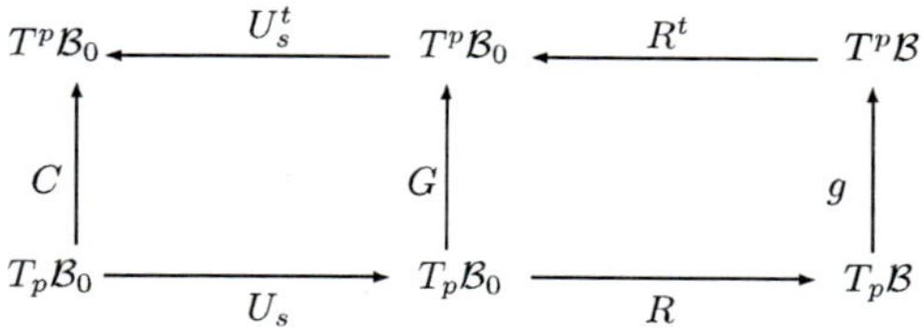

Fig. 7.4 Polar decomposition of F

2.6 *Stress Tensors*

A surface in the body is defined by a function $f : \mathcal{B} \to \Re$. The unit normal to this surface may be considered to be the local one-form $df = \sum_i \frac{\partial f}{\partial p_i} dp_i$. Note that this one-form corresponds to the gradient ∇f under the usual isomorphism between the tangent and cotangent space to a point in a finite dimensional manifold.

The Cauchy stress, σ, is defined on the cotangent bundle, $T^*\mathcal{B}$. The Cauchy stress, σ, is defined from the balance of linear momentum and Cauchy's theorem to be the unique contravariant tensor on the cotangent space such that the stress $\boldsymbol{t} = \sigma \mathbf{n}$ where $\boldsymbol{n}$ is a unit normal to a plane in the current state of the body. The domain is therefore the sphere bundle associated with the cotangent bundle, T^*B.

The second Piola stress, $S = JF^{-1}\sigma F^{-t}$, where $J = \det(F)$, is a contravariant tensor on the reference cotangent space. It is the pull-back of the contravariant Kirchhoff stress tensor $J\sigma$ (Fig. 7.5). The first Piola stress, $P = FS$, is a tensor on the reference cotangent space at X to the tangent space at $\chi(X,t)$. If it is desired to have $P : T^*\mathcal{B}_0 \to T^*\mathcal{B}$ so that P is dual to F, then one must define $P = gFS$, where g is the metric in the current configuration (Fig. 7.6).

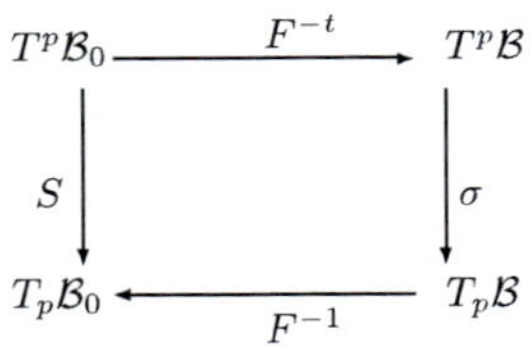

Fig. 7.5 Second Piola stress S

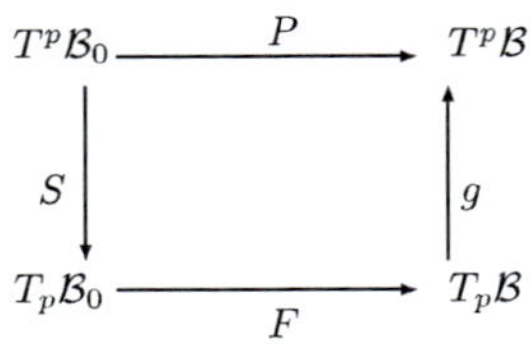

Fig. 7.6 First Piola stress P

2.7 Thermodynamic Variables

Each thermodynamic variable is either a scalar, vector or tensor field on the configuration $\mathcal{B}(t)$. Each field is interpreted as a section into the ith component of the range, $w_i : \mathcal{B} \to \mathcal{N} \equiv (\mathcal{B}\times\Re) \bigoplus T_*\mathcal{B} \bigoplus T^*\mathcal{B} \bigoplus (T_*\mathcal{B}\otimes T_*\mathcal{B}) \bigoplus (T^*\mathcal{B}\otimes T^*\mathcal{B}) \bigoplus (T_*\mathcal{B}_o\otimes T_*\mathcal{B})$, the Whitney sum of fiber bundles on $\mathcal{B}$. The Whitney sum $\xi_1 \bigoplus \xi_2$ of two fiber bundles ξ_1 and ξ_2 over $\mathcal{B}$ is the fiber bundle formed by taking the pointwise direct sum of the fibers for each point of $\mathcal{B}$. Note that the fiber of $\mathcal{N}$ is a subspace of the tensor algebra.

Order the m thermodynamic variables and use the local representation of each variable to define a chart $\psi : U \subset \mathcal{N} \to \Re^3 \times \Re^N$, where N is the total number of components of all the thermodynamic variables, as $\psi(\boldsymbol{u}) = (p_1, p_2, p_3, \psi_1 w_1, \ldots, \psi_{i_m} w_m)$, where $(p_1, p_2, p_3) \in \Re^3$ and $\boldsymbol{u} \in U$. Here, $\psi_i,\ i = 1, \ldots 6$, denotes the chart for the ith summand of the Whitney sum $\mathcal{N}$. The local chart, ψ_i, for each summand is given by an ordered tuple of the components of the tensor. The ordering for second order tensors may be done using the vec operator. The vec operation, also called the stacking operator, maps a matrix to a column vector by concatenating the columns of the matrix. It therefore may be applied to tensors written in Cartesian coordinates because a matrix may represent such tensors. For example, if the matrix $A = (a_{ij})$ is 2×2,

$$\begin{bmatrix} a_{11} & a_{12} \\ a_{21} & a_{22} \end{bmatrix} \text{ then } vec(A) = \begin{bmatrix} a_{11} \\ a_{21} \\ a_{12} \\ a_{22} \end{bmatrix}. \tag{6}$$

For example, if a variable is the second order tensor $T_{ij}\mathbf{g}_i \otimes \mathbf{g}_j$, then $\psi w_i(\boldsymbol{p}) = (T_{11}, \ldots, T_{33})$, a sextuple if the tensor is symmetric, where $\boldsymbol{p} \in \mathcal{B}$. If the variable is a vector, then the image of the chart is in $\Re^3$. Locally each thermodynamic variable has the form $\psi_i w_i : \mathcal{B} \to \Re^k$, where ψ_i is the appropriate chart and $k = 1, 3, 6, 9$. The local energy density is a function $\varphi : \Re^3 \times \Re^n \to \Re$ defined on the image of the chart.

Having estabilished that a chart may be defined, in the following the components of the thermodynamic variables will be divided into control variable $\mathbf{y} = (y_1, \ldots, y_n)$ and state variables $\mathbf{x} = (x_1, \ldots, x_n)$, as in Chapter 3.

3 Contact Structures and Thermostatics

In the Gibbs (1873) relationship between the entropy, volume and total internal energy of a fluid in equilibrium represented by a surface in three dimensional space, the temperature, θ, and the pressure, $\mathcal{P}$, are represented by the partials of $\mathcal{U}$ with respect to $\mathcal{S}$ and $\mathcal{V}$. These partials determine the normal to, and thus, the tangent plane to the energy surface, the graph of $\mathcal{U}$, at the point with coordinates $(\mathcal{S}, \mathcal{V}, \mathcal{U}(\mathcal{S}, \mathcal{V}))$. Therefore there is a correspondence between state variables and tangent hyperplanes.

To generalize Gibbs surface, the thermodynamic state variables at equilibrium are determined by the tangent hyperplane to the equilibrium energy surface in $\Re^{n+1}$, $(n+1)$-dimensional real space, where n is the number of extensive variables. The role of the hyperplanes suggests that the system of all possible equilibrium states is a submanifold of a bundle of contact elements. The tangent hyperplane to a submanifold of a manifold, M, is a subspace of the fiber of the tangent bundle, T_*M. The map of tangent bundles induced by a manifold map, $\phi : M \to M'$, is denoted by $T\phi : T_*M \to T_*M'$. Two submanifolds which have the same tangent hyperplanes at a common point model the identical behavior locally.

Definition 19 Let M be a manifold of dimension m. Two n-dimensional submanifolds $\phi : N \to M$ and $\phi' : N' \to M$ have first order contact at $p = \phi(n) = \phi'(n')$ if $T\phi(T_n(N)) = T\phi'(T_{n'}(N'))$, where the equality indicates equality of sets.

The equivalence class of such submanifolds under first order contact is an n-contact element at p. There is a one-to-one correspondence between contact elements at p and n-dimensional subspaces of the tangent space, T_pM, to M at p.

Definition 20 Let M be an m-dimensional manifold. The bundle of n-contact elements in $M, C(M, n)$, is the manifold of pairs consisting of points, p, of M and n-contact elements $[N]$ of M at p. $C(M, n)$ is a fiber bundle with base space M and fiber map $\pi : C(M, n) \to M$ defined by $\pi(p, [N]) = p$. The fiber at p in M is $\pi^{-1}(p)$.

The proper setting for Gibbs thermostatics of a closed system consisting of a single substance is in the contact bundle, $\pi : C(\Re^3, 2) \to \Re^3$. The fiber of $C(\Re^3, 2)$ consists of classes represented by tangent planes to a surface in $\Re^3$. There are as many of these planes to all possible surfaces as there are lines normal to the surface or equivalently as many as there are lines in $\Re^3$ through the origin. The lines can be distinguished by their direction angles and so the set is two-dimensional. The bundle $C(\Re^3, 2)$ over $\Re^3$ is five-dimensional, and its fiber is two-dimensional real projective space, $RP(2)$. The normal to the tangent plane at a point of the Gibbs energy surface, $(\mathcal{S}, \mathcal{V}, \mathcal{U}(\mathcal{S}, \mathcal{V}))$, is determined by $-\partial\mathcal{U}/\partial\mathcal{V}$ and $\partial\mathcal{U}/\partial\mathcal{S}$, where $\mathcal{S}$ and $\mathcal{V}$ are the control variables. The contact element, the equivalence class of tangent planes at a point on the energy surface, is given coordinates the components of the normal direction, $(-\mathcal{P}, \theta)$, called the state variables. The set of all equilibrium states, a submanifold of $C(\Re^3, 2)$, is shown below to be mapped onto the graph of $\mathcal{U}$ by the fiber map $\pi : C(\Re^3, 2) \to \Re^3$.

To represent the thermodynamics for an arbitrary number of thermodynamic variables, each hyperplane in the fiber of the tangent bundle at a point of the body is assigned state coordinates, that are unique up to multiplication by a non-zero scalar, by the n components of the oriented unit normal to the hyperplane. Since each n-contact equivalence class can be represented by a hyperplane, each class has the coordinates, $(x_1, \ldots, x_n)$. For example, the fiber of the n-contact bundle, $C(\Re^{n+1}, n)$, therefore has dimension n, and the local coordinates are $(y_1, \ldots, y_n, z, x_1, \ldots, x_n)$, where the n coordinates, $(x_1, \ldots, x_n)$, of the fiber of $C(\Re^{n+1}, n)$ are called state variables.

3.1 Lift of the Energy Surface in $\Re^{n+1}$ in the Contact Bundle

An energy function on thermostatic states, $\varphi : \Re^n \to \Re$, is placed in the context of a contact bundle by considering the graph of φ in $\Re^{n+1}$, denoted gr(φ), as the Gibbs energy surface and lifting it into the contact bundle of n-contact elements $C(\Re^{n+1}, n)$ since the tangent planes to the graph are n-dimensional hyperplanes in $\Re^{n+1}$. In the adapted local coordinates of $C(\Re^{n+1}, n)$, $(y_1, \ldots, y_n, z, x_1, \ldots, x_n)$, where the n coordinates, $\boldsymbol{x} = (x_1, \ldots, x_n)$, of the fiber are the state variables and the $\boldsymbol{y} = (y_1, \ldots, y_n)$ of the base are controls, the n-dimensional submanifold which is the lift of the graph of φ can be described by n equations of state,

$$x_i = f_i(y_1, \ldots, y_n), \qquad i = 1, \ldots, n.$$

These equations are traditionally given to define a thermostatic system. A section of the bundle $\pi : C(\Re^{n+1}, n) \to \Re^{n+1}$, is a map $\beta : \Re^{n+1} \to C(\Re^{n+1}, n)$ such that $\beta\pi = 1$, The section β defined in terms of local coordinates by $\beta(y_1, \ldots, y_n, z) = (y_1, \ldots, y_n, z, x_1(\boldsymbol{y}), \ldots, x_n(\boldsymbol{y}))$ yields a lift of the n-dimensional gr(φ), the image

$\beta(\text{gr}(\varphi))$, that is an n-dimensional submanifold of $C(\Re^{n+1}, n)$ described by the n equations of state $x_i = f_i(y_1, \ldots, y_n), i = 1, \ldots, n$. This structure improves the Gibbs energy surface since it explicitly involves the equations of state in the submanifold lifting $\text{gr}(\varphi)$. The graphs of possible energy functions are therefore sections of the fiber bundle, $\pi : C(\Re^{n+1}, n) \to \Re^{n+1}$.

3.2 Thermostatics in a Contact Manifold

The geometric setting for the thermostatics of a closed system consisting of a single substance described by n thermodynamic control variables, $(y_1, \ldots, y_n)$, is in the contact bundle, $C(\Re^{n+1}, n)$, which is assigned a contact one-form that accounts for the second law of thermodynamics. Recall that the Gibbs analysis was based on the combined representation of the first and second laws of thermodynamics in the form $d\mathcal{U} = \theta d\mathcal{S} - \mathcal{P} d\mathcal{V}$.

A description of the equilibrium submanifold is introduced through the Gibbs contact form on $C(\Re^{n+1}, n)$ that was discussed in Chapter 3, Section 4.5.

Definition 21 In local coordinates, $(y_1, \ldots, y_n, z, x_1, \ldots, x_n)$, where the y_i are the control and the x_i are the state variables, the Gibbs one-form on $C(\Re^{n+1}, n)$ is $\omega = dz - \sum_i x_i dy_i$.

The thermostatic submanifold is an integral submanifold since the Gibbs contact form is zero on equilibrium states.

Definition 22 Let M have a contact structure defined by the one form, ω. A connected submanifold $\phi : N \to M$ is an integral submanifold of M if the pullback of ω on cotangent bundles is $\phi^*(\omega) = 0$.

A Legendre submanifold is an integral submanifold of maximal dimension, n. Following Hermann (1973, p. 264) and Mrugala et al. (1991),

Definition 23 Let M be a $2n + 1$ dimensional manifold with a contact structure given by ω and $\phi : N \to M$ be an n-dimensional submanifold. The submanifold N is a thermostatic system if ϕ is a Legendre submanifold for M.

The lift of the graph of an energy function, representing a homogeneous thermostatic system, is therefore a Legendre submanifold of the contact manifold of contact elements on $\Re^{n+1}$, $C(\Re^{n+1}, n)$, with the Gibbs form.

An arbitrary thermostatic system was viewed as a Legendre submanifold of a contact manifold by Hermann (1973), by Burke (1985) and by Arnold (1990). In the contact model, at each equilibrium state, the Gibbs contact form is zero and defines a hyperplane which is equivalent to the Gibbs tangent plane. Mrugala et al. (1991) extended this model of thermostatics by showing that a real function on the contact manifold induces a vectorfield which either preserves the Legendre submanifolds or transforms one into another. They interpreted this field as a thermostatic process in the case that the vectorfield induces a map of the thermostatic Legendre submanifold to itself.

3.3 Interpretations of $C(\Re^{n+1}, n)$

Other geometric descriptions of the contact bundle are available. The n-contact bundle on an $(n+1)$-dimensional manifold has one dimension less than the corresponding tangent or cotangent bundle since one degree of freedom is ignored, the magnitude of the vector normal to the tangent plane. Each fiber of the contact bundle can therefore be viewed as a projective space, so that the manifold of n-contact elements, $C(M, n)$, on a manifold M of dimension $n+1$, is a projective bundle. To be more precise, let $\pi : T^*M \to M$ be the cotangent bundle and $q : T_o^*M \to M$ be the open subbundle with the zero section removed. The multiplicative group of non-zero real numbers, $\Re_o$, acts on T_o^*M by dilation. This action $h : \Re_o \times T_o^*M \to T_o^*M$ is free since $h(\lambda, \xi) \neq \xi$ if $\lambda \neq 1$, $\forall \xi \in T_o^*M$ and regular since the map, π, from T_o^*M onto the quotient space, PT^*M, is a submersion (Fig. 7.7).

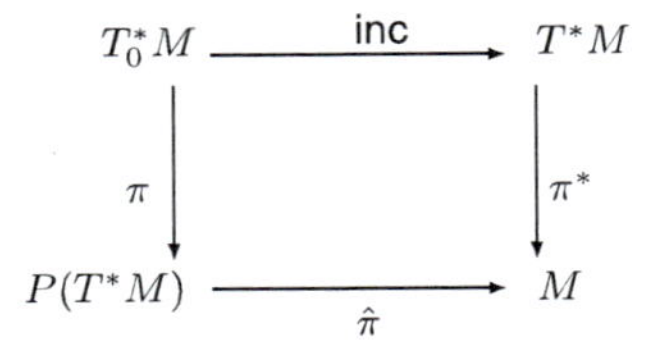

Fig. 7.7 Projective bundle from the group action ($\pi^* = \hat{\pi}\pi$)

The projective bundle over M, $\hat{\pi} : PT^*M \to M$, is isomorphic to $C(M, n)$ under the map which takes an element $[\xi] \in PT^*M$, where the element ξ is a one-form, to the subspace $\ker(\xi)$ in T_*M; $\ker(\xi)$ is the tangent hyperplane to an n-dimensional submanifold of M. For example in the homogeneous case, the contact bundle, $C(\Re^{n+1}, n)$, is the projective bundle associated with the cotangent bundle of $\Re^{n+1}$ (see Libermann and Marle, 1987, p. 283, or Arnold, 1989, p. 354). Therefore, $C(\Re^{n+1}, n)$ is diffeomorphic to $\Re^{n+1} \times RP(n)$, where $RP(n)$ is n-dimensional real projective space.

Mrugala et al. (1991) point out that the contact form induces a principle bundle structure with group, $\Re$, the additive group of real numbers, induced by the Reeb vectorfield (Libermann and Marle, 1987, p. 291). In the homogeneous case, the principal $\Re$-bundle is $C(\Re^{n+1}, n) \to \Re^n \times RP(n)$.

In another viewpoint, the bundle $C(\Re^{n+1}, n) \to \Re^{n+1}$ is a jet bundle since the base space is itself a fiber bundle $\Re^{n+1} \to \Re^n$.

3.4 Symplectic Representation of the Thermostatic Manifold in $\Re^{2n}$

The fundamental concept of a generalized thermodynamic energy function, $\varphi^* : \Re^{2n} \to \Re$, may be defined on a symplectic domain, whose two-form, $\Omega = \sum_i dx_i \wedge dy_i$, distinguishes the conjugate variables. Dynamical systems have often been described in terms of a symplectic structure. An alternative to the contact structure is to show that the manifold, M_e, forms a Lagrange submanifold of $\Re^{2n}$, an n-dimensional submanifold on which the symplectic form is zero.

Definition 24 A subspace W of a vector space V is isotropic if the symplectic form restricted to W, $\Omega_W = 0$, i.e. W is contained in orth$(W) = \{\mathbf{v} \in V | \Omega(\mathbf{v}, \mathbf{w}) = 0, \forall \mathbf{w} \in W\}$. W is coisotropic if $\Omega_{\text{orth}(W)} = 0$, i.e. orth$(W)$ is contained in W. The subspace W is Lagrangian if it is both isotropic and coisotropic so that $W = \text{orth}(W)$.

Definition 25 If $f : N \to M$ is a differentiable map into the symplectic manifold (M, Ω), such that $T_p N$ is a Lagrangian subspace of $(T_{f(p)} M, \Omega_{f(p)})$ for all $p \in N$, then N is a Lagrangian submanifold of M. In this case, Ω is zero on N.

Lemma 26 *The thermostatic manifold, M_e, is a Lagrange submanifold of M^{2n}.*

Proof By (9), on M_e, $y_i = -\partial\hat{\varphi}/\partial x_i \equiv g_i(x_1, \ldots, x_n)$, and so $dy_i = \sum_j (\partial g_i/\partial x_j) dx_j$. The proof is complete if, on M_e,

$$\Omega = \sum_{i=1}^{n} dx_i \wedge dy_i = \sum_{i=1}^{n} \{dx_i \wedge \sum_{j=1}^{n} [(\partial g_i/\partial x_j) dx_j]\} = 0.$$

But $dx_j \wedge (\partial g_i/\partial x_j) dx_i + dx_i \wedge (\partial g_j/\partial x_i) dx_j = 0$ because of the facts that the exterior product is antisymmetric and $\partial g_i/\partial x_j = \partial g_j/\partial x_i = -\partial^2 \varphi/\partial x_i \partial x_j$.

To relate the thermostatic manifold viewed as a Lagrange submanifold or as a Legendre submanifold, define the embedding $\phi' : \Re^{2n} \to T_o^* \Re^{n+1}$, the subbundle of the cotangent bundle with the zero section removed, by

$$\phi'(y_1, \ldots, y_n, x_1, \ldots, x_n) = (y_1, \ldots, y_n, \varphi^*(\mathbf{y}, \mathbf{x}), dz - x_1 dy_1 + \cdots + x_n dy_n),$$

where $\{y_1, \ldots, y_n, z\}$ are the coordinate maps associated with the vector basis $\{\mathbf{e}_i\}$ for $\Re^{n+1}$. This map is completely defined by φ^* since $\partial\varphi^*/\partial y_i = x_i$, for $i = 1, \ldots, n$. The element in the fiber of $T_o^* \Re^{n+1}$ is the Gibbs contact element on $C(\Re^{n+1}, n)$, $dz - x_1 dy_1 + \ldots + x_n dy_n$.

Since $C(\Re^{n+1}, n)$ is the projective bundle associated with $T_o^* \Re^{n+1}$, define the map $\phi : \Re^{2n} \to C(\Re^{n+1}, n)$ to be that induced by $\phi' : \Re^{2n} \to T_o^* \Re^{n+1}$. The map $\phi : \Re^{2n} \to C(\Re^{n+1}, n)$ can be also expressed as $\phi(\mathbf{y}, \mathbf{x}) = (\mathbf{y}, \varphi^*(\mathbf{y}, \mathbf{x}), [N])$, where $[N]$ is the contact class of the unique hyperplane plane in $\Re^{n+1}$ perpendicular to $-x_1\mathbf{e}_1 - \ldots - x_n\mathbf{e}_n + \mathbf{e}_{n+1}$. When $x_i = \partial\varphi^*/\partial y_i$, N is tangent to the surface formed by the graph of φ in $\Re^{n+1}$.

The map, ϕ, takes the thermostatic submanifold of the domain to a Legendre submanifold of $C(\Re^{n+1}, n)$.

Lemma 27 *The image of ϕ restricted to the thermostatic manifold M_e, $\phi(M_e)$, is a Legendre submanifold of $(C(\Re^{n+1}, n), \omega)$.*

Proof $\phi(M_e)$ has dimension n since M_e does. The pullback $\phi^*(\omega) = 0$ since on M_e, $\partial\varphi^*/\partial x_i = 0$ and

$$d\varphi^* = \sum_{i=1}^{n}(\partial\varphi^*/\partial x_i)dx_i + \sum_{i=1}^{n}(\partial\varphi^*/\partial y_i)dy_i = \sum_{i=1}^{n} x_i dy_i,$$

and so on the cotangent space, T^*M_e, $\phi^*(\omega) = \phi^*(d\varphi^* - \sum_{i=1}^{n} x_i dy_i) = 0$. Since ϕ is an immersion, $\omega = 0$ on $T^*\phi M_e$.

The relationship between $M_e \subset \Re^{2n}$ and the Legendre submanifold of $C(\Re^{n+1}, n)$ holds because $C(\Re^{n+1}, n)$ is the contactification (Arnold 1989, p. 368) of the symplectic structure on $\Re^{2n}$.

To summarize the construction of this section, a full thermodynamic system is locally a $2n$-dimensional submanifold of the $2n + 1$ dimensional contact bundle on $\Re^{n+1}$. The restriction of the embedding ϕ to the thermostatic manifold of the generalized energy function φ^* is the Legendre submanifold of $C(\Re^{n+1}, n)$ representing the associated thermostatic system, i.e. the lift of the graph of the classical thermostatic function φ of the controls which is equal to the restriction of φ^* to the thermostatic submanifold.

4 Geometry of Maximum Dissipation Non-equilibrium Thermodynamics for Small Displacements

To model time-dependent non-equilibrium behavior, a generalized thermodynamic function of all thermodynamic variables is defined on all possible states, equilibrium and non-equilibrium (Chapter 1, Section 3 Definition 2). Further, the thermostatic manifold is obtained from a zero gradient condition on the generalized thermodynamic function. All non-equilibrium processes lie on the manifold that is the graph of the generalized energy function.

The lift of the graph of a classical thermostatic energy density function, $\hat{\varphi}(y_1, \ldots, y_n)$, to the graph of φ^* is associated to a first order partial differential equation on $\Re^n$, that is defined by Libermann and Marle (1987, p. 478) as a function $F = 0$, where $F : T^*(\Re^n) \times \Re \to \Re$ (This F is not the deformation gradient). $F^{-1}(0)$ defines a $2n$-dimensional submanifold, W, of $T^*(\Re^n) \times \Re$. If $T^*(\Re^n) \times \Re$ has coordinates $(y_1, \ldots, y_n, x_1, \ldots, x_n, z)$, then

$$F.j^1 f = F(y_1, \ldots, y_n, \partial f/\partial y_1, \ldots, \partial f/\partial y_n, f(y_1, \ldots, y_n)) = 0, \qquad (7)$$

where $f(y_1, \ldots, y_n)$ is a solution of the differential equation, and $j^1 f$ is its one jet.

Put $x_i = \partial\hat{\varphi}/\partial y_i$, and let $\varphi(x_1, \ldots, x_n)$ be the Legendre transform of $\hat{\varphi}(y_1, \ldots, y_n)$. Then on the thermostatic manifold,

$$\hat{\varphi}(y_1, \ldots, y_n) = \varphi(x_1, \ldots, x_n) + \sum_i y_i x_i.$$

Therefore $\hat{\varphi}(y_1, \ldots, y_n)$ is a solution at the thermostatic states of the partial differential equation,

$$F(y_1, \ldots, y_n, x_1, \ldots, x_n) = \varphi(x_1, \ldots, x_n) + \sum_i y_i x_i - \hat{\varphi}(y_1, \ldots, y_n) = 0. \quad (8)$$

Guided by this result, define the local generalized energy associated with $\hat{\varphi}(y_1, \ldots, y_n)$ to be

$$\varphi^*(x_1, \ldots, x_n, y_1, \ldots, y_n) \equiv \varphi(x_1, \ldots, x_n) + \sum_i y_i x_i \quad (9)$$

because setting its gradient with respect to the coordinates $x_1, \ldots, x_n$ equal to zero also defines the thermostatic manifold through the equations of state $y_i = -\partial\varphi/\partial x_i = 0$ for $i = 1, \ldots, n$. The fact that $F = 0$ ensures that the generalized energy equals the classical energy density, $\hat{\varphi}$, on the thermostatic manifold.

Conversely, such a submanifold W can be obtained from $\varphi^* : \Re^{2n} \to \Re$, where $\Re^{2n}$ is thought of as $T^*(\Re^n)$, since $\partial\varphi^*/\partial y_i = x_i$. Define the map $F : T^*(\Re^n) \times \Re \to \Re$ by $F(\mathbf{y}, \mathbf{x}, z) = \varphi^*(\mathbf{y}, \mathbf{x}) - z$. W is represented by the map $\hat{\psi} : \Re^{2n} \to T^*(\Re^n) \times \Re$, where $\hat{\psi}(\mathbf{y}, \mathbf{x}) = (\mathbf{y}, \mathbf{x}, \varphi^*(\mathbf{y}, \mathbf{x}))$, which can be viewed as the graph of φ^*. The solution to the first order partial differential equation is the associated thermostatic energy function $\hat{\varphi}(\mathbf{y}) : \Re^n \to \Re$, where the $\mathbf{y}$ are the control variables.

Example As discussed in the example of Chapter 3, section 3, $\varphi^*(y, x) = -x^3/3 + xy$ is a generalized thermodynamic energy function, where y is the control and x is the state variable. Its thermostatic manifold is defined by $x = y^{1/2}$, and its thermostatic function is $\hat{\varphi}(y) = (2/3)y^{3/2}$. The associated first order partial differential equation is $-(1/3)(df/dy)^3 + (df/dy)y - (2/3)y^{3/2} = 0$, which has solution $\hat{\varphi}(y) = (2/3)y^{3/2}$.

The definition of the generalized thermodynamic function in terms of local coordinates (Chapter 3, Section 3) is supported by the previous discussion.

4.1 Morse Family Formulation of the Generalized Energy Function

The generalized energy function can also be represented as a Morse family in the sense defined and used by L. Hormander, A. Weinstein, W. Tulczyjew and S. Benenti. Their original works are cited in Libermann and Marle (1987), whose notation is used here. The natural projection $\pi : B = \Re^{2n} \to \Re^n$ defined by $\pi(\boldsymbol{y}, \boldsymbol{x}) = \boldsymbol{y}$ is a surjection from the space of state variables to the index space (or from space of state and control variables to the space of control variables). The energy function is defined on B, $\varphi^* : \Re^{2n} \to \Re$.

Definition 28 Suppose $\pi : B \to N$ is a submersion. Then the map $\varphi^* : B \to \Re$ is a Morse family iff the image $d\varphi^*$ in T^*B and the bundle N_π of $T^*(B)$ are transverse, where N_π, the bundle conormal to the fibers of π, is the annihilator of the kernal of the tangent bundle map, $T\pi$.

The subspace $d\varphi^*$ of $T^*(B)$ is defined locally in the adapted coordinates for T^*B,

$$(y_1, \ldots, y_n, x_1, \ldots, x_n, p_1, \ldots, p_n, q_1, \ldots, q_n),$$

by

$$\left(y_1, \ldots, y_n, x_1, \ldots, x_n, \frac{\partial \varphi^*}{\partial y_1}, \ldots, \frac{\partial \varphi^*}{\partial y_n}, \frac{\partial \varphi^*}{\partial x_1}, \ldots, \frac{\partial \varphi^*}{\partial x_n}\right).$$

The family is called a Morse family because there are no singularities if N_π and $d\varphi^*$ are transverse. The control coordinates, $\boldsymbol{y} = (y_1, \ldots, y_n)$, serve to index the functions $\varphi^*(\boldsymbol{x}; \boldsymbol{y})$ defined on the fibers of π.

This representation is equivalent to that in the contact manifold in the absence of singularities of φ^*. Suppose that the cotangent space has adapted coordinates $(y_1, \ldots, y_n, x_1, \ldots, x_n, p_1, \ldots, p_n, q_1, \ldots, q_n)$. The equilibrium submanifold $d\varphi^* \cap N_\pi$ is defined by

$$\begin{aligned}
\frac{\partial \varphi^*}{\partial y_i} &= p_i \\
\frac{\partial \varphi^*}{\partial x_i} &= 0 \\
q_i &= 0.
\end{aligned}$$

Under the projection, $q_\pi : T^*\Re^{2n} \to \Re^{2n}$, $q_\pi(d\varphi^* \cap N_\pi)$ is defined by

$$\frac{\partial \varphi^*}{\partial x_i} = 0.$$

These are exactly the equilibrium conditions for φ^*. Therefore the function, ϕ : $\Re^{2n} \to C(\Re^{n+1}, n)$, defined by

$$\phi(\boldsymbol{x}, \boldsymbol{y}) = \left(\boldsymbol{y}, \varphi^*(\boldsymbol{x}; \boldsymbol{y}), [N]\right).$$

takes this submanifold onto the graph of φ, the equilibrium submanifold.

The formulation in terms of Morse families and the formulation by a generalized thermodynamic function in a contact manifold give equivalent representations of the equilibria of a thermodynamic system. However, the second law of thermodynamics can easily be represented in the contact structure, which is therefore more advantageous for the study of non-equilibrium processes. This result appears in Haslach (1997).

4.2 Legendre Transformations

The important transformations, or symmetries, in thermodynamics are those, called Legendre transformations, which preserve the thermostatic surface. The Legendre transformation allows the transfer from one set of controls chosen from the pairs of conjugate variable to another set of control variables. When thermodynamics is placed in a contact setting, the classical Legendre transformation of thermodynamic energy functions becomes a map of contact structures.

Definition 29 Suppose (M, ω) and (M', ω') are contact manifolds of the same dimension $n + 1$. A contact transformation is a diffeomorphism $L : M \to M'$ such that $L^*(\omega') = \omega$.

A contact transformation maps Legendre submanifolds to Legendre submanifolds since the contact structure is preserved. The diffeomorphism, L, carries the field of contact hyperplanes to itself.

The classical Legendre transforms, internal energy, enthaply, etc., are related by special contact maps of $C(\Re^{n+1}, n)$, called Legendre involutions (Arnold, 1989, p. 366). The composition of the involution with itself yields the identity transformation.

For example, suppose M has local coordinates $(x_1, \ldots, x_n, y_1, \ldots, y_n, z)$ and $\omega = dz - \sum_{i=1}^{n} x_i dy_i$. M' has local coordinates $(x_1', \ldots, x_n', y'1, \ldots, y_n', z'$ and $\omega' = \sum_{i=1}^{n} x_i' dy'i - dz'$ Then the Legendre involution, $L : M \to M'$, is defined by

$$\begin{aligned} x_i' &= L_i(x_1, \ldots, x_n, y_1, \ldots, y_n, z) = y_i \;\; i = 1, \ldots, n; \\ y_i' &= L_{i+n}(x_1, \ldots, x_n, y_1, \ldots, y_n, z) = x_i \;\; i = 1, \ldots, n; \\ z' &= L_{2n+1}(x_1, \ldots, x_n, y_1, \ldots, y_n, z) = \sum_{i=1}^{n} x_i y_i - z. \end{aligned}$$

This map is a discrete, as opposed to a continuous, or infinitesimal, contact transformation. L is a contact map because $L^*(\omega') = L^*(\sum_{i=1}^{n} x_i' dy_i' - dz' = \sum_{i=1}^{n} y_i dx_i - \sum_{i=1}^{n}(x_i dy_i + y_i dx_i) + dz = dz - \sum_{i=1}^{n} x_i dy_i = \omega$.

If a Legendre submanifold in $C(\Re^{n+1}, n)$ that is the lift of a graph of $\hat{\varphi}$ in $\Re^{n+1}$ is mapped by a Legendre involution to another Legendre submanifold in $C(\Re^{n+1}, n)$ that is the lift of a graph of $\hat{\varphi}'$ in $\Re^{n+1}$, the functions, $\hat{\varphi}$ and $\hat{\varphi}'$, are called Legendre transforms. In general if N is a Legendre submanifold of $C(\Re^{n+1}, n)$, then the projection of N into $\Re^{n+1}$ may have singularities. However, if N is the lift of the graph of a convex function, such as the internal energy, then the projection of its transform under the Legendre involution is also the graph of a function (Arnold, 1989, p. 367).

The graph of $\hat{\varphi}'$ is the tangential coordinate description of the graph of $\hat{\varphi}$. The tangent hyperplane to the graph of $\hat{\varphi}$ at $(\{y_i^o\}, \hat{\varphi}^o)$ is

$$\hat{\varphi} - \hat{\varphi}^o - \sum_{i=1}^{n} \left[\frac{\partial \hat{\varphi}}{\partial y_i}\bigg|_p (y_i - y_i^o) \right] = 0,$$

where p denotes $\{y_i^o\}$. This expression can be rewritten as

$$\hat{\varphi} - \sum_{i=1}^{n} \left[\frac{\partial \hat{\varphi}}{\partial y_i}\bigg|_p y_i \right] + \left(\sum_{i=1}^{n} \left[\frac{\partial \hat{\varphi}}{\partial y_i}\bigg|_p y_i^o \right] - \hat{\varphi}^o \right) = 0.$$

So the tangential coordinates of the graph of $\hat{\varphi}$ at $(\{y_i^o\}, \hat{\varphi}^o)$ are

$$\left(\frac{\partial \hat{\varphi}}{\partial y_1}\bigg|_p, \ldots, \frac{\partial \hat{\varphi}}{\partial y_n}\bigg|_p, -\sum_{i=1}^{n} \left[\frac{\partial \hat{\varphi}}{\partial y_i}\bigg|_p y_i^o \right] + \hat{\varphi}^o \right)$$

But these are just $(x_1, \ldots, x_n, \hat{\varphi}')$ if $\hat{\varphi}' = -\sum_{i=1}^{n} y_i x_i + \hat{\varphi}$.

Geometrically, visualize the intersection of the tangent plane to the graph of $\hat{\varphi}$ at $\{y_i^o\}$ with the z-axis. The z-coordinate of this intersection is the value of $\hat{\varphi}'$ at the corresponding $\{x_i^o\} = \{\partial \hat{\varphi}/\partial y_i)|_p\}$. This value exists only if the tangent plane is not vertical; in other words, if $\hat{\varphi}$ has no singularities. A Legendre transformation exists if $x_i = \partial \hat{\varphi}/\partial y_i$ for $i = 1, \ldots, n$ can be solved for the y_i. This is possible if the Hessian of $\hat{\varphi}$ is non-singular.

A thermodynamic problem can sometimes be analyzed more efficiently if a different subset of the thermodynamic variables is chosen as control variables. Two generalized energy functions will be called Legendre transforms if they define the same thermodynamic system, but use different selections of control variables from the fixed pairs of conjugate thermodynamic variables.

Definition 30 A Legendre transformation of a generalized energy density function, $E^* : \Re^{2n} \to \Re$, is a generalized energy density function, $F^* : \Re^{2n} \to \Re$, having the same thermostatic submanifold as E^*, but with different control variables, selected one from each conjugate pair defined by the symplectic form $\Omega = \sum_i dy_i \wedge dx_i$ on the domain, $\Re^{2n}$, of E^*.

The generalized energy density function $E^* : \Re^{2n} \to \Re$ has controls $\mathbf{y}$ and state variables $\mathbf{x}$. Suppose that the Legendre transform, $F^* : \Re^{2n} \to \Re$ is to have controls y_j for $j \in J$ and x_i for $i \in I$, where I is a subset of $\{1, \ldots, n\}$ and J is its complement. The symplectic form defining the conjugate pairs in $\Re^{2n}$, $\Omega = \sum_i dy_i \wedge dx_i$, can be rewritten as $\Omega' = \sum_{j \in J} dy_j \wedge dx_j - \sum_{i \in I} dx_i \wedge dy_i$. The contact form in $C(\Re^{n+1}, n)$ associated with Ω is $\omega = dz - \sum_i x_i dy_i$ and with Ω' is $\omega' = dz' - \sum_{j \in J} x_j dy_j + \sum_{i \in I} y_i dx_i$. A contact map $L : C(\Re^{n+1}, n) \to C(\Re^{n+1}, n)$ such that $L^*(\omega') = \omega$ is defined by $L(\mathbf{y}, \mathbf{x}, z) = (y_J, x_I, x_J, -y_I, z - \sum_{i \in I} y_i x_i)$. Then

$$L^*(\omega') = dz - \sum_{i \in I} x_i dy_i - \sum_{i \in I} y_i dx_i - \sum_{j \in J} x_j dy_j + \sum_{i \in I} y_i dx_i = dz - \sum_{i} x_i dy_i = \omega.$$

Such contact maps are Legendre involutions.

A two-step calculation produces the Legendre transformation of a generalized energy function. The equations for the thermostatic manifold are solved for the desired new controls in terms of the desired state variables as $x_i = \varphi_i(y_I, x_J)$, for $i \in I$ and $j \in J$. These equations are substituted into the expression $E^*(\mathbf{y}, \mathbf{x}) = \varphi(\mathbf{x}) + \sum x_k y_k$ to obtain $E'(y_I, x_J) = \varphi'(y_I, x_J) + \sum_{j \in J} x_j y_j$. Finally the term $\sum_{i \in I} x_i y_i$ is subtracted to produce $F^* = E' y_I, x_J) - \sum_{i \in I} x_i y_i$. The functions E^* and F^* have identical thermostatic submanifolds since they have equivalent equations of state.

An inversion of states and controls within pairs of conjugate variables in the symplectic space $(\Re^{2n}, \Omega)$, where $\Omega = \sum_i dy_i \wedge dx_i$, corresponds to a Legendre involution on $C(\Re^{n+1}, n)$. Each also corresponds to a possible contact structure, ω', on $C(\Re^{n+1}, n)$ consistent with Ω, in the sense that the pullback $\phi^*(d\omega') = \Omega$, where $\phi : \Re^{2n} \to C(\Re^{n+1}, n)$.

5 Compound Systems and Chemical Reactions

A closed thermodynamic system is a system which reversibly exchanges heat and volume change work with its surroundings, but which admits no mass transfer across its boundary. An open system is one which allows mass transfer across its boundary. A compound system is a closed system composed of open subsystems.

Definition 31 A compound system is closed but is composed of several substances between which mass transfer may occur.

The Gibbs form $\omega = d\mathcal{U} - \theta d\mathcal{S} + \mathcal{P} d\mathcal{V}$ defines the equilibrium states of closed systems, where $\mathcal{U}$ is the total internal energy, $\mathcal{S}$ is the total entropy, $\mathcal{P}$ is the pressure, and $\mathcal{V}$ is the volume. In an open system, such as that for a single substance at equilibrium in a compound system,

$$d\mathcal{U} = \frac{\partial \mathcal{U}}{\partial \mathcal{S}} d\mathcal{S} + \frac{\partial \mathcal{U}}{\partial \mathcal{V}} d\mathcal{V} + \frac{\partial \mathcal{U}}{\partial \mathcal{N}} d\mathcal{N} = \theta d\mathcal{S} - \mathcal{P} d\mathcal{V} + \mu d\mathcal{N}, \tag{10}$$

where μ is the chemical potential, and $\mathcal{N}$ is the number of moles of the substance. Then an equilibrium condition must be obtained for the total closed compound system.

Let (M_i, Ω_i) be the symplectic manifold defining the ith substance in the compound system and $\varphi_i^* : M_i \to \Re$ be the generalized energy function for the i_{th} substance, where each $M_i = \Re^{2n_i}$. The product of the manifolds, M_i, defines a symplectic manifold, $M = \times_i M_i$, with symplectic form $\Omega = \sum \Omega_i$. The total generalized energy function $\varphi* : M \to \Re$ is defined by $\varphi^*(p_1, \ldots, p_m) = \sum \varphi_i^*(p_i)$, for $p_i \in M_i$. Let $n = \sum n_i$ and ω be the contact form on $T^*(\Re^n) \times \Re$ defined by the choices of control variables for each M_i. A submanifold, defined by the function, $\phi : N \to \Re^{2n} \to T^*(\Re^n) \times \Re$, is an equilibrium submanifold if the pullback $\phi^*(\omega) = 0$.

Hermann (1973) defines interactions between thermostatic systems in terms of a product of the contact structures.

Lemma 32 *Let* $M = \times_i M_i$ *be the domain for a compound system in which the temperature,* θ*, and the pressure,* $\mathcal{P}$*, is constant and which is composed of m substances. Then* $\phi : N \to M$ *is an equilibrium submanifold if* $\phi^*(\sum \mu_i d\mathcal{N}_i) = 0$.

Proof Each $d\mathcal{U}_i = \theta d\mathcal{S}_i - \mathcal{P} d\mathcal{V}_i + \mu_i d\mathcal{N}_i$ since each substance is in equilibrium. For the compound system, $d\mathcal{U} = \theta \sum_i d\mathcal{S}_i + \mathcal{P} \sum_i d\mathcal{V}_i$. Therefore the pull-back is $\phi^*(\omega) = \phi^*(\sum \mu_i d\mathcal{N}_i) = 0$.

At constant temperature and pressure, two phases are in equilibrium if $\mu_1 = \mu_2$ since $d\mathcal{N}_1 = -d\mathcal{N}_2$. For a given $\mathcal{N}_1$, the equation $\mu_1 = \mu_2$ determines $\mathcal{N}_2$.

5.1 Chemical Reactions

A possible reversible chemical reaction is described by

$$\nu_1 B_1 + \ldots + \nu_n B_n \Leftrightarrow \nu_1' B_1' + \ldots + \nu_m' B_m', \tag{11}$$

where the B_i are the chemical symbols of the reactants and the B_i' are the chemical symbols of the products (Yeremin, 1981, p. 180). The stoichiometric coefficients, ν_i and the ν_i', define the number of moles of each substance in the possible reaction. In order for there to be atomic balance, whether the system is in equilibrium or not, it must be true that

$$d\mathcal{N}_i/\nu_i = d\mathcal{N}_j/\nu_j,$$

for all i, j and for $\mathcal{N}_i$ and $\mathcal{N}_j$ either reactants or products.

The reaction coordinate, ξ, is defined by putting $d\mathcal{N}_i/\nu_i = d\xi$, for all i. Then for $\mathcal{N}_i^o$, the original number of moles of substance i,

$$\mathcal{N}_i = \mathcal{N}_i^o + \nu_i \xi.$$

Because each $d\mathcal{N}_i = \nu_i d\xi$, $\sum \mu_i d\mathcal{N}_i = \sum (\mu_i \nu_i) d\xi$ and for equilibrium $\phi^*(\omega) = \phi^*(\sum \mu_i d\mathcal{N}_i) = 0$ if and only if $\sum \mu_i \nu_i = 0$. The reaction coordinate, ξ, measures the extent of the reaction and is zero for no reaction at the given control variables. The reaction coordinate defines a path in M, and the timewise mass transfer as a possible reaction occurs is a non-equilibrium process.

The ith component of the compound system has controls, θ, $\mathcal{P}$, and μ_i. Its generalized Gibbs energy function for fixed temperature and pressure is $G_i^* = G(\theta, \mathcal{P}, \mathcal{N}_i) - \mu_i \mathcal{N}_i$. At equilibrium, $\partial G_i^*/\partial \mathcal{N}_i = \partial G_i/\partial \mathcal{N}_i - \mu_i = 0$ recovers the equation of state. The maximum dissipation evolution equation describes the non-equilibrium evolution of $\mathcal{N}_i$ as the reaction occurs. Since $N_i = N_i^o + \nu_i \xi$, the Gibbs energy for each substance is rewritten in terms of the reaction coordinate, and after a short calculation, the maximum dissipation evolution of the state variable, ξ, is

$$\dot{\xi} = -k \left(\frac{d^2 G_i}{d\xi^2}\right)^{-2} \left(\frac{dG_i}{d\xi} - \nu_i \mu_i\right). \tag{12}$$

Therefore, at equilibrium, where $\dot{\xi} = 0$, $dG_i/d\xi = \nu_i \mu_i$.

However, in the classical analysis of reversible chemical reactions, each substance is assumed to be in self-equilibrium during each instant of the reaction, even though the total system is not in equilibrium. The total compound system is closed in the sense that there is no mass transfer across its boundaries. Since each component substance is in self-equilibrium, the generalized Gibbs energy for the total system is the sum of the equilibrium Gibbs energies of the substances, $G^* = \sum_i G(\theta, \mathcal{P}, \mathcal{N}_i)$. At constant temperature, θ, and pressure, $\mathcal{P}$, the affinity for the total system is, by the chain rule,

$$X = \frac{dG^*}{d\xi} = \sum_i \frac{\partial G^*}{\partial \mathcal{N}_i} \frac{\partial \mathcal{N}_i}{\partial \xi} = \sum_i \nu_i \mu_i. \tag{13}$$

At equilibrium where the affinity $X = 0$, the classical equilibrium condition that $\sum_i \mu_i \nu_i = 0$ is recovered for G^*.

The traditional non-equilibrium analysis of a chemical reaction is closely related to the contact form associated with G^*. The Clausius uncompensated heat, Q', is defined by the relation for the variation of heat δQ,

$$\delta Q = dG - \mathcal{P} d\mathcal{V} = \mathcal{S} d\theta - \delta Q',$$

so that the contact form for G^* is $\omega = -\delta Q'$. Because $\omega \leq 0$ for an admissible process, $\delta Q' \geq 0$ as required.

Because the equilibrium condition is $\sum_i \mu_i \nu_i = 0$, the chemical affinity, A, for the reaction has been classically defined by $A = \sum_i \mu_i \nu_i$. The chemical affinity was defined by de Donder (see Yeremin, 1981, p. 319) in terms of the uncompensated heat by the relation $\delta Q' = A d\xi \geq 0$. Therefore the contact form is $\omega = -(\sum_i \mu_i \nu_i) d\xi \leq 0$.

References

V. I. Arnold (1989). *Mathematical Methods of Classical Mechanics*, Springer- Verlag, New York.

V. I. Arnold (1990). Contact Geometry: the geometrical method of Gibbs's thermodynamics. In *Proceedings of the Gibbs Symposium*, ed. D. G. Caldi and G. D. Mostow, American Mathematical Society, American Institute of Physics, pp. 163–180.

W. L. Burke (1985). *Applied Differential Geometry*, Cambridge University Press, Cambridge.

L. Conlon (1993). *Differentiable Manifolds*, Birkhäuser, Boston.

J. W. Gibbs (1873). A method of geometrical representation of the thermodynamic properties of substances by means of surfaces. *Transactions of the Connecticut Academy* II, 382–404. Also in *The Collected Works*, Vol. 1, Yale University Press, New Haven 1948, 33–54.

H. W. Haslach, Jr. (1997). Geometrical structure of the non-equilibrium thermodynamics of homogeneous systems. *Reports on Mathematical Physics* **39**, 147–162.

R. Hermann (1973). *Geometry, Physics, and Systems*, Marcel Dekker, New York.

G. A. Holzapfel (2000). *Nonlinear Solid Mechanics*, 2005 reprinting. Wiley, Chichester.

P. Libermann and C-M. Marle (1987). *Symplectic Geometry and Analytical Mechanics*, D. Reidel Pub. Co., Dordrecht.

S. MacLane (1963). *Homology*. Springer-Verlag, Berlin.

L. E. Malvern (1969). *Introduction to the Mechanics of a Continuous Medium*, Prentice-Hall, Englewood Cliffs, NJ.

J. E. Marsden and T. J. R. Hughes (1994). *Mathematical Foundations of Elasticity*, Dover, New York.

R. Mrugala, J. D. Nulton, J. C. Schon, and P. Salaman (1991). Contact structure in thermodynamic theory, *Reports on Mathematical Physics* **29**, 109–121.

J.C. Simo and K. S. Phister (1984). Remarks on rate constitutive equations for finite deformation problems: computational implications. *Computer Methods in Applied Mechanics and Engineering* **46**, 201–215.

E. N. Yeremin (1981). *Fundamentals of Chemical Thermodynamics*, Mir Publishers, Moscow.

Chapter 8
Bifurcations in the Generalized Energy Function

1 Introduction

The generalized thermodynamic function defined in Chapter 3 simplifies the thermodynamic analysis of physical systems. An advantage of the generalized thermodynamic functions is that the resulting model can account for bifurcations and singularities by methods related to energy methods. The earliest published examples of generalized theromodynamic functions were invented to study singularities by catastrophe theory. Gilmore (1981) defined generalized energy functions on the full set of thermodynamic variables which are only equivalent on thermostatic states to those presented here, but did not produce a model for non-equilibrium behavior. Other generalized functions have been used by Lavis and Bell (1977) and by Poston and Stewart (1978) to describe the neighborhood of the singularity point of a van der Waals fluid as a cusp catastrophe, a description not possible with classical thermostatic energy functions. Their analysis requires a reduction to a single essential state variable to permit the application of elementary catastrophe theory.

The long-term quasi-static energy density function must be determined experimentally in order to construct the maximum dissipation non-equilibrium evolution equation. Small errors are likely to arise in this determination. Therefore, an important question is whether or not the thermostatic function deduced from experiment changes properties under small perturbations. The analysis depends on the particular loading applied to the boundary of the body. Small changes in response when the load is perturbed slightly do not affect the form of the function determined, but the changes are qualitatively significant if the generalized energy exhibits bifurcations when the load is considered as a parameter. One may characterize all possible perturbations using catastrophe theory.

The equilibrium behavior of a physical system under varying conditions is often described by the equilibrium set of a family of thermodynamic energy density functions, φ. The equilibrium points are not critical points of these energy functions. A critical point is a solution to the set of equations $\partial\varphi/\partial x_i = 0$, for $i = 1, \ldots, n$. The direct determination of the stability of a thermostatic function without relating it to a generalized thermodynamic function is a complex problem (see Section 2). Mathematical analysis is made simpler by using the associated generalized energy function

H.W. Haslach Jr., *Maximum Dissipation Non-Equilibrium Thermodynamics and its Geometric Structure*, DOI 10.1007/978-1-4419-7765-6_8,

for which the equilibria are critical points. Then, physically complex responses, such as the initiation of buckling or phase changes with respect to some parameter, are predicted by the model if at least one memeber of the family indexed by the parameter has a singularity. In fact, most classical thermodynamic functions are assumed to be convex so that all states are stable and no singularity is possible. Avoiding this restriction is another advantage of the generalized thermodynamic function.

Rank one singularities and the resulting bifurcations in isotropically symmetric planar families of functions, indexed by the thermodynamic control variables and perhaps by additional material or structural parameters, appear in many important mechanics problems. If the functions represent the energy of a nonlinear elastic body parametrized by load, a bifurcation occurs at that load at which the body no longer has a unique equilibrium deformation response. Such functions appear in modeling the behavior of sheets or membranes under loads. A catastrophe theory analysis is applied to the example of a membrane since many substructures in biomechanics are membranes. The example sheet is made of rubber, but no suggestion that rubber models are valid for biological tissue is implied. The history of the search for a constitutive model for rubber again shows that many functional forms can fit the same data if a sufficient number of empirical constants are adjusted properly. Some hyperelastic energy functions for rubber were mentioned in Chapter 2. The associated generalized energy functions are analyzed for singularities. However well they fit other data, all of the classical constitutive models for rubber discussed here fail to produce enough bifurcations or disjoint equilibria to predict Treloar's experimental measurements. The models built from statistical chain representations of the rubber molecules are the least successful. The results on the singularities of the energy function for an equibiaxially loaded rubber sheet first appeared in Haslach (2000).

After examples of an analysis to determine bifurcations and singularities, an investigation is begun into the influence of the equilibrium states bifurcation with respect to a parameter on the behavior of the non-equilibrium maximum dissipation evolution process.

1.1 Lavis and Bell Generalized Thermodynamic Function: Van der Waals Fluid

Catastrophe theory is a natural technique to apply to the study of phase changes in a material because a phase change involves a singularity at the critical point at which the solid, liquid and gas phases meet in the phase diagram. The first order phase transitions bifurcate from the thermodynamic critical point. In order to apply catastrophe theory to the phase change of a van der Waals fluid, Lavis and Bell (1977) assumed the existence of a potential of the form of the generalized energy defined in Chapter 3. So their work can be thought of as one of the earliest published applications of a generalized thermodynamic energy function with a bifurcation.

Lavis and Bell reduced the behavior to a single state variable, which allowed an elementary catastrophe theory analysis. The number of state variables is reduced in a generalized energy function $\varphi^* : \Re^{2n} \rightarrow \Re$ by elimination. To remove the state variables x_i for $i \in I$, where I is a subset of $\{1, 2, \ldots n\}$ and J is its complement, it must be possible to solve the equations of state $x_i = f_i(\boldsymbol{y})$ to obtain $x_i = h_i(\boldsymbol{x}_J, \boldsymbol{x}_I)$ for each state variable x_i, for $i \in I$. When the resulting expressions for the x_i are substituted, the generalized energy function no longer has the form $\varphi^*(x_1, \ldots, x_n; y_1, \ldots, y_n) = \varphi(x_1, \ldots, x_n) + \sum_{i=1}^{n} x_i y_i$, but is a function of only the x_i, for $i \in I$.

The Lavis and Bell result is recovered in terms of the generalized energy function defined in Chapter 3. The equation of state, $P = NkT/(V - bN) - aN^2/V^2$ with material constants a and b, for a van der Waals fluid can represent either the gas or liquid phase. A global generalized internal energy function, U^*, whose state variables are the temperature T and the pressure P and control variables are U and S does not exist since the equation of state cannot be inverted to obtain V globally. However, a global generalized Gibbs function whose control variables are the temperature T and the pressure P can be obtained from the equation for P and T in terms of the state variables: entropy, number of moles and volume, S, N and V.

The constant volume heat capacity is assumed to be $(\partial U/\partial T)_V = 3NR/2$. The equilibrium internal energy is $U = 3NRT/2 + f(V)$. Since $dU - TdS + PdV = 0$ at equilibrium,

$$(3NR/2)[(\partial T/\partial S)dS + (\partial T/\partial V)dV)] + (df/dV)dV - TdS$$
$$+[NRT/(V - bN) - aN^2/V^2]dV = 0.$$

If it is assumed that $df/dV = aN^2/V^2$ so that $f(V) = -aN^2/V$, solution of the resulting pair of differential equations yields

$$T = (2N^{2/3}/3R)(V - bN)^{-2/3} \exp(2S/3NR); \tag{1}$$

$$P = (2N^{5/3}/3)(V - bN)^{-5/3} \exp(2S/3NR) - aN^2/V^2. \tag{2}$$

The coefficient of T was chosen so that $U(cS, cV, cN) = cU(S, V, N)$ for any constant, c. This determines the equilibrium internal energy, U, for fixed N, in state variables S and V. The generalized Gibbs function, $G^* = U(S, V, N) + PV - TS$, up to a constant, is

$$G^*(S, V, N; T, P) = N^{5/3}(V-bN)^{-2/3} \exp(2S/3NR) - aN^2/V + \beta(N) + PV - TS. \tag{3}$$

The equation of Lavis and Bell for G^* in the single state variable, V, is recovered for a constant in N by substituting into (3), the entropy obtained from (1),

$$S = (3NR/2)[\ln T + (2/3)\ln(V - bN) - \ln(2N^{2/3}/3R)].$$

Lavis and Bell instead use a Legendre transform to eliminate S by writing $P = f(T, V.N) = NkT/(V - bN) - aN^2/V^2$ and obtain the generalized function as

$$G^* = PV - \int f(T, V, N)dV + \alpha(T, N). \tag{4}$$

The function $\alpha(T, N)$ is determined from the heat capacity, $2Nk/2$, and the Sacker-Tetrode equation giving the asymptotic behavior of the entropy.

The generalized energy function, G^*, in (3) has a singularity at the classical critical point for a van der Waals fluid, $V_c = 3bN$, $P_c = a/27b^2$, $T_c = 8a/27bR$, which is the end of the liquid-gas coexistence curve in the $P - V$ diagram (Callen, 1985, p. 241). The Hessian of G^* at this point

$$\begin{pmatrix} \frac{4a}{81b^3N} & \frac{-8a}{81b^2NR} \\ \frac{-8a}{81b^2NR} & \frac{16a}{81bNR^2} \end{pmatrix}$$

has determinant equal to zero. The rank of the Hessian is 1, and therefore the function of the two variables V and S has corank 1. Since the corank is one, only one of the variables is essential.

The van der Waals equation $(P + \alpha/V^2)(V - \beta) = RT$ is reduced by translating the bifurcation point to (1, 1, 1) and normalizing to obtain an energy function of the form $x^2 + ax + b$ in the version of this analysis given by Poston and Stewart (1978, p. 329). A function with this form is shown below to be a cusp catastrophe.

1.2 Rubber Sheet Under Biaxial Loading

The use of the generalized energy function to analyze singularities is illustrated in the remainder of the chapter by a biaxially loaded rubber sheet. Nonlinear elastic bodies, in which the deformations are non-infinitesimal, do not always have unique equilibrium states for a given load. Rivlin (1974) first demonstrated this fact for an isothermal, incompressible, neo-Hookean cube under equal and opposite normal surface tractions. Experiments on a rubber sheet under equal biaxial in-plane tensile loads show that unequal stable equilibrium stretches are possible. Experiments such as those by Treloar (1948) and by Alexander (1968) show that a rubber sheet under equal in-plane biaxial tensile loads can be in stable equilibrium with unequal, as well as equal, in-plane principal stretches. To model these unequal stretches, the sheet strain energy function, when parametrized by the load, must have several bifurcations in the equilibrium set or must have paths of equilibria disjoint from the equal stretches equilibria path. A Liapunov-Schmidt reduction for the equilibria of a class of isotropically symmetric energy functions and elementary catastrophe theory are used to classify the degenerate singularity behavior. The classical empirical constitutive models proposed for rubber-like, isothermal, incompressible nonlinear elastic materials are shown by this analysis to fail to generate enough bifurcations or disjoint equilibria paths to represent the experimental rubber sheet behavior under equal biaxial loads. Based on a full description of the equilibria behavior of any Ogden strain invariant near a singularity, a model which has three degenerate singularities and which reproduces the qualitative structure of Treloar's sheet data is constructed from linear combinations of three of Ogden's strain invariants. Errors in making the two in-plane tensions equal are represented by an imperfection parameter in the catastrophe universal unfolding of the energy function. In the rubber sheet, the isothermal energy density is $\varphi(\lambda_1, \lambda_2)$, where the in-plane principal stretches λ_1 and λ_2 are the state variables. The control variables are the in-plane principal stresses T_1 and T_2 (Fig. 8.1). The equilibrium stresses are calculated from $\partial\varphi/\partial\lambda_1 = T_1$ and $\partial\varphi/\partial\lambda_2 = T_2$. The function, $\varphi(\lambda_1, \lambda_2)$, for a isotropic sheet has symmetry under the operation of switching λ_1 and λ_2 so that $\varphi(\lambda_1, \lambda_2) = \varphi(\lambda_2, \lambda_1)$.

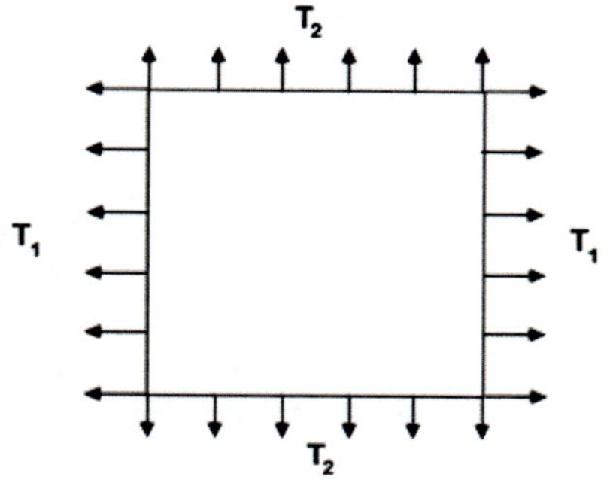

Fig. 8.1 Sheet subject to an in-plane biaxial load

Definition 1 A function, φ, indexed by parameters $a_1, \ldots, a_m$, is isotropically symmetric in the state variables if

$$\varphi(\lambda_1, \lambda_2, a_1, \ldots, a_m) = \varphi(\lambda_2, \lambda_1, a_1, \ldots, a_m).$$

If $T_1 = T_2$, there is a single control variable. The sheet is equibiaxially loaded. An important question is whether or not the equibiaxially loaded sheet can have unequal equilibrium stretches. This entails discovering whether or not the function can have a bifurcation when $\lambda_1 = \lambda_2$.

To facilitate an analysis of the stability of the equilibria of the function, φ, it is desirable for the equilibria to be extrema of a function. The associated generalized energy density is defined to this end,

$$\varphi^*(\lambda_1, \lambda_2, a_1, \ldots, a_m) = \varphi(\lambda_1, \lambda_2, a_1, \ldots, a_m) - T_1\lambda_1 - T_2\lambda_2. \tag{5}$$

Recall that the two functions, φ^* and φ, have the same equilibria state variables, λ_1 and λ_2, at a given value of T_1 and T_2. A bifurcation in the family, φ^*, occurs at a critical point where the Hessian of a member of φ^* is singular.

2 Stability in Energy Density Functions for Which the Equilibria Are Not Critical Points

If one does not use the associated generalized energy function, then the stability analysis is very complex. Supple considered planar function families with a different symmetry, indexed by a single parameter, p, and containing a member with singularity of rank zero Hessian. Supple's analysis determines the stability behavior near the bimodal singularity point and applies it to bimodal post buckling paths, with respect to p. Supple (1967) analyzes systems in two state variables, u_1 and u_2, and one control variable, p, for which the potential energy, $V(u_1, u_2; p)$, has the symmetry,

$$V(u_1, u_2; p) = V(-u_1, u_2; p) = V(u_1, -u_2; p).$$

The function V is an energy function similar to the function φ^* of Eq. (5). He restricts his attention to those functions linear in p, as is the case for many physical situations in which p is directly related to the load on a structure.

Supple shows that, in addition to the fundamental equilibrium path $u_1 = u_2 = 0$, there are at most four equilibrium paths parametrized by p. This is a consequence of the fourth order approximation to the function about (0, 0) not involving any u_1u_2 terms so that each derivative is a cubic with no constant term. There are two paths as p varies, one in each coordinate plane $p - u_i$, which Supple calls the uncoupled paths. The remaining paths, at most two, lying out of the coordinate planes, are called the coupled paths by Supple.

Supple's theorems determine when the points on the coupled paths are stable or unstable equilibria. Assume that the uncoupled paths are rising; in other words, the equilibria on the paths are stable. Subscripts on functions refer to partial derivatives. Then if $V_{1111}V_{2222} - 3V_{1122}^2 < 0$, the coupled paths are falling, and all the equilibria on them are unstable. Conversely, if $V_{1111}V_{2222} - 3V_{1122}^2 > 0$, the coupled paths are rising, and all the equilibria on them are stable. If at least one of the uncoupled paths is falling, then the coupled paths must be falling, and all the equilibria on them are unstable.

For example, these theorems describe the post-buckling behavior of a linear orthotropic elastic plate under in-plane biaxial compressive loads. The uncoupled paths for a linear orthotropic plate are rising and $V_{1111}V_{2222} - 3V_{1122}^2 > 0$; therefore the coincident buckling point and the post-buckling equilibria are stable. Details are given in Haslach (1986).

The other case occurs in the Augusti model (Thompson and Hunt, 1973, p. 243), a rigid rod axially loaded in compression on one end and supported on the other end by two perpendicular torsional springs and a ball and socket. The coalescence of stable buckling modes in a double cusp produces unstable post-buckling behavior since the uncoupled paths are rising but $V_{1111}V_{2222} - 3V_{1122}^2 < 0$.

The equilibrium set in λ_1–λ_2 space for the sheet under the loading $T_1 = T_2$ is all points (λ_1, λ_2) satisfying $\varphi_1 = \varphi_2$, where the subscripts on the function φ

refer to partial derivatives, e. g. $\varphi_i = \partial\varphi/\partial\lambda_i$, etc. By the symmetry, $\varphi_1(\lambda, \lambda) = \varphi_2(\lambda, \lambda)$, so that $\lambda_1 = \lambda_2$ is always a solution of $\varphi_1 - \varphi_2 = 0$. The function $f(\lambda_1, \lambda_2) = \varphi_1(\lambda_1, \lambda_2) - \varphi_2(\lambda_1, \lambda_2)$ is anti-symmetric, $f(\lambda_1, \lambda_2) = -f(\lambda_2, \lambda_1)$, because $\varphi_1(\lambda_1, \lambda_2) = \varphi_2(\lambda_2, \lambda_1)$ and $\varphi_2(\lambda_1, \lambda_2) = \varphi_1(\lambda_2, \lambda_1)$. Consequently, f can be factored in the form $f(\lambda_1, \lambda_2) = \hat{f}(\lambda_1, \lambda_2)(\lambda_1 - \lambda_2)$, where the function $\hat{f}(\lambda_1, \lambda_2)$ is isotropically symmetric. The fundamental path of equilibria as the load is varied is the set of points, $\lambda_1 = \lambda_2$.

Conditions are available (Ogden 1985, 1987) to identify the bifurcations from the path of equal stretches in the equilibrium set of an isothermal strain energy density function, $\varphi(\lambda_1, \lambda_2)$, for a sheet. Sufficient conditions for a sheet, made of an incompressible material and which is in plane stress, to be in an infinitesimal stable equilibrium state under arbitrary homogeneous in-plane tensile stresses are that the energy function satisfies

$$\varphi_{11} > 0 \tag{6}$$

$$\varphi_{11}\varphi_{22} - \varphi_{12}^2 > 0 \tag{7}$$

$$\frac{\varphi_1 - \varphi_2}{\lambda_1 - \lambda_2} > 0. \tag{8}$$

The latter inequality arises from the Baker and Ericksen (1954) inequality for stable equilibria.

These stability conditions reduce to a simpler form on the fundamental path, $\lambda_1 = \lambda_2$, in stretch space. The symmetry of the strain energy function, $\varphi(\lambda_1, \lambda_2) = \varphi(\lambda_2, \lambda_1)$, implies $\varphi_{11} = \varphi_{22}$ there, so that

$$\varphi_{11} > 0 \tag{9}$$

$$\varphi_{11} - \varphi_{12} > 0 \tag{10}$$

$$\varphi_{11} + \varphi_{12} > 0, \tag{11}$$

The second inequality is a consequence of the Baker and Ericksen inequality by taking the limit as λ_1 approaches λ_2. The third is derived from the second by factoring Eq. (7) under the condition that $\varphi_{11} = \varphi_{22}$ when $\lambda_1 = \lambda_2$.

If the loads are equal, the points on an equilibrium path bifurcating from the fundamental path satisfy $\varphi_1(\lambda_1, \lambda_2) = \varphi_2(\lambda_1, \lambda_2)$ for $\lambda_1 \neq \lambda_2$. Then $\varphi_1 - \varphi_2 = 0$ implies that

$$\frac{\varphi_1 - \varphi_2}{\lambda_1 - \lambda_2} = 0.$$

At the intersection of the bifurcating path with the fundamental path, found by letting λ_1 tend to λ_2, the requirement

$$\varphi_{11} - \varphi_{12} = 0 \tag{12}$$

becomes a necessary condition for a bifurcation from the fundamental path to occur. Possible bifurcations can be sought by first locating the points on the fundamental path where Eq. (12) holds. Such points are also singularities of the energy function. However, not all singularities induce bifurcations. Singularities on the fundamental path with $\varphi_{11} + \varphi_{12} = 0$ or $\varphi_{11} = 0$ need not be considered, because no bifurcation can develop at such singularities when the loads are equal.

Equilibrium points for equal biaxial loads cannot satisfy the stability condition, $(\varphi_1 - \varphi_2)/(\lambda_1 - \lambda_2) > 0$, because the equilibrium condition is $\varphi_1 - \varphi_2 = 0$. Under these stability criteria, each point is said to have neutral stability. However, it is possible that $\varphi_1 - \varphi_2 = 0$, but the equilibrium state is stable, as can be shown by other methods which involve higher order derivatives of the energy function. This analysis is much more complex than an energy methods based analysis of the associated generalized energy function. Catastrophe theory is used below to identify both the bifurcation points of the associated generalized energy function and the stability behavior along the fundamental path and the stability of any equilibrium points with $\lambda_1 \neq \lambda_2$ near the fundamental path.

3 Stability, Equivalence and Unfoldings

Catastrophe theory can define what it means so say that two functions with a degenerate bifurcation, with respect to the load viewed as a parameter in the sheet, are essentially the same under perturbations. For some values of the load parameter, the member of the family may have a bifurcation at a critical point that is a singularity. Since it is impossible to perfectly repeat an experiment, mathematical descriptions of any experiment must be essentially the same under small perturbations, a property called structural stability.

Definition 2 Two families of functions, $f, g : \Re^n \times \Re^r \to \Re$ depending on r parameters are equivalent if there exists

(a) a local diffeomorphism of the parameters, $e : \Re^r \to \Re^r$;
(b) a local change of coordinates, which depends smoothly on the parameters, in the sense that for each parameter $c \in \Re^r$, the function $y_c : \Re^n \to \Re^n$, defined by $(x, c) \to (y_1(x, c), \ldots, y_n(x, c))$ is a diffeomorphism in a neighborhood of the origin;
(c) a shear map, $\gamma : \Re^r \to \Re$ that varies smoothly with $s \in \Re^r$,

so that $g(x, s) = f(y_s(x), e(s)) + \gamma(s)$ for all (x, s) in a neighborhood.

Following Poston and Stewart (1978, p. 92),

Definition 3 A function $f : \Re^n \times \Re^r \to \Re$ is structurally stable if it is equivalent to $f + p : \Re^n \times \Re^r \to \Re$, where p is a sufficiently small family $p : \Re^n \times \Re^r \to \Re$.

Structural stability and the qualitative response of the system may be influenced by the existence of singularities in the long-term energy density function.

Definition 4 The function $f : M \to N$, where M and N are differentiable manifolds, has a singularity at $x \in M$ if rank $df(x) : T_*M \to T_*N$ is less than the smaller of the dimensions of M and N, where T_*M and T_*N are the tangent bundles.

Definition 5 The point x is a non-degenerate critical point of $f : \Re^n \to \Re$ if $df(x) = 0$ and if the determinant of the Hessian $\det(\partial^2 f/\partial x_i \partial x_j) \neq 0$. If the determinant of the Hessian is zero, then the critical point is degenerate.

The degeneracy of the singularity may be measured by the corank of the Hessian. If the Hessian is $n \times n$ and r is the rank of the Hessian, then the corank of the Hessian is $n-r$. The corank gives the number of independent directions in which the system is degenerate. This idea is clearer if the Hessian is put in diagonal form. The nature of the critical points in a member of the family can change only by passing through a degenerate critical point at some value of the parameters because of the Morse Lemma.

Lemma 6 *(Morse) Let $u \in \Re^n$ be a nondegenerate critical point of the smooth function $f : \Re^n \to \Re$. Then there is a local coordinate system $(x_1, \ldots, x_n)$ in a neighborhood U of u with $x_i(u) = 0$ for all i such that $f(x_1, \ldots, x_n) = f(u) - x_1^2 - \ldots - x_i^2 + x_{i+1}^2 + \ldots + x_n^2$ for all $(x_1, \ldots, x_n) \in U$ (Poston and Stewart, 1978, p. 54).*

Such a nondegenerate critical point is also called a Morse critical point by some.

A critical point is called hyperbolic if the Hessian of φ has no eigenvalues with zero real part. The Hessian of a real-valued function is symmetric and therefore its eigenvalues are real. So, a critical point of a real-valued function is hyperbolic if none of the eigenvalues are zero. Hyperbolic critical points are locally structurally stable, in other words a perturbation does not change the structure. Singularities that are not hyperbolic are called degenerate singularities because they can have bifurcations, and so are not locally structurally stable.

3.1 Equivalence, Unfoldings, and Perturbations of Real Valued Functions

The families of perturbations of real valued functions with a degenerate singularity are given by families of truncated Taylor series (Poston and Stewart, 1978; Castrigiano and Hayes, 1993). The identification of which, if any, of the truncated Taylor series of a real valued function captures its full qualitative behavior and the enumeration of all possible perturbations near a degenerate singularity of such a real valued function is the program of catastrophe theory. A coordinate translation moves a critical point of interest to the origin. Poston and Stewart (1978, p. 125) make the definition

Definition 7 Two functions $f, g : \Re^n \to \Re$ are equivalent at $x = 0$ if there is a local diffeomorphism on a neighborhood U of $0 \in \Re^n$, $\phi : U \to \Re^n$ and a constant

γ such that for $x \in U$, $g(x) = f(\phi(x)) + \gamma$. The two functions f, g are said to be strongly equivalent if the diffeomorphism ϕ has derivative which is the identity at the origin.

Another type of equivalence determines when the Taylor series of a function can be truncated.

Definition 8 Two functions are k-equivalent if they have the same Taylor series out to the kth order.

Definition 9 A function is k-determined if it is equivalent to any function with which it is k-equivalent. The function is strongly k-determined if the change of coordinate diffeomorphism has derivative which is the identity at the origin.

Therefore, any k-determined function is equivalent to its Taylor series out to the kth order, its k-jet. The terms of order greater than order k of the Taylor series of such a function can be removed by a coordinate transformation. The k-jet, in this case, contains all qualitative information about the function.

The perturbations of a real valued C^∞ function are captured in an unfolding of the function. Let $\Re^n$ be the space of state variables, $x = (x_1, \ldots, x_n)$, and $\Re^r$ be the space of perturbation (or unfolding) parameters, $c = (c_1, \ldots, c_r)$. Identify $\Re^n \times \Re^r$ with $\Re^{n+r}$.

Definition 10 An r-unfolding of $f : \Re^n \rightarrow \Re$ is a function $F : \Re^{n+r} \rightarrow \Re$ such that $F(x_1, \ldots, x_n, 0, \ldots, 0) = f(x_1, \ldots, x_n)$.

An unfolding of f has exactly the same number of state variables as f. A relation is defined among unfoldings.

Definition 11 An s-unfolding of $f : \Re^n \rightarrow \Re$, $G : \Re^{n+s} \rightarrow \Re$, is induced from the r-unfolding, F, of f if there exists

(a) a smooth function of the unfolding parameters, $e : \Re^s \rightarrow \Re^r$;
(b) a change of coordinates, which depends on the unfolding parameters, in the sense that for each parameter $c \in \Re^s$, the function $y_c : \Re^n \rightarrow \Re^n$, defined by $(x, c) \rightarrow (y_1(x, c), \ldots, y_n(x, c))$ is a diffeomorphism in a neighborhood of the origin;
(c) a shear map, $\gamma : \Re^s \rightarrow \Re$;
and $G(x, c) = F(y_c(x), e(c)) + \gamma(c)$ in a neighborhood of the origin.

The idea is that G can contain no more information about the degenerate singularity of f than F does. G is a reparametrization of F.

Definition 12 An unfolding is versal if all other unfoldings of f can be induced from it. It is called universal if it involves the least number of perturbation parameters.

A universal unfolding is the smallest versal family containing all possible perturbations.

3.2 The Simple Catastrophes of One State Variable

Elementary catastrophe theory can be used to describe the possible degenerate singularities of the reduction, $G(x, t)$. One accomplishment of elementary catastrophe theory was to classify all degenerate singularities of functions of one variable and their possible perturbations. Assume that the singular critical point occurs at $x = 0$ and that the function value is zero at $x = 0$. A translation preserving the singularity structure is always possible to obtain this condition. '

Definition 13 A function of one variable with Taylor series $f(x) = \sum_{q=2}^{\infty} c_q x^q$ about the critical point $x = 0$ has a degenerate singularity of order k at $x = 0$ if $c_q = 0$ for $q = 2, \ldots, k-1$ and $k > 2$.

A real valued function of one variable, $f(x)$, is equivalent to the first non-zero term of its Taylor series, $(1/k!)(d^k f/dx^k)x^k = \alpha x^k$, i.e. its lowest order non-zero derivative at the critical point $x = 0$. The appropriate coordinate transformation is (Poston and Stewart, 1978, p. 57)

$$y = \phi(x) = x \left| \frac{f(x)}{\alpha x^k} \right|^{1/k}. \tag{13}$$

Then $f(x)$ is equivalent to αy^k in the sense of Definition (7). The coordinate transformation, ϕ, is a local diffeomorphism, whose derivative at the origin is the identity.

The family of possible perturbations of a function with a degenerate singularity is given by its universal unfolding. A function, $f(x)$, of a single variable with degenerate singularity which is equivalent to $b_k x^k$ for some k has universal unfolding, $F(x; c_i) = c_1 x + \cdots + c_{k-2} x^{k-2} + b_k x^k$, (Poston and Stewart, 1978, p. 121; Castrigiano and Hayes, 1993). In other words, the Taylor series can be truncated at order k and preserve all qualitative information about the function. Because $k-2$ parameters are required for the universal unfolding, the function, $b_k x^k$, has codimension $k-2$ in the family of all real valued functions of one variable which are zero at the critical point $x = 0$.

The possible types of low order families with a degenerate singularity at $x = 0$ are the universal unfoldings of the functions $f(x) = x^k$,

$$\text{Fold:} \quad F(x, c_1) = x^3 + c_1 x \tag{14}$$

$$\text{Cusp:} \quad F(x, c_1, c_2) = x^4 + c_2 x^2 + c_1 x \tag{15}$$

$$\text{Dual Cusp:} \quad F(x, c_1, c_2) = -x^4 + c_2 x^2 + c_1 x \tag{16}$$

$$\text{Swallowtail:} \quad F(x, c_1, c_2, c_3) = x^5 + c_3 x^3 + c_2 x^2 + c_1 x \tag{17}$$

$$\text{Butterfly:} \quad F(x, c_1, c_2, c_3, c_4) = x^6 + c_4 x^4 + c_3 x^3 + c_2 x^2 + c_1 x, \tag{18}$$

in Thom's classification (Thom, 1975, p. 61ff). The coefficients c_i can be functions of other parameters. However, the c_i must vary independently to produce the full universal unfolding.

Definition 14 A symmetric catastrophe is a subfamily of a catastrophe in which all coefficients of odd order polynomial terms are zero.

For example, the symmetric cusp or dual cusp has as its equilibrium set the classical pitchfork bifurcation. The pitchfork is the subset of the equilibrium surface for the cusp defined by setting the imperfection, $c_1 = 0$. There are four possible sets of equilibria for a function linearly dependent on the parameter t and of the form, $h(x, t) = c_2(t)x^2 + c_4(t)x^4$, if $c_2(t_c) = 0$ for some t_c. These pitchforks are determined by all combinations of $c_2(t)$ increasing or decreasing in t and c_4 positive or negative (Fig. 8.2). In the standard cusp or dual cusp, $c_2(t)$ is decreasing in t near t_c.

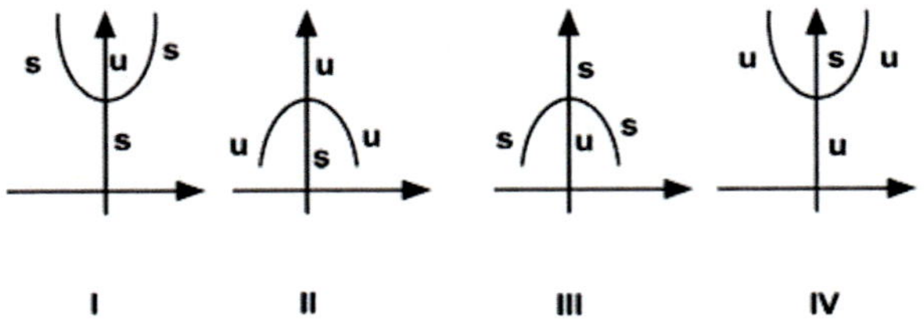

Fig. 8.2 Pitchfork types from the positive and negative cusp and dual cusp

For example, the parametrized symmetric cusp, with $c_1 = 0$, has as its equilibrium set the solution to $F_x = 4x^3 + 2c_2(t)x = 0$. The point $x = 0$ is always a solution. The bifurcation occurs at $c_2(t_c) = 0$ for some t_c. There are two other real solutions, $x = \pm[-c_2(t)/2]^{1/2}$ for $t > t_c$. The equilibrium set is the standard supercritical pitchfork. On the other hand, the equilibrium set of the symmetric dual cusp is the solution to $F_x = -4x^3 + 2c_2(t)x = 0$. In addition to $x = 0$, the two other real solutions are $x = \pm[-c_2(t)/2]^{1/2}$ for $t < t_c$. The equilibrium set is the standard subcritical pitchfork. In both cases, if $c_2(t)$ is decreasing in t as happens in many physical applications, then for $t < t_c$, the equilibria with $x = 0$ are stable since $c_2(t) > 0$ and unstable for $t > t_c$ since $c_2(t) < 0$. The converse situation occurs if $c_2(t)$ is increasing in t. These four possibilities are shown in Fig. 8.2. These pitchforks are the slice of the equilibrium surface for the cusp or dual cusp respectively defined by setting the imperfection, $c_1 = 0$.

The function, $F(x, c_2, c_4) = x^6 + c_4x^4 + c_2x^2$ has a singularity at $x = 0$ if $c_2 = 0$. If also $c_4 = 0$, then this is a symmetric butterfly degenerate singularity. It is a cusp or dual cusp if $c_4 \neq 0$ and $c_2 = 0$. The bifurcation set and sketches of the equilibrium set for the full butterfly are given by Poston and Stewart (1978, p. 178). For example, a sheet of cusp bifurcations is obtained when $c_4 > 0$ is fixed as c_2 passes through zero from positive to negative.

4 Asymmetric Deformations in Experiments on a Rubber Sheet

Treloar's experiments showed that asymmetric equilibrium stretches in a rectangular rubber sheet are possible when it is subjected to equal in-plane biaxial loads applied perpendicular to the sheet edges (Treloar, 1948, Table 1). The rubber tested by

Treloar was vulcanized with 2% sulfur. He performed parallel experiments on rubber which was swollen to about twice its original volume in medicinal paraffin. The volume fraction of rubber in the swollen state was 0.525. The shear modulus for the dry rubber was measured as $2.90\,\text{kg/cm}^2$ and for the swollen, $G = 2.09\,\text{kg/cm}^2$, by plotting $\sigma_1 - \sigma_2 = G(\lambda_1^2 - \lambda_2^2)$. The responses to equal loading along the two in-plane edges of the sheet are summarized in Table 8.1, which presents Treloar's 1948) data for the principal stretches, λ_1 and λ_2, of an unswollen rubber sheet due to in-plane biaxial loads f_1 and f_2. At biaxial loads of 400 and 600 gm, the two stretches differ by 7.5% and 12.4% respectively so that these experimental stretches are unequal under equal loads. Notice that at the intermediate load of 500 gm there is a state with nearly equal stretches. All must represent stable equilibrium states; otherwise they could not have been experimentally measured.

Table 8.1 Treloar (1948) data for a rubber sheet

f_1 (gm)	f_2 (gm)	λ_1	λ_2
200	200	1.16	1.20
300	300	1.35	1.37
400	400	1.60	1.72
500	500	2.07	2.10
600	600	2.34	2.63

This data shows that there is a change in the stretch response from equal stretches to an asymmetrical stable state between 300 and 400 gm loads. Another transition in which the response returns to symmetrical stretches occurs at a higher load around 500 gm. A third change to an asymmetric response occurs before 600 gm. The jumps between symmetric and asymmetric stretches as the load is discontinuously increased may be a consequence of several possible equilibrium path structures. The asymmetric stable equilibria measured by Treloar could lie on primary stable equilibrium paths which bifurcate from the path of equal stretches or on secondary paths of equilibrium states which may bifurcate from a primary path. It is also possible that there are paths of equilibria disjoint from the path of equal stretches to which the sheet could jump.

Treloar discussed, in both his original paper and in the 1949 edition of his book, the possibility that the unequal stretches are spurious effects. He reported in (Treloar, 1948) that the order of application of the forces slightly affected the stretches, but not the qualitative response. Later, in the 1949 edition of his book (pp. 118–120), Treloar flatly stated, while discounting the possible influence of stress relaxation, that the results were reproducible and independent of the order of force application. Relaxation was negligible since the results were the same for the swollen rubber, in which any relaxation effects would be accelerated. Specimen anisotropy and lack of strain uniformity were also dismissed as a cause of the unequal stretches. Treloar pointed out that the neo-Hookean model cannot fit this data and that the Mooney-Rivlin model is an improvement.

Further experimental support for the existence of an asymmetric response to equal principal stresses in a rubber sheet appears in the inflation of a spherical balloon, which produces an equal biaxial load on a rectangular section of the balloon material. Two bifurcations were found in the continuous and monotonic inflation of neoprene (polychloroprene) balloons by Alexander (1971). Initially spherical balloons become asymmetric after a certain inflation pressure. At a higher inflation pressure, the balloon again becomes spherical. The first bifurcation occurred at the stretch $R/R_o = 1.6$ and the second at $R/R_o = 4.3$, where R_o is the original balloon radius and R is the inflated radius. A stretch of 1.6 is very close to the first transition in natural rubber obtained by Treloar. In the asymmetric region, Alexander observed two stable and one unstable equilibrium states, similar to that in snap through buckling. His balloon did not produce a third bifurcation to an aspherical shape at an even higher inflation pressure. He also reported that if the natural rubber is stiffened sufficiently, say by crosslinking with 8% sulfur, the balloon does not show the bifurcation to asymmetry. He stated that materials with low moduli are more likely to exhibit the asymmetric behavior. The additional transition in Treloar's experiment may therefore have occurred because natural rubber (isoprene) is less stiff than neoprene. Alexander conjectured that the bifurcations are related to crystallization of the material due to the increasing stress.

Alexander's observation of bifurcations for continuously inflated balloons suggests that Treloar's experimental findings are due to the bifurcation of a primary equilibrium path from the path of equal stretches, not due to secondary bifurcations or disjoint paths of equilibria. Further experiments with continuous loading of flat sheets would be required to confirm this possibility.

5 Incompressible Elastic Energy Functions

Hyperelastic models for isotropic, isothermal, incompressible materials are local, current elasticity models. No viscous behavior is considered, nor is any dependence on the past history of the loading. Deformations of the form

$$\mathbf{x}(\mathbf{X}) = (\lambda_1 X_1, \lambda_2 X_2, \lambda_3 X_3), \tag{19}$$

for λ_i constant over the body are called purely homogeneous, where X_i are the reference Cartesian coordinates and x_i are the current Cartesian coordinates. The constant λ_i is the stretch in the X_i direction. The deformations are homogeneous in the sense that the deformation gradient, $F = (\partial x_i/\partial X_j)$, is constant throughout the body. Ball and Schaeffer (1983, Theorem 2.1) showed that the study of arbitrary homogeneous deformations can be reduced to the study of purely homogeneous deformations.

Various constitutive modelers of rubber-like materials have viewed the isothermal strain energy as either a function of the strain invariants or of the stretches, λ_i. The principle of material frame indifference allows the transformation of the strain energy from a function of the Cauchy strain to a function of the left Cauchy-Green tensor, $B = FF^t$. The corresponding strain energy function, $\varphi(B)$, can be rewritten as a function of the strain invariants I_B, II_B, and III_B. In the sheet, λ_1 and λ_2 are the in-plane principal stretches, and λ_3 is the stretch in the thickness direction.

Since the current density is $\rho = \rho_0 \det(F)$, $\det(F) = \lambda_1\lambda_2\lambda_3 = 1$ when the material is incompressible, and $\lambda_3 = 1/\lambda_1\lambda_2$. For a purely homogeneous deformation, the invariants of B are $III_B = \det(B) = 1$ and

$$\begin{aligned} I_B &= \lambda_1^2 + \lambda_2^2 + \lambda_1^{-2}\lambda_2^{-2} \\ II_B &= \lambda_1^2\lambda_2^2 + \lambda_2^{-2} + \lambda_1^{-2}. \end{aligned} \tag{20}$$

The internal energy is then a function of just the first two strain invariants, $\varphi = \varphi(I_B, II_B)$. In the past, much experimental data was given in terms of $\varphi_I = \partial\varphi/\partial I_B$ and $\varphi_{II} = \partial\varphi/\partial II_B$.

Experiments by Rivlin and Saunders (1951) showed that, over a large range of controlled stretches, the derivatives $\partial\varphi/\partial I_B$ and $\partial\varphi/\partial II_B$ are functions of I_B and of II_B. The term, $\partial\varphi/\partial I_B$, is positive and seems to be constant in I_B and almost parabolic in II_B. The change with respect to II_B is very small so that one could approximate $\partial\varphi/\partial I_B$ as a constant for small stretches. The partial, $\partial\varphi/\partial II_B$, which is positive, is approximately linearly decreasing in I_B and decreases asymptotically to zero in II_B.

The conditions on the derivative behavior define the particular constitutive model. In terms of the stretches, one constraint is that the derivative evaluated at $\lambda_1 = \lambda_2 = 1$, $\partial^2\varphi/\partial\lambda_1\partial\lambda_2 = 2\mu$, where μ is the shear modulus. Rivlin and Saunders assumed that $\partial\varphi/\partial I_B$ is constant and $\partial\varphi/\partial II_B$ is a function of only II_B. Beatty (1987) verified analytically that the requirement that φ_I and φ_{II} both be functions of only II_B is sufficient to produce the Rivlin-Saunders constitutive model,

$$\varphi(I_B, II_B) = c_1(I_B - 3) + h(II_B - 3), \tag{21}$$

where c_1 is a positive constant and the function $h(II_B - 3)$ is arbitrary with $h(0) = 0$. The -3 terms have been added to make the strain energy zero when the body is undeformed. This model includes the neo-Hookean model when h is constant, the Mooney-Rivlin model when $h = c_2(II_B - 3)$ and c_2 is a positive constant, and the Gent-Thomas model when $h(II_B - 3)$ is inversely proportional to II_B. The experimental data of Rivlin and Saunders is not sufficient to generate the properties needed to produce bifurcations in the sheet model.

5.1 A Bifurcation Condition

A simple case is that in which the strain energy function depends only on the first strain invariant. To connect the bifurcation analysis to the classical rubber experiments which are expressed in terms of the strain invariants, the second derivatives of the strain invariants are required.

$$\frac{\partial^2 I_B}{\partial \lambda_1^2} = 2 + 6\lambda_1^{-4}\lambda_2^{-2} \quad \text{and} \quad \frac{\partial^2 I_B}{\partial \lambda_1 \partial \lambda_2} = 4\lambda_1^{-3}\lambda_2^{-3}$$

$$\frac{\partial^2 II_B}{\partial \lambda_1^2} = 6\lambda_1^{-4} + 2\lambda_2^2 \quad \text{and} \quad \frac{\partial^2 II_B}{\partial \lambda_1 \partial \lambda_2} = 4\lambda_1\lambda_2.$$

Proposition 15 *No model based on a strain energy function $\varphi(I_B, II_B)$ which is a function of only I_B and which meets the experimental requirement (Rivlin and Saunders, 1951) that $\partial\varphi/\partial I_B > 0$ can reproduce a sheet bifurcation from the fundamental path. Furthermore, the only equilibria for such a model are on the fundamental path and all are stable.*

Proof The equilibria satisfy

$$\varphi_1 - \varphi_2 = \frac{\partial \varphi}{\partial I_B}\left(\frac{\partial I_B}{\partial \lambda_1} - \frac{\partial I_B}{\partial \lambda_2}\right) + \frac{\partial \varphi}{\partial II_B}\left(\frac{\partial II_B}{\partial \lambda_1} - \frac{\partial II_B}{\partial \lambda_2}\right) = 0, \tag{22}$$

where

$$\frac{\partial I_B}{\partial \lambda_1} - \frac{\partial I_B}{\partial \lambda_2} = 2\lambda_1^{-3}\lambda_2^{-3}(\lambda_1 - \lambda_2)(\lambda_1^3\lambda_2^3 + 1),$$

$$\frac{\partial II_B}{\partial \lambda_1} - \frac{\partial II_B}{\partial \lambda_2} = 2(\lambda_1 - \lambda_2)[\lambda_1^{-3}\lambda_2^{-3}(\lambda_1^2 + \lambda_1\lambda_2 + \lambda_2^2) - \lambda_1\lambda_2].$$

When $\partial\varphi/\partial II_B = 0$,

$$\varphi_1 - \varphi_2 = \frac{\partial \varphi}{\partial I_B}\left[2\lambda_1^{-3}\lambda_2^{-3}(\lambda_1 - \lambda_2)(\lambda_1^3\lambda_2^3 + 1)\right] = 0. \tag{23}$$

Therefore any state with $\lambda_1 = \lambda_2$ is an equilibrium state. Because $\varphi_1 - \varphi_2$ factors into $\lambda_1 - \lambda_2$ and a positive function, these are the only equilibria.

That no bifurcations exist at any $\lambda_1 = \lambda_2$ is verified by the fact that the expression

$$\begin{aligned}\varphi_{11} - \varphi_{12} &= \frac{\partial^2 \varphi}{\partial I_B^2}\frac{\partial I_B}{\partial \lambda_1}\left(\frac{\partial I_B}{\partial \lambda_1} - \frac{\partial I_B}{\partial \lambda_2}\right) + \frac{\partial \varphi}{\partial I_B}\left(\frac{\partial^2 I_B}{\partial \lambda_1^2} - \frac{\partial^2 I_B}{\partial \lambda_1 \partial \lambda_2}\right) \\ &\quad + \frac{\partial^2 \varphi}{\partial II_B^2}\frac{\partial II_B}{\partial \lambda_1}\left(\frac{\partial II_B}{\partial \lambda_1} - \frac{\partial II_B}{\partial \lambda_2}\right) + \frac{\partial \varphi}{\partial II_B}\left(\frac{\partial^2 II_B}{\partial \lambda_1^2} - \frac{\partial^2 II_B}{\partial \lambda_1 \partial \lambda_2}\right) \\ &= \frac{\partial \varphi}{\partial I_B}(2 + 2\lambda^{-6}) + \frac{\partial \varphi}{\partial II_B}(6\lambda^{-4} - 2\lambda^2)\end{aligned} \tag{24}$$

because the first and third terms are zero when $\lambda_1 = \lambda_2 = \lambda$. If $\partial\varphi/\partial II_B = 0$, no bifurcation is possible. In this case, the sign of $\varphi_{11} - \varphi_{12}$ at $\lambda_1 = \lambda_2$ is determined by the sign of $\partial\varphi/\partial I_B$. Because $\varphi_{11} - \varphi_{12}$ is positive, all equilibria on the fundamental path are stable.

Many of the statistical chain models such as the neo-Hookean or the Arruda-Boyce models result in strain energy functions which depend only on I_B. These models cannot represent the bifurcation behavior of a rubber sheet.

6 Bifurcation Types

The bifurcation types possible in the response of a sheet are more easily obtained by a coordinate transformation of the associated generalized energy function. Once a bifurcation point is located on the fundamental path, it remains to describe the local bifurcation structure as the applied tensile equal biaxial loads are increased, including the number of equilibria at each load and their stability. The coordinate transformation of the potential energy and the Liapunov-Schmidt reduction of its equilibria permit the use of one dimensional elementary catastrophe theory to classify the behavior.

The $\lambda_1 - \lambda_2$ coordinate system is rotated 45° to x-y coordinates by requiring that the condition $\lambda_1 = \lambda_2$ occur at $x = 0$. This coordinate transformation produces a diagonalized Hessian, which greatly simplifies the Liapunov-Schmidt reduction and its analysis. Make the diffeomorphic coordinate change

$$\begin{aligned} x &= \frac{1}{2}(\lambda_1 - \lambda_2), \\ y &= \frac{1}{2}(\lambda_1 + \lambda_2), \\ t_x &= T_1 - T_2, \\ t_y &= T_1 + T_2. \end{aligned} \tag{25}$$

That the stretch, λ_i, is positive by definition implies y is greater than zero and also $y > x$. Furthermore, in tension $y > 1$ since the λ_i are also greater than 1. The body is undeformed when $x = 0$ and $y = 1$. The stretches are equal when $x = 0$. The biaxial loads are equal when $t_x = 0$, i.e. $T_1 = T_2$. Define

$$V(x, y) = \varphi\big(\lambda_1(x, y), \lambda_2(x, y)\big), \tag{26}$$

where

$$\begin{aligned} \lambda_1(x, y) &= x + y \\ \lambda_2(x, y) &= y - x. \end{aligned} \tag{27}$$

The potential energy function in the new coordinates is

$$\varphi^*(x, y, t_x, t_y) = V(x, y) - t_x x - t_y y. \tag{28}$$

The transformed function remains denoted as φ^* by an abuse of notation. This function can be viewed as a family of potential energy functions parametrized by t_x and t_y. The equilibrium states are determined by $\partial\varphi^*/\partial x = 0$ and $\partial\varphi^*/\partial y = 0$, so that at equilibrium,

$$t_x = \frac{\partial V}{\partial x} \quad \text{and} \quad t_y = \frac{\partial V}{\partial y}, \tag{29}$$

respectively. The equilibria are defined as critical points of φ^* rather than as solutions to the more cumbersome expressions, $\partial\varphi/\partial\lambda_i = T_i$. The requirement that the in-plane stresses in the sheet be equal ($T_1 = T_2$) is that $t_x = 0$. Imperfections in the loading are represented by small values of the parameter t_x. Of course, $t_y = \partial V/\partial y \neq 0$ as long as there is any stress on the sheet.

All odd order derivatives with respect to x of V at $x = 0$ are zero, because $V(x, y) = V(-x, y)$. As a result of the isotropic symmetry of φ,

$$V(x, y) = \varphi(x + y, y - x) = \varphi(y - x, y + x) = V(-x, y).$$

Again, in the following, the subscripts involving x and y on the function V or involving 1 and 2 on the function φ refer to partial derivatives, so that $V_x = \partial V/\partial x$, $\varphi_{11} = \partial\varphi^2/\partial\lambda_1^2$, etc. By the chain rule,

$$T_1 = \varphi_1 = (V_x + V_y)/2 \tag{30}$$

$$T_2 = \varphi_2 = (-V_x + V_y)/2. \tag{31}$$

By an additional application of the chain rule,

$$\begin{aligned} \varphi_{11} &= (V_{xx} + 2V_{xy} + V_{yy})/4 \\ \varphi_{22} &= (V_{xx} - 2V_{xy} + V_{yy})/4 \\ \varphi_{12} &= (-V_{xx} + V_{yy})/4. \end{aligned} \tag{32}$$

Therefore

$$\varphi_{11}\varphi_{22} - \varphi_{12}^2 = (V_{xx}V_{yy} - V_{xy}^2)/4. \tag{33}$$

The change in coordinates is a rotation and a uniform stretch; it preserves the positive definiteness of the function.

Lemma 16 *The Hessian of the function φ is positive definite if and only if the Hessian of V is positive definite.*

Proof Equation (33) shows that $\varphi_{11}\varphi_{22} - \varphi_{12}^2 > 0$ if and only if $V_{xx}V_{yy} - V_{xy}^2 > 0$.

Assume that $\varphi_{11} > 0$ and that $\varphi_{11}\varphi_{22} - \varphi_{12}^2 > 0$ and show that V_{xx} must be positive. By the chain rule and substituting $\varphi_{22} > \varphi_{12}^2/\varphi_{11}$,

$$
\begin{aligned}
V_{xx} &= \varphi_{11} - 2\varphi_{12} + \varphi_{22} \\
&> \varphi_{11} - 2\varphi_{12} + \varphi_{12}^2/\varphi_{11} \\
&= \frac{1}{\varphi_{11}}(\varphi_{11}^2 - 2\varphi_{12}\varphi_{11} + \varphi_{12}^2) \\
&= \frac{1}{\varphi_{11}}(\varphi_{11} - \varphi_{12})^2 > 0.
\end{aligned}
$$

A similar calculation verifies the converse that $V_{xx} > 0$ implies $\varphi_{11} > 0$.

The stability conditions for the equilibria defined by φ easily translate into conditions on the critical points, $x = 0$, of the energy function, φ^*, defined by Eq. (28) since $\varphi^*_{xx} = V_{xx}$, $\varphi^*_{yy} = V_{yy}$ and $\varphi^*_{xy} = V_{xy}$. The symmetry of the strain energy function, $\varphi(\lambda_1, \lambda_2) = \varphi(\lambda_2, \lambda_1)$, implies that whenever $\lambda_1 = \lambda_2$, it must be true that $\varphi_{11} = \varphi_{22}$. But by Eq. (32),

$$V_{xy} = \varphi_{11} - \varphi_{22}.$$

Therefore, when $x = 0$,

$$V_{xy}(0, y) = 0, \tag{34}$$

for all y.

The Hessian of φ at $\lambda_1 = \lambda_2$ has eigenvalues $\varphi_{11} - \varphi_{12}$ and $\varphi_{11} + \varphi_{12}$. The first corresponds to $V_{xx}/2$ and the second to $V_{yy}/2$ by Eq. (32) because $V_{xx} = 2(\varphi_{11} - \varphi_{12})$ and $V_{yy} = 2(\varphi_{11} + \varphi_{12})$ at $x = 0$. Ogden's condition $\varphi_{11} + \varphi_{12} > 0$ is equivalent to assuming that $V_{yy} > 0$. The Hessian of φ^* is singular if $V_{xx}V_{yy} = 0$ since $V_{xy} = 0$. In the case that $V_{xx} = 0$ and $V_{yy} > 0$, the Hessian has rank one for all $x = 0$ and y arbitrary. The classical constitutive models for rubber all satisfy $V_{yy} > 0$ for all y when $x = 0$. For energy functions with the proper symmetry, when $\varphi^*_{yy} = V_{yy} > 0$ and $\varphi^*_{xy} = V_{xy} = 0$, Ogden's conditions on the fundamental path, $x = 0$, are equivalent, by Lemma (16), to the statement that an equilibrium is stable if $\varphi^*_{xx} = V_{xx} > 0$, unstable if $\varphi^*_{xx} = V_{xx} < 0$, and a singularity if $\varphi^*_{xx} = V_{xx} = 0$.

6.1 The Liapunov-Schmidt Reduction for the Equilibria

The equilibrium states defined by V are in one-to-one correspondence with the critical points of $\varphi^*(x, y, t_x, t_y) = V(x, y) - t_x x - t_y y$. The equilibrium conditions, $\varphi^*_x = 0$ and $\varphi^*_y = 0$, form a system of two equations which define a map $\Phi : \Re^2 \times \Re^2 \to \Re^2$ whose zeros are the equilibria,

$$\Phi(x, y, t_x, t_y) = \big(\varphi^*_x(x, y, t_x, t_y), \varphi^*_y(x, y, t_x, t_y)\big).$$

The Jacobian of Φ with respect to x and y is the Hessian of φ^*. If the Hessian of φ^* has rank 2 at an equilibrium state, the state can be determined locally by solving the system using the implicit function theorem. The Hessian of φ^* is identical to that of V because of the linearity of the additional terms in φ^*.

Singularities are investigated when $\lambda_1 = \lambda_2$ so that $x = 0$. The analysis is initially restricted to the case, $T_1 = T_2$, i.e. $t_x = 0$. The Hessian is forced to have at least rank one by requiring that $V_{yy} > 0$ at $x = 0$. The assumed isotropic symmetry of φ implies that $V_{xy} = 0$ and so the Hessian, H, of φ^* is diagonal with eigenvalues V_{xx} and V_{yy} when $x = 0$. A singularity for φ^* under the assumptions occurs at a point $(0, y_c, t_c)$ where $V_{xx} = 0$. In a neighborhood of a rank one singularity, the system, $\varphi^*_x = 0$ and $\varphi^*_y = 0$, Eq. (29), has a Liapunov-Schmidt reduction $g(x, t_y)$ whose zeros are in one-to-one correspondence with the zeros of the system.

To simplify notation, set $t_y = t$. Because $V_{yy} \neq 0$, the implicit function theorem ensures that the relation $V_y(x, y) - t = 0$ can be solved for y in terms of x and t to give on a neighborhood of $(0, y_c, t_c)$,

$$y = f(x, t). \tag{35}$$

The function, $f(x, t)$, is used to eliminate y from $\varphi^*_x(x, y, t) = 0$ to define a relation involving one state variable,

$$g(x, t) = V_x(x, f(x, t), t) = 0.$$

There is a one-to-one correspondence between the equilibria of φ^* and the zeros of g in the sense that $g(x, t) = 0$ iff $\varphi^*_x(x, f(x, t), t) = 0$ and $\varphi^*_y(x, f(x, t), t) = 0$.

The singularity of φ^* is classified by thinking of the function $g(x, t)$ as the first derivative of a function $G(x, t)$ with respect to x. In this case, since $t_x = 0$

$$G(x, t) = \varphi^*(x, f(x, t), t) \tag{36}$$

To verify this, the first derivative is $G_x = \varphi^*_x + \varphi^*_y f_x = V_x + (V_y - t) f_x = V_x$ when $y = f(x, t)$, the solution to $V_y - t = 0$ is substituted. There is then a one-to-one correspondence between the equilibria of G and φ^*.

To compare the stability behavior of corresponding equilibria, the second derivative of G is needed. By definition,

$$G_x(x, t) = V_x\big(x, f(x, t)\big).$$

By the chain rule,

$$G_{xx}(x, t) = V_{xx} + V_{xy} f_x. \tag{37}$$

The key to using the stability behavior of the equilibria of G to study those of φ^* is the following fact.

Lemma 17 *Corresponding equilibria of G and of φ^* at which $V_{yy} > 0$ have the same stability behavior.*

Proof By the definition of φ^*, $\varphi^*_{xx}\varphi^*_{yy} - \varphi^{*2}_{xy} = V_{xx}V_{yy} - V^2_{xy}$. Also

$$G_{xx}(x,t) = V_{xx} + V_{xy}f_x = V_{xx} - \frac{V_{xy}V_{yx}}{V_{yy}} = \frac{1}{V_{yy}}(V_{xx}V_{yy} - V^2_{xy}). \tag{38}$$

Therefore, G_{xx} and $\varphi^*_{xx}\varphi^*_{yy} - \varphi^{*2}_{xy}$ have the same sign. The critical point of φ^* is stable iff $G_{xx} > 0$ and is unstable iff $G_{xx} < 0$.

In other situations, extra conditions are required to determine how the Liapunov-Schmidt reduction affects the stability of the corresponding equilibria. Any bifurcation point when $x = 0$ occurs for both G and φ^* at the same value of the parameter t where $V_{xx}(0, y) = 0$, because the second derivative $G_{xx}(0,t) = V_{xx}(0, f(0,t)) = \varphi^*_{xx}(0, f(0,t), t)$ when $x = 0$.

The stability behavior of G near a degenerate bifurcation can be classified by elementary catastrophe theory since G is a function of the single state variable, x. The type of the degenerate singularity is determined from the Taylor series expansion of $G(x,t)$ about $x = 0$, the singularity. Therefore, higher order derivatives of $G(x,t)$ with respect to x evaluated at $x = 0$ are needed.

The symmetry of V with respect to x produces $V_y(x,y) = V_y(-x,y)$ so that $f(x,t) = f(-x,t)$. The function G has the symmetry $G(x,t) = G(-x,t)$ because $f(x,t) = f(-x,t)$ and $\varphi^*(x,y,t) = \varphi^*(-x,y,t)$ when $t_x = 0$. This implies that all odd order derivatives of G with respect to x are zero at $x = 0$.

By the chain rule, applied to $G_x(x,t) = V_x(x, f(x,t), t)$,

$$G_{xx}(x,t) = V_{xx} + V_{xy}f_x; \tag{39}$$

$$G_{xxx}(x,t) = V_{xxx} + 2V_{xxy}f_x + V_{xyy}f_x^2 + V_{xy}f_{xx}; \tag{40}$$

$$\begin{aligned} G_{xxxx}(x,t) = {} & V_{xxxx} + 3V_{xxxy}f_x + 3V_{xxyy}f_x^2 + 3V_{xxy}f_{xx} \\ & + V_{xyyy}f_x^3 + 3V_{xyy}f_xf_{xx} + V_{xy}f_{xxx}; \end{aligned} \tag{41}$$

$$\begin{aligned} G_{xxxxx}(x,t) = {} & V_{xxxxx} + 4V_{xxxxy}f_x + 6V_{xxxyy}f_x^2 + 6V_{xxxy}f_{xx} \\ & + 4V_{xxyyy}f_x^3 + 12V_{xxyy}f_xf_{xx} + 4V_{xxy}f_x^3 + V_{xyyyy}f_x^4 \\ & + 6V_{xyyy}f_x^2f_{xx} + 4V_{xyy}f_xf_{xxx} + 3V_{xyy}f_{xx}^2 + V_{xy}f_{xxxx}; \end{aligned} \tag{42}$$

$$\begin{aligned} G_{xxxxxx}(x,t) = {} & V_{xxxxxx} + 5V_{xxxxxy}f_x + 10V_{xxxxyy}f_x^2 + 10V_{xxxxy}f_{xx} \\ & + 10V_{xxxyyy}f_x^3 + 30V_{xxxyy}f_xf_{xx} + 10V_{xxxy}f_{xxx} \\ & + 5V_{xxyyyy}f_x^4 + 30V_{xxyyy}f_x^2f_{xx} + 20V_{xxyy}f_xf_{xxx} \\ & + 15V_{xxyy}f_{xx}^2 + 5V_{xxy}f_{xxxx} + V_{xyyyyy}f_x^5 + 10V_{xyyyy}f_x^3f_{xx} \\ & + 10V_{xyyy}f_x^2f_{xxx} + 15V_{xyyy}f_xf_{xx}^2 + 5V_{xyy}f_xf_{xxxx} \\ & + 10V_{xyy}f_{xx}f_{xxx} + V_{xy}f_{xxxxx}. \end{aligned} \tag{43}$$

Now $V_y(x, f(x,t), t) = t$ implies that $V_{xy} + V_{yy}f_x = 0$ so that

$$f_x = -\frac{V_{xy}}{V_{yy}}. \tag{44}$$

Therefore, $f_x = 0$ at $x = 0$ because $V_{xy} = 0$ at $x = 0$, by Eq. (34). Likewise, differentiating $V_{xy} + V_{yy} f_x = 0$ repeatedly yields

$$\begin{aligned}
f_{xx} &= -\frac{1}{V_{yy}}(V_{xxy} + 2V_{xyy} f_x + V_{yyy} f_x^2);\\
f_{xxx} &= -\frac{1}{V_{yy}}(V_{xxxy} + 3V_{xxyy} f_x + 3V_{xyyy} f_x^2 + V_{yyyy} f_x^3 + 3V_{xyy} f_{xx}\\
&\quad +3V_{yyy} f_x f_{xx});\\
f_{xxxx} &= -\frac{1}{V_{yy}}(V_{xxxxy} + 4V_{xxxyy} f_x + 6V_{xxyyy} f_x^2 + 6V_{xxyy} f_{xx} + 4V_{xyyyy} f_x^3\\
&\quad +12V_{xyyy} f_x f_{xx} + 4V_{xyy} f_{xxx} + V_{yyyyy} f_x^4 + 6V_{yyyy} f_x^2 f_{xx}\\
&\quad +4V_{yyy} f_x f_{xxx} + 3V_{yyy} f_{xx}^2).
\end{aligned} \tag{45}$$

These expressions can be substituted into the expressions for the derivatives of G to write the derivatives in terms of powers of f_x.

The isotropic symmetry implies $f_x = 0$ at $x = 0$ so that the fourth derivative of $G(x, t)$ at $x = 0$ becomes, by Eqs. (41) and (45),

$$G_{xxxx}(0, t) = V_{xxxx} - 3\frac{V_{xxy}^2}{V_{yy}}. \tag{46}$$

Since $V_{xyy} = 0$ and $V_{xxxy} = 0$ at $x = 0$, the sixth derivative of $G(x, t)$ at $x = 0$ is

$$G_{xxxxxx}(0, t) = V_{xxxxxx} - 15\frac{V_{yyy} V_{xxy}^3}{V_{yy}^3} + 45\frac{V_{xxy}^2 V_{xxyy}}{V_{yy}^2} - 15\frac{V_{xxy} V_{xxxxy}}{V_{yy}}, \tag{47}$$

by Eqs. (43) and (45).

6.2 Bifurcation with Respect to the Load Parameter

The equilibrium set of the potential energy function, φ^*, and the stability of these equilibria can be predicted from knowledge of the types of bifurcation behavior possible in families of functions of one state variable and the constructions of the previous sections. Any singularity at $x = 0$ leading to a symmetric bifurcation must be degenerate of order at least four. The following theorem characterizes the equilibria stability of φ near the simplest degenerate singularity.

Theorem 18 *Let $\varphi(\lambda_1, \lambda_2)$ be a function with isotropic symmetry and with $\varphi_i = T_i$ so that $V(x, y)$ is independent of any parameters. Assume that there is a value*

$\lambda = \lambda_1 = \lambda_2$, where $\varphi_{11} + \varphi_{12} > 0$, such that $\varphi_{11}(\lambda, \lambda) - \varphi_{12}(\lambda, \lambda) = 0$. Then the transformed family, $\varphi^*(x, y, t_x, t_y) = V(x, y) - t_x x - t_y y$, has a rank one degenerate singularity at $(0, \lambda, 0, t_y)$ for some value $t_y = 2T_1 = 2T_2$.

If V_{xx} is decreasing in y, i.e. $V_{xxy} < 0$, if $V_{xx}(0, y_c) = 0$ for a value of y_c and if at the singularity $(0, y_c)$

$$V_{xxxx} - 3\frac{V_{xxy}^2}{V_{yy}} > 0, \tag{48}$$

then the equilibrium states of φ^* form a distorted pitchfork bifurcation in x-y-t_y – space and the projection in x-t_y – space is the pitchfork, the equilibria of the symmetric cusp catastrophe. If the expression of Eq. (48) is negative, the singularity is a symmetric dual cusp. When it is zero, the singularity is degenerate of order at least six.

Proof After the coordinate transformation of Eq. (27), $\varphi^*(x, y, t_x, t_y) = V(x, y) - t_x x - t_y y$, and $t_x = 0$ since $T_1 = T_2$. Also at $x = 0$, $\varphi_{11} + \varphi_{12} > 0$ implies that $V_{yy} > 0$, and $V_{xx} = 0$ iff $\varphi_{11} - \varphi_{12} = 0$. Put $t = t_y$. The Liapunov-Schmidt construction of the previous section applies.

If $f(x, t)$ is the function defined from $\varphi^*_y = 0$ (Eq. (35)), the mapping $(x, t) \to (x, f(x, t), t)$ defines a one-to-one relation between the equilibria of the function, $G(x, t)$, associated with the Liapunov-Schmidt reduction of φ^* and the equilibria of φ^* near the singularity. As a consequence of Lemma 17, corresponding nonsingular equilibria have the same stability behavior.

The stability behavior of the equilibria of φ^* is then determined by the catastrophe type of the function, $G(x, t)$. All the odd order derivatives of $G(x, t) = \varphi^*(x, f(x, t), t)$ with respect to x at $x = 0$ are zero because $G(x, t) = G(-x, t)$. The Taylor series expansion of G is $G(x, t) = G(0, t) + \sum_q c_{2q}(t)x^{2q}$. A singularity exists if the coefficient of x^2 is zero for some value t_c of t.

The condition, $V_{xxy} < 0$, guarantees that as t increases the equilibrium behavior changes from stable to unstable at the bifurcation point, because at $x = 0$,

$$G_{xxt} = V_{xxy}\frac{dy}{dt}. \tag{49}$$

When $x = 0$, G_{xxt} and V_{xxy} have the same sign because, as a consequence of $V_{yy} > 0$, t is an increasing function of y. Therefore $G_{xxt} < 0$.

Condition (48) determines the sign of the coefficient of x^4 and therefore when the catastrophe is a cusp, dual cusp, or degenerate singularity of higher order.

This theorem is a local result, valid in a neighborhood of the degenerate singularity. More general results involving higher order degenerate singularities are given in Haslach (1999). Furthermore, the family, $G(x, t)$, is stable under perturbations which preserve the isotropic symmetry, but not under arbitrary perturbations. The expression Eq. (48) follows a pattern similar to Supple's criterion for functions with additional symmetry and linearly dependent on a load parameter.

To apply the criterion, use a computer algebra program to transform $\varphi(\lambda_1, \lambda_2)$ to $V(x, y)$. Then, in the program, write the Taylor series of $V(x, y)$ about $x = 0$. Determine the values of y which make the coefficient of x^2 zero to locate any bifurcation points $(0, y_c)$. Obtain the corresponding value of the control variable, t, from $t = V_y(0, y_c)$. The criterion of Eq. (48) determines if the degenerate singularity is of fourth order or of higher order.

This criterion saves the work of writing the projected function in closed form; a task which may be impossible if the relation $\varphi^*_y = 0$ cannot be solved explicitly for the function $y = f(x, t)$. One can determine the behavior in a neighborhood of a degenerate singularity directly from the coefficients of the desired Taylor series by evaluating the derivatives of the original functions at the bifurcation values of $x = 0$ and y_c.

6.2.1 Bifurcation with Respect to a Load Parameter and Involving Multiple Parameters

The family of functions considered in the previous section was indexed by the load parameter, t. Such families may arise from constitutive models which depend on additional empirical or material parameters or from models of structures involving dimensional parameters. The presence and variation of the additional parameters can affect the behavior of the degenerate singularities of the generalized energy function associated with the family, $\varphi(\lambda_1, \lambda_2, a_1, \ldots, a_m)$.

Theorem 19 *Let $\varphi(\lambda_1, \lambda_2, a_1, \ldots, a_m)$ be a function parametrized by $a_1, \ldots, a_m$, with conjugate parameters $\varphi_i = T_i$ and with isotropic symmetry. Assume that of those points, (λ_1, λ_2) where $\varphi_1 = \varphi_2$ and where $\varphi_{11} + \varphi_{12} > 0$, there is a value $\lambda = \lambda_1 = \lambda_2$ such that $\varphi_{11}(\lambda, \lambda) - \varphi_{12}(\lambda, \lambda) = 0$. Then, the transformed family, $\varphi^*(x, y, t_x, t_y, a_1, \ldots, a_m) = V(x, y, a_1, \ldots, a_m) - t_x x - t_y y$, has a rank one singularity at $(0, \lambda, 0, t_y)$ for some value $t_y = 2T_1 = 2T_2$. The equilibrium states of φ^* in a neighborhood of the singularity at $x = 0$ in $x - y - t_y - a_1 - \ldots - a_m -$ space are the image of a stability preserving one-to-one map of the equilibrium set in $x - t_y - a_1 - \ldots - a_m -$ space of an even order symmetric catastrophe family of functions of one variable.*

Proof The beginning of the proof of Theorem (18) is repeated to obtain the function, $G(x, t, a_1, \ldots, a_m)$, associated with the Liapunov-Schmidt reduction of φ^* and the equilibria of φ^* near the singularity. The stability behavior of the equilibria of φ^* is then determined by the catastrophe type of the function, $G(x, t, a_1, \ldots, a_m)$. All the odd order derivatives of $G(x, t, a_1, \ldots, a_m) = \varphi^*(x, f(x, t), t, a_1, \ldots, a_m)$ with respect to x at $x = 0$ are zero because $G(x, t, a_1, \ldots, a_m) = G(-x, t, a_1, \ldots, a_m)$. The Taylor series expansion of G is $G(x, t, a_1, \ldots, a_m) = G(0, t, a_1, \ldots, a_m) + \sum_q c_{2q}(t)x^{2q}$. A singularity exists if the coefficient of x^2 is zero for some value t_c of t. This singularity is degenerate of order $2n$ if the coefficients of x^{2q} are zero for $q = 2, \ldots, n - 1$ at t_c and $c_{2n}(t_c) \neq 0$.

This is a local result, valid in a neighborhood of the degenerate singularity. Furthermore, the family, $G(x, t, a_1, \ldots, a_m)$, is stable under perturbations which preserve the isotropic symmetry, but not under arbitrary perturbations.

The number of parameters which can qualitatively affect the response increases with the order of the singularity. The simplest even order degenerate singularity when the family, φ^*, depends on no parameters, a_i, other than the conjugate parameters t_y and possibly t_x, is a subfamily of the cusp or dual cusp catastrophe. Alternatively, this family can be produced if any such parameters a_i which exist are held fixed.

6.2.2 Bifurcations with Respect to One Additional Parameter

A more specific result may be obtained if the family of energies, φ, depends linearly on a single additional parameter, a. The function, φ, is transformed to $V(x, y, a)$. The equilibrium surface of the family φ^* is locally two-dimensional in $x - y - t - a$ space. If the zero of the coefficient of x^2 depends on the parameter, a, then each member of the family may have a different singularity point, as a is varied.

Theorem 20 *Under the hypotheses of Theorem 19, if $V(x, y, a)$ depends on a single parameter a, if $V_{xx}(0, y_c, a_c) = 0$ for a value of y_c and of a_c, if at the singularity $(0, y_c, a_c)$*

$$V_{xxxx} - 3\frac{V_{xxy}^2}{V_{yy}} = 0, \tag{50}$$

and if at the singularity $(0, y_c, a_c)$

$$V_{xxxxxx} - 15\frac{V_{yyy}V_{xxy}^3}{V_{yy}^3} + 45\frac{V_{xxy}^2 V_{xxyy}}{V_{yy}^2} - 15\frac{V_{xxy}V_{xxxxy}}{V_{yy}} > 0, \tag{51}$$

then the equilibrium set of φ^* is the image of the equilibria of a symmetric butterfly in $x - t - a$ – space.

Proof The Taylor series of the reduced function, $G(x, t, a)$, contains only even order terms by Theorem 19. Equation (50) implies that the coefficient of the fourth order term is zero and Eq. (51) implies that the coefficient of the sixth order term is positive. Then by the standard form Eq. (18), the equilibria of $G(x, t, a)$ are those of a symmetric butterfly catastrophe. The function $(x, t, a) \to \big(x, f(x, t, a), t, a\big)$ is one-to-one from the equilibria of $G(x, t, a)$ to the equilibria of φ^* in a neighborhood of the butterfly degenerate singularity point.

Example Consider the constitutive model for the strain energy, φ, which depends on an additional parameter, a.

$$\varphi(\lambda_1, \lambda_2, a) = 2 + \frac{1}{16}(\lambda_1 - \lambda_2)^4 - \frac{3}{2}(\lambda_1 + \lambda_2) + \frac{1}{4}(\lambda_1 + \lambda_2)^2 + \frac{1}{4}\left[(\lambda_1 - \lambda_2)^2\left(2 - \frac{1}{2}(\lambda_1 + \lambda_2)\right)a\right]. \tag{52}$$

It is isotropically symmetric since $\varphi(\lambda_1, \lambda_2, a) = \varphi(\lambda_2, \lambda_1, a)$. Assume $t_x = 0$ and $t_y = t$. Apply Eq. (25) to obtain the family of energies, $V(x, y, a)$, indexed by the parameter, a,

$$V(x, y, a) = 2 - 3y + y^2 + a(2 - y)x^2 + x^4. \tag{53}$$

The structure of the equilibrium surface of the generalized energy family

$$\varphi^*(x, y, t, a) = V(x, y, a) - ty = 2 - 3y + y^2 + a(2 - y)x^2 + x^4 - ty \tag{54}$$

can be obtained from Theorem 18, for fixed a. V is already in the form of the Taylor series about $x = 0$. $V_{xx} = 2a(2 - y)$ implies that a singularity exists at $y = 2$ for arbitrary $a \neq 0$. The value of t at the singularity of $x = 0$ and $y = 2$ is easily calculated from $V_y = t$ to be $t = 1$. Another singularity occurs at $x = 0$ when $a = 0$ and y is arbitrary. Since the third derivative of G is zero at $x = 0$, both these singularities are degenerate. A calculation of the fourth derivative of G determines whether or not the degenerate singularity is a cusp, a dual cusp, or higher order degenerate, including flat.

$$V_{xxxx} - 3\frac{V_{xxy}^2}{V_{yy}} = 24 - 3\frac{(-2a)^2}{2} = 24 - 6a^2. \tag{55}$$

The fourth derivative, G_{xxxx} (Eq. (55)), is positive if $|a| < 2$ and negative if $|a| > 2$. Furthermore, $V_{yy} = 2$ is positive for all (x, y). Therefore, G_{xxt} has the same sign as the derivative $V_{xxy}(0, y) = -2a$, which changes sign as a does. All four types of pitchforks illustrated in Fig. 8.2 occur as a is varied. When $a < -2$, the equilibria are of type IV; when $-2 < a < 0$, of type III; when $0 < a < 2$, of type I from the cusp; when $2 < a$, of type II from the dual cusp. At $|a| = 2$ and $t = 1$, the singularity is either degenerate of higher order or the reduction is flat. The reduction has equilibria at all x if $|a| = 2$.

Alternatively, the analysis of the Liapunov-Schmidt reduction, $G(x, t)$, can be carried out explicitly for this example by writing the expression for $G(x, t)$ in closed form, for fixed a. The system is reduced by substituting in φ^* the expression $y = f(x, t) = (t + 3 + ax^2)/2$ obtained from $\varphi^*_y = 0$ to produce the family indexed by the two parameters, t and a,

$$G(x, t, a) = -\frac{1}{4} - \frac{3}{2}t - \frac{1}{4}t^2 + \frac{a}{2}(1 - t)x^2 + \left(1 - \frac{a^2}{4}\right)x^4. \tag{56}$$

The fact that the bifurcation point occurs at $x = 0$ when $t = 1$ or when $a = 0$ is easily verified by setting the coefficient of x^2 to be zero. The bifurcation value of t does not change for any $a \neq 0$. The coefficient of x^4 matches the value calculated from $G_{xxxx}(0, t)/24$ using Eq. (55). Since $G_{xxt} = -a$ changes signs when a does, the series of pitchfork types examined in the previous paragraph is reproduced. When $t = 1$ and $a = \pm 2$, $G(x, t, a) = -2$ and is flat because the non-constant terms of the Taylor series are zero.

The family $G(x, t, a)$ is a symmetric subfamily of a cusp catastrophe with respect to the control parameter, a, and state variable, x, when $t < 1$ because $G_{xxa} = 1 - t$ is negative and the coefficient of x^4 is positive. The cusp singularity at $x = 0$ occurs when $a = 0$. The equilibria are the pitchfork of Type IV (Fig. 8.2) when $t > 1$.

The stability results for the equilibria of φ^* implied by Theorem 18 in this simple example are verified explicitly from φ^* using its 2×2 Hessian. The function φ^* has equilibria at those points in $x - y - t$ – space satisfying

$$\varphi_x^* = 2a(2 - y)x + 4x^3 = 0 \tag{57}$$

$$\varphi_y^* = -3 + 2y - ax^2 - t = 0. \tag{58}$$

Since there are two equilibrium equations, the equilibria form a locally one-dimensional subset of $\Re^3$, $x - y - t$ – space. One solution for given t is the (x, y) such that $x = 0$ and $t = -3+2y$. The other points correspond to $x \neq 0$. Elimination of the variable x from these two equations gives the load parameter, t, as a function of y at equilibrium. If $x \neq 0$, the first equation becomes $2a(2-y)+4x^2 = 0$. Solve for x^2 and substitute into the second equation to obtain

$$t(y) = -3 + a^2 + \left(2 - \frac{1}{2}a^2\right) y. \tag{59}$$

Therefore, on the equilibrium states, $t(y)$ is increasing when $|a| < 2$, constant if $|a| = 2$, and decreasing if $|a| > 2$.

The second derivatives of φ^* are

$$\varphi_{xx}^* = 2a(2 - y) + 12x^2 \tag{60}$$

$$\varphi_{xy}^* = -2ax \tag{61}$$

$$\varphi_{yy}^* = 2. \tag{62}$$

The Hessian has a singularity if $\varphi_{xx}^*\varphi_{yy}^* - \varphi_{xy}^{*2} = 4a(2 - y) + 24x^2 - 4a^2x^2 = 0$ at some equilibrium point. For the equilibria with $x = 0$ forming the middle stem of the pitchfork, the singularity is at $y = 2$ and so $t = 1$ or at $a = 0$ in which case φ^* has no x^2 term. Along this stem, the equilibria change from stable to unstable as y increases through 2 when a is positive and vice versa when a is negative. The non-zero equilibrium value of x, defining the outer prongs of the pitchfork of equilibria for fixed a is $x^2 = -a(2-y)/2$. When $a > 0$, then $y > 2$ and conversely when $a < 0$, then $y < 2$. Substitution produces the stability criterion

$$\varphi^*_{xx}\varphi^*_{yy} - \varphi^{*2}_{xy} = 4a(2-y) + 24x^2 - 4a^2x^2 = (-8a + 2a^3)(2-y). \quad (63)$$

This expression is zero for $x \neq 0$, when $a = 2$ and y is arbitrary. Since φ^*_{yy} is always positive, the Hessian is positive definite for $|a| < 2$ and the equilibria $x \neq 0$ are stable. The equilibria $x \neq 0$ are unstable if $|a| > 2$. The stability of the equilibria on the outer prongs changes at $a = \pm 2$.

The polynomial, $G(x, t, a)$, is not equivalent to the even order subfamily of the butterfly catastrophe, Eq. (18),

$$G'(x, t, a) = -\frac{1}{4} - \frac{3}{2}t - \frac{1}{4}t^2 + \frac{a}{2}(1-t)x^2 + \left(1 - \frac{a^2}{4}\right)x^4 + x^6,$$

because all derivatives of order higher than four are all zero at $x = 0$.

7 Rubber Constitutive Models Without a Bifurcation

Proposed isothermal, incompressible, elastic, isotropically symmetric, constitutive models for rubber-like materials can be validated by whether or not they generate the experimentally observed non-unique equilibrium states of a rubber sheet under equal in-plane biaxial tension. Any model which is a function of only the first strain invariant, I_B, cannot produce a bifurcation in the deformation response of a sheet by Proposition (15). This eliminates the classical neo-Hookean and the Arruda-Boyce models, which are both based on statistical chain models for the rubber molecules. These are presented for completeness and to show that these models, as required, have stable equilibria at $x = 0$ because $V_{xx} > 0$. Two other models which fail to produce the required bifurcations are described because other models, producing bifurcations, have been built from these models.

7.1 The Neo-Hookean Model

A neo-Hookean material is one for which $\varphi_I = \partial\varphi/\partial I_B = c_1$, a positive constant, and $\varphi_{II} = \partial\varphi/\partial II_B = 0$.

$$\begin{aligned}\varphi(\lambda_1, \lambda_2) &= c_1(I_B - 3) \\ &= c_1(\lambda_1^2 + \lambda_2^2 + 1/\lambda_1^2\lambda_2^2 - 3).\end{aligned} \quad (64)$$

The function, $V(x, y)$, is after making the transformation of Eq. (27),

$$V(x, y) = c_1[(x+y)^2 + (y-x)^2 + (y^2 - x^2)^{-2} - 3]. \quad (65)$$

The neo-Hookean energy function for a sheet has the equilibrium positions for equal loads, $t_x = 0$, given by points in x-y-t–space with $x = 0$ and y, t satisfying

$V_y(0, y) = t$. Both $V(x, y)$ and its derivatives are undefined at $x = y$ for each value of y, but these points are not in the physically acceptable region $y > x$. Also

$$V_{yy} = 4 + 24y^2(-x^2 + y^2)^{-4} - 4(-x^2 + y^2)^{-3} \tag{66}$$

is positive if $x = 0$. Therefore, by Lemma (17), for each y, the equilibrium $x = 0$ $(\lambda_1 = \lambda_2)$ is stable because

$$V_{xx} = c_1[4(y^2 - x^2)^{-4}][(y^2 - x^2)^4 + y^2 - 5x^2] \tag{67}$$

is positive if $x = 0$.

7.2 The Arruda-Boyce Model

Arruda and Boyce (1993) observed that few phenomenological models have a direct connection to the physical structure of the material. They developed a rubber model which is based on an eight chain network. The model has the advantage that one experiment is sufficient to determine the required empirical coefficients.

They proposed a thermoelastic model in terms of the stresses,

$$\sigma_1 - \sigma_2 = \frac{Nnk\theta}{3}\mathcal{L}^{-1}\left(\frac{\Gamma}{\sqrt{N}}\right)\frac{\lambda_1^2 - \lambda_2^2}{\Gamma}, \tag{68}$$

where $\Gamma = (\lambda_1^2 + \lambda_2^2 + \lambda_3^2)^{1/2}/\sqrt{3}$. The Langevin function $\mathcal{L}(x) = \coth(x) - x^{-1}$, k is the Boltzmann constant, N is a measure of the number of links in the long chain molecule, and n is the chain density. The temperature is θ. This expression is integrated to obtain the strain energy,

$$\varphi = nk\theta\left[\frac{1}{2}(I_B - 3) + \frac{1}{20N}(I_B^2 - 9) + \frac{11}{1050N^2}(I_B^3 - 27) + \frac{19}{7000N^3}(I_B^4 - 81) + \ldots\right]. \tag{69}$$

After the coordinate change of Eq. (27), using the fact that

$$I_B = 2(y^2 + x^2) + (y^2 - x^2)^{-2} - 3,$$

a calculation shows that $V_{yy} > 0$ for $y > 1$ and $x = 0$ and $V_{xy} = 0$ when $x = 0$. However, the Arruda - Boyce model has no bifurcation when applied to a sheet. The only equilibria are at $x = 0$ and are stable because the truncated term

$$V_{xx} = nk\theta\{0.5(2+2y^{-6}) + [(2+2y^{-6})(y^{-4}+2y^2)]/10N + [11(2+2y^{-6})(y^{-4}+2y^2)^2]/(350N^2) + [19(2+2y^{-6})(y^{-4}+2y^2)^3]/(1750N^3)\} \quad (70)$$

is positive at $x = 0$ for all $y > 1$. The equilibrium behavior of the model under equal loads is qualitatively equivalent to the neo-Hookean.

7.3 The Valanis-Landel Hypothesis and Model

For isotropic materials, Valanis and Landel (1967) postulated a function $w(\lambda)$ such that

$$\varphi(\lambda_1, \lambda_2, \lambda_3) = w(\lambda_1) + w(\lambda_2) + w(\lambda_3). \quad (71)$$

The stress per unit of deformed area in terms of derivatives of the function $w(\lambda)$ is written as a difference,

$$T_{11} - T_{22} = \lambda_1 w'(\lambda_1) - \lambda_2 w'(\lambda_2), \quad (72)$$

where $w'(\lambda) = dw/d\lambda$. The function $w(\lambda)$ is determined from a shear test by $w'(\lambda_1) = (T_{11} - T_{22})/\lambda_1$.

From pure shear tests with $\lambda_2 = 1$, Valanis and Landel, (1967, Eq. 53) produced a logarithmic model, which Treloar, (1975, p. 240) later modified by adding c/λ. The principal stress per unit reference area is

$$T = \frac{dw(\lambda)}{d\lambda} = 2\mu \ln(\lambda) + c/\lambda,$$

for $1 \le \lambda \le 2.5$ and $\mu = 4.1$ kg/cm^3. Integration yields up to a constant

$$\varphi(\lambda_1, \lambda_2, \lambda_3) = \sum_{i=1}^{3} \left[2\mu[\lambda_i \ln(\lambda_i) - \lambda_i] + c \ln(\lambda_i)\right].$$

For an incompressible material, since $\lambda_1\lambda_2\lambda_3 = 1$, $\sum_{i=1}^{3} c \ln(\lambda_i) = 0$, and c is arbitrary and inconsequential in φ, even though Treloar found it must be non-zero to fit the stress data.

Making the coordinate transformation of Eq. (27),

$$V(x, y) = 2\mu[(x+y)\ln(x+y) + (y-x)\ln(y-x) - (y^2-x^2)^{-1} \ln(y^2-x^2) - 2y - (y^2-x^2)^{-1}]. \quad (73)$$

The equilibrium state at which $t_x = 0$ satisfies $V_y = t$ and

$$V_x = \ln(x+y) - \ln(y-x) - 2x(y^2 - x^2)^{-2}\ln(y^2 - x^2) = 0. \qquad (74)$$

The only solutions to the equilibrium equation have $x = 0$, i.e. $\lambda_1 = \lambda_2$. Again by symmetry, $V_{xy} = 0$. Also

$$V_{yy}/2\mu = 4y^{-4} + 2y^{-1} - 6y^{-4}\ln(y^2)$$
$$V_{xx}/2\mu = 2y^{-1} - 2y^{-4}\ln(y^2)$$

Both are positive for $y \geq 1$, so that this energy function does not exhibit a bifurcation. The equilibrium set of the Valanis-Landel logarithmic model is qualitatively equivalent to the neo-Hookean in tension.

7.4 The Gent-Thomas Model

The Gent-Thomas model (1958) takes the form of the Rivlin-Saunders model (Eq. (21)) with $\partial\varphi/\partial II_B$ proportional to $1/II_B$ to account for the fact that $\partial\varphi/\partial II_B$ was experimentally found to be positive and asymptotic to zero for increasing II_B. Their model is $\varphi(I_B, II_B) = c_1(I_B - 3) + c_2\ln(II_B/3)$. The resulting energy function after the transformation of Eq. (27) and rescaling is

$$\begin{aligned} V(x, y) = {} & K'[2(y^2 + x^2) + (y^2 - x^2)^{-2} - 3] \\ & + \ln[(y+x)^{-2} + (y-x)^{-2} + (y^2 - x^2)^2], \end{aligned} \qquad (75)$$

where K' is a material constant. A numerical solution at equal loads, $t_x = 0$, of $V_x = 0$ where

$$\begin{aligned} V_x = {} & 20K'x[1 + (-x^2 + y^2)^{-3}] + [2(-x+y)^{-3} - 2(x+y)^{-3} - 4x(-x^2 \\ & + y^2)]/[(-x+y)^{-2} + (x+y)^{-2} + (-x^2 + y^2)^2] \end{aligned} \qquad (76)$$

shows this model has only equilibrium states $x = 0$, for a sheet in tension. At $x = 0$, $V_{xy} = 0$ by symmetry and

$$\begin{aligned} V_{yy} = {} & K'(4 + 20y^{-6}) - (-4y^{-3} + 4y^3)^2(2y^{-2} + y^4)^{-2} \\ & + (12y^{-4} + 12y^2)(2y^{-2} + y^4)^{-1} > 0, \end{aligned} \qquad (77)$$

for positive y if $K' \geq 0.166$. Also, there are no bifurcations and the equilibria are all stable since

$$V_{xx} = K'(4 + 4y^{-6}) + (12y^{-4} - 4y^2)(2y^{-2} + y^4)^{-1} > 0, \qquad (78)$$

for all y if $K' \geq 0.235$. Gent and Thomas show that their constant is related to the Mooney-Rivlin constant by $K' = (K + 0.247)/2.18$. Since K is always greater than

0.265 to fit any rubber, there are no bifurcations for a rubber sheet represented by the Gent-Thomas model. It is interesting to note that for smaller moduli, $0.166 \leq K' \leq 0.235$, there are two singularities on the line $x = 0$ near $y = 2.2$. A numerical calculation shows that no disjoint equilibrium paths exist.

8 Rubber Constitutive Models that Produce a Bifurcation

Treloar's (1948) experiments on sheets and Alexander's on balloons (1971) show that equal in-plane biaxial tensile loads on a rubber sheet can produce unequal elastic stable stretches in the direction of the loads for large enough loads. Modeling such behavior requires more than one bifurcation must exist in the family of nonlinear generalized energy functions, $\varphi^*(x, y, t_x, t_y) = V(x, y) - t_x x - t_y y$, for a biaxially loaded sheet as the load is increased. The Mooney-Rivlin model for an in-plane biaxially loaded sheet produces only a single bifurcation point and provides an example in which Theorem 18 is needed because the projected Liapunov-Schmidt reduction function cannot be calculated in closed form. The Alexander model produces two and the Ogden model can produce three bifurcation points when $x = 0$.

8.1 The Mooney-Rivlin Model

The Mooney-Rivlin model for an in-plane biaxially loaded sheet provides an example in which Theorem 18 is needed because the projected Liapunov-Schmidt reduction function cannot be calculated in closed form. A higher order degenerate singularity is obtained if the family is regarded as indexed by the material constant as well as by the equibiaxial load.

A Mooney-Rivlin material is one in which $\varphi_I = \partial\varphi/\partial I_B = c_1$ and $\varphi_{II} = \partial\varphi/\partial II_B = c_2$ are positive constants. The Mooney-Rivlin strain energy for the isothermal equilibrium states of a rubber sheet is, in terms of the in-plane stretches,

$$\varphi(\lambda_1, \lambda_2) = c_1(\lambda_1^2 + \lambda_2^2 + \lambda_1^{-2}\lambda_2^{-2} - 3) + c_2(\lambda_1^{-2} + \lambda_2^{-2} + \lambda_1^2\lambda_2^2 - 3) \quad (79)$$

Since the derivative $\partial^2\varphi/\partial\lambda_1\partial\lambda_2 = 2\mu$, where μ is the shear modulus, when evaluated at $\lambda_1 = \lambda_2 = 1$, it must be true that $2(c_1 + c_2) = \mu$. The two positive material constants, c_1 and c_2 can be reduced to one by rescaling φ to the equivalent form, using $K = c_1/c_2$, so that $K > 0$.

$$\varphi(\lambda_1, \lambda_2) = K(\lambda_1^2 + \lambda_2^2 + \lambda_1^{-2}\lambda_2^{-2} - 3) + \lambda_1^{-2} + \lambda_2^{-2} + \lambda_1^2\lambda_2^2 - 3 \quad (80)$$

The function φ is isotropically symmetric. The associated generalized energy function is,

$$\begin{aligned}\varphi^*(\lambda_1, \lambda_2; T_1, T_2) &= \varphi(\lambda_1, \lambda_2) - \lambda_1 T_1 - \lambda_2 T_2 \\ &= K(\lambda_1^2 + \lambda_2^2 + \lambda_1^{-2}\lambda_2^{-2} - 3) + (\lambda_1^{-2} + \lambda_2^{-2} + \lambda_1^2\lambda_2^2 - 3) \\ &\quad -\lambda_1 T_1 - \lambda_2 T_2, \end{aligned} \tag{81}$$

where λ_1 and λ_2 are the state variables describing the response to changes in the control variables, T_1 and T_2. The transformed family of energy functions has the form

$$\begin{aligned}\varphi^*(x, y, t_x, t_y) &= V(x, y) - t_x x - t_y y \\ &= K\left[2(x^2 + y^2) + (y^2 - x^2)^{-2} - 3\right] + (x + y)^{-2} \\ &\quad +(y - x)^{-2} + (y^2 - x^2)^2 - 3 - x t_x - y t_y. \end{aligned} \tag{82}$$

The equilibrium states are determined by $\varphi_x^* = 0$ and $\varphi_y^* = 0$, so that at equilibrium,

$$V_x = 4x\left\{K\left[1 + (y^2 - x^2)^{-3}\right] + (x^2 + 3y^2)(y^2 - x^2)^{-3} - (y^2 - x^2)\right\} = 0. \tag{83}$$

$$V_y = 4y\left\{K[1 - (y^2 - x^2)^{-3}] - (3x^2 + y^2)(y^2 - x^2)^{-3} + (y^2 - x^2)\right\} = t_y. \tag{84}$$

The expressions will be simplified again by writing $t_y = t$. Clearly, $y = f(x, t)$ of Eq. (35) cannot be obtained in closed form from $V_y = t$ so that $G(x, t)$ (Eq. (36)) also cannot be written in closed form. While $V_x = 0$ when $t_x = 0$, $V_y \neq 0$ as long as there is any stress on the sheet. The equilibria are not critical points of V.

The conditions of Theorem 18 hold for this family of functions. At $x = 0$ when the stretches are equal, the Hessian of φ^* is diagonal because $\varphi_{xy}^* = V_{xy} = 0$. The second derivative is

$$\begin{aligned}V_{yy} &= 8y^2 + 6(-x + y)^{-4} + 6(x + y)^{-4} + 4(-x^2 + y^2) \\ &\quad +K[4 + 24y^2(-x^2 + y^2)^{-4} - 4(-x^2 + y^2)^{-3}]. \end{aligned} \tag{85}$$

At $x = 0$ when the stretches are equal, the Hessian of φ^* has at least rank one because

$$V_{yy}(0, t) = 12y^2 + 12y^{-4} + K(4 + 20y^{-6}) \tag{86}$$

is positive for any positive K. Also, at $x = 0$,

$$V_{xx} = K(2 + 2y^{-6}) + 6y^{-4} - 2y^2. \tag{87}$$

The singularity occurs when $\varphi_{xx}^* = V_{xx} = 0$. In terms of K, the critical values, y_c, are solutions to

$$K = \frac{y^2(y^6 - 3)}{y^6 + 1}. \tag{88}$$

This result agrees with Kearsley's (1986) equation (2.12) and with Ericksen's (1991) equation (6.2.11). Note that $y_c = 3^{1/6} = 1.2009$ when $K = 0$. As $K \to \infty$, the Mooney-Rivlin model approximates the neo-Hookean model and $y_c \to \infty$, which verifies that the Neo-Hookean model exhibits no singularity at $x = 0$. Each positive value of K produces a unique real root of Eq. (88). Therefore, for each value of K, the Mooney-Rivlin model produces a unique singularity at $x = 0$. The singularity is due to the behavior of the second invariant, II_B, of the tensor $B = FF^t$, where F is the deformation gradient, when added to the neo-Hookean model.

Applying the condition Eq. (48) to the Mooney-Rivlin model at $x = 0$ yields for fixed K,

$$V_{xxxx} - 3\frac{V_{xxy}^2}{V_{yy}} = 24(1 + 3Ky^{-8} + 10y^{-6}) - \frac{12(-12Ky^{-7} - 24y^{-5} - 4y)^2}{K(4 + 20y^{-6}) + 12y^{-4} + 12y^2}. \tag{89}$$

Substituting the expression for K given by Eq. (88) and plotting the result shows that G_{xxxx} is positive for the critical values $y_c > 1.2171$. Experimental values of K are greater than 4 at which $y_c > 2.5$.

In order for G to be a cusp catastrophe, the coefficient of x^2 must be decreasing in the parameter, t, at the singularity. In such a case, the equal stretches state is stable for small stretches. The derivative at $x = 0$,

$$V_{xxy} = -24Ky^{-7} - 48y^{-5} - 8y \tag{90}$$

is negative for all positive y and positive K. At $x = 0$, as a consequence of $V_{yy} > 0$, t is a monotonically increasing function of y,

$$t = 4y[K(1 - y^{-6}) - y^{-4} + y^2]. \tag{91}$$

Therefore, when V_{xxy} is negative, G_{xxt} is negative by Eq. (49).

Then, by Theorem 18, the equilibria set for the family φ^* based on the Mooney-Rivlin model for fixed $K > 0$ contains a rank one degenerate singularity and is an image of the pitchfork equilibria of the subfamily of the symmetric cusp catastrophe, $G(x, t)$. The equilibrium curves of $\varphi^*(x, y, t)$ form a distorted pitchfork bifurcation in x-y-t – space for fixed K in which the middle prong is curved in the y-t - plane, much like the equilibria of the example of Section 6.2.2 above for $0 < a < 2$. This is further supported by the fact that a numerical calculation shows that the projection, for fixed K, of the equilibria for φ^* into x-y – space is a standard pitchfork. The equilibria of φ^* on the central prong, $x = 0$, are minima for $t < t_c$, and maximum for $t > t_c$ and therefore unstable since as a function of t the coefficient of x^2 in G changes from positive to negative at t_c. The coefficient of x^4 in G is positive at t_c.

The equilibrium states on the two outer prongs of the pitchfork are minima of φ^* and are stable.

If K is taken as a parameter in $V(x, y, K)$, then as K increases from zero, the degenerate singularity at $(0, y_c, t_c)$ moves in $x - y - t$ space and the subfamily for each fixed K changes from a dual cusp to a cusp indexed by t at about $K = 0.08732$ which corresponds to $y_c = 1.2171$. This contrasts with the example of Section 6.2.2 in which the singularity point did not move in $x - y - t$ space with variations of the parameter a. Experimentally measured values of K are usually between 4 and 12 for rubber so that the butterfly point is not observed in rubber tests.

The family parametrized by K and t which is generated by the reduction is a symmetric subfamily of the butterfly catastrophe,

$$G(x, t, K) = \frac{1}{2}G_{xx}(0, t, K)x^2 + \frac{1}{24}G_{xxxx}(0, t, K)x^4 + \frac{1}{720}G_{xxxxxx}(0, t, K)x^6.$$

The butterfly point at $x = 0$ occurs when the coefficients $G_{xx}(0, t, K)$ and $G_{xxxx}(0, t, K)$ are simultaneously zero at $K = 0.08732$ and $t = 5.2874$ so that $y = 1.2171$. At this point, $G_{xxxxxx} = 40.078$, by Eq. (51) of Theorem 20 and so the coefficient of x^6 is positive.

As Kearsley, Ogden and others have shown, the Mooney-Rivlin model admits a single unsymmetric equilibrium path defined by

$$K(\lambda_1^3\lambda_2^3 + 1) + \lambda_1^4\lambda_2^4 - (\lambda_1^2 + \lambda_1\lambda_2 + \lambda_2^2) = 0. \tag{92}$$

This path bifurcates from the line $\lambda_1 = \lambda_2$ and admits no secondary bifurcations. Clearly, then there is also no equilibrium path which is disjoint from the line $\lambda_1 = \lambda_2$. Ogden pointed out in (1987), as verified above, that these equilibria are stable if K is large and are unstable if $K > 0$ is near 0.

Treloar obtained a fit of the Mooney-Rivlin model to his data by taking the constants in equation (10) of (Treloar, 1948) to be $2c_1 = 1.0$ (his $G = 2c_1$) and $2c_2 = 0.1$ (his $K = 2c_2$). Therefore in Eq. (88) $K = c_1/c_2 = 10$, which produces a critical point $y_c = 3.16854$. Unfortunately, the values of y corresponding to Treloar's asymmetric stretches are $y = 1.66$ and $y = 2.49$, both of which are less than the computed y_c. This is further evidence that the Mooney-Rivlin model is unsatisfactory or that constants other than those used by Treloar to fit the data are required.

8.1.1 Imperfection Behavior

Experimenters may find it very difficult to create exactly equal in-plane tensions on the rubber sheet. Therefore, the predictions of the model when the loads are slightly unequal, if $t_x = T_1 - T_2 \neq 0$, must be investigated. To model the rubber sheet experiments, the energy function should predict an equilibrium state for small t_x that is only slightly different from the perfect case $t_x = 0$. In this case, the imperfection t_x introduces a perturbation of the equilibrium set and destroys the singularity at $x =$

0. The universal unfoldings of catastrophe theory can help identify the predicted behavior (Poston and Stewart, 1978; Castrigiano and Hayes, 1993).

For the pitchfork type of bifurcation which appears in sheet models, when $t_x \neq 0$, the function associated with the Liapunov-Schmidt reduction is of the form $G(x, t) = x^4 + c_2(t)x^2 - t_x x$. This family is a subfamily of the cusp catastrophe locally near $(0, t_c)$. As an illustration of how the imperfection parameter, t_x, perturbs the equilibrium states, consider the Mooney-Rivlin model with an imperfection $t_x = 1$. The equilibrium set is shown in Fig. 8.3 as it projects into x-y – space. For small enough y, the stretch response is unique. For y larger than the minimum on the left branch, there are still three possible equilibrium states. This minimum is a singularity; the equilibrium is stable on the outer portion of this left branch and unstable on the inner portion. The minimum in this branch occurs at a larger value of y than the bifurcation point, y_c, in the perfect case. If the sheet were increasingly loaded from zero while maintaining the difference, t_x, the right hand equilibrium path would be followed; these equilibria are all stable. Once the minimum on the left branch is exceeded, snap through is possible. The inability of the Mooney-Rivlin model to reproduce Treloar's sheet data is not due to any failure by Treloar to achieve $T_1 = T_2$ exactly.

Any catastrophe function of one variable, x, admits a term of the form, cx, in its universal unfolding which accounts for the perturbations of the original function

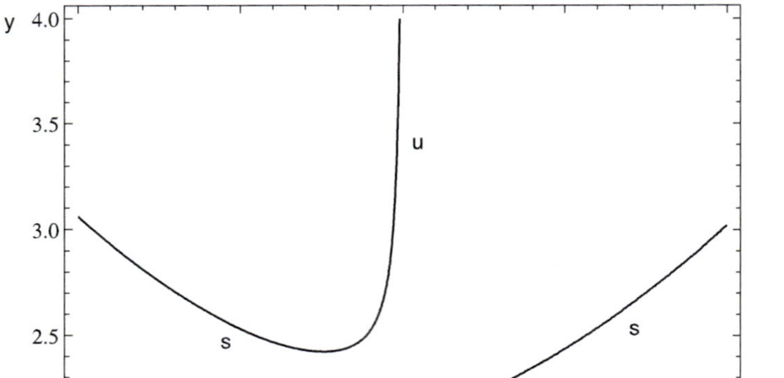

Fig. 8.3 Imperfection $t_x = 1$ for a Mooney-Rivlin material, $K = 5$ (s = stable and u = unstable equilibria)

due to slightly unequal loads. Therefore, the strategy for viewing the reduction of the equilibria of the strain energy as the equilibria of a catastrophe function in x can model small imbalances in the loading on the sheet.

8.2 Alexander Model

To predict the two bifurcations he observed in the continuous inflation of a neoprene (polychloroprene) balloon, Alexander (1968, 1971) proposed the model,

$$\frac{\partial\varphi}{\partial I_B} = c_1 \exp[k(I_B - 3)^2] \quad \text{and} \quad \frac{\partial\varphi}{\partial II_B} = \frac{c_2}{II_B - 3 + \gamma} + c_3, \tag{93}$$

where $c_1 = 17.0$, $c_2 = 19.85$, $c_3 = 1.0$, $k = 0.00015$, and $\gamma = 0.735$. So there are no cross terms involving both strain invariants. $\partial\varphi/\partial II_B$ is monotonically decreasing when II_B increases as required by experiment. Here it decreases asymptotically to c_3; in the Gent-Thomas model, $\partial\varphi/\partial II_B$ decreases asymptotically to zero.

This expression for $\partial\varphi/\partial I_B$ was originally developed by Hart-Smith (1966) and (Hart-Smith and Crisp, 1967) in an attempt to extend the statistical models to large deformations. He sought a function which is nearly constant for small stretches, as experimentally observed by Rivlin and Saunders, but increases sharply for large stretches to model $\partial\varphi/\partial I_B$. The expression for $\partial\varphi/\partial II_B$ is a slight generalization of the form taken by Gent and Thomas and by Hart-Smith in which the constant, γ, is introduced so that this energy derivative is defined in the undeformed state. The Alexander energy function cannot be obtained in closed form because of the presence of $\exp[(I_B - 3)^2]$. However, chain rule calculations produce the necessary derivatives to write the Taylor series of the reduction, $G(x, t)$.

$$\begin{aligned} V_{yy} = {} & 17\exp[0.00015(-3 + y^{-4} + 2y^2)^2](4 + 20y^{-6}) \\ & +0.0051\exp[0.00015(-3 + y^{-4} + 2y^2)^2](-4y^{-5} + 4y)^2(-3 + y^{-4} + 2y^2) \\ & -19.85(-4y^{-3} + 4y^3)^2(-2.265 + 2y^{-2} + y^4)^{-2} \\ & +(12y^{-4} + 12y^2)[1 + 19.85(-2.265 + 2y^{-2} + y^4)^{-1}] \end{aligned} \tag{94}$$

is positive for all y at $x = 0$. Therefore there is a one-to-one relationship between t and y, when $x = 0$,

$$\begin{aligned} t = {} & 17\exp[0.00015(-3 + y^{-4} + 2y^2)^2](-4y^{-5} + 4y) \\ & +(-4y^{-3} + 4y^3)[1 + 19.85(-2.265 + 2y^{-2} + y^4)^{-1}], \end{aligned} \tag{95}$$

a rapidly increasing function. The conditions of Theorem (18) hold.

$$\begin{aligned} V_{xx} = {} & 17\exp[0.00015(-3 + y^{-4} + 2y^2)^2](4 + 4y^{-6}) \\ & +(12y^{-4} - 4y^2)[1 + 19.85(-2.265 + 2y^{-2} + y^4)^{-1}]. \end{aligned} \tag{96}$$

The model produces two singularities at which $V_{xx} = 0$, approximately $y = 4.4$ and $y = 6.3$, when $x = 0$ so that there are two bifurcation points with respect to t. The function V_{xx} is negative at $x = 0$ for $4.4 < y < 6.3$ and positive otherwise. So along the fundamental path $x = 0$ in x-y-t – space the equilibria are unstable when $4.4 < y < 6.3$ and stable elsewhere.

Numerical calculations show that the reduced derivative, G_{xxxx}, of Eq. (48) is positive at both degenerate singularities. The derivative V_{xxy} at $x = 0$ is negative at the lower bifurcation point, $y = 4.4$, and positive at the higher point, $y = 6.3$. Therefore there are stable equilibria for $x \neq 0$ lying on the joined outer prongs of a supercritical pitchfork with singularity at $y = 4.4$ and a subcritical pitchfork with singularity at $y = 6.3$. To help visualize the equilibrium set, Fig. 8.4 shows its projection into x-y – space. The equilibrium set of φ^* in x-y-t – space is a curved version of this projection. These are the only equilibrium states in the range of the Treloar data, $1 \leq \lambda \leq 3$.

The Alexander model has a single equilibrium path other than the line $\lambda_1 = \lambda_2$. No secondary bifurcations from a primary path or disjoint paths of equilibria exist. The Alexander model is defined in terms of the strain invariants, so that by the chain rule,

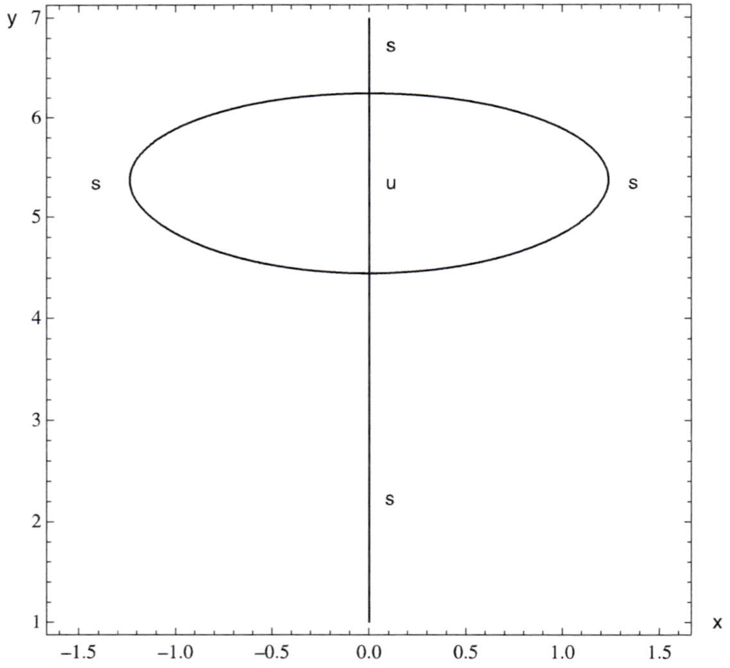

Fig. 8.4 The projection of the equilibrium set for the Alexander model into x-y – space (s = stable and u = unstable equilibria)

$$\varphi_1 - \varphi_2 = \frac{\partial \varphi}{\partial I_B}\left(\frac{\partial I_B}{\partial \lambda_1} - \frac{\partial I_B}{\partial \lambda_2}\right) + \frac{\partial \varphi}{\partial II_B}\left(\frac{\partial II_B}{\partial \lambda_1} - \frac{\partial II_B}{\partial \lambda_2}\right) \tag{97}$$
$$= \Big[c_1 \exp[k(I_B - 3)^2](1 + \lambda_1^{-3}\lambda_2^{-3})$$
$$+ \left(\frac{c_2}{II_B - 3 + \gamma} + c_3\right)\lambda_1^{-3}\lambda_2^{-3}\,(\lambda_1^2 + \lambda_1\lambda_2 + \lambda_2^2) - \lambda_1\lambda_2\Big]\,2(\lambda_1 - \lambda_2).$$

The equilibrium path of unequal stretches is obtained by setting this expression equal to zero after dividing by the non-zero term, $2(\lambda_1 - \lambda_2)$. This path forms a closed curve symmetric about the line $\lambda_1 = \lambda_2$. It corresponds to the closed path shown in Fig. 8.4.

8.3 The Ogden Models

Ogden (1972) proposed a functional form for the strain energy which would both satisfy the Valanis-Landel hypothesis and fit the stress difference $T_1 - T_2$ versus λ_1 data from biaxial tests of sheets with unique large tensile and large compressive stretch responses to varying loads. He generalized the first and second strain invariants of B to define strain invariants of the form

$$\varphi(\alpha) = \lambda_1^\alpha + \lambda_2^\alpha + \lambda_3^\alpha - 3,$$

where α is any fixed real number. Such an expression clearly satisfies the Valanis-Landel hypothesis by taking $w(\lambda) = \lambda^\alpha - 1$.

Ogden built constitutive models for rubber from linear combinations of the $\varphi(\alpha)$ so that the form of the strain energy is

$$\varphi(\lambda_1, \lambda_2, \lambda_3) = \sum_{p=1}^{N} \frac{\mu_p}{\alpha_p}\varphi(\alpha_p), \tag{98}$$

for empirical constants μ_p and α_p. Since the derivative $\partial^2\varphi(\alpha)/\partial\lambda_1\partial\lambda_2 = \alpha^2$ when evaluated at $\lambda_1 = \lambda_2 = 1$, the coefficients are constrained by $\sum_{p=1}^{N} \mu_p\alpha_p = 2\mu$, where μ is the shear modulus. This function satisfies the Valanis-Landel hypothesis since each summand does. Ogden (1972) pointed out that a sufficient condition for the model to satisfy Hill's inequality for admissible constitutive models for incompressible materials is that in each summand $\mu_p\alpha_p$ be positive.

The bifurcation behavior of such models for an incompressible sheet can be studied after first setting $\lambda_3 = 1/(\lambda_1\lambda_2)$ and making the transformation of Eq. (27). The Ogden invariant then has the form

$$\varphi(\alpha) = V(x, y) = (x + y)^\alpha + (y - x)^\alpha + (y^2 - x^2)^{-\alpha}.$$

The following derivatives at $x = 0$ are needed to describe the behavior of the invariant when viewed as an energy function.

$$V_{xx} = \alpha\left[2y^{-2-2\alpha} + 2(-1+\alpha)y^{-2+\alpha}\right];$$
$$V_{yy} = 2\alpha y^{-2-2\alpha}\left[1 + 2\alpha + (\alpha - 1)y^{3\alpha}\right];$$
$$V_{xxy} = \alpha\left[4(-1-\alpha)y^{-3-2\alpha} + 2(-2+\alpha)(-1+\alpha)y^{-3+\alpha}\right];$$
$$V_{xxxx} = \alpha\left[12(1+\alpha)y^{-4-2\alpha} + 2(-3+\alpha)(-2+\alpha)(-1+\alpha)y^{-4+\alpha}\right].$$

A bifurcation for $x = 0$ is possible when the energy, $\varphi(\alpha)$, has a single term so that $N = 1$. The derivative $V_{xx} = 0$ if $y_c^{3\alpha} = (1-\alpha)^{-1}$. This recovers the result of Ogden (1987, Eq. 3.15) because $y = \lambda_1 = \lambda_2$ at $x = 0$. Since the sheet is in tension, only critical points $y_c > 1$ are of interest and so one only considers $-\infty < \alpha < 1$. The second derivative, V_{yy}, at $x = 0$ is positive for all $-\infty < \alpha < 1$ and all $y_c > 1$. Therefore Theorem (18) can be used to classify this degenerate singularity. In this range of α, at the singularity, $y_c^{3\alpha} = (1-\alpha)^{-1}$, for $x = 0$,

$$V_{yy} = 4\alpha^2(1-\alpha)^{2/3+2/(3\alpha)};$$
$$V_{xxy} = 2\alpha(1-\alpha)^{1/\alpha+2/3}\left[-2 + 2(1-\alpha)^{-1} - 2\alpha - 3\alpha(1-\alpha)^{-1} + \alpha^2(1-\alpha)^{-1}\right];$$
$$V_{xxxx} = \alpha[12(1+\alpha)(1-\alpha)^{2/3-4/(3\alpha)} +2(-3+\alpha)(-2+\alpha)(-1+\alpha)(1-a)^{-1/3+4/(3\alpha)}].$$

At the singularity, the expression

$$V_{xxxx} - 3\frac{V_{xxy}^2}{V_{yy}} = [-5(1-\alpha)^{4/(3\alpha)}\alpha^2 + 3(1-\alpha)^{4/(3\alpha)}\alpha^3 + 2(1-\alpha)^{4/(3\alpha)}\alpha^4](1-\alpha)^{-1/3} \tag{99}$$

is shown numerically to be negative for $-\infty < \alpha < 1$, as is the term V_{xxy}. Therefore this degenerate singularity is a subcritical pitchfork if $-\infty < \alpha < 1$.

The linear combination of two of the Ogden strain invariants can also produce a bifurcation from the fundamental path. The Mooney-Rivlin model is a linear combination of the two Ogden invariants, one for $\alpha = 2$ with no bifurcation in the range $y > 0$ and one for $\alpha = -2$ with a subcritical pitchfork singularity at the sixth root of 3. As shown above, the sum forming the Mooney-Rivlin model produces a supercritical pitchfork singularity at a point $(0, y_c)$ which depends on the coefficients chosen for the two functions.

Not all sums of three Ogden invariants produce a bifurcating model. A good fit was obtained by Jones and Treloar (1975) to their data representing a unique stretch response to loads by taking $N = 3$ and $\alpha_1 = 1.3$, $\alpha_2 = 4.0$, and $\alpha_3 = -2.0$, while

$\mu_1 = 0.69$, $\mu_2 = 0.01$, and $\mu_3 = -0.0122$. The α_1 term fits the data for large compressive stretches and the α_3 term fits the data for the large tensile stretches. At $x = 0$, again $V_{xy} = 0$ and

$$V_{yy} = 0.5308(9.36y^{-4.6} + 0.78y^{-0.7}) + 0.0061(12y^{-4} + 12y^2) + 0.0025(72y^{-10} + 24y^2); \tag{100}$$

$$V_{xx} = 0.5308(2.6y^{-4.6} + 0.78y^{-0.7}) + 0.0061(12y^{-4} - 4y^2) + 0.0025(8y^{-10} + 24y^2). \tag{101}$$

Both V_{yy} and V_{xx} are positive for all $y \geq 1$ so that this model has no bifurcation from the fundamental path in tension. This Ogden model produces the same qualitative equilibrium behavior for a sheet biaxially loaded under equal in-plane stresses as the neo-Hookean model. A similar result is obtained for the choice of the six empirical constants given in equation (36) of (Ogden, 1986).

Choices of the material coefficients exist which create a secondary bifurcation off a primary bifurcating path (Ogden, 1987). It will be shown below that it is possible to construct models with several bifurcations from the fundamental path using three Ogden invariant functions.

9 Rubber Constitutive Model with a Three Bifurcation Points Structure

The rubber models examined do not produce sufficient bifurcations to obtain Treloar's result. But the analysis of the Ogden strain invariants makes it possible to speculate on a possible form for a successful model.

Treloar's (1948) data for the 400, 500 and 600 gm loads suggests that a square rubber sheet stretched under equal in-plane tensions applied to the edges of the sheet can have both symmetric and asymmetric stable equilibria. The fundamental path $x = 0$, i.e. $\lambda_1 = \lambda_2$, must always be an equilibrium solution of the model. The data at 200 and 300 gm in Table 8.1 show that the fundamental path is initially stable. At the next higher load of 400 gm there are asymmetric stable stretches. This indicates that a bifurcation point has been passed through and that the corresponding states on the fundamental path are unstable. But the 500 gm load produces stable equal stretches. Therefore, another bifurcation point has been passed through and the fundamental path is again stable. Finally, at the 600 gm load, there are unequal stable stretches so that the fundamental path is again unstable. This suggests that a simple model to reproduce Treloar's data would be a function with three bifurcations on the fundamental path. In the simplest form, each of these is locally one of the pitchfork types.

As the stress is increased on the sheet, the value of y increases if the response remains on the fundamental path. It is possible to build a model with the three singularities occurring at appropriate values of y from a linear combination of three

Ogden invariants. The idea is to make the graph of $V_{xx}(0, y)$ look qualitatively like that of a cubic polynomial which intersects the y-axis in three appropriate points. If $0 < \alpha < 1$, the graph of $\varphi(\alpha)_{xx}$ is decreasing and intersects the y-axis once at some $1 < y < 3$, has a minimum at a somewhat larger value of y and then is asymptotic from below to the y-axis as y increases. Furthermore when α is negative, the graph of $\varphi(\alpha)_{xx}$ is decreasing for $y > 1$. The sum of $\varphi(0.5)$ and $\varphi(-2)$ has a second derivative which looks like that of a cubic function since at larger values of y the negative contribution from $\varphi(-2)_{xx}$ bends down the graph of $\varphi(0.5)_{xx}$. This curve only intersects the y-axis once. To lift the graph, a small, relatively constant, positive function is added. Such a function is a multiple of $\varphi(2)_{xx}$, which is positive and nearly constant when $y > 1$. The candidate strain energy function resulting from some trial and error choices of the coefficients is the following three bifurcation model built from three Ogden invariant functions

$$\begin{aligned} V(x, y) = {} & .00045\left[(x+y)^{-2} + (y-x)^{-2} + (y^2 - x^2)^2\right] \\ & +1.5\left[(x+y)^{0.5} + (y-x)^{0.5} + (y^2 - x^2)^{-0.5}\right] \\ & +.026\left[(x+y)^2 + (y-x)^2 + (y^2 - x^2)^{-2}\right]. \end{aligned} \tag{102}$$

Ogden's requirement that $\sum_{p=1}^{N} \mu_p \alpha_p = 2\mu$ can be satisfied if one multiplies $V(x, y)$ by the appropriate scaling coefficient.

This model is, in a sense, a generalization of the Mooney-Rivlin model since two of the Ogden strain invariants used are those appearing in the Mooney-Rivlin model. For this model in tension when $y > 1$, the derivative $V_{yy}(0, y)$ is positive where

$$V_{yy} = 0.026(4 + 20y^{-6}) + 1.5(2y^{-3} - 0.5y^{-1.5}) + 0.00045(12y^{-4} + 12y^2) \tag{103}$$

and the graph of $V_{xx}(0, y)$ is qualitatively similar to a cubic crossing the axes at three points, where

$$V_{xx} = 0.026(4 + 4y^{-6}) + 1.5(y^{-3} - 0.5y^{-1.5}) + 0.00045(12y^{-4} - 4y^2) \tag{104}$$

Recall that the equilibria of the Liapunov-Schmidt reduction have the same stability as the corresponding equilibria of φ^* when $V_{yy} > 0$ (Lemma 17) and that the degenerate singularities of both the reduction and φ^* are derived from $G_{xx} = V_{xx} = 0$. The bifurcation points, $V_{xx} = 0$, when $x = 0$ occur at approximately, $y = 2.4$, $y = 3.2$, and $y = 5.6$. The one-to-one increasing relationship between y and the load t, since $V_{yy} > 0$ at $x = 0$, is obtained from $\varphi^*_y = 0$ so that

$$t = 0.026(-4y^{-5} + 4y) + 0.00045(-4y^{-3} + 4y^3) + 1.5(y^{-0.5} - y^{-2}). \tag{105}$$

The bifurcation points can be moved to the left, if necessary to fit a particular rubber, by reducing the exponent in the third term to 1.99 or 1.98 since this lowers the graph of the second derivative of the other two terms.

Theorem (18) and the analysis preceding it describes the nature of the bifurcation points and leads to a description of the equilibrium set in x-y-t – space. The derivative at $x = 0$,

$$V_{xxy} = 0.00045(-48y^{-5} - 8y) - 0.624y^{-7} + 1.5(0.75y^{-2.5} - 3y^{-4}),$$

is negative if $2.72 < y < 4.7$ and positive otherwise. Also, at $x = 0$,

$$V_{xxxx} = 0.00045(24 + 240y^{-6}) + 1.872y^{-8} + 1.5(-1.875y^{-3.5} + 9y^{-5}).$$

The fourth derivative at $x = 0$ of the function, $G(x, t)$, associated with the reduction

$$G_{xxxx} = V_{xxxx} - 3\frac{V_{xxy}^2}{V_{yy}}$$

is always positive. The second derivative, $G_{xx} = V_{xx}$, is positive if $1 < y < 2.4$ and $3.2 < y < 5.6$ and negative otherwise. Since $V_{yy} > 0$, $dy/dt > 0$, then by Eq. (49) V_{xxy} and G_{xxt} have the same sign. The singularities at $y = 2.4$ and $y = 5.6$ are pitchforks of Type I (Fig. 8.2) because $G_{xxt} < 0$ and $G_{xxxx} > 0$. The singularity at $y = 3.2$ is a pitchfork of type III (Fig. 8.2) since $G_{xxt} > 0$ and $G_{xxxx} > 0$ there.

The qualitative shape of the equilibrium set of φ^* in x-y-t – space can be deduced from the behavior of G. Along the fundamental path, $x = 0$ where y and t are related by Eq. (105), the equilibria are stable for $1 < y < 2.4$ and $3.2 < y < 5.6$ and unstable otherwise. For $2.4 < y < 3.2$, there are stable equilibria for $x \neq 0$ lying on the joined outer prongs of a supercritical pitchfork with singularity when $y = 2.4$ at $(0, 2.4, 0.98)$ and a subcritical pitchfork with singularity when $y = 3.2$ at $(0, 3.2, 1.08)$. The point when $y = 5.6$ at $(0, 5.6, 1.48)$ is the singularity for a supercritical pitchfork for which the equilibria on the outer prongs are stable. To aid in visualizing the equilibrium set, the projection of the equilibrium states for the model of Eq. (102) into (x, y) space is shown in Fig. 8.5. The equilibrium set in x-y-t – space for φ^* is a distorted version of this set.

The effect of the two applied loads being slightly unequal, so that $t_x = \epsilon$, is modeled by the term, $t_x x$, in the energy function, which acts as an imperfection term. Fig. 8.6 shows the equilibrium set when $\epsilon = -0.0001$. This is a global perturbation as opposed to the local perturbations often considered in singularity analyses.

This example demonstrates the technique. For a given rubber, the shear modulus and non-unique stretch data would have to be obtained from which to construct the constitutive model. A valid model must reproduce the Treloar sheet experiment, all unique stretch response uniaxial and biaxial test data, as well as data showing the dependence on the variation of the strain invariants. Validation would require a full array of such tests on the same type of rubber.

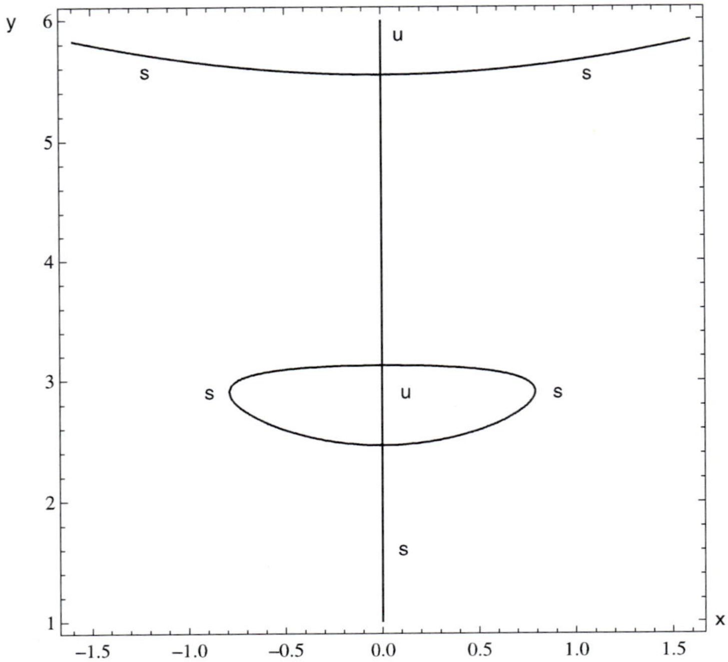

Fig. 8.5 The projection into x-y – space of the equilibrium set for the proposed model when $t_x = 0$ (s = stable and u = unstable equilibria)

10 Influence of Bifurcations on Maximum Dissipation Non-equilibrium Evolution Processes

The presence of a bifurcation influences the behavior of the maximum dissipation evolution equations. One question is whether the response is stable or unstable. Again, a structurally stable evolution equation is desired so that small experimental errors do not influence the qualitative response. To make this idea more precise, follow Hale and Koçak (1991, pp. 62–63),

Definition 21 Two scalar differential equations $\dot{x} = f(x)$ and $\dot{x} = g(x)$ are topologically equivalent if there is a homeomorphism $h : \Re \rightarrow \Re$ such that h takes the orbits of one differential equation to the orbits of the other and preserves the direction in time.

Definition 22 Let $\dot{x} = F(a, x)$ be a vectorfield that depends on k parameters $a = (a_1, \ldots, a_k)$. For a fixed set of parameter values, $\bar{a}$, the vectorfield $\dot{x} = F(\bar{a}, x)$ is structurally stable if there is an ϵ such that $\dot{x} = F(a, x)$ is topologically equivalent to $\dot{x} = F(\bar{a}, x)$ for all sets of parameters a such that $||a - \bar{a}|| < \epsilon$.

Conditions that guarantee generic behavior have been developed from the work of Guckenheimer (1973) and codified by Hale and Koçak (1991, p. 438),

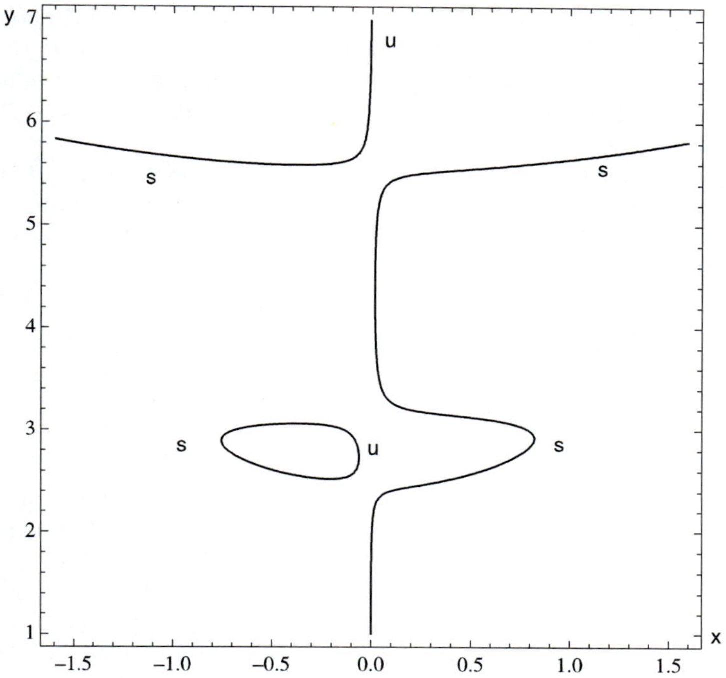

Fig. 8.6 Global perturbation of the equilibria of the proposed model by an imperfection of $t_1 = -0.0001$ (s = stable and u = unstable equilibria)

Definition 23 A gradient vectorfield $\dot{x} = -\nabla F$ is generic if

(i) F has a finite number of critical points;
(ii) each critical point is nondegenerate;
(iii) there are no saddle connections;
(iv) $|F(x)| \to \infty$ as $||x|| \to \infty$.

Such generic vectorfields are dense in the set of all gradient vectorfields and so are structurally stable.

Definition 24 (e.g. Hale and Koçak; 1991, p. 303) A planar differential equation $\dot{x} = f(x)$ has a saddle connection between two saddle points $x^{(1)}$ and $x^{(2)}$ if the intersection of the unstable manifold for point $x^{(1)}$, $W^u(x^{(1)})$, and the stable manifold for $x^{(2)}$, $W^s(x^{(2)})$, is not empty. If the saddle points coincide, the orbit is homoclinic; if not the orbit is heteroclinic.

A planar gradient dynamical system, such as describes the response of a rubber sheet, cannot have a homoclinic orbit because, except at critical points, the evolution path cross level sets orthogonally and point inward. For the same reason, it cannot have a periodic orbit.

Saddle-node degenerate bifurcations (one eigenvalue is zero) and heteroclinic saddle connections are possible. The saddle-node bifurcation results from the collision of a saddle and a node (of a stable and unstable node) as the bifurcation parameter varies. For example $\dot{x} = x^2+\alpha$, where α is a parameter, has a saddle-node bifurcation when $\alpha = 0$.

Global bifurcations may not correspond to bifurcations in the potential function. An example of a heteroclinic saddle connection is given by the following gradient vectorfield, which is based on Poston and Stewart (1978, p. 237). The potential function is $F(x, y, \epsilon) = (x^2 - 1)y + \epsilon x$, parametrized by ϵ.

$$\begin{aligned} \dot{x} &= \frac{\partial F}{\partial x} = 2xy + \epsilon \\ \dot{y} &= \frac{\partial F}{\partial y} = x^2 - 1. \end{aligned} \tag{106}$$

There are two saddles, one at $(1, -\epsilon/2)$ and the other at $(-1, \epsilon/2)$, which are connected by a heteroclinic orbit when $\epsilon = 0$. This saddle connection is not preserved when ϵ is varied and there is no associated bifurcation in the family $F(x, y, \epsilon)$.

The saddle connection perturbs to break the connection and such a vectorfield is a structurally stable perturbation of the original gradient field of $f(x, y)$ which is not equivalent to the gradient of any function in the unfolding of $f(x, y)$ (Guckenheimer, 1973).

10.1 Dynamic Behavior in "Snap-Through"

Static analyses of bifurcations assume that the system instantaneously jumps from one state to a stable equilibrium. Of course, in reality this dynamic process takes time. The maximum dissipation evolution equation predicts how this process takes place in a sheet.

Let $\bar{\varphi}^* = \varphi(\lambda_1, \lambda_2) - T_1\lambda_1 - T_2\lambda_2$ and let $\varphi^* = V(x, y) - t_x x - t_y y$. Since the transformation (25) maps $-t_x x - t_y y$ to $-T_1\lambda_1 - T_2\lambda_2$, the transformation maps $\bar{\varphi}^*$ to φ^*. Now consider the transformation of affinities. Let X_i be the affinities associated with $\bar{\varphi}^*$ and X_i' be the affinities associated with φ^*. The transformation produces the following relations

$$\begin{aligned} \varphi_1 - T_1 &= 0.5(V_x + V_y) - 0.5(t_x + t_y) \\ \varphi_2 - T_2 &= 0.5(-V_x + V_y) - 0.5(-t_x + t_y). \end{aligned}$$

Therefore, by combining these relations, $X_1 + X_2$ transforms to X_y' and $X_1 - X_2$ transforms to X_x'.

In the Liapunov-Schmidt reduction, $V_y - t = 0$, where $t = t_y$. This equation solves by the implicit function theorem, since $V_{yy} \neq 0$ is assumed, to give

$y = f(x, t)$. In this case, it is always true that $X'_y = 0$, so that the system is one-dimensional.

$G(x, t)$ is the reduction of the generalized energy function $\varphi^*(x, y, t)$ by $G(x, t) = \varphi^*(x, f(x, t), t)$; it is not itself a generalized energy function. The reduced transformed system is obtained by also reducing $X'_x(x, f(x, t), t)$, $X'_y = 0$, and the evolution equation, which becomes

$$\frac{dx}{dt} = -k\left(\frac{d\varphi^*(x, f(x, t), t)}{dx^2}\right)^{-2}[V_x(x, f(x, t), t) - t]. \tag{107}$$

So the Hessian term in the reduced evolution equation is a function of t. Further, the evolution expression is the product of a negative quantity times the affinity $X'_x = V_x(x, f(x, t), t) - t$.

If the function, $G(x, t)$, has both stable and unstable equilibria, the behavior of the maximum dissipation evolution depends on the type of bifurcation. If a stable equilibrium exists for a larger value of the state variable than that corresponding to the unstable equilibrium and if the perturbation is close to the unstable equilibrium, the evolution is repelled from the unstable equilibrium until it enters the basin of attraction for the stable equilibrium. In this sense, the dynamic response is stable. The response is unstable if there is no stable equilibrium to the right of the unstable equilibrium. Any time that the perturbation is in the basin of attraction of a stable equilibrium, the response is asymptotically stable.

The affinities are zero at any equilibrium. Therefore the affinity as a function of the state variables is strictly positive between a stable and unstable equilibrium and strictly negative between an unstable and a stable equilibrium. The boundary of the basin of attraction occurs at $\partial X'_x/\partial x = 0$ because the affinity, X'_x, changes from increasing to decreasing or vice versa. The boundary of the basin of attraction is at

$$dX'_x(x, f(x, t), t)/dx = V_{xx}(x, f(x, t), t) = 0, \tag{108}$$

but $V_x(x, f(x, t), t) \neq 0$. Therefore it is not an equilibrium point.

For example, assume that an unstable equilibrium is surrounded by two stable equilibria (they must alternate). $X'_x(x)$ decreases from a positive value at the boundary of the basin of attraction of the left-hand stable equilibrium to a negative value at the boundary of the basin of attraction of the right-hand stable equilibrium; it passes through zero at the unstable equilibrium (Fig. 8.7).

Example Suppose that the Liapunov-Schmidt reduction yields a cusp catastrophe $G(x, t) = x^4 + c(t)x^2 + t_x x$, where $t_x = T_1 - T_2$ is the perturbation, and $t = T_1 + T_2$. The function $c(t)$ is negative, and its magnitude is an increasing function of t. Assume no perturbation; $t_x = 0$. Then the equilibria for a given value of t greater than the critical load are at $x = 0$, which is unstable, and at $x = \pm\sqrt{-c(t)}$, which are stable. The boundaries of the basin of attraction for the stable equilibria are at $d^2X'_x/dx^2 = \pm\sqrt{-c(t)/3}$. Even though the unstable equilibrium exists, the dynamic response is asymptotically stable. Eventually the process approaches

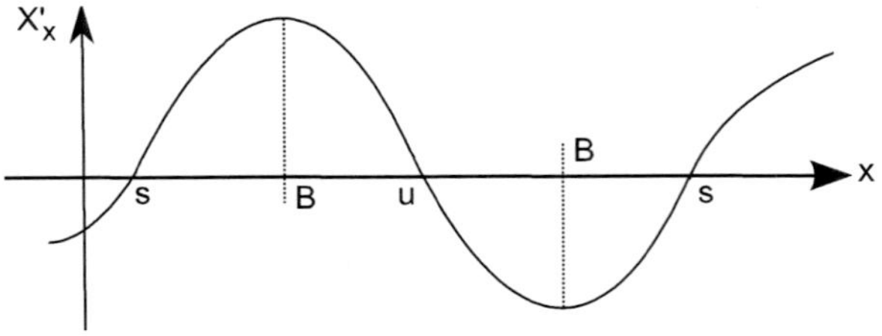

Fig. 8.7 Behavior of the affinity $X'_x(x)$ between the equilibria of a pitchfork bifurcation. The lines, B, indicate the boundary of the basin of attraction for the stable equilibria. (s = stable and u = unstable equilibria)

one of the stable equilibria corresponding to the chosen value of t. The maximum dissipation process gives the dynamical path taken by the snap-through predicted by a static analysis.

References

H. Alexander (1968). A constitutive relation for rubberlike materials. *International Journal of Engineering Science* **6**, 549–562.

H. Alexander (1971). Tensile instability of initially spherical balloons. *International Journal of Engineering Science* **9**, 151–162.

E. Arruda and M. Boyce (1993). A three-dimensional constitutive model for the large stretch behavior of rubber elastic materials. *Journal of the Mechanics and Physics of Solids* **41**, 389–412.

M. Baker and J. L. Ericksen (1954). Inequalities restricting the form of the stress-deformation relations for isotropic elastic solids and Reiner-Rivlin fluids. *Journal of the Washington Academy of Sciences* **44**, 33–35.

M. F. Beatty (1987). Topics in finite elasticity: Hyperelasticity of rubber, elastomers, and biological tissues - with examples. *Applied Mechanics Reviews* **40**, 1699–1734.

J. M. Ball and D. G. Schaeffer (1983). Bifurcation and stability of homogeneous equilibrium configurations of an elastic body under dead-load tractions. *Mathematical Proceedings of the Cambridge Philosophical Society* **94**, 315–339.

D. P. L. Castrigiano and S. A. Hayes (1993). *Catastrophe Theory*, Addison-Wesley, Reading MA.

J. L. Ericksen (1991). *Introduction to the Thermodynamics of Solids*, Chapman and Hall, London.

A. N. Gent and A. G. Thomas (1958). Forms for the stored (strain) energy function for vulcanized rubber. *Journal of Polymer Science* **28**, 625–628.

R. Gilmore (1981). *Catastrophe Theory for Scientists and Engineers*, Wiley, New York.

J. Guckenheimer (1973). Bifurcation and catastrophe. In *Dynamic Systems*, ed. M. M. Peixoto, Academic Press, New York, 95–109.

J. Hale and H. Koçak (1991). *Dynamics and Bifurcations*, Springer, New York.

L. J. Hart-Smith (1966). Elasticity parameters for finite deformations of rubber-like materials. *Zeitschrift für Angewandte Mathematik und Physik* **17**, 608–625.

L. J. Hart-Smith and J. D. C. Crisp (1967). Large elastic deformations of thin rubber membranes. *International Journal of Engineering Science* **5**, 1–24.

H. W. Haslach, Jr. (1986). Post-buckling stability of orthotropic, linear elastic, rectangular plates under biaxial loads. *International Journal of Mechanical Sciences* **28**, 739–756.

H. W. Haslach, Jr. (1999), Dynamical effects of degenerate singularities in the potential for mechanical systems. In *Nonlinear Techniques, Structural Dynamics Systems, Computational*

Techniques, and Optimization, ed. C. T. Leondes, Vol. 15, International Series in Engineering, Technology and Applied Science, Gordon and Breach Science Publishers, NY, 273–330.

H. W. Haslach, Jr. (2000). Constitutive models and singularity types for an elastic biaxially loaded rubber sheet. *Mathematics and Mechanics of Solids* **5**, 41–73.

D. F. Jones and L. R. G. Treloar (1975). The properties of rubber in pure homogeneous strain. *Journal of Physics D: Applied Physics* **8**, 1285–1304.

E. A. Kearsley (1986). Asymmetric stretching of a symmetrically loaded elastic sheet. *International Journal of Solids and Structures* **22**, 111–119.

D. A. Lavis and G. M. Bell (1977). Thermodynamic phase changes and catastrophe theory. *Bulletin of the Institute of Mathematics and its Application* **13**, 34–42.

R. W. Ogden (1972). Large deformation isotropic elasticity - on the correlation of theory and experiment for incompressible rubberlike solids. *Proceedings of the Royal Society of London A*, **326**, 565–584.

R. W. Ogden (1985). Local and global bifurcation phenomena in plain-strain finite elasticity. *International Journal of Solids and Structures* **21**, 121–132.

R. W. Ogden (1986). Recent advances in the phenomenological theory of rubber elasticity. *Rubber Chemistry and Technology* **59**, 361–383.

R. W. Ogden (1987). On the stability of asymmetric deformations of a symmetrically-tensioned elastic sheet. *International Journal of Engineering Science* **25**, 1305–1314.

T. Poston and I. Stewart (1978). *Catastrophe Theory and its Applications*, Pitman, London.

R. S. Rivlin (1974). Stability of pure homogeneous deformations of an elastic cube under dead loading. *Quarterly of Applied Math* **32**, 265–271.

R. S. Rivlin and D. W. Saunders (1951). Large elastic deformations of isotropic materials VII. Experiments on the deformation of rubber. *Philosophical Transactions of the Royal Society of London A* **243**, 251–285.

W. J. Supple (1967). Coupled branching configurations in the elastic buckling of symmetric structural systems International Journal of Mechanical Science **9**, 97–112.

R. Thom (1975), *Structural Stability and Morphogenesis*, W. A. Benjamin, Inc., Reading, MA.

J. M. T. Thompson and G. W. Hunt (1973). *A General Theory of Elastic Stability*, Wiley, London.

L. R. G. Treloar (1948). Stresses and birefringence in rubber subjected to general homogeneous strain. *Proceedings Physics Society* **60**, 135–144.

L. R. G. Treloar (1949). *The Physics of Rubber Elasticity*, 1st ed., Clarendon Press, Oxford, UK.

L. R. G. Treloar (1975). *The Physics of Rubber Elasticity*, 3rd ed., Clarendon Press, Oxford, UK.

K. C. Valanis and R. F. Landel (1967). The strain-energy function of a hyperelastic material in terms of the extension ratios. *Journal of Applied Physics* **38**, 2997–3002.

Chapter 9
Maximum Dissipation Evolution Construction for Non-homogeneous Thermodynamic Systems

1 Introduction

A body is thermodynamically non-homogenous if the thermodynamic variables in the homogeneous construction vary from point to point in the body. The non-homogeneous non-equilibrium thermodynamic model is an extension of the homogeneous non-equilibrium construction obtained by adding fluxes such as the heat flux as variables and a generalized entropy function that depends on these fluxes. The geometric construction requires locally a second contact manifold in which the fluxes operate.

The entropy production should be given by a constitutive relation that extends the older idea that the entropy flux is equal to the heat flux divided by temperature (Müller, 1967). Likewise internal entropy production is represented by a constitutive relation. If the thermodynamic state varies from point to point in the body, then thermodynamic fluxes such as the heat, mass, and stress fluxes, become thermodynamic variables. Müller and Ruggeri (1998) obtain a system of non-linear hyperbolic first order evolution equations by minimizing their entropy production subject to the constraints given by the balance laws. Similarly, de Groot and Mazur (1984) use a variational statement to show that their entropy production has such a minimum at the stationary states. Stationary means that the state variables are independent of time. Their linear irreversible thermodynamics results only in linear constitutive models for the stationary states because of the use of the Onsager relations. Common linear stationary constitutive equations include the Fourier relation for heat conduction, Fick's law for diffusion, and Ohm's law for electromagnetism. A body for which the flux satisfies these relations responds instantly to variations of the boundary conditions because the relations involve no time derivatives. The stationary manifold at each point of the body is defined at each time by a relation for each flux in terms of the control variables. In contrast to the case in which φ is the hyperelastic strain energy density in the definition of a generalized thermodynamic function (Section 3 in Chapter 3) and M_e is the manifold of equilibrium states, when φ is entropy production per unit volume, then M_e is the manifold of stationary states.

In the construction of a generalized entropy production function, the product of pairs of conjugate fluxes and gradients, one a state variable and the other a control variable, is required to have units of specific entropy production, in contrast to the

H.W. Haslach Jr., *Maximum Dissipation Non-Equilibrium Thermodynamics and its Geometric Structure*, DOI 10.1007/978-1-4419-7765-6_9,

conjugate variables of the homogeneous model that must multiply to give specific energy. A generalized entropy production function is produced from empirical constitutive equations for stationary states by requiring that the stationary manifold is defined by setting the gradient of the generalized entropy production function with respect to the state variables equal to zero. The steady state manifold plays a role for generalized entropy production analogous to that the thermostatic manifold plays for generalized energy density. A relaxation process involving gradients of the thermodynamic variables evolves to the steady state manifold.

As in the homogeneous case, this system is placed in the context of a contact manifold whose contact one-form is negative on an admissible process. A particular admissible non-equilibrium evolution of the state variables is defined by a maximum change in the generalized entropy production, as in Chapter 3. This construction gives a criterion to predict which of the many possible processes consistent with the second law of thermodynamics is taken. This strategy also allows the study of the stability behavior of systems not in stationary states. The construction produces a system of flux evolution equations so that transfer occurs in finite time rather than instantaneously. The process involving a change in gradients also generates a path on the graph of the generalized energy density function. As a relaxation process evolves to the thermostatic manifold, the fluxes tend to zero on the stationary manifold so that the construction provides a geometric relation between stationary processes and equilibrium states. The thermostatic manifold could be viewed as a submanifold of the steady state manifold. The current control variable in a gradient relaxation process derived from the generalized energy function is the value in the equilibrium manifold determined from the generalized entropy production function. Therefore, in a non-homogeneous non-equilibrium system, the evolution equations derived from generalized energy interact with those derived from the generalized entropy function under the constraints of the balance laws to define the non-equilibrium response. The construction of the generalized entropy production parallels that of the generalized energy, except in the units chosen. Therefore all constructions deduced from the generalized thermodynamic function in Chapter 3 carry over to the generalized entropy flux.

The geometry is more complex than that for the homogeneous system which could be represented in $\mathcal{B} \times C(\Re^{n+1}, n)$ in which each point has associated with it a contact manifold for the thermodynamic variables. To account for the gradients, an additional contact manifold is associated with each point of the body. The balance laws connect the thermodynamics at one point with those of the other points of the body. These are expressed as differential forms so that they may act on tangents to the process path in thermodynamic space. Portions of this chapter originally appeared in Haslach (2009).

2 Generalized Entropy Production and Flux Evolution

The stationary manifold at each point of $\mathcal{B}$ is defined at each time by a relation for each flux in terms of the control variables. In analogy with the generalized energy density from which the thermostatic manifold is obtained by a zero gradient condition with respect to the state variables, a generalized entropy production is defined that gives the stationary manifold from a zero gradient condition with respect to the fluxes.

Denote the vector fluxes by $\boldsymbol{f}_i$ for $i = 1, \ldots, r$ and their corresponding control vectorfields by $\boldsymbol{c}_i$. A scalar product, which has units of specific entropy production, is defined for each conjugate pair. A flux vector is objective if $\boldsymbol{f}'_i = Q\boldsymbol{f}_i$, where the prime indicates the second observer. In many cases, the control vectors associated with flux vectors are gradients of scalar functions and so are objective. When modeling fluids, it is convenient to think of the spatial rate of deformation second-order tensor, D, as a flux, so that a flux may be a second order tensor with a conjugate control second order tensor, such as the Cauchy stress, σ. A local chart is given by an ordered listing of the components of the fluxes and controls, $(p_1, p_2, p_3, f_1, \ldots, f_m, c_1, \ldots c_m) \in \Re^3 \times \Re^{2m}$.

The generalized entropy production is defined to satisfy a condition in addition to Definition 2 of Section 3 in Chapter 3 that gives the stationary manifold from a zero gradient condition with respect to the fluxes.

Definition 1 A local generalized entropy production function $\Psi^* : \Re^{2m} \to \Re$ satisfies at each point (p_1, p_2, p_3), (1) $\partial\Psi^*/\partial f_i = 0$ for $i = 1, \ldots, m$, defines the stationary manifold; (2) $\partial\Psi^*/\partial c_i = f_i$; and (3) If all fluxes $\boldsymbol{f}_i = \boldsymbol{O}$, then $\Psi^* = 0$.

The third condition was not assumed in the definition of a generalized thermodynamic function because some thermodynamic variables cannot take the value zero. The units of Ψ^* are specific energy per degree per second. As for the generalized energy density function, the form of Ψ^* is $\Psi^* = \Psi(f_1, \ldots, f_m) + \sum_{i=1}^{m} f_i c_i$, where $\partial\Psi/\partial f_i = -c_i$. The generalized entropy production is objective if for each flux vector, $\boldsymbol{f}_i$, $\Psi^*(\boldsymbol{f}_i) = \Psi^*(Q\boldsymbol{f}_i)$. In practice the generalized entropy production is computed by integrating, with respect to the state variable, the relation for each flux in terms of the control variables.

A thermodynamic process involves the evolution of both the thermodynamic state variables appearing in the generalized energy density function and the fluxes. Let $\boldsymbol{p}^0$ be a point in the reference state of the body and $\boldsymbol{p} = \chi(\boldsymbol{p}^0, t)$ in the current state of the body.

Definition 2 A thermodynamic process is locally a function $\gamma : \Re^3 \times \Re \to \Re^3 \times \Re^{2n+1} \times \Re^{2m+1}$.

$$\gamma(\boldsymbol{p}^0, t) = \big(\boldsymbol{p}, \boldsymbol{x}(\boldsymbol{p}, t), \boldsymbol{y}(\boldsymbol{p}, t), \varphi^*(\boldsymbol{p}, \boldsymbol{x}, \boldsymbol{y}, t), \boldsymbol{f}(\boldsymbol{p}, t), \boldsymbol{c}(\boldsymbol{p}, t), \Psi^*(\boldsymbol{p}, \boldsymbol{f}, \boldsymbol{c}, t)\big), \tag{1}$$

where $\boldsymbol{x} = (x_1 \ldots, x_n)$, $\boldsymbol{y} = (y_1 \ldots, y_n)$, $\boldsymbol{f} = (f_1 \ldots, f_m)$, and $\boldsymbol{c} = (c_1 \ldots, c_m)$.

In geometric terms, a thermodynamic process is locally a function $\gamma : \Re^3 \times \Re \to \Re^3 \times C(\Re^{n+1}, n) \times C(\Re^{m+1}, m)$. The process is locally a time-dependent curve lying on the graphs of φ^* and Ψ^* in $\Re^3 \times \Re^{2n+1} \times \Re^{2m+1}$.

The definition of a gradient relaxation process for the fluxes parallels that for a thermodynamically homogeneous system. Affinities are associated with each flux component by $X_i = \partial\Psi^*/\partial f_i, \ i = 1, \ldots, m$. Transform the function Ψ^* to a function Ψ'^* of the affinities and fixed controls as in the thermodynamically homogeneous case. Recall that $\mathcal{L}_{\mathbf{v}}X_i$ is the Lie time derivative of X_i.

Definition 3 A thermodynamically non-homogenous gradient relaxation process for the fluxes is defined by the system of equations, for affinities $X_i, \ i = 1, \ldots, m$, at each point of the body, as

$$\mathcal{L}_{\mathbf{v}}X_i = -k_i \frac{\partial \Psi'^*}{\partial X_i}. \tag{2}$$

The analogous relation to that for the evolution in terms of the state variables of Chapter 3 gives the evolution in terms of each f_i. All results for the generalized thermodynamic function given in Chapter 3 carry over for the generalized entropy production function.

The constructions in Chapter 3 of an associated one-form and the definition of admissible processes apply to the generalized entropy production function. For example, the entropy production one-form at each point of $\mathcal{B}$ is defined to be locally

$$\bar{\omega} = dz - \sum_{i=1}^{m} f_i dc_i, \tag{3}$$

where z is the entropy production coordinate, and $\bar{\omega} = 0$ on the stationary manifold. This one-form acts on $T_*\big(\Re^3 \times C(\Re^{n+1}, n) \times C(\Re^{m+1}, m)\big)$. The entropy production in relaxation decreases over time (as the entropy approaches a maximum) because the one-form acting on a tangent to an admissible process path with fixed controls is negative.

The stationary manifold defined only in terms of the control coordinates, c_i, is associated with an entropy production $\Psi : \Re^m \to \Re$ whose graph in $\Re^{m+1}$ lifts to the graph of Ψ^* in the contact bundle of m-contact elements $C(\Re^{m+1}, m)$ since the tangent planes to the graph of Ψ are m-dimensional hyperplanes in $\Re^{m+1}$ by 1) of Def. 1. The submanifold which is the lift of the graph of Ψ is described by m equations $f_i = f_i(c_1, \ldots, c_m), \ i = 1, \ldots, m$, the equations of state for the stationary manifold. In this contact manifold, $\bar{\omega} = 0$ on the stationary manifold since by condition 1) of Definition (1),

$$\bar{\omega} = d\Psi^* - \sum_{i=1}^{m} f_i dc_i = d\Psi + \sum_{i=1}^{m} c_i df_i = \sum_{i=1}^{m} \left(\frac{\partial \Psi}{\partial f_i} + c_i\right) df_i = 0.$$

2.1 Relaxation Towards Equilibrium

The stationary manifold organizes the evolution of the fluxes, while the thermostatic manifold organizes the evolution of the other thermodynamic variables. The long-term behavior depends on the boundary conditions. The boundary conditions may force the process to the thermostatic manifold in which the thermodynamic variables remain constant over all points of the body, and the fluxes are all zero over all points of the body, as in the homogeneous case. Alternatively, the boundary conditions may imply that the long-term behavior is a time-dependent steady state in which some fluxes are non-zero.

If no forcing function is applied to the system and if one of the thermodynamic variables is perturbed so that the system is thermodynamically non-homogenous, then the evolution under a gradient relaxation process should approach an equilibrium state determined by the perturbed control variable and in which all the fluxes are zero.

Let the generalized entropy production be $\Psi^* = \Psi + \sum_{i=1}^{m} f_i c_i$. The requirement that the one-form, $\bar{\omega} = dz - \sum_i f_i dc_i$, be negative on an admissible non-equilibrium process, for fixed controls, implies that as for the generalized energy

$$\frac{d\Psi^*}{dt} = \frac{d\Psi}{dt} + \sum_{i=1}^{m} \frac{df_i}{dt} c_i < 0. \tag{4}$$

Therefore Ψ^* is decreasing for any gradient relaxation process. Note that this derivative is zero when the fluxes are zero.

Proposition 4 *As a process approaches equilibrium, all fluxes, f_i, tend to zero.*

Proof As the process approaches equilibrium, the rate of change of the internal energy density $\dot{u}$ tends to zero. Therefore $\operatorname{div} \boldsymbol{f}_i$ also tends to zero, so that $\boldsymbol{f}_i$ tends to a constant for all points in B. But the control tends to its equilibrium value, which is constant over all points of B. Therefore, the flux $\boldsymbol{f}_i$ approaches to zero.

3 Examples of Stationary Manifolds and Evolution of Fluxes

All of the stationary manifolds discussed below are composed of points that occur at minima of the generalized entropy production, Ψ^*, in $\Re^{2m}$. The stationary submanifolds of thermodynamic space act as attractors for the gradient relaxation process evolution equations. The flux paths approach the stationary manifold, but never catch up. Further, the control variables also evolve due to the gradient relaxation process defined from the generalized energy. The following examples exhibit the equations for the evolution of various fluxes under fixed controls; these thermodynamically based equations produce finite velocity transport processes.

3.1 Thermal Gradients and Fluxes

The Fourier law prediction of instantaneous response to disturbances has long been known to be physically unrealistic. Let $\boldsymbol{q}$ be the heat flux. The modification,

$\boldsymbol{q} + \tau\dot{\boldsymbol{q}} = -K\nabla\theta$, of the Fourier relation was proposed by Cattaneo in 1948 to produce a finite velocity of propagation of heat. See Müller and Ruggeri (1998, pp. 12–14) for a discussion of Cattaneo's reasoning. Bogy and Naghdi (1970) say this is an empirical equation that Vernotte hypothesized in 1958. A derivation in terms of extended irreversible thermodynamics is given in Section 3.3.1 in Chapter 1. The Bowen (1989) continuum thermodynamics derivation of the Maxwell-Cattaneo equation in a rigid heat conductor restricts it to small perturbations from equilibrium. The rigid body assumption eliminates questions about objectivity. The energy is taken to be a function of $\boldsymbol{q}$ by the usual continuum thermodynamic argument based on the Clausius-Duhem inequality, and $\tilde{\sigma}$ is the affinity corresponding to $\boldsymbol{q}$. The Bowen derivation is only valid near equilibrium because it essentially involves taking a quadratic Taylor series approximation of the specific energy and a linear Taylor series approximation of the assumed function $\dot{\boldsymbol{q}}(\theta, \nabla\theta, \tilde{\sigma})$ (Bowen 1989, Sect. 5.2). This argument breaks down if one does not permit the specific energy to be a function of $\boldsymbol{q}$. In the development presented here the specific entropy production, not the specific energy, is taken to be a function of $\boldsymbol{q}$, a strategy which seems more natural. Specific energy is never taken as a function of a rate in the maximum dissipation non-equilibrium construction. This choice avoids the problems encountered by Green and Naghdi (1977) that were discussed in Chapter 1, Section 3.4.

Requiring the entropy production, not the energy, to be a function of the heat flux is consistent with Müller and Ruggeri (1998, p. 11). They also point out that the classical Maxwell-Cattaneo equation is not frame invariant. Their method of accounting for this problem differs from the maximum dissipation construction which employs the gradient relaxation process that puts the Cattaneo equation on a consistent thermodynamic foundation.

The conjugate thermodynamic variables are taken to be the gradient, $\nabla(1/\theta)$, as control, and as state, the heat flux vector, $\boldsymbol{q}$, which points in the direction of the energy flow. To guarantee that heat energy flows from hot to cold, it is required that $\boldsymbol{q} \cdot \nabla\theta < 0$. The generalized entropy production function, $\Psi^*(\boldsymbol{p}, \boldsymbol{q}, \nabla(1/\theta))$, for $\boldsymbol{p}$ a point in $\mathcal{B}$, is chosen so that $\partial\Psi^*/\partial\boldsymbol{q} = 0$ gives the emprical Fourier relation defining the Fourier stationary submanifold, $\boldsymbol{q} = -K\nabla\theta$. The tensor K does not depend on time, but may depend on temperature, strain, or the point in $\mathcal{B}$. The local generalized entropy production function is, at each point in $\mathcal{B}$,

$$\Psi^* = \left(\frac{1}{2\theta^2}\right) K^{-1}\boldsymbol{q} \cdot \boldsymbol{q} - \boldsymbol{q} \cdot \nabla\left(\frac{1}{\theta}\right). \tag{5}$$

The control $\nabla(1/\theta)$ is fixed and $\dot{\theta} = 0$. Note that Ψ^* has units of energy per volume per second per degree. As desired, the zero gradient condition on Ψ^*, $\partial\Psi^*/\partial\boldsymbol{q} = 0$, produces the Fourier relation. The stationary manifold contains minima of the generalized entropy production function since the Hessian of Ψ^* with respect to state variables is positive.

The Objective Maxwell-Cattaneo Equation Assume that the heat conduction tensor is isotropic of the form KI and that $\nabla(1/\theta)$ is fixed. From Eq. (5), the affinity is $X = \left(\frac{1}{\theta^2}\right) K^{-1}\boldsymbol{q} + \frac{1}{\theta^2}\nabla\theta$. Then

$$\Psi'^{*} = \left(\frac{1}{2}\theta^2\right)(X \cdot KX - \nabla\theta \cdot K\nabla\theta).$$

The gradient relaxation process $\mathcal{L}_{\mathbf{v}}X = -k(\partial\Psi'^{*}/\partial X)$, for $k = (1/\theta^2)K^{-1}\tau^{-1}$ with τ diagonal, since $\mathcal{L}_{\mathbf{v}}X = (1/\theta^2)K^{-1}\mathcal{L}_{\mathbf{v}}\boldsymbol{q}$ yields

$$\boldsymbol{q} + \tau\mathcal{L}_{\mathbf{v}}\boldsymbol{q} + K\nabla\theta = 0. \tag{6}$$

This relation will be called the objective Maxwell-Cattaneo equation.

The Maxwell-Cattaneo relation in the reference coordinates is obtained by pulling back (6) to generalize Holzapfel (2000, 41–46). The term $\text{grad}(\theta) = \nabla\theta$ is covariant and pulls back by F^t so that $F^t\text{grad}(\theta) = \text{Grad}(\theta)$. Likewise, since $\boldsymbol{q}$ is contravariant, $JF^{-1}\boldsymbol{q} = \boldsymbol{Q}$, the heat flow in the reference configuration. Also $JK = K_o$ and $J\tau = \tau_o$. Substitution in (6) yields

$$JF\boldsymbol{Q} + \tau F\frac{D}{Dt}(JF^{-1}\boldsymbol{q}) + KF^{-t}\text{Grad}(\theta) = 0,$$

so that dividing by J,

$$\boldsymbol{Q} + \tau_o\frac{D\boldsymbol{Q}}{Dt} + F^{-1}K_oF^{-t}\text{Grad}(\theta) = 0.$$

One can view $F^{-1}K_oF^{-t}$ in this expression as a strain tensor by interpreting K_o^{-1} as a metric tensor.

In the case of small displacements and heat conduction tensor KI, the gradient relaxation process, for some relaxation coefficient k, defining the flux evolution is, by writing the relaxation process in terms of the states variables (Chapter 3),

$$\dot{\boldsymbol{q}} = -k\left(\frac{1}{K\theta^2}\right)^{-2}\left[\left(\frac{1}{K\theta^2}\right)\boldsymbol{q} - \nabla\left(\frac{1}{\theta}\right)\right]. \tag{7}$$

The evolution (7) for $k = 1/(K\theta^2\tau)$ with τ constant simplifies to $\boldsymbol{q} + \tau\dot{\boldsymbol{q}} = -K\nabla\theta$, which is the classical Maxwell-Cattaneo evolution equation. The evolution of the heat flux approaches the submanifold of solutions to the Fourier equation in the long term as $\dot{\boldsymbol{q}}$ tends to zero. The evolution defined by (7) satisfies the condition $\boldsymbol{q} \cdot \nabla\theta < 0$ at all times during the evolution when $\nabla\theta$ is held constant.

Example: *Heat conduction and radiation in a uniaxially loaded neo-Hookean rod* A rod of length, ℓ, is initially in a steady state with temperature $\theta(x) = \theta_1 + (\theta_2 - \theta_1)x/\ell$, $0 \leq x \leq \ell$. The heat flux in the rod is assumed to obey the Fourier relation in steady state. In the Fourier stationary relation, let $K(\theta) = K_o[1 - \xi(\theta - \theta_o)]$, where θ_o is the reference temperature and ξ is a softening parameter (e.g. Holzapfel, 2000, p. 342). Therefore, initially $q(x) = K(\theta(x))(\theta_2 - \theta_1)/\ell$, which is not homogeneous.

The rod is suddenly placed in a bath of temperature θ_B and a stress P_1 is suddenly applied at the two ends of the rod, and evolution equations are sought describing the relaxation to equilibrium defined by $\theta(x) = \theta_B$ and $q(x) = 0$. The temperature $\theta(x)$ and the stretch $\lambda(x)$ are taken as the state variables at each point x. The equilibrium stretch $\lambda(x)$ is unknown a priori. The rod is not immediately in force equilibrium so that the Piola stress at points along the rod at each time, $P(x, t)$, remains to be determined.

Assume entropic elasticity in which the mechanical deformation is incompressible. Then $J = \exp[3\alpha(\theta - \theta_o)]$. Assume $\lambda_1 = \lambda$, and $\lambda_2 = \lambda_3 = (J/\lambda)^{1/2}$. As in Holzapfel (2000, Eq. 7.107), or in Haslach and Zeng, (1999), the long-term Helmholtz free energy is

$$\varphi(\lambda, \theta) = \frac{\theta}{\theta_o}(\lambda^2 + 2(J/\lambda) - 3) + c_o[(\theta - \theta_o) - \theta \ln(\theta/\theta_o)], \tag{8}$$

where c_o is the specific heat capacity at the reference temperature. The associated generalized energy is $\varphi^*(\lambda, \theta; P, \eta) = \varphi(\lambda, \theta) + \theta\eta - P\lambda$. The evolution equations for θ and λ are obtained from the construction applied to the generalized energy, φ^*. Compatibility between the evolution of the thermodynamic variables at each point, x, of the rod is enforced by the balance laws. The affinity driving the temperature is that obtained by imagining the specific entropy perturbed to its value at equilibrium; i.e. that value, η_B, corresponding to θ_B. If θ_o is the reference temperature, then $\eta_B = (-1/\theta_o)[\lambda + (2 + 3\alpha\theta_B(J(\theta_B)/\lambda) - 3] + c_o \ln(\theta_B/\theta_o)$.

The evolution of the temperature at point, x, of the rod is

$$\dot{\theta} = -k_\theta \left(\frac{\partial^2\varphi}{\partial\theta^2}\right)^{-2} \left(\frac{\partial\varphi}{\partial\theta} + \eta_B\right). \tag{9}$$

The evolution of the stretch, λ, is given at each point, x, of the rod by

$$\dot{\lambda} = -k_\lambda \left(\frac{\partial^2\varphi}{\partial\lambda^2}\right)^{-2} \left(\frac{\partial\varphi}{\partial\lambda} - P(x, t)\right). \tag{10}$$

No Lie derivative appears in these equations because the state variables are scalars.

As θ evolves, the control $\nabla(1/\theta)$ in the evolution of the scalar heat flux, q, also changes. The evolution of the heat flux is obtained from the entropy production function, Ψ^*, which yields the one-dimensional version of (7),

$$\dot{q} = -k_q \left(\frac{1}{K\theta^2}\right)^{-2} \left[\left(\frac{1}{K\theta^2}\right) q - \nabla\left(\frac{1}{\theta}\right)\right]. \tag{11}$$

One could choose k_q to produce the Maxwell-Cattaneo evolution with the initial condition of the heat flux given above.

In general, energy balance $\dot{u} = \sigma : D - \nabla \boldsymbol{q} + r$ yields a relation for $P(x, t)$. Assume that the radiation term $r = \alpha_1(\theta_B - \theta)$ is Newton's law of cooling, which is a maximum dissipation process by Haslach (1997). In terms of the Helmholtz energy,

$$\dot{\varphi} + \dot{\theta}\eta_B = (J^{-1}P(x,t)F^t) : (\dot{F}F^{-1}) - \nabla \boldsymbol{q} + r.$$

Then the one-dimensional energy balance assuming the constant perturbed entropy, η_B, is

$$\dot{\varphi} + \dot{\theta}\eta_B = J^{-1}P(x,t)\lambda\dot{\lambda}\lambda^{-1} - \frac{\partial q}{\partial x} + \alpha_1(\theta - \theta_B). \tag{12}$$

This equation produces $P(x, t)$ because evolution equations are available for all other terms.

The transient behavior as the rod relaxes to equilibrium is obtained by simultaneously solving the four equations for $\dot{\theta}$, $\dot{\lambda}$, $\dot{q}$, and the energy balance subject to the initial conditions for θ, q, and $\lambda = 1$.

3.2 Non-steady Transport in Porous Biological Membranes

Many studies of fluid transport through biological membranes assume steady behavior in the combined form called Starling's law (Fung, 1990, p. 291) for membrane filtration, $\mathcal{Q} = K(\Delta p - \hat{\sigma}\Delta\Pi)$, where $\mathcal{Q}$ is the volume rate of flow per area (and so has units of velocity), K is the permeability coefficient, $\hat{\sigma}$ is the reflection coefficient, $\Delta\Pi$ is the osmotic pressure which forces flow from the side of higher vapor pressure to the lower, and Δp is an externally applied pressure. The opposite sign indicates that a large enough externally applied pressure can force flow from the low vapor pressure side of the membrane. Fung develops these relations from the controversial linear Onsager relations. Here, these relations are taken instead as empirically determined models for steady behavior. But it is well recognized that in vivo these transport processes are not in fact steady. The flux evolution construction described above, based on constitutive equations for the steady behavior, provides a technique to model realistic non-steady behavior. For example, a simple model assumes that the two processes are uncoupled and that the relaxation modulus in either case is a constant.

The generalized entropy production for the fluid is

$$\Psi^*(\mathcal{Q}_1, p) = \frac{1}{2\mathcal{K}\theta}\mathcal{Q}_1 \cdot \mathcal{Q}_1 + \mathcal{Q}_1 \cdot \nabla\left(\frac{p}{\theta}\right),$$

where $\mathcal{Q}_1$ is the pressure induced volume flux per area per second so that it has units of velocity, p is the hydrostatic pressure, and $\mathcal{K}$ is the hydraulic permeability. The steady state is then Darcy's law, under the isothermal assumption that $\nabla\theta = 0$, given by

$$0 = \frac{d\Psi^*(\mathcal{Q}_1, p)}{d\mathcal{Q}_1} = \frac{1}{\mathcal{K}\theta}\mathcal{Q}_1 + \nabla\left(\frac{p}{\theta}\right).$$

The maximum change in entropy production, in the same manner as for the heat flux, yields the unsteady equation, for one choice of the relaxation modulus $k = 1/(\mathcal{K}\theta\tau_1)$, where τ_1 is a constant,

$$\tau_1\dot{\mathcal{Q}}_1 + \mathcal{Q}_1 = -\mathcal{K}\nabla p.$$

Such a transport process induced by the external pressure may compete with diffusion of the water solvent under osmotic transfer. In a multi-substance system, the chemical potential, μ_i, and the concentration, $C_i = \rho_i/\rho$ where ρ is the density of the composite and ρ_i is the density of the ith component substance, are conjugate thermodynamic variables. The Gibbs one form, ω, includes the term $-\sum_i \mu_i dC_i$ in this case.

The osmotic pressure induced flow of water per area per second is $\mathcal{Q}_2$. The control variable is $\nabla(\mu/\theta)$. The generalized entropy production is, for D_m a constant,

$$\Psi^* = \frac{1}{2D_m\theta}\mathcal{Q}_2 \cdot \mathcal{Q}_2 + \mathcal{Q}_2 \cdot \nabla\left(\frac{\mu}{\theta}\right). \tag{13}$$

By a calculation from Eq. (13) similar to that for the heat flux, the evolution equation for fixed control variable, $\nabla(\mu/\theta)$, is

$$\mathcal{Q}_2 + \tau_2\dot{\mathcal{Q}}_2 + D_m\nabla\mu = 0, \tag{14}$$

where the constant τ_2 has units of time and $k = 1/(D_m\theta\tau)$ is the gradient relaxation coefficient. As before, the general evolution is obtained by letting $\nabla\mu$ vary with time in (14).

The net unsteady flow is obtained by combining these results.

$$\Psi^* = \frac{1}{2\mathcal{K}\theta}\mathcal{Q}_1 \cdot \mathcal{Q}_1 + \frac{1}{2D_m\theta}\mathcal{Q}_2 \cdot \mathcal{Q}_2 + \mathcal{Q}_1 \cdot \nabla\left(\frac{p}{\theta}\right) + \mathcal{Q}_2 \cdot \nabla\left(\frac{\mu}{\theta}\right). \tag{15}$$

Therefore, the net unsteady flow $\mathcal{Q} = \mathcal{Q}_1 - \mathcal{Q}_2$ is obtained from the differential equation

$$\mathcal{Q} + \tau_1\dot{\mathcal{Q}}_1 - \tau_2\dot{\mathcal{Q}}_2 + \mathcal{K}\nabla p - D_m\nabla\mu = 0. \tag{16}$$

Water moves from high to lower vapor pressure under osmotic pressure and so passes from pure water to a solution. It is not the same as diffusion of the solute. Starling's law shows that while osmotic pressure forces flow from the high to the low vapor pressure, an externally applied pressure can reverse the flow. The flow can reverse direction on each cycle of the hemodynamic pressure. This process can compete with diffusion.

3.3 Electromagnetic Fluxes

Ohm's law, as is well known, is an empirical relation describing a steady state (e.g. Fabrizio and Morro, 2003). The flux, $\mathbf{J_e}$, is the free current density in amperes per square meter. Ohm's law is $\mathbf{J_e} = \bar{\sigma}\mathbf{E}$, where $\mathbf{E}$ is the electrical field in volts per meter and $\bar{\sigma}$ is the conductivity. In steady state $\mathbf{E} = -\nabla\phi$, where ϕ is the scalar-valued electric potential. Application of the construction produces a relation for the transient behavior as the system approaches the stationary state described by Ohm's law.

The state variable is $\mathbf{J_e}$ and the control variable is $\nabla(\phi/\theta)$. The generalized entropy production is

$$\Psi^* = \frac{\mathbf{J_e} \cdot \mathbf{J_e}}{2\bar{\sigma}\theta} + \mathbf{J_e} \cdot \nabla\left(\frac{\phi}{\theta}\right), \tag{17}$$

The isothermal zero gradient condition gives Ohm's law as the stationary state since $\nabla\theta = 0$. The associated gradient relaxation process if small displacements are assumed is

$$\dot{\mathbf{J}}_\mathbf{e} = -k(\bar{\sigma}\theta)^2\left[\frac{\mathbf{J_e}}{\bar{\sigma}\theta} + \nabla\left(\frac{\phi}{\theta}\right)\right]. \tag{18}$$

Setting $k = 1/\tilde{L}\bar{\sigma}^2\theta$, where $\tilde{L}$ is the inductance, yields Ohm's law modified to account for inductance (e.g. Ulbrich, 1961, Eq. (7)),

$$\tilde{L}\dot{\mathbf{J}}_\mathbf{e} = -\left(\frac{\mathbf{J_e}}{\bar{\sigma}} + \nabla\phi\right). \tag{19}$$

This equation predicts finite velocity propagation of the current in contrast to the infinite velocity prediction of Ohm's law. Subjecting this maximum dissipation relation to conservation of charge, $\nabla\mathbf{J_e} = -\tilde{C}(d\phi/dt)$, where $\tilde{C}$ is the capacitance and restricting to one dimension recovers the classical differential equation of telegraphy. The gradient relaxation evolution equation gives a class of models indexed by k for the transient to the steady state, which is typically quite rapid in circuits.

3.4 Fluids

The standard stress-strain constitutive equations for fluids are for stationary states, since the equilibrium state is stress free. Therefore the constitutive equations for fluids may be obtained from the zero gradient condition on a generalized entropy production function.

For example, the constitutive equation for an isothermal Newtonian fluid arises from the generalized entropy production function,

$$\Psi^* = \frac{1}{\theta}[-pD + \lambda \mathbf{Tr}(D)D + \mu D : D - \sigma : D], \tag{20}$$

where here D is the spatial rate of deformation tensor, σ is the Cauchy stress tensor, and λ and μ are positive material constants with units of stress-time. The zero gradient condition for state variable, D, and control variable, σ, is

$$\frac{\partial \Psi^*}{\partial D} = \frac{1}{\theta}[-pI + \lambda \mathbf{Tr}(D)I + 2\mu D - \sigma] = 0,$$

which yields the Newtonian fluid equation. If a fluid were at rest and a stress, σ, suddenly is applied, the gradient relaxation process predicts the evolution of rate of deformation to the stationary state defined by the constitutive equation.

Example A Maxwell model consisting of a spring and dashpot in series is most suited to fluids since it can model stress relaxation but not creep. The classical uniaxial Maxwell model with spring constant E_1 and dashpot constant c is obtained as a gradient relaxation process from the generalized entropy production in which the flux, $\dot{\epsilon}$, is the control variable,

$$\Psi^* = \frac{\sigma^2}{2c\theta} - \frac{\sigma\dot{\epsilon}}{\theta}.$$

The gradient relaxation process with $k = E_1/c^2\theta$ yields the uniaxial Maxwell model,

$$\dot{\sigma} = -k\left(\frac{d^2\Psi^*}{d\sigma^2}\right)^{-2}\left[\frac{d\Psi^*}{d\sigma}\right] = -\frac{E_1}{c^2\theta}(c^2\theta^2)\left(\frac{\sigma}{c\theta} - \frac{\dot{\epsilon}}{\theta}\right) = -E_1\frac{\sigma}{c} + E_1\dot{\epsilon}. \tag{21}$$

The stationary manifold $\sigma = c\dot{\epsilon}$, is an attractor since $d^2\Psi^*/d\sigma^2 = 1/c\theta > 0$.

4 Admissible Non-homogeneous Processes

In a thermodynamically homogeneous system, no fluxes exist. The requirement that the Gibbs form be non-positive on a homogeneous process cannot guarantee for a thermodynamically non-homogeneous process that heat flows from hot regions to cold, or that diffusion moves particles from regions of high concentration to low concentration, or that electrical current flows from regions of high voltage potential to regions with lower voltage potential, etc. Additional criteria are needed to ensure that admissible processes obey these physical observations.

Here the proper direction is ensured by a condition on the scalar product, denoted by $\langle \cdot, \cdot \rangle$. The scalar product induces the heat flux one-form $\omega_q = \langle \cdot, (\nabla\theta)/\theta \rangle$, the diffusion one-form $\omega_d = \langle \cdot, (\nabla\mu)/\theta \rangle$ for mass flux $\mathbf{J}$, and the electromagnetic one-form $\omega_e = \langle \cdot, (\nabla\phi)/\theta \rangle$. The flow direction for a process is defined by

$$\begin{aligned} \omega_q(\mathbf{q}) &= \langle \mathbf{q}, (\nabla\theta)/\theta \rangle < 0; \\ \omega_d(\mathbf{J}) &= \langle \mathbf{J}, (\nabla\mu)/\theta \rangle < 0; \\ \omega_e(\mathbf{J}_e) &= \langle \mathbf{J}_e, (\nabla\phi)/\theta \rangle < 0. \end{aligned} \tag{22}$$

The non-homogeneous local Gibbs form, ω_G, for a body made of a single substance and subject to no electromagnetic effects is, at each point $\boldsymbol{p}$ of $\mathcal{B}$, the sum of the Gibbs form for the homogeneous case and the heat flux one form.

$$\omega_G = dz - \sum_i x_i dy_i + \omega_q(\boldsymbol{q}). \tag{23}$$

Definition 5 A thermodynamically admissible non-equilibrium process in the thermodynamic system defined by the generalized energy function, φ^*, and the generalized entropy production function, Ψ^*, is a process for which $\omega < 0$ and $\bar{\omega} < 0$ and the transport one-form relations (22) are satisfied, and the balance laws are obeyed at each point along the path.

This definition requires that $\omega_G(\boldsymbol{t}_p, q) < 0$ for a process on a body made of one substance and subject to no electromagnetic effects.

4.1 Relation to the Clausius-Duhem Inequality

For a thermodynamically non-homogeneous system in a body made of one substance and subject to no electromagnetic effects, the second law in the form of the Clausius-Duhem inequality, from continuum thermodynamics (e.g. Holzapfel, 2000), is in the reference configuration,

$$0 \geq \frac{du}{dt} - \theta\frac{d\eta}{dt} - P_{ij}\frac{dF_{ij}}{dt} - A_i\frac{da_i}{dt} + \boldsymbol{q} \cdot \left(\frac{\nabla\theta}{\theta}\right), \tag{24}$$

where u is the internal energy density, a_i and A_i are conjugate internal variables, and P is the first Piola stress. Because the Clausius-Duhem inequality pulls back to the reference state up to the factor $J > 0$, since all energies are per unit volume, it suffices to work in the reference configuration. The Gibbs form must be non-positive on any admissible process. The Clausius-Duhem inequality of continuum thermodynamics is implied by this requirement. Let $\boldsymbol{t}_p$ be the tangent to the path of a process (Section 4.5 in Chapter 3).

Proposition 6 *For any process, the Clausius-Duhem inequality is equivalent to* $\omega_G(\boldsymbol{t}_p, \boldsymbol{q}) \leq 0$.

Proof The free, or complementary, energy required in the construction of the generalized internal energy, $U^*(\theta, P_{ij}, A_i; \eta, F_{ij}, a_i) = \psi(\theta, P_{ij}, A_i) + \theta\eta + P_{ij}F_{ij} + A_i a_i$, is the Legendre transform $\psi(\theta, P_{ij}, A_i) = u - \theta\eta - P_{ij}F_{ij} - A_i a_i$ where η, F_{ij}, and a_i are functions of θ, P_{ij}, and A_i. Therefore by substituting for u in (24),

$$0 \geq \frac{d\psi}{dt} + \eta\frac{d\theta}{dt} + F_{ij}\frac{dP_{ij}}{dt} + a_i\frac{dA_i}{dt} + \boldsymbol{q} \cdot \left(\frac{\nabla\theta}{\theta}\right).$$

For any process on the graph of U^* with tangent to the path $\boldsymbol{t}_p$, the action of the Gibbs form is given by the calculation in Section 4.6 in Chapter 3), $\omega(\boldsymbol{t}_p) = d\psi/dt + \sum_{i=1}^{n} y_i(dx_i/dt)$,

$$\omega_G(\boldsymbol{t}_p, \boldsymbol{q}) = \frac{d\psi}{dt} + \eta\frac{d\theta}{dt} + F_{ij}\frac{dP_{ij}}{dt} + a_i\frac{dA_i}{dt} + \boldsymbol{q} \cdot \left(\frac{\nabla\theta}{\theta}\right).$$

Therefore $\omega_G(\boldsymbol{t}_p, \boldsymbol{q}) \leq 0$ iff the Clausius-Duhem inequality holds.

4.2 The Balance Laws as Differential Forms

The balance laws are given here for the case of a body made of a single substance and subject to no electromagnetic effects. The balance laws are represented by differential one-forms that act on the tangent bundle, $T_*(\mathcal{N} \times \Re)$, where the factor $\Re$ is time, and in particular act on the tangents to the paths representing processes. Recall from Chapter 7 that $\mathcal{N} \equiv (\mathcal{B} \times \Re) \bigoplus T_*\mathcal{B} \bigoplus T^*\mathcal{B} \bigoplus (T_*\mathcal{B} \otimes T_*\mathcal{B}) \bigoplus (T^*\mathcal{B} \otimes T^*\mathcal{B}) \bigoplus (T_*\mathcal{B}_o \otimes T_*\mathcal{B})$.

Local balance of linear momentum gives the equation of motion. The velocity and body force per unit volume have the components v_{p_i} and b_{p_i} respectively, and σ_{ij} is the Cauchy stress. To obtain the equation of motion in the p_1-direction, define the 1-form $\rho v_{p_1} dt + \sigma_{11} dp_1 + \sigma_{21} dp_2 + \sigma_{31} dp_3$, where ρ is the mass density. Apply the Hodge star operator (Darling, 1994) to map this form to the 3-form,

$$\begin{aligned}\beta_{p_1} = {} & -\rho v_{p_1} dp_1 \wedge dp_2 \wedge dp_3 - \sigma_{11} dp_2 \wedge dp_3 \wedge dt \\ & + \sigma_{21} dp_1 \wedge dp_3 \wedge dt - \sigma_{31} dp_1 \wedge dp_2 \wedge dt.\end{aligned} \tag{25}$$

Then, since $d\beta_{p_1} = (\rho \dot{v}_{p_1} - \partial\sigma_{11}/\partial p_1 - \partial\sigma_{21}/\partial p_2 - \partial\sigma_{31}/\partial p_3) dp_1 \wedge dp_2 \wedge dp_3 \wedge dt$, the process obeys the equation of motion in the p_1-direction if $d\beta_{p_1} - b_{p_1} dp_1 \wedge dp_2 \wedge dp_3 \wedge dt = 0$. The analogous construction gives the equations of motion for the p_2 and p_3-directions.

The conservation of energy form is constructed in a similar manner. Denote the internal energy per unit volume by u and the p_i component of $\sigma \boldsymbol{v}$ by $(\sigma \boldsymbol{v})_{p_i}$ and likewise for the heat flux vector $\boldsymbol{q}$. Define the 1-form, $(0.5\rho\boldsymbol{v}\cdot\boldsymbol{v}+u)dt+[-q_{p_1}+(\sigma\boldsymbol{v})_{p_1}]dp_1+[-q_{p_2}+\sigma\boldsymbol{v})_{p_2}]dp_2+[-q_{p_3}+(\sigma\boldsymbol{v})_{p_3}]dp_3$. Apply the Hodge star operator to obtain the 3-form,

$$\begin{aligned}\alpha &= -(0.5\rho\boldsymbol{v}\cdot\boldsymbol{v}+u)dp_1\wedge dp_2\wedge dp_3-(-q_{p_1}+(\sigma\boldsymbol{v})_{p_1})dp_2\wedge dp_3\wedge dt\\ &\quad +(-q_{p_2}+(\sigma\boldsymbol{v})_{p_2})dp_1\wedge dp_3\wedge dt-(-q_{p_3}+(\sigma\boldsymbol{v})_{p_3})dp_1\wedge dp_2\wedge dt.\end{aligned} \tag{26}$$

Let r denote the heat radiation term. Conservation of energy is satisfied on the process if $d\alpha-(\boldsymbol{b}\cdot\boldsymbol{v}+r)dp_1\wedge dp_2\wedge dp_3\wedge dt=0$ because

$$d\alpha=[\rho\boldsymbol{v}\cdot\dot{\boldsymbol{v}}+\dot{u}-\mathbf{div}(\sigma\cdot\boldsymbol{v})+\nabla\cdot\boldsymbol{q}]dp_1\wedge dp_2\wedge dp_3\wedge dt.$$

This expression reduces to the standard result, $\dot{u}=\sigma:D-\nabla\cdot\boldsymbol{q}+r$, using the balance of linear momentum form and using the relation $\sigma:\nabla\boldsymbol{v}=\sigma:D$, where D is the spatial rate of deformation tensor. The term $\boldsymbol{b}\cdot\boldsymbol{v}+r$ is neglected in many problems.

The conservation of mass form is obtained by applying the Hodge star operator to the 1-form, $\rho dt+\rho v_1dp_1+\rho v_2dp_2+\rho v_3dp_3$ to obtain

$$\begin{aligned}\alpha_m &= -\rho dp_1\wedge dp_2\wedge dp_3-\rho v_1dp_2\wedge dp_3\wedge dt\\ &\quad +\rho v_2dp_1\wedge dp_3\wedge dt-\rho v_3dp_1\wedge dp_2\wedge dt.\end{aligned} \tag{27}$$

Conservation of mass is satisfied by a process if

$$d\alpha_m=\left(\frac{\partial\rho}{\partial t}-\mathbf{div}(\rho\boldsymbol{v})\right)dp_1\wedge dp_2\wedge dp_3\wedge dt=0. \tag{28}$$

4.3 Non-homogeneous Examples

Example An initially unstrained, straight bar of uniform specific weight, ρ, and length, L, is mounted to a ceiling so that it is loaded by its own weight. The goal is to determine the time evolution of the deformation gradient as it approaches equilibrium. The conjugate thermodynamic state variables are the uniaxial first Piola-Kirchhoff stress, P, and the uniaxial deformation gradient, $F=\partial\chi(x,t)/\partial x$, where x is the material reference coordinate measured down from the ceiling (Fig. 9.1). The long-term stress-strain relation is assumed linear, $P=E_1(F-I)$, where E_1 is a constant modulus and where I is the identity tensor. The control variable $P(x,t)$ varies with both time and position. The generalized energy density function is

$$\varphi^*=0.5E_1(F-I)\cdot(F-I)-P\cdot F.$$

Fig. 9.1 Hanging Bar

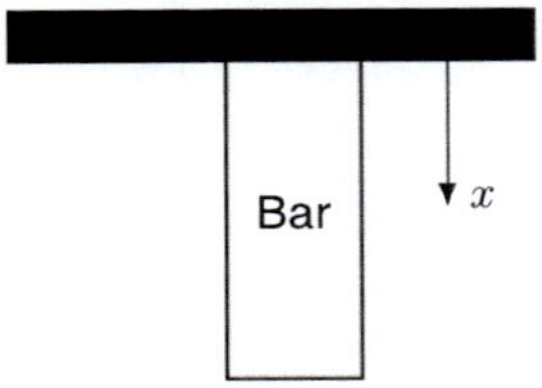

On an admissible process, the balance of linear momentum holds in material coordinates, where $m = \rho AL/g$ is the mass, and gives the variation of $P(x, t)$.

$$\frac{\partial P(x,t)}{\partial x} + \rho = m\frac{\partial^2 \chi(x,t)}{\partial t^2} \tag{29}$$

The evolution equation for a gradient process for a long-term behavior $P = E_1(F - I)$ is

$$\dot{F} = -kE_1^{-2}[E_1(F - I) - P(x,t)] \tag{30}$$

Differentiating (29) once with respect to x and (30) twice with respect to x and combining the results to eliminate $\partial^2 P/\partial x^2$ yields the governing equation for F,

$$m\frac{\partial^2 F}{\partial t^2} = E_1\frac{\partial^2 F}{\partial x^2} + \frac{E_1^2}{k}\frac{\partial^3 F}{\partial t \partial x^2}. \tag{31}$$

Note that it was not necessary to use conservation of energy.

The simplified boundary conditions are $F(L, t) = 1$ since $P(L, t) = 0$, and $F(x, 0) \equiv 1$. Assume that $F(x, t) = w(x, t) + a(x) + c(t)$, where $w(x, t)$ satisfies (31) and has homogeneous boundary conditions. To satisfy the boundary conditions and (31), take $a(x) = \rho(L - x)/E_1$ and $c(t) \equiv 1$. Then $F(L, t) = 1$. The function $w(x, t) = f(x)g(t)$ is determined by separation of variables. The homogeneous boundary conditions $f(0) = 0$ and $f(L) = 0$ imply that for constants B_n, $n = 1, 2, 3, \ldots$,

$$f_n(x) = B_n \sin\left(\frac{n\pi x}{2L}\right).$$

Also for constants C_{n_1}, $n = 1, 2, 3, \ldots$,

$$g_n(t) = \exp\left(\frac{-n^2\pi^2 E_1^2 t}{8L^2 mk}\right)[C_{n_1}\exp(r_1 t\iota) + C_{n_2}\exp(r_2 t\iota)],$$

where

$$r_i = \pm\sqrt{\frac{n^2\pi^2 E_1}{mL^2} - \left(\frac{n^2\pi^2 E_1^2}{8mL^2k}\right)^2}.$$

Note that $\lim_{t\to\infty} g_n(t) = 0$. The solution $F(x, t)$ is therefore

$$F(x,t) = a(x) + \sum_{n=1}^{\infty} f_n(x)g_n(t) + 1. \tag{32}$$

To satisfy the condition that $F(x, 0) \equiv 1$, the coefficients in the Fourier series for $f(x)$ are chosen so that the summand at $t = 0$ is $-\rho(L - x)/E_1$. The long-term behavior is $\lim_{t\to\infty} F(x, t) = a(x) + 1$, which is the equilibrium solution obtained in linear elasticity. The equilibrium values of the Piola stress are $P(x) = \rho(L - x)$ for $0 \le x \le L$.

Example Consider a one-dimensional rigid rod of length L initially in equilibrium at temperature θ_o and entropy η_o. Perturb the temperature along the rod to $\theta(x, 0)$. This makes $\theta(x, t)$ the control variable. The energy equation relates flux evolution to the time-dependent variation of the control in the homogeneous contact manifold. Assume the constitutive relation that $\dot{u} = c_v\dot{\theta}$. Assume that the generalized energy gives a linear relation between temperature and entropy so that $\varphi^* = 0.5b\eta^2 - \eta\theta$. The generalized entropy production is $\Psi^* = \left(\frac{1}{2\lambda\theta^2}\right)\boldsymbol{q}\cdot\boldsymbol{q} - \boldsymbol{q}\cdot\nabla\left(\frac{1}{\theta}\right)$. Choose the relaxation modulus to produce the Maxwell-Cattaneo evolution for the heat flux.

$$c_v\dot{\theta} = \frac{dq}{dx}; \tag{33}$$

$$q + \tau\dot{q} = -\lambda\frac{d\theta}{dx}; \tag{34}$$

$$\dot{\eta}(x,t) = -kb^{-2}\big(b\eta(x,t) - \theta(x,t)\big). \tag{35}$$

The latter equation shows the entropy evolution at each point $0 \le x \le L$.

The heat flux equation after differentiating by x is written in terms of the control variable, $\theta(x, t)$, by means of energy balance to obtain

$$c_v\dot{\theta} + \tau c_v\ddot{\theta} = -\lambda\frac{d^2\theta}{dx^2}. \tag{36}$$

The solution to this partial differential equation gives the instantaneous value of the control variable at each point along the rod.

As the system approaches equilibrium, the temperature gradient $\nabla\theta$ and the heat flux must both tend to zero to give a uniform equilibrium temperature θ_e. As they tend to this limit, the stationary state manifold in the flux contact space tends to a point at some value, η_e, of the entropy. The uniform equilibrium temperature is given by $\theta_e = b\eta_e$.

References

D. B. Bogy and P. M. Naghdi (1970). On heat conduction and wave propagation in rigid solids. *Journal of Mathematical Physics* **11**, 917–923.

R. M. Bowen (1989). *Introduction to Continuum Mechanics for Engineers*, Plenum Press, New York, NY.

R. W. R. Darling (1994). *Differential Forms and Connections*, Cambridge University Press, Cambridge.

M. Fabrizio and A. Morro (2003). *Electromagnetism of Continuous Media*, Oxford University Press, Oxford, UK.

Y. C. Fung (1990). *Biomechanics: Motion, Flow, Stress, and Growth*, Springer, New York, NY.

A. E. Green and P. M. Naghdi (1977). On thermodynamics and the nature of the second law. *Proceedings of the Royal Society of London, Series A* **357**, 253–270.

S. R. de Groot and P. Mazur (1984). *Non-equilibrium Thermodynamics*, Dover, New York, NY.

H. W. Haslach, Jr. (1997). Geometrical structure of the non-equilibrium thermodynamics of homogeneous systems. *Reports on Mathematical Physics* **39**, 147–162.

H. W. Haslach, Jr. (2009). Thermodynamically consistent, maximum dissipation, time-dependent models for non-equilibrium behavior. *International Journal of Solids and Structures* **46**, 3964–3976. DOI: 10.1016/j.ijsolstr.2009.07.017

H. W. Haslach, Jr. and N-N. Zeng (1999). Maximum dissipation evolution equations for nonlinear thermoviscoelasticity. *International Journal of Non-linear Mechanics* **34**(2), 361–385.

G. A. Holzapfel (2000). *Nonlinear Solid Mechanics*, 2005 reprinting. Wiley, Chichester, UK.

I. Müller (1967). On the entropy inequality. *Archive for Rational Mechanics and Analysis* **26**, 118–141.

I. Müller and T. Ruggeri (1998). *Rational Extended Thermodynamics*, 2nd, ed. Springer, New York, NY.

C. W. Ulbrich (1961). Exact electrical analogy to the Vernotte hypothesis. *Physical Review* **123**, 2001–2002.

Chapter 10
Electromagnetism and Joule Heating

1 Introduction

Joule heating was one of the first non-equilibrium processes studied as thermodynamics developed in the first half of the nineteen century. The Carnot caloric theory that heat is a substance dominated thought in 1850, but Joule argued that heat is vibration and so is a dynamic process. Joule described the heating effect of a current in a copper wire conductor immersed in water as showing that for a fixed current the quantity of "heat evolved in a given time is proportional to the resistance multiplied by the square of the electric intensity" (Joule, 1841). Joule described the heat generated per unit volume as $\mathbf{q} = (1/\hat{\sigma})\mathbf{J} \cdot \mathbf{J}$, where $\mathbf{J}$ is the conduction current or free current density in amperes per square meter per second, the heat flux is $\mathbf{q}$, and $\hat{\sigma}$ is the conductivity. Joule heating (ohmic heating) results from the transfer of momentum during impact of the moving charged particles. The change in momentum changes the kinetic energy of the free particles. In general the impact is elastic; if the structure of the material permanently changes, the process is called electromigration. So, Joule heating is due to the resistance of the material. Joule heating is a dissipative steady process that requires forcing by an electromotive force to produce the steady state.

A thermoelectric process involves both heat and an electric current. The Thomson effect is the change in heat of a conductor of electricity in the presence of a temperature gradient as experimentally observed by Thomson. Thomson (Lord Kelvin) agreed with Joule's viewpoint about the dynamic nature of heat and rejected the idea that heat is a substance. Thomson (1851) showed experimentally and by an argument by contradiction that heat is either released or absorbed, depending on the material, as a current passes from a region of one temperature to a region of another temperature. The total heat produced includes that from Joule heating as well as that from the Thomson effect, as pointed out by Thomson.

$$\mathbf{q} = (1/\hat{\sigma})\mathbf{J} \cdot \mathbf{J} - \mu \mathbf{J} \cdot \nabla\theta, \tag{1}$$

where μ is the Thomson coefficient. The Thomson effect occurs in a single material, but the Seebeck-Pelter effects occur at the junction of two materials. A current is

H.W. Haslach Jr., *Maximum Dissipation Non-Equilibrium Thermodynamics and its Geometric Structure*, DOI 10.1007/978-1-4419-7765-6_10,

induced by a temperature difference across the junction of two materials, called the Seebeck effect as observed in 1821. This thermoelectric effect is the basis of thermocouples. Conversely, the Peltier effect observed in 1834 is that applying a current across the junction of two materials produces heat. The combination is often called the Seebeck-Peltier effect.

Current flow of charge and electromagnetic processes are non-equilibrium processes. Transient electromagnetic forces may induce Joule heating in an initially isothermal conductor. But the time-dependent temperature variation is not included in the classical analyses which use the Fourier principle for heat conduction. This omission is significant because the change in temperature modifies the resistance of a conductor and thus the Joule heating.

The thermoelectric interaction is often taken as an example of applicability of the Onsager linear relations. The electromagnetic models proposed by de Groot and Mazur (1984) restrict their attention to linear steady constitutive relations, in this case Ohm's law $\mathbf{J} = \hat{\sigma}\mathbf{E}$, where $\hat{\sigma}$ is the electrical conductivity, and use the linear Onsager reciprocal relations to develop the interaction of various non-equilibrium effects. As mentioned in Chapter 9, Ohm's law is a steady relationship and is analogous to the Fourier law $\mathbf{q} = k\nabla\theta$ since $\mathbf{E} = -\nabla\phi$, where ϕ is the electric potential.

The maximum dissipation thermodynamic model for the time-dependent response caused by thermoelectric effects in an electro-magnetic system gives the increase in temperature from Joule heating. A strategy using the generalized entropy production function is employed in which the charge flux is treated like other fluxes in a manner that instantaneous flow is not assumed in the model. This finite time flow created is analogous in heat transfer to the replacement of the Fourier law by the Maxwell-Cattaneo relation. In this construction, Ohm's law is the constitutive equation for steady states. A model is given for the heating of a wire that shows the approach to steady state after a current is switched on. The general geometric thermodynamic model includes the Maxwell equations as one-forms. The goal is not to fully describe electromagnetism, or thermoelectric processes, but to show how they fit into the maximum dissipation non-equilibrium thermodynamic model.

2 Constitutive Models

Constitutive models previously proposed for electromagnetism range from the application of the Onsager relations to the Coleman (1964) model of simple materials with memory taken from viscoelasticity (Chapter 4). The fundamental thermodynamic variables of electromagnetism are vectors. The electric intensity, $\mathbf{E}$, is the electrical field in volts per meter, the electric induction, $\mathbf{D} = \epsilon_o\mathbf{E} + \mathbf{P}$ is the electric displacement field, where $\mathbf{P}$ is the electric polarization. The magnetic intensity, $\mathbf{H}$, is the magnetic field strength. The magnetic induction is $\mathbf{B}$ is the magnetic field (magnetic flux density) in webers per square meter. By definition $\mathbf{H} = \mathbf{B} - 4\pi\mathbf{M}$. The magnetic polarization vector, $\mathbf{M}$, in amperes/m, is the analogue to $\mathbf{P}$. The conjugate flux pairs are $\mathbf{E}$ and $\mathbf{D}$ as well as $\mathbf{H}$ and $\mathbf{B}$. Another conjugate energy pair is temperature and entropy per volume. One can control **E**, the electric field but one cannot control **D**. **B** is not easily controlled, but **H** is (Purcell, 1985, p. 433).

In a linear material, $\mathbf{D} = \epsilon \mathbf{E}$, where ϵ is the electric permittivity (dielectric coefficient), and $\mathbf{B} = \mu \mathbf{H}$, where μ is the magnetic permeability. These coefficients are frequency dependent, but may be taken as constants if the time variation is slow or if the frequency interval is very narrow.

The Introduction of the classical book by de Groot and Mazur (1984) begins by crediting Thomson as the first to apply thermodynamics to non-equilibrium processes in his study of the relationship between the Seebeck-Peltier and the Thomson thermoelectric effects. They go on to recover the two Thomson relations described in the 1854 paper from the Onsager relations. While some say these thermoelectric processes are reversible, they are not equilibrium processes because the temperature changes.

The Onsager analysis of de Groot and Mazur (1984) chooses **E**, **B** as state variables for their isothermal theory. They base their analysis on an entropy production with terms for Fourier heat transfer and a Peltier term. The steady flux **E** is given with an Ohmic term and a term $\alpha \nabla \theta$, where α is the differential thermoelectric power. From these two constitutive equations, they recover through the Onsager relations that $\theta \eta = -\pi$, where π / θ is the entropy flux per unit current in a uniform temperature. Further manipulations lead to the first and second Thomson relations.

Coleman and Dill (1971a, 1971b) chose the electromagnetic state variables **E** and **H**, along with θ and $\nabla \theta$ as the fundamental quantities for a continuum mechanics analysis. The enthalpy density given by Coleman and Dill is

$$\psi = u - \theta \eta - \mathbf{H} \cdot \mathbf{B} - \mathbf{E} \cdot \mathbf{D}. \tag{2}$$

The rate of production of entropy is assumed to be

$$\dot{\eta} + \nabla \cdot \left(\frac{\mathbf{q}}{\theta} \right).$$

Their version of the second law is that the production of entropy is non-negative for all processes; they call this inequality the dissipation principle. They use this principle, as is the usual strategy of those working in continuum mechanics, to develop constraints on the constitutive equations.

An entropy balance may replace the entropy inequality when representing electromagnetic effects (Green and Naghdi, 1984). The usual balance laws are supplemented with

$$\rho \dot{\eta} = \rho (s + \bar{\xi}) - \mathrm{div}(\bar{\mathbf{p}}), \tag{3}$$

where ρ is the density, the external volume rate of entropy supply is $s = r/\theta$ for r the volume rate of external heat supply, $\bar{\xi}$ is the internal rate of production of entropy, and $\bar{\mathbf{p}}$ is the entropy flux vector.

Fabrizio and Morro (2000, 2003), for example, establish a thermodynamic model for electromagnetic behavior in terms of the integrals of simple materials with memory. As mentioned in Chapter 4, such integral models require the difficult task of experimentally fitting the empirical time-dependent functions that appear in the integrands. Following Coleman and Dill, Fabrizio and Morro (2003, p. 204) also take **E** and **H** as controls for the enthalpy, ψ, in their electromagnetic model based on materials with memory. The enthalpy for a conductor has partials $\mathbf{D} = -\partial\psi/\partial\mathbf{E}$ and $\mathbf{B} = -\partial\psi/\partial\mathbf{H}$, verifying the conjugate pairs of variables.

In the case that $\mathbf{B} = 0$, the flux constitutive equations for a conductor given by Fabrizio and Morro (2003, p. 292) are

$$\mathbf{J} = \hat{\sigma}(\mathbf{E} - \mu\nabla\theta); \tag{4}$$

$$\mathbf{q} = \lambda\mathbf{J} - \mathcal{K}\nabla\theta, \tag{5}$$

where $\hat{\sigma}$ is the conductivity, μ is the Thomson coefficient, λ is the Peltier coefficient, and $\mathcal{K}$ is the Fourier coefficient. The second Thomson relation says that the Peltier coefficient satisfies $\mathbf{q} = \lambda\mathbf{J}$. Their constitutive models for the fluxes are chosen to be linear as in the Onsager representation and to force **q** and **J** to be zero when **E** and $\nabla\theta$ are zero. The expression for **J** generalizes the Ohm law and the expression for **q** generalizes the Fourier law. The dissipation inequality, from the Clausius-Duhem inequality is

$$\mathbf{E}\cdot\mathbf{J} + \frac{1}{\theta}\mathbf{q}\cdot\nabla\theta \geq 0. \tag{6}$$

Fabrizio and Morro (2003, p. 295) take $\lambda = \phi$, the electric potential, for thermoelectric processes. The flux constitutive equations for thermoelectric processes replace (5) by another expression for the heat flux,

$$\mathbf{q} = \phi\mathbf{J} + \beta\mathbf{E} - \gamma\nabla\theta, \tag{7}$$

where the coefficients γ and β are constants to be determined. The Onsager relations require that $\beta = \alpha\theta\hat{\sigma}$. Note that the assumption for **J** is the same in both cases.

2.1 Electromagnetic Relations and the Maxwell Equations

The vector potential **A** (e.g. Corson and Lorrain, 1962, p. 186) gives the relations

$$\mathbf{E} = -\frac{\partial\mathbf{A}}{\partial t} - \nabla\phi \text{ and } \mathbf{B} = \nabla\times\mathbf{A}. \tag{8}$$

The term $-\nabla\phi$ accounts for the accumulation of electric charge. The term $-\partial\mathbf{A}/\partial t$ accounts for changing magnetic fields (Corson and Lorrain, 1962, p. 439). So only in steady state does $\mathbf{E} = -\nabla\phi$ for steady currents (**A** is constant).

Maxwell's equations apply in any medium at rest with the chosen coordinate system.

(1) Gauss law: $\nabla \cdot \mathbf{D} = \rho$;
(2) Gauss law for magnetism: $\nabla \cdot \mathbf{B} = 0$;
(3) Faraday law of induction: $\nabla \times \mathbf{E} = -\frac{\partial \mathbf{B}}{\partial t}$;
(4) Ampere law: $\nabla \times \mathbf{H} = \mathbf{J} + \frac{\partial \mathbf{D}}{\partial t}$.

The minus sign in the Faraday law represents the Lenz law, which says that when a magnetic field change induces a current, the current flow opposes the change causing the flow. The system attempts to relax to an equilibrium state. If $\mathbf{E}$ is steady so that $\mathbf{E} = -\nabla\phi$, then since curl(grad ϕ) = $\nabla \times \nabla\phi = 0$, the Faraday equation implies that $\dot{\mathbf{B}} = 0$.

2.2 Energy Balance

The system must obey the various balance laws of continuum mechanics in addition to satisfying Maxwell's laws. Several different forms of energy balance are given in the literature for steady or for linear steady electromagnetic systems.

The energy flux at a point is often written in terms of the Poynting vector, which is the vector, $\mathbf{S} = \mathbf{E} \times \mathbf{H}$. The Poynting vector is normal to both $\mathbf{E}$ and $\mathbf{H}$. By substitution of Maxwell's equations,

$$\nabla(\mathbf{E} \times \mathbf{H}) = -\mathbf{H}\frac{\partial \mathbf{B}}{\partial t} - \mathbf{E}\left(\frac{\partial \mathbf{D}}{\partial t} + \mathbf{J}\right).$$

For a linear isotropic material, this reduces to (Corson and Lorrain, 1962, p. 334)

$$\nabla(\mathbf{E} \times \mathbf{H}) = -\frac{\partial}{\partial t}\left(\frac{1}{2}\mu H^2 + \frac{1}{2}\epsilon E^2\right) - \mathbf{E} \cdot \mathbf{J}. \tag{9}$$

The integral of $\nabla(\mathbf{E} \times \mathbf{H})$ over a surface gives the energy flux through the surface, S, by the divergence theorem and (9),

$$\int_S \nabla(\mathbf{E} \times \mathbf{H}) dS = \int_{vol} (\mathbf{E} \times \mathbf{H}) dV = \int_{vol} -\frac{\partial}{\partial t}\left[\left(\frac{1}{2}\mu H^2 + \frac{1}{2}\epsilon E^2\right) - \mathbf{E} \cdot \mathbf{J}\right] dV. \tag{10}$$

As is often done in continuum mechanics, let the volume tend to zero to show that the Poynting vector $\mathbf{E} \times \mathbf{H}$ can represent the energy flux at a point.

Ignoring the heat flux in a linear theory, the rate of acceptance of energy per unit volume of a conductor is $-\nabla \cdot \mathbf{S}$ (Carter, 1968). Carter uses the fact that $\mathbf{E}$, $\mathbf{D}$ and $\mathbf{J}$ are parallel and that $\mathbf{B}$ and $\mathbf{H}$ are parallel to write the energy sum as a dot product of vectors, which transforms to $-\nabla \cdot \mathbf{S}$ using the Ampere and Faraday relations.

Conservation of energy in a linear material (Di Bartolo, 1991, p. 136) is given by setting the internal energy $U = \frac{1}{8\pi}(\mathbf{B} \cdot \mathbf{H} + \mathbf{E} \cdot \mathbf{D})$. Then conservation of energy is for Poynting vector **S**,

$$\frac{\partial U}{\partial t} + \nabla \cdot \mathbf{S} = -\frac{\partial W}{\partial t}. \tag{11}$$

The energy supplied to a conductor when the current density changes from zero to **J** (Corson and Lorrain, 1962, p. 239) is, by Eq. (8),

$$\frac{dW}{dt} = (-\nabla\phi) \cdot \mathbf{J} = \left(\mathbf{E} + \frac{\partial \mathbf{A}}{\partial t}\right) \cdot \mathbf{J}.$$

The first term is the Ohmic loss (Joule heating). The second term is the work done by the source against the induced electromotance, the self-inductance due to a change in current. The magnitude of the electromotance (energy or work per unit charge around a path) is proportional to the rate of change. Therefore the energy stored in the magnetic field is

$$\frac{dW}{dt} = \int_{vol} \frac{\partial \mathbf{A}}{\partial t} \cdot \mathbf{J} d(vol). \tag{12}$$

Fabrizio and Morro (2003, p. 204) give the expression for specific energy u

$$\dot{u} = -\nabla \cdot \mathbf{q} + r + \mathbf{H} \cdot \dot{\mathbf{B}} + \mathbf{E} \cdot \dot{\mathbf{D}} + \mathbf{E} \cdot \mathbf{J}. \tag{13}$$

Coleman and Dill (1971a) also give (13). To derive it, they combine the Faraday and Ampere laws with conservation of energy, $-\dot{u} = \nabla \cdot (\mathbf{q} + \mathbf{E} \times \mathbf{H})$, where $\mathbf{q}$ is the heat flux and $\mathbf{E} \times \mathbf{H}$ is the Poynting vector, to obtain Eq. (13) using $\mathrm{div}(\mathbf{E} \times \mathbf{H}) = \mathrm{curl}(\mathbf{E}) \cdot \mathbf{H} + \mathbf{E} \cdot \mathrm{curl}(\mathbf{H})$. They do not use the Gauss laws, but the partials of the Helmholtz energy must satisfy the Gauss laws so that constraints are placed on the Helmholtz energy. This seems to be the reverse of the statement by Carter.

2.3 The Maxwell Equations as Differential Forms

The geometry of Section 4.2 in Chapter 9 may be extended to electromagnetism. Misner et al. (1973, Sec. 4.5) give the Maxwell equations as differential forms. The electromagnetic two-form, **F**, the Faraday, is axisymmetric.

$$\mathbf{F} = -E_x dt \wedge dx - E_y dt \wedge dy - E_z dt \wedge dz + B_x dy \wedge dz + B_y dz \wedge dx + B_z dx \wedge dy.$$

As is common with balance laws, setting the exterior derivative of **F** equal to zero yields the second and third Maxwell equations (see Misner et al., 1973, p. 113).

$$
\begin{aligned}
d\mathbf{F} = & \left(\frac{\partial B_x}{\partial x} + \frac{\partial B_y}{\partial y} + \frac{\partial B_z}{\partial z}\right) dx \wedge dy \wedge dz \\
& + \left(\frac{\partial B_x}{\partial t} + \frac{\partial E_x}{\partial y} - \frac{\partial E_y}{\partial z}\right) dt \wedge dy \wedge dz \\
& + \left(\frac{\partial B_y}{\partial t} + \frac{\partial E_x}{\partial z} - \frac{\partial E_z}{\partial x}\right) dt \wedge dz \wedge dx \\
& + \left(\frac{\partial B_z}{\partial t} + \frac{\partial E_y}{\partial x} - \frac{\partial E_x}{\partial y}\right) dt \wedge dx \wedge dy = 0.
\end{aligned} \tag{14}
$$

The Maxwell dual two-form is the Hodge star operator acting on the Faraday two-form, $\mathbf{F}$.

$$
{}^*\mathbf{F} = B_x dt \wedge dx + B_y dt \wedge dy + B_z dt \wedge dz + E_x dy \wedge dz + E_y dz \wedge dx + E_z dx \wedge dy.
$$

The remaining two Maxwell laws are described by $d^*\mathbf{F} = 4\pi {}^*\mathbf{J}$, where the charge three-form is ${}^*\mathbf{J} = \rho dx \wedge dy \wedge dz - J_x dt \wedge dy \wedge dz - J_y dt \wedge dz \wedge dx - J_z dt \wedge dx \wedge dy$.

2.4 Joule Heating

The expression, RJ^2, where R is the resistance and J is the magnitude of the current, for Joule heating given by Joule (1841) is for steady state. Joule heating is often interpreted as the energy dissipation during steady current flow (Purcell, 1985, p. 153). For unsteady Joule heating, others define Joule heating as $\int_{vol} \mathbf{E} \cdot \mathbf{J}$ so that heating equals $J^2 R$ only if Ohm's law holds (e.g. Corson and Lorrain, 1962, p. 334).

3 Unsteady Thermoelectric and Electromagnetic Evolution

The maximum dissipation non-equilibrium evolution construction allows the study of transient effects rather than restricting one's view to steady processes and their interactions as in the use of the Onsager equations.

3.1 Unsteady Ohm's Law

The derivation of an unsteady version of Ohm's law given in Section 3.3 in Chapter 9 is based on the generalized entropy production

$$
\Psi^* = \frac{\mathbf{J} \cdot \mathbf{J}}{2\hat{\sigma}\theta} + \mathbf{J} \cdot \nabla\left(\frac{\phi}{\theta}\right), \tag{15}
$$

Applying the zero gradient condition gives Ohm's law as the stationary state since system is assumed isothermal, $\nabla\theta = 0$. The unsteady Ohm's law is the associated gradient relaxation process for small displacements

$$\dot{\mathbf{J}} = -k(\hat{\sigma}\theta)^2 \left[\frac{\mathbf{J}}{\hat{\sigma}\theta} + \nabla\left(\frac{\phi}{\theta}\right)\right]. \tag{16}$$

3.2 Classical Joule Heating with the Maxwell-Cattaneo Heat Flux

A current is suddenly applied to a wire of length ℓ that is not thermally insulated. The wire radiates heat into the ambient fluid. Assume that the current flow is steady and is described by Ohm's Law. The temperature is initially non-uniform. The maximum dissipation evolution equations in conjunction with energy balance describe how, corresponding to the steady current, the temperature of the wire approaches the equilibrium value, θ_o, with the ambient fluid. Since $\mathbf{E}$ is assumed constant, initial dissipation in the system is required.

The rate of change of energy $\dot{u} = r - \nabla\mathbf{q}$. The radiation term is given by Newton's law of cooling, $r = \alpha(\theta - \theta_o)$, at each point in the wire. Let $\bar{\theta}$ be the temperature from the reference state so that $r = \alpha\bar{\theta}$. Therefore,

$$\dot{u} = \alpha\bar{\theta} - \nabla\mathbf{q}. \tag{17}$$

Further, assume that the resistance is a function of temperature $R(\bar{\theta}) = a\bar{\theta}$. An alternative is the Steinhart-Hart equation, $1/\theta = A + B\ln(R) + C(\ln(R))^3$. Assume that the current is in a steady state so that the magnitude of $\mathbf{J}$, denoted J, is constant and that $\dot{u} = R(\theta)J^2$. Substitution into Eq. (17) and differentiation with respect to time yields

$$\dot{\mathbf{q}}_x = -(J^2 a - \alpha)\dot{\bar{\theta}}. \tag{18}$$

The maximum dissipation evolution for the heat is the Maxwell-Cattaneo equation (Section 3.1 in Chapter 9).

$$\mathbf{q} + \tau\dot{\mathbf{q}} = -\mathcal{K}\nabla\theta. \tag{19}$$

Let $0 \le x \le \ell$ be the spatial coordinate of the wire. Then differentiation of Eq. (19) with respect to x yields

$$-\tau\dot{\mathbf{q}}_x = \mathbf{q}_x + \mathcal{K}\bar{\theta}_{xx}. \tag{20}$$

Combining the two Eqs. (18) and (20) and using Eq. (17) with $\dot{u} = R(\bar{\theta})J^2$ produces

$$\tau(J^2a-\alpha)\dot{\bar{\theta}}+[a\bar{\theta}J^2-\alpha\bar{\theta}]=\mathcal{K}\bar{\theta}_{xx}. \tag{21}$$

This partial differential equation in $\bar{\theta}$ may be solved by separation of variables. Rewrite the equation as

$$\beta\dot{\bar{\theta}}+\gamma\bar{\theta}=\mathcal{K}\bar{\theta}_{xx}, \tag{22}$$

where $\beta=\tau(J^2a-\alpha)$ and $\gamma=aJ^2-\alpha$.

Assume $\bar{\theta}(x,t)=f(x)g(t)$. Then (22) becomes

$$(\beta\dot{g}+\gamma g)/g=\mathcal{K}f_{xx}/f.$$

Therefore each side equals some constant K. The boundary conditions determine the sign of K. The function $g(t)$ satisfies $(K-\gamma)g-\beta\dot{g}=0$. Therefore $g(t)=g(0)\exp[(K-\gamma)t/\beta]$. Note that if the radiation is large enough or if $K<0$, then $g(t)$ decreases since the coefficient of t becomes negative and so $g(t)$ tends to zero. The spatial function satisfies $\mathcal{K}f_{xx}-Kf=0$. The solution for $f(x)$ is a Fourier series that depends on the initial temperature distribution. The amplitude of the solution, $\bar{\theta}(x,t)$, decreases to zero over time so that the temperature becomes a constant θ_o.

3.3 Transient Model of Joule Heating

The steady state constitutive relations (4) and (5) may be derived from a generalized entropy production function, for state variable **J** and control variable $\nabla(\phi/\theta)$,

$$\Psi_J^*=\frac{\mathbf{J}\cdot\mathbf{J}}{2\hat{\sigma}\theta}+\left(\frac{\phi}{\theta^2}-\frac{d\alpha}{d\theta}\right)\mathbf{J}\cdot\nabla\theta+\mathbf{J}\cdot\nabla\left(\frac{\phi}{\theta}\right), \tag{23}$$

where α is absolute thermoelectric power (volts/K^o), also known as the Seebeck coefficient. The resulting zero gradient condition is

$$\frac{\mathbf{J}}{\hat{\sigma}\theta}+\left(\frac{\phi}{\theta^2}-\frac{d\alpha}{d\theta}\right)\nabla\theta+\frac{1}{\theta}\nabla\phi-\frac{\phi}{\theta^2}\nabla\theta=0, \tag{24}$$

which reproduces (4) since the Thomson coefficient, μ, by the first Thomson relationship is $\mu/\theta=d\alpha/d\theta$. The second term accounts for the interaction of the Peltier and Thomson effects.

The heat flux described by (5) is accounted for in the generalized entropy production function, for state variable **q** and control variable $\nabla(1/\theta)$,

$$\Psi_q^*=\left(\frac{1}{2\theta^2}\right)\mathcal{K}^{-1}\boldsymbol{q}\cdot\boldsymbol{q}-\frac{\mathcal{K}^{-1}}{\theta^2}\phi\mathbf{J}\cdot\mathbf{q}-\frac{\lambda}{\theta^2}\mathbf{J}\cdot\boldsymbol{q}-\boldsymbol{q}\cdot\nabla\left(\frac{1}{\theta}\right). \tag{25}$$

Then the zero gradient condition produces (5) from

$$\left(\frac{1}{\theta^2}\right) K^{-1}\boldsymbol{q} - \frac{\lambda\mathcal{K}^{-1}}{\theta^2}\mathbf{J} - \frac{1}{\theta^2}\nabla\theta = 0. \tag{26}$$

The total entropy production is $\Psi^* = \Psi_J^* + \Psi_q^*$. Therefore the maximum dissipation evolution equations for the fluxes have cross terms and are

$$\dot{\mathbf{J}} = -k_J(\hat{\sigma}\theta)^2\left[\frac{\mathbf{J}}{\hat{\sigma}\theta} + \left(\frac{\phi}{\theta^2} - \frac{d\alpha}{d\theta}\right)\nabla\theta - \frac{\mathcal{K}^{-1}}{\theta^2}\phi\mathbf{q} - \frac{\lambda}{\theta^2}\boldsymbol{q} + \nabla\left(\frac{\phi}{\theta}\right)\right]; \tag{27}$$

$$\dot{\mathbf{J}}_q = -k_q\left(\theta^4\mathcal{K}^2\right)\left[\left(\frac{1}{\theta^2}\right)\mathcal{K}^{-1}\boldsymbol{q} - \frac{\mathcal{K}^{-1}}{\theta^2}\phi\mathbf{J} - \frac{\lambda}{\theta^2}\mathbf{J} - \nabla\left(\frac{1}{\theta}\right)\right]. \tag{28}$$

Example Suppose that a wire at equilibrium and with no current flow has a fixed voltage, **E**, applied suddenly. The initial temperature may vary from point to point in the body. A constant deformation, ϵ_o, is applied at the same time as the fixed voltage. The maximum dissipation evolution equations predict how Joule heating influences stress relaxation. The maximum dissipation criterion produces the transient process to steady state. The current comes to a steady state but the temperature tends to a uniform equilibrium greater than the initial temperature θ_r. As a simple example, assume that the material is linearly elastic with elastic modulus E_1.

The maximum dissipation evolution equations for the heat flux and for Ohm's law require the current temperature as a control variable. The additional equation required is conservation of energy for the unsteady system. Since the only power applied is $\mathbf{E}\cdot\mathbf{J}$,

$$c\theta\dot{\theta} + \frac{1}{E_1(\theta)}\sigma\dot{\sigma} = \mathbf{E}\cdot\mathbf{J} - \nabla\cdot\mathbf{q} + r + \mathbf{E}\cdot\dot{\mathbf{D}} \tag{29}$$

where c is a constant related to the heat capacity and r may be given by Newton cooling. Assume $\dot{\mathbf{D}} = 0$ since $\dot{\mathbf{E}} = 0$. But the current value of the stress is also required in this equation and may be obtained from the maximum dissipation evolution derived from the energy density function.

For stress relaxation, the state variables are stress, σ, and temperature, θ, so that the generalized energy density function at each point of the body is

$$\varphi^* = 0.5c\theta^2 + \frac{1}{2E_1(\theta)}\sigma^2 - \alpha_t\bar{\theta}\sigma - \epsilon\sigma - \eta\theta. \tag{30}$$

If the stress is not zero, and the wire is in stress relaxation, then the maximum dissipation evolution equation for a fixed strain, ϵ_o is required.

$$\dot{\sigma} = -k_\sigma E_1(\theta)^2\left(\frac{\sigma}{E_1(\theta)} - \alpha_t\bar{\theta} - \epsilon_o\right), \tag{31}$$

where α_t is the linear coefficient of thermal expansion and $\bar{\theta} = \theta - \theta_r$.

To predict the behavior of bodies subjected to Joule heating and stress relaxation with an initial temperature variation, solve simultaneously the Eqs. (27), (28), (29) and (31).

If desired, the current value of the entropy, $\eta(t)$, at each point is obtained from the maximum dissipation evolution equation for temperature

$$\dot{\theta} = -k_\theta c^2 \left[c\theta - \frac{1}{2E_1(\theta)^2} \frac{dE_1(\theta)}{d\theta} \sigma^2 - \alpha_t \sigma - \eta(t) \right]. \tag{32}$$

If creep rather than stress relaxation is to be modeled, the generalized energy density must have strain as a state variable. The generalized energy density function for state variables strain, ϵ, and temperature, θ, is

$$\varphi^* = 0.5c\theta^2 + 0.5E_1(\theta)\epsilon^2 - E_1(\theta)\alpha_t\bar{\theta} - \epsilon\sigma - \eta\theta. \tag{33}$$

This example shows the interaction of the generalized energy density function, the generalized entropy production function and energy balance during relaxation by maximum dissipation processes. The voltage is a forcing function which eventually produces a steady current, while the temperature relaxes to equilibrium.

References

B. Di Bartolo (1991). *Classical Theory of Electromagnetism*. Prentice-Hall, Englewood Cliffs, NJ.

G. W. Carter (1968). *The Electromagnetic Field in its Engineering Aspects*, American Elsevier, New York, NY.

B. D. Coleman (1964), Thermodynamics of materials with memory. *Archive for Rational Mechanics and Analysis* **17**, 1–46.

B. D. Coleman and E. H. Dill (1971a). Thermodynamic restrictions on the constitutive equations of electromagnetic theory. *Zeitschrift für angewandte Mathematik und Physik (ZAMP)* **22**, 691–702.

B. D. Coleman and E. H. Dill (1971b). On the thermodynamics of electromagnetic fields in materials with memory. *Archive for Rational Mechanics and Analysis* **41**, 132–162.

D. R. Corson and P. Lorrain (1962). *Introduction to Electromagnetic Fields and Waves*, W. H. Freeman and Co., San Francisco, CA.

M. Fabrizio and A. Morro (2000). Dissipativity and irreversibility of electromagnetic systems. *Mathematical Models and Methods in Applied Sciences* **10**(2), 217–246.

M. Fabrizio and A. Morro (2003). *Electromagnetism of Continuous Media*. Oxford University Press, Oxford, UK.

A. E. Green and P. M. Naghdi (1984). Aspects of the second law of thermodynamics in the presence of electromagnetic effects. *The Quarterly Journal of Mechanics and Applied Mathematics* **37**, 179–193.

S. R. de Groot and P. Mazur (1984). *Non-equilibrium Thermodynamics*, Dover, New York, NY.

J. P. Joule (1841). On the heat evolved by metallic conductors of electricity, and in the cells of a battery during electrolysis. *Philosophical Magazine* **19**, 260, Also in J. P. Joule (1963). *The Scientific Papers of James Prescott Joule*, Dawsons of Pall Mall, London.

C. Misner, K. Thorne, and J. Wheeler (1973). *Gravitation*, W. H. Freeman, San Francisco, CA.
E. M. Purcell (1985). *Electricity and Magnetism*, 2nd ed., McGraw-Hill, New York, NY.
W. Thomson (1851). On a mechanical theory of thermoelectric currents, *Proceedings of the Royal Society of Edinburgh* **3**, 91–98.
W. Thomson (1854). *Proceedings of the Royal Society of Edinburgh* **3**, 225. *Transactions of the Royal Society of Edinburgh* **21** Part I.

Chapter 11
Fracture

1 Introduction

Non-steady crack propagation is a non-equilibrium thermodynamic process. The application of the maximum dissipation construction to fracture mechanics provides an example in which the critical manifold represents neither thermostatic nor steady states. In this case, it is a set of states satisfying a critical condition or event, and the set may be either an attractor or a repeller depending on whether the crack propagation is stable or unstable. In unstable propagation, the non-equilibrium process is repelled from the quasi-static critical manifold. If the initial state is stable, then the crack growth process approaches the quasi-static critical manifold and eventually the crack is arrested. The maximum dissipation non-equilibrium evolution model describes the non-steady, crack propagation rate for both brittle fracture and for viscoplastic behavior at the crack tip. The class of models produced includes the classical Freund model and a modification that is consistent with the experimental maximum crack velocity. Further, the thermodynamic relaxation modulus for brittle fracture has a physical interpretation in terms of the microscopic response of the material. An application of the construction gives the craze growth in PMMA. A simple viscoplastic model for metals predicts the change in temperature at the crack tip as the crack grows.

In fracture mechanics, the non-equilibrium thermodynamic evolution equations should predict the rate of crack propagation and should capture the transient crack propagation, the stability or instability of the crack growth, the influence of the plastic zone at the crack tip, and the temperature change at the crack tip. The thermodynamic maximum dissipation construction generates a new class of evolution equations for the non-steady rate of crack propagation that are based on the chosen quasi-static crack propagation criterion.

Much current research on fracture either focuses on the microstructural mechanisms of crack propagation or numerical techniques based on lattice structures (Ravi-Chandar, 2004), but fewer attempts have been made to develop a thermodynamic viewpoint (e.g. Gurtin, 1979). Rice (1978) considered the thermodynamics of Griffith cracks, ones which grow in a lattice without generation or motion of dislocations. The growth is quasi-static, which requires that the kinetic energy and

H.W. Haslach Jr., *Maximum Dissipation Non-Equilibrium Thermodynamics and its Geometric Structure*, DOI 10.1007/978-1-4419-7765-6_11,

the change in temperature are negligible. The new thermodynamic variables are the crack length, l, and the crack driving force at the crack tip, G. The work required to reversibly separate the surfaces is 2γ so that $(G - 2\gamma)\dot{l} \geq 0$ is the condition for non-negative entropy production. A kinetic rate law for quasi-static crack growth in which $\dot{l}$ is a function of G must give $\dot{l} = 0$ at $G = 2\gamma$ to be thermodynamically consistent. This condition is met by the critical manifold defined in the thermodynamic construction presented here.

In their description of the thermodynamics of crack propagation, Lemaitre and Chaboche (1990) took l and G as thermodynamic variables. A key difference with the maximum dissipation construction is that they assume a separate dissipation potential, $\phi(\dot{l})$, such that $\partial\phi/\partial\dot{l} = G$ and such that the crack propagation rate $\dot{l} = \partial\psi/\partial G$, where ψ is the Legendre transform of ϕ. Such an assumption, which is similar to that made in rate-independent plasticity theories, is tantamount to assuming the evolution laws rather than deriving them from a thermodynamic principle. Maugin (1992) put fracture in a thermodynamic setting by computing the dissipation, ϕ, as the difference of the applied forces and the stored energy under the assumption of a moving coordinate system with origin at the crack tip. This computation leads to $\phi = G\dot{l}$, where G has the form of the J-integral. This expression is formally the same as the rate-independent plasticity flow law with $\dot{l}$ corresponding to the plastic strain rate, $\dot{\epsilon}^p$, and so is a quasi-static fracture theory. Maugin proposed the potential $\phi(G) = G^2/2\eta$ for materials, such as polymers or rubber, with viscosity, η, to define $\dot{l} = d\phi/dG$.

A non-equilibrium thermodynamic development for the propagation velocity has been presented by Berezovski and Maugin (2007), who write the driving force as the sum of the equilibrium stress and the contact stress from the excess energy over the equilibrium energy. The contact stress plays an analogous role to the affinity defined in the model presented here. The splitting of the strain energy density into equilibrium and excess summands, as in the Muschik and Berezovski (2004) model for lumped non-equilibrium systems, gives a non-equilibrium model viewing the crack front as a phase-transition. The jump in linear momentum at the discontinuity provides an estimate of the crack tip velocity. The velocity is written in terms of the contact stress between the two systems divided by the crack front, but no means is available to compute this contact stress. The assumption that the contact and equilibrium stresses are proportional produces an expression for the crack velocity in terms of a limiting velocity, a characteristic length, and the stress intensity factor. A choice of the first two parameters recovers an experimental crack velocity versus stress intensity factor relation for Homolite-100. The rate of crack propagation is often viewed as a function of the stress intensity factor because the energy release rate is the crack driving force. However, the question of the uniqueness of the stress intensity factor - crack velocity relationship, raised over twenty years ago by Dally et al. (1985) for example, remains unresolved.

The continuum thermodynamics approach of Gurtin (2000) consolidates all behavior near the tip into a single configurational force, $\mathbf{g}_{tip}$, so that his configuration force balance applies. The dissipation at the crack tip is defined to be $-\mathbf{g}_{tip} \cdot \mathbf{v}$ and as in continuum thermodynamics the behavior is required to satisfy a dissipation inequality. Unfortunately, the theory cannot predict the crack velocity, $\mathbf{v}$, since it requires the assumption of a constitutive equation for the crack velocity as a function of the crack angle and forcing function. No examples of such a constitutive function are given by Gurtin. Other models assume a constant crack tip velocity (e.g. Gao, 1996). Empirical equations based on a power law to reflect the Paris law have been proposed (Christensen and Miyano, 2007), but have not been shown to be thermodynamically consistent.

Bui et al. (1980) construct the temperature field near the crack tip in the thermoelastic case and find that it has a mathematical singularity. The field therefore predicts that the temperature at the crack tip rises significantly as the crack propagates. Such temperature rises have been observed experimentally by Fuller et al. (1975), Weichert and Schönert (1974, 1978) and Döll (1984).

The general instantaneous maximum dissipation non-equilibrium thermodynamic model is applied to develop thermodynamically consistent evolution equations for the rate of non-steady dynamic crack propagation. Crack propagation under a constant load is viewed as a relaxation process that moves in thermodynamic state space in the direction of maximum decrease of the generalized energy at a speed given by a relaxation modulus. In this model, the evolution of the internal state variables is not derived from a dissipation potential separate from the energy, as in traditional quasi-static thermodynamic models.

The construction of the evolution equations is illustrated in several examples, some for dynamic loading and some for quasi-static loading. Each constitutive model requires choosing a critical manifold and a thermodynamic relaxation modulus. Application of the model to brittle crack growth in a linearly elastic material with a lattice structure reproduces the classical dynamic fracture model of Freund (1990) if a particular thermodynamic relaxation modulus is chosen based on the Rayleigh wave speed, showing that the non-equilibrium instantaneous maximum dissipation construction includes traditional models. A different choice of the relaxation modulus based on the shear wave speed yields an evolution equation that predicts a maximum crack tip velocity in better agreement with experiment. The stability of the crack growth is shown to depend on the manifold of quasi-static fracture states, the critical manifold, defined by the zero gradient condition on the generalized function. Again, the critical manifold may be an attractor or a repeller; this viewpoint is similar to that of nonlinear dynamics. The role of the critical manifold in determining whether or not the crack propagation is stable is exhibited in the well-known quasi-static loading example of the split beam, and the transient crack growth prior to arrest is captured. Experimental results of Döll et al. (1981) on the growth of the craze region in PMMA under creep are also successfully modeled by this construction. Finally the influence of the plastic zone in a metal is examined by combining the crack growth model with an instantaneous maximum dissipation model for viscoplasticity (Chapter 5; Haslach, 2002). The latter in conjunction with an energy balance equation, which accounts for the fact that the process including the evolution of temperature is non-homogeneous, is applied to predict the change in temperature at the crack tip. The non-equilibrium construction using an instantaneous maximum dissipation principle gives a class of evolution equations for non-steady crack growth, in contrast to classical models making the unrealistic assumption of constant speed.

2 Construction of the Model for the Non-equilibrium Thermodynamics of Fracture

The conjugate pair (l, G) is composed of the crack driving force, G, which is the control variable and the crack length, l, which is the state variable. Other conjugate pairs are the stress and strain, and the specific entropy, η, and temperature, θ. In the crack growth model, the organizing manifold, M_e, is the critical submanifold whose coordinates include the set of all (l, G) such that, in the quasi-static case, G corresponds to l. These manifolds, M_e, which define the crack propagation criterion are directly related to the thermostatic functions, $\hat{\varphi}(y_1, \ldots, y_n)$, called F for fracture function in this chapter. The path evolution may also depend on internal state variables and their conjugates to represent effects such as hardening.

For the case of crack growth with the fracture generalized energy, F^*, the zero gradient condition, $\partial F^*/\partial l = 0$, gives the constitutive relation in the chosen quasi-static model for the crack driving force G as a function of the other thermodynamic variables that defines the incipient crack growth criterion. F^* is still called a generalized energy function to be consistent with the other applications of the construction of non-equilibrium evolution equations even though F may or may not be a well-known energy function. For crack propagation, if the force G is perturbed to $\bar{G}$ so that the pair $(l, \bar{G})$ is no longer on the critical manifold, the process relaxes by increasing the crack length. The non-equilibrium behavior is forced by the affinity associated to l. Further, no threshold fracture stress is required. The path may be either attracted or repelled from the critical manifold, depending on whether the critical manifold is composed of minima or maxima of the generalized energy. If the critical manifold is an attractor, then the crack propagation is eventually arrested and is stable.

As described in Chapter 3, the maximum dissipation relaxation process, as a system of first-order nonlinear ordinary differential equations in terms of the state variables, x_i, and the control variables, y_i, is equivalently

$$\begin{bmatrix} \frac{dx_1}{dt} \\ \frac{dx_2}{dt} \\ \cdots \\ \frac{dx_n}{dt} \end{bmatrix} = -k \begin{bmatrix} \frac{\partial^2\varphi}{\partial x_1^2} & \frac{\partial^2\varphi}{\partial x_1 \partial x_2} & \cdots & \frac{\partial^2\varphi}{\partial x_1 \partial x_n} \\ \frac{\partial^2\varphi}{\partial x_2 \partial x_1} & \frac{\partial^2\varphi}{\partial x_2^2} & \cdots & \frac{\partial^2\varphi}{\partial x_2 \partial x_n} \\ \cdots & \cdots & \cdots & \cdots \\ \frac{\partial^2\varphi}{\partial x_n \partial x_1} & \frac{\partial^2\varphi}{\partial x_n \partial x_2} & \cdots & \frac{\partial^2\varphi}{\partial x_n^2} \end{bmatrix}^{-2} \begin{bmatrix} y_1 + \frac{\partial\varphi}{\partial x_1} \\ y_2 + \frac{\partial\varphi}{\partial x_2} \\ \cdots \\ y_n + \frac{\partial\varphi}{\partial x_n} \end{bmatrix}. \quad (1)$$

Relaxation in the case of crack growth means that the evolution equation (1) for the rate of crack growth is obeyed. Relaxation in a body with a crack requires that G be held fixed; the response is then measured by the time-dependent variation of the crack length, l. A forced process again is a non-equilibrium process in which the control variables vary with time.

This construction begins with a criterion for the incipient growth of a crack, such as that of Griffith. Assume a crack propagation criterion $G = f(l, u_1, \ldots, u_m)$, where the criterion may depend on other parameters, $u_1, \ldots, u_m$, such as material parameters. The external load normal to a crack face may also appear as a parameter in the criterion. Such a choice of parameter is not a thermodynamic state or control variable at each point of the body in the sense used in the construction of non-equilibrium evolution equations. Such a choice does not imply that the theory is one-dimensional in stress or strain. The criterion depends on a single pair (G, l) of thermodynamic variables. This could be the two-dimensional Griffith criterion, for example. Given $G = f(l, u_1, \ldots, u_m)$, the function, F, describing the quasi-static states is obtained by integration, and the associated generalized energy, F^*, is well defined and exists because its equation is explicitly expressed by $F^*(\boldsymbol{x}; \boldsymbol{y}) = F(\boldsymbol{x}) + \sum_{i=1}^{n} x_i y_i$ (see Chapter 3).

3 Linear Elastic Instantaneous Maximum Dissipation Crack Propagation

The generalized energy function, F^*, for brittle fracture obtained from the Griffith criterion coincidently agrees with the potential energy in the Griffith analysis (1921, 1924). The potential energy required to derive the Griffith equilibrium condition for incipient crack propagation is already in the form of a generalized energy function. Griffith described his quasi-static fracture theory in terms of the equilibria of a potential energy function, Π, which may be written as

$$\Pi = 4wl\gamma - \Delta U = 4wl\gamma - \frac{\pi l^2 \sigma^2 w}{E'}. \qquad (2)$$

Here γ is the surface energy per unit area, w is the width of the crack face, σ is the external applied stress, and in plane strain $E' = E/(1-\nu^2)$, where E is the elastic modulus. According to Griffith, the equilibria of Π obtained by setting $d\Pi/dl = 0$ determine the critical states for crack propagation. This potential function is related to the generalized energy function, F^*, below by letting $G = 2\gamma$. This construction also represents the Irwin theory because $K_I^2 = E'G$.

The Griffith criterion is based on the diminution of potential energy as the external load is applied and the crack prepares to grow. The total diminution of the potential energy is the increase in strain energy due to the crack formation W minus the potential energy of the crack surface U (which increases as the crack grows). Just after the tractions are annihilated, the system is not in equilibrium. Griffith (1921) hypothesizes that the crack propagation begins when the decrease in potential energy due to crack formation is at an equilibrium. Note that the Griffith expression $W - U$ is defined on non-equilibrium states.

It seems that the fact that the generalized energy function coincides with the Griffith potential energy is an accidental consequence of the manner in which Griffith justified his criterion. No such comparison is true for the particular generalized energies used to obtain viscoelastic or viscoplastic non-equilibrium evolution equations using this general thermodynamic construction.

3.1 Freund Equation of Motion as a Maximum Dissipation Evolution Equation

The evolution equation construction (1), in the case of the Griffith criterion for brittle fracture of a linearly elastic material, recovers the Freund Mode I equation of motion for one choice of k. Another choice of k produces a model that better agrees with experiment. Freund (1990) writes the dynamic energy release rate as

$$G(t, l, \dot{l}) = \frac{1-\nu^2}{E}[A_I K_I^2 + A_{II} K_{II}^2] + \frac{1}{\mu} A_{III} K_{III}^2, \tag{3}$$

where μ is the shear modulus, and, the intensities K are functions of time, l and $\dot{l}$, the A_I, A_{II}, A_{II} depend on the crack propagation speed, but are independent of the loading (Freund, 1990, p. 234). Therefore in Freund's theory, a crack growth criterion, called the equation of motion for the crack tip, must be added to the governing equations. The classical model of Freund for the unsteady Mode I crack growth speed is an equation of motion developed from the Griffith energy balance and from modifying the Irwin stress intensities to be functions of the crack propagation velocity. Freund uses the dilatational solution to the wave equation obtained from the Navier equation of motion to show that the mode I dynamic intensity is a product of a function of the crack tip velocity and the quasi-static intensity, $K_I(t, l, 0)$, for the crack of initial length l so that $K_I(t, l, \dot{l}) = k(\dot{l})K_I(t, l, 0)$, where $k(\dot{l}) \sim (1 - \dot{l}/c_R)/\sqrt{1 - \dot{l}/c_d}$. The elastic dilatational wave speed is $c_d = ((\lambda + 2\mu)/\rho)^{1/2}$ and the shear wave speed is $c_s = \sqrt{\mu/\rho}$, using the Lamé constants, λ and the shear modulus, μ, and the density, ρ, of the material. These are body waves while the Rayleigh wave is a surface wave. The Rayleigh wave speed, c_R, is approximately (Freund, 1990, p. 83)

$$c_R \sim \frac{0.862 + 1.14\nu}{1+\nu} c_s. \tag{4}$$

Here, the static stress intensity factor depends on the loading and on the geometry. Presumably,

$$K_I(t, l, 0) = \sigma(t)\sqrt{\pi l(t)},$$

for shape factor, $Y = 1$.

To obtain Freund's equation of motion for mode I crack opening, the quasi-static value of G is perturbed to the value of the resistance to crack growth, Γ, a material property that may depend on l or could also be a function of crack speed. The experimentally determined parameter Γ may be assumed to be a constant in the simplest cases. Freund assumes that growth occurs if $G(t, l, \dot{l}) = \Gamma$. In this case, setting G from (3) equal to this critical value, Γ, gives a differential equation for the crack velocity. The general crack growth expression for mode I opening in plane strain given by Freund (1990, p. 397) is

$$\frac{E\Gamma}{(1-\nu^2)K_I(t,l,0)^2} = 1 - \frac{\dot{l}}{c_R}. \tag{5}$$

3.1.1 The Maximum Dissipation Construction

The Freund crack tip equation of motion (5) is recovered from the generalized energy function, F^*, in plane strain,

$$F^* = -\frac{(1-\nu^2)}{2E}\sigma^2\pi l^2 + Gl, \tag{6}$$

where E is the elastic modulus, $w = 1$, and ν is the Poisson ratio. This is half the change in potential energy due to the crack formation as calculated by Griffith (1921, 1924). Let $\bar{G}$ be a perturbation, which is larger, of the critical value corresponding to the initial crack length l_o. The affinity, $X = \partial F^*/\partial l = -(1-\nu^2)\sigma^2\pi l/E + \bar{G}$, drives the crack growth.

The evolution of the crack length induced by (1) is

$$\dot{l} = k_l\left(\frac{E^2}{(1-\nu^2)^2\sigma^4\pi^2}\right)\left[\frac{(1-\nu^2)}{E}\sigma^2\pi l - \bar{G}\right], \tag{7}$$

where $K_I = \sigma\sqrt{\pi l}$. Also $\bar{G} = \Gamma$ may be assumed constant, as would be true in pure cleavage. The choice of k_l is part of the constitutive model.

A suddenly applied load generates waves that radiate from the load point. Assume that the stress waves cause the critical deformation leading to rupture. Depending on the choice of constitutive model, the stress waves may be either surface or body waves. The critical stress waves have multiple frequencies even though all have the same wave speed, v, defined as the wavelength times the frequency. Let l be the distance from the point of load application to the crack tip along the crack face. Such a critical wave reaches the crack tip at l/v seconds after the application of the load. Assume that in thermodynamic space, the speed of relaxation, the relaxation modulus, is the wave power of such waves. The wave power, P, at the crack tip for stress waves $\sigma\sin(\omega t)$, in a linearly elastic material, is by definition the area under the curve $\epsilon\sigma(\omega l/v)$ for $0 \le \omega \le 2\pi v/l$, over all critical waves, indexed by frequency ω, with wavelength greater than l. Set $\epsilon = (1-\nu^2)\sigma/E$.

$$P = \int_0^{2\pi v/l} \frac{(1-\nu^2)\sigma^2}{E}\sin^2(\omega l/c_R)d\omega = v\frac{(1-\nu^2)\sigma^2\pi}{El}. \tag{8}$$

The relaxation modulus k for this crack propagation model under dynamic loading is taken to be the wave power. This rate of change increases with an increase in applied stress at fixed l and decreases with an increase in the instantaneous crack length, l at a fixed stress. In contrast, the thermodynamic space relaxation modulus is taken to be a constant in quasi-static loading. The modulus is large for metals in the linearly elastic model and small for polymers because of their viscosity.

The Freund model is based on the transport of Rayleigh waves along the crack surface. The Freund model is obtained if Rayleigh waves, surface waves, are taken to be critical, so that $v = c_R$. A Rayleigh wave reaches the crack tip at time $t = l/c_R$. The thermodynamic relaxation modulus is the wave power, the time rate of change of the work done on the crack tip by all Rayleigh waves of wavelength longer than l. In this model, k_l is chosen from (8) to be

$$k_l = c_R \frac{(1-\nu^2)\sigma^2\pi}{El},$$

which has units of energy density per second, then

$$\dot{l} = c_R \left[1 - \frac{E\Gamma}{(1-\nu^2)K_I^2} \right]. \tag{9}$$

The class of instantaneous maximum dissipation models proposed here therefore includes the Freund theory (5), with this choice of k_l.

The relations (5) and (9) imply that c_R is the upper limit on the crack tip speed, $\dot{l}$. However, the maximum crack tip speed experimentally observed in any material is much less than the Rayleigh wave speed, c_R. The limiting speeds collected by Ravi-Chandar and Knauss (1984) indicate that the speeds range from about $0.4c_s$–$0.6c_s$. The lattice model of Gao (1996) indicates a limiting speed of $c_s\sqrt{\sigma_{max}/\mu}$, where σ_{max} is the maximum cohesive stress, i. e. the equi-biaxial stress at the crack tip, due to a local wave that competes with the global Rayleigh wave. To estimate the coefficient of c_s in metals, recall that the maximum cohesive stress is approximately $E/10$ from the classical atomic bonding energy estimate. The shear modulus is about $\mu = E/2.5$. Therefore $\sqrt{\sigma_{max}/\mu} \sim 0.5$.

The construction proposed here requires the maximum dissipation possible at each instant. Slepyan (1993) determined the limiting speed by computing the velocity associated with the maximum of all the instantaneous dissipations and found that the limiting velocity in an elastic body is about half of the shear wave speed, c_s. The Slepyan computation closely approximates the experimental measurements in glass ($\nu = 0.22$) and in PMMA ($\nu = 0.35$).

A shear wave model is obtained when the microstructure or local stress state affects the shear wave speed. The relaxation modulus is obtained by substituting the Gao (1996) velocity $v = c_s\sqrt{\sigma_{max}/\mu}$ in Eq. (8)

$$k_l = c_s \sqrt{\frac{\sigma_{max}}{\mu}} \frac{(1-\nu^2)\sigma^2\pi}{El}, \tag{10}$$

so that

$$\dot{l} = c_s \sqrt{\frac{\sigma_{max}}{\mu}} \left[1 - \frac{E\Gamma}{(1-\nu^2)K_I^2} \right]. \tag{11}$$

This relaxation modulus, k_l, better accounts for the experimentally measured maximum crack velocity than the classical Freund model.

Figure 11.1 compares the crack propagation response predicted by (9) and (11) in PMMA, a quasi-brittle material. Döll (1976) measured $c_R = 1,242\,\text{m/s}$ and $\nu = 0.36$. Therefore by (4), $c_s = 1,327.5\,\text{m/s}$. Eq. (11) is computed under the assumption that $\sqrt{\sigma_{max}/\mu} = 0.5$. The initial crack length is $l = 0.005\,\text{m}$, $\Gamma = 100\,\text{N/m}$, $E = 3\,\text{GPa}$, and the applied stress is $\sigma = 5\,\text{MPa}$.

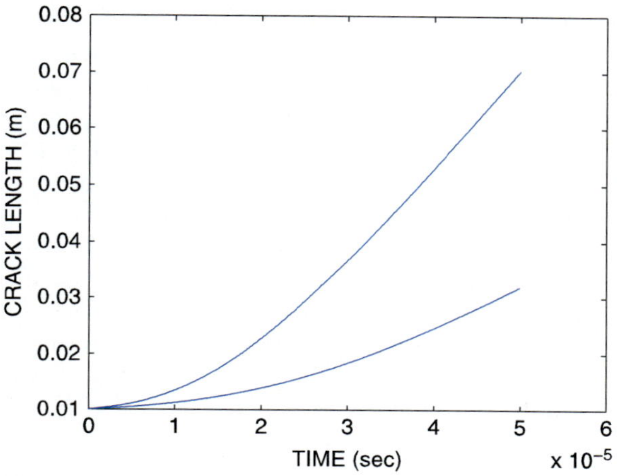

Fig. 11.1 Comparison of the crack propagation in PMMA predicted by the Freund model (9) and by the modification (11). The upper curve is the Freund model prediction

Other choices of the relaxation modulus, k_l, for the process speed in thermodynamic space might be made to represent experimentally observed micromechanical mechanisms that influence the transmission of shear waves and thus crack propagation.

3.2 Stability in the Griffith-Irwin Theory Viewed as Maximum Dissipation Fracture

An advantage of the maximum dissipation thermodynamic model is its ability to easily capture the stability of crack propagation. The role of the critical manifold contained in thermodynamic space in determining the stability of the predicted crack propagation is simply illustrated by the classical split beam model. The response also depends on whether the loading is in load or deformation control because the choice of the quasi-static fracture criterion determines whether the critical manifold is a set of minima or maxima for the generalized energy.

Crack propagation is said to be stable if the crack does not continue to grow. Stability in the Griffith theory depends on the behavior of the potential energy function at the point of critical crack growth initiation. The propagation is stable if the equilibrium is a minimum and unstable if the equilibrium is a maximum of the change in potential energy function.

The stability of crack propagation is defined in the context of the non-equilibrium thermodynamic model proposed for crack propagation by the behavior of the process initiated by perturbing the value of G so that the state (l, G) is no longer on the critical manifold. If, while all parameters are then held fixed, the state is repelled from the critical manifold, then the crack is said to be unstable. If it is attracted to the critical manifold, the state is said to be stable. Crack arrest may occur if the dynamical system response approaches the critical manifold.

As in the classical Griffith-Irwin analysis, different stability results may be obtained in deformation or load control. The split, or double, cantilever beam that has a longitudinal crack starting at the free end separated by a wedge is a traditional example of a stable fracture because the crack is arrested. Let d be the height of each cantilever section. Let $2h$ be the fixed separation between cantilever sections, so that h is the deflection of the free end of each cantilevered section, and l be the length of the cantilever sections, viewed as the crack length. This is an example of quasi-static deformation control because once the wedge is moved into the split to create the separation h, the wedge does not move further so that h is held fixed. The model predicts the resulting crack growth when the wedge is moved into the split and held in a new position, as well as the eventual crack arrest. So it captures the transient behavior of the crack growth.

Let U be the change in strain energy for the structure. Then $G = \frac{1}{w}\frac{dU}{dl}$ for constant crack face width, w. The generalized energy, F^*, is constructed to preserve this relation on the critical manifold. When G is perturbed to $\bar{G}$, then the system relaxes to a value of l that corresponds to $\bar{G}$ if the system is stable.

The strain energy is $U = (wEd^3h^2)/(4l^3)$, where h is the load. Put, with $w = 1$,

$$F^*(G; l) = \frac{Ed^3h^2}{4l^3} + Gl \tag{12}$$

Then $\partial^2 F^*/\partial^2 l > 0$ so that the critical manifold is stable. The non-equilibrium evolution (1) becomes

$$\dot{l} = -k\left(\frac{2l^5}{3Ed^3h^2}\right)^2\left[-\frac{3Ed^3h^2}{4l^4} + \bar{G}\right] \tag{13}$$

for fixed $\bar{G}$. As a numerical example, let $E = 70\,\text{GPa}$, $h = 0.0001\,\text{m}$, and $d = 0.01\,\text{m}$. Assume $k = 10^8$. The initial crack length is $l_o = 0.01\,\text{m}$ with corresponding equilibrium value of G = 2,6250 N/m. The perturbed value of the control variable G is $\bar{G} = 3,360\,\text{N/m}$. The crack length grows but asymptotically approaches the equilibrium length, $l = 0.0167185$ corresponding to $\bar{G}$ so that the behavior is stable in the sense of this thermodynamic model (Fig. 11.2). The non-equilibrium evolution path, starting from the point $(l, G) = (0.01, 3, 360)$ off the critical manifold, is attracted back to the critical manifold.

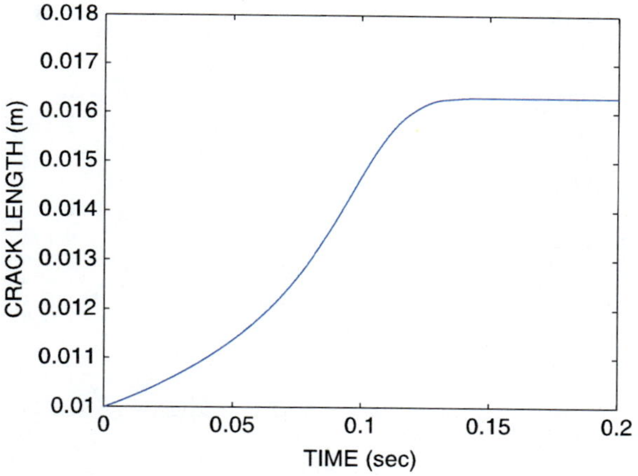

Fig. 11.2 Crack growth in the split cantilever beam under deformation control showing crack arrest

The double cantilever beam subjected to a fixed opening force, P, is in load control. In this case, the strain energy is $U = (4P^2l^3)/(wEd^3)$. The corresponding generalized energy is

$$F^* = -\frac{4P^2l^3}{Ed^3w^2} + Gl, \tag{14}$$

where w is the beam width, $2d$ is the beam height, and l is the crack length. Then $\partial^2 F^*/\partial^2 l < 0$ so that the equilibrium manifold is unstable. The non-equilibrium evolution (1) becomes

$$\dot{l} = -k\left(\frac{Ed^3w^2}{24P^2l}\right)^2\left[-\frac{12P^2l^2}{Ed^3w^2} + \bar{G}\right] \tag{15}$$

for fixed $\bar{G}$. The crack evolution for the same material parameters used in the wedge example is given in Fig. 11.3. Assume $E = 70\,\text{GPa}$, $w = 0.01$ m, and $d = 0.01$ m. Assume $k = 10^8$. The initial crack length is again $l_o = 0.01$ m and the perturbed value of the control variable is $\bar{G} = 3,360$ N/m. Take $P = 4,500$ N, slightly greater than the critical load of $P = 4,427.2$ N corresponding to the initial crack length of 0.01 m. The growth of the crack is unbounded and so is unstable.

This is not the same problem as described by Slepyan (1993) in which the wedge is continuously driven into the crack, resulting in a extremal speed of $\dot{l} = \sqrt{E/3\rho}$ where ρ is the material mass density. This maximum velocity value is equal to $\sqrt{2(1+\nu)/3}c_s$ so that it is larger than the experimental values reported by Ravi-Chandar and Knauss (1984). Here the wedge is moved some distance and the transient crack growth is predicted.

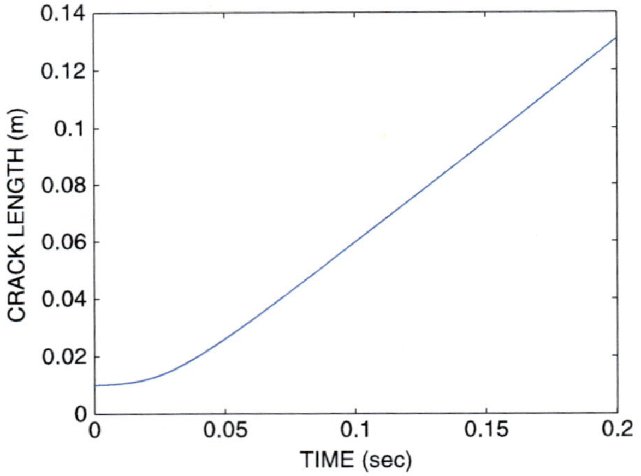

Fig. 11.3 Crack growth in the split cantilever beam under force control showing unstable crack growth

3.3 Craze Growth in PMMA Under Creep

Döll et al. (1981) give data for the growth of the craze region in PMMA 233 under a constant load applied by a tensile testing apparatus. Let s be the craze length, $\sigma_Y = 70\,\text{MPa}$ be the yield strength, and $K_I = 19.98\,\text{N/mm}^{3/2}$ be the stress intensity factor. They represented the time-dependent behavior of the craze length in terms of the Dugdale model for the extent of a plastic zone,

$$s = \frac{\pi}{8}\frac{K_I^2}{\sigma_Y^2}(1-\nu^2) \tag{16}$$

Döll et al., (1981) assume that K_I is a constant during the process. The craze length growth is obtained from the evolution equation (1) by assuming that the quasi-static behavior is given by Eq. (16). The conjugate variables in the associated generalized energy, F^*, are the state variable, s, and the control variable, K_I. The generalized energy is, after rescaling by multiplying by the constant E/K_I,

$$F^*(s, K_I) = -\frac{2}{3}\sigma_Y a s^{3/2} + K_I s, \tag{17}$$

where $a = \sqrt{8/[\pi(1-\nu^2)]} = 1.67282$ if $\nu = 0.3$. Then, from Eq. (1), the evolution of s is given by

$$\dot{s} = -k\left(-2s + \frac{2K_I}{\sigma_Y a}s^{1/2}\right). \tag{18}$$

This evolution equation closely matches the experimental curve of Döll et al. (1981) when the relaxation modulus is chosen as $k = 1.8 \times 10^{-7}$ (Fig. 11.4). Döll et al. tried to account for the growth as a quasi-static process by assuming that the yield stress decreases as computed in their Fig. 8. However in the dynamic model proposed here, the assumption that the yield stress changes during the process is not needed.

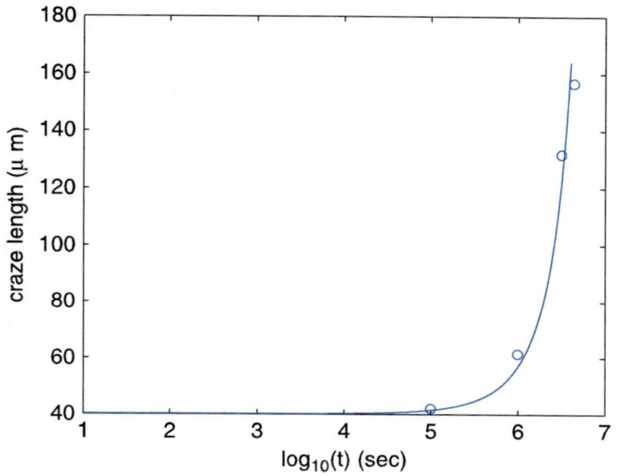

Fig. 11.4 The evolution model for craze growth in PMMA 233. Circles are the experimental data

4 Temperature at the Crack Tip

The temperature, θ, at the tip of a rapidly moving crack increases significantly. Fuller et al. (1975) measured a rise of about 500K in PMMA in the crack velocity range 200–650 m/s and about 300K in polystyrene. Glass exhibits a higher temperature change of about 900K at a crack speed of 20 m/s (Weichert and Schönert, 1974). Most of the energy deforming the plastic region at the crack tip is expected to convert to heat. While some models have assumed that all of the energy release becomes heat, others find that 90% of the energy is consumed in heat in metals and about 70% in polymers (Weichert and Schönert, 1978). The heat increase in PMMA is related to craze formation (Döll, 1984). These measurements are all maximum temperature changes. None of these experimentalists reported the change in temperature as a function of crack length while the crack grows.

Williams (1965) estimates the temperature increase by assuming the temperature change, $\Delta\theta$, in a cylindrical region around the tip is obtained from the difference between the dissipation and the heat conducted through the boundary. Other models of the temperature near the crack tip have been based on the heat equation under the assumption of a constant crack tip velocity (Weichert and Schönert, 1974, 1978; Döll, 1976). The result strongly depends on the size assumed for the plastic zone (e.g. Weichert and Schönert, 1974, 1978).

The maximum dissipation non-equilibrium model (1) in conjunction with the energy balance equation produces a model for the crack tip temperature along the path of the crack tip under non-constant crack tip velocity. The coordinate system is assumed to move with the crack tip so that $x + l_o = l(t)$, where l_o is the initial crack length.

The balance of energy equation is $\dot{u} = \mathbf{T} : \mathbf{D} - \text{div}(\mathbf{q})$, where u is the internal energy density, $\mathbf{T}$ is the Cauchy stress tensor, $\mathbf{D}$ is the rate of deformation tensor, and $\mathbf{q}$ is the heat flux. Following Bui et al., (1980) and Rice (1978), $\dot{u} = (G-2\gamma)\dot{l}+c_v\dot{\theta}$, where γ is the surface energy per length of the crack face, c_v is the specific heat. Assuming that c_v is constant, that Fourier heat conduction $q = -k_f(d\theta/dx)$ holds, that $\mathbf{T} : \mathbf{D} = \sigma\dot{\epsilon}^p$, where ϵ^p is the plastic strain at the crack tip, the energy balance is, since θ is a function of x and $\dot{l} = \dot{x}$,

$$(G - 2\gamma)\dot{l} + c_v\frac{d\theta}{dx}\dot{l} = \sigma\dot{\epsilon}^p + k_f\frac{d^2\theta}{dx^2}. \tag{19}$$

The value of the quasi-static mode I stress intensity must account for temperature since a rise in temperature increases the critical value of K_I. The critical mode I intensity varies nearly linearly with temperature for some materials and nearly exponentially for others (Hellan, 1984, p. 150). Here, assume a temperature dependent mode I stress intensity of the form

$$K_I = \sigma\sqrt{\pi l}g(\theta). \tag{20}$$

A simple linear temperature function is $g(\theta) = \theta/293$.

A viscoplastic model with zero yield stress, Y, is assumed because some dislocation motion starts with the application of the load so that a small plastic strain of up to 10μ may occur in the so-called elastic region (Haslach, 2002). Therefore, the question of the variation of the yield stress with temperature does not arise. The applicable generalized energy is constructed by adding a term for G and l to the quadratic generalized energy used in the viscoplastic analysis presented in Chapter 5. The back stress, B, accounts for kinematic hardening, and the drag stress, D, accounts for isotropic hardening. Corresponding conjugate variables must be defined in this thermodynamic model. The control variable, the back strain b, is set to be $b = LH$, where H is a constant and L is the limiting value of B because its energy is assumed quadratic. The relaxation modulus for the viscoplastic variables in thermodynamic space is the magnitude of the plastic strain rate in order to account for the change in the material structure due to dislocation multiplication, $k = ||\dot{\boldsymbol{\epsilon}}^p||$. A Norton type model is assumed for the norm of the plastic strain rate, but other choices are possible,

$$\dot{\epsilon}^p = \frac{nA}{D}\exp\left(\frac{-Q}{R\theta}\right)\left(\frac{||\sigma - B||}{D}\right)^{n-1}. \tag{21}$$

As the crack grows both the plastic strain rate and the back stress at the crack tip depend on the temperature, θ, at the crack tip.

The generalized energy for crack propagation is

$$F^*(l, \theta, B) = -\frac{(1-\nu^2)}{2E}\sigma^2 \pi l^2 g(\theta)^2 + \frac{1}{2}H_\theta \theta^2 + \frac{1}{2}HB^2 + Gl + \eta\theta + Bb, \quad (22)$$

The state variables are l, B, and θ. The conjugate controls are respectively G, the back strain b, and specific entropy η. The process is assumed isentropic. For the purposes of this illustrative computation, assume that the evolution of the crack length is given by the revised Freund model (11) additionally modified for temperature.

The system to be solved with Eq. (21) is then, after substituting the expression for k,

$$\dot{B} = -\frac{nA}{D}\exp\left(\frac{-Q}{R\theta}\right)\left(\frac{||\sigma - B||}{D}\right)^{n-1}\left(\frac{B}{H} - \frac{b}{H^2}\right);$$
$$\dot{l} = 0.5c_s\left[1 - \frac{E\Gamma}{(1-\nu^2)K_I^2}\right]. \quad (23)$$

The relaxation equation for the temperature θ is not given because it describes the temperature evolution with time around the crack tip at a fixed crack length; that evolution is not of interest in this analysis of the temperature change with position at the moving crack tip.

The model for the change in temperature at the propagating crack tip is given by the energy equation (19) obtained by substituting (23) for $\dot{l}$ and (20) for K_I. Here γ is ignored as in Bui et al. (1980).

$$0.5c_s\left(\frac{\theta^2\sigma^2\pi l(1-\nu^2)}{(293^2 E)} - \Gamma\right) + 0.5c_s c_v\left[1 - \frac{E\Gamma}{(1-\nu^2)K_I^2}\right]\frac{d\theta}{dx}$$
$$= \sigma \exp\left(\frac{Q}{R\theta}\right)\left(\frac{A}{D}\right)n\left(\frac{(\sigma - B)}{D}\right)^{n-1} + k_f\frac{d^2\theta}{dx^2} \quad (24)$$

As a numerical example, the model is applied to copper. Since the shear modulus is $\mu = 34\,\text{GPa}$ and the density is $\rho = 9,000\text{kg/m}^3$, the shear wave velocity is $c_s = \sqrt{\mu/\rho} = 2,185.8\,\text{m/s}$. For comparison, $c_R = 2,062\,\text{m/s}$. Also the surface energy at room temperature is $\gamma = 1.8\,\text{J/m}$; this value decreases slightly with temperature increase to about 1.5 J/m at 1,310 K. The other parameters are $k_f = 130\,\text{W/mK}$; $c_v = 0.4$; $Q = 3\,\text{W}$; $R = 8.31$; $\nu = 0.34$; $E = 117\,\text{GPa}$; $A = 2.3785 \times 10^{-2}$; $n = 4.5$; $H = 1/32,000$; $L = 20\,\text{kPa}$; $D = 13\,\text{MPa}$. For an applied stress of $\sigma = 12\,\text{MPa}$ and the control values $\Gamma = 10\,\text{N/m}$ and $b = LH = 0.625$, the change of temperature as a function of crack length obtained by solving (24) is given in Fig. 11.5. The initial conditions are that the crack length is 0.001 m, $B = 5\,\text{MPa}$, $\theta = 293\,\text{K}$ and $d\theta/dx = 0$. The Griffith critical stress for the

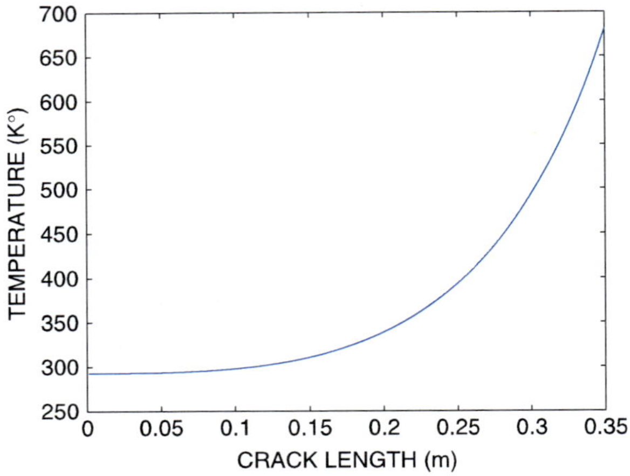

Fig. 11.5 Variation of the temperature at the crack tip as the crack propagates. Example of copper

initial crack is $\sigma_c = 10.7$ MPa. The constant applied stress becomes increasingly larger than the decreasing Griffith quasi-static critical stress as the crack grows. The curve shows that the temperature at the crack tip monotonically increases as the crack propagates and in this example reaches nearly 700 K at a crack length of about 0.35 m. This length is shorter than the 700 mm long plates used by Weichert and Schönert (1978). In this example, only a 5 K temperature increase occurs up to a crack length of 0.1 m, but the temperature blows up at about a crack length of 0.58 m perhaps because radiation is neglected. The non-equilibrium process model for fracture in conjunction with the energy balance relation qualitatively predicts the expected growth of the temperature at the moving crack tip.

References

A. Berezovski and G. A. Maugin (2007). On the propagation velocity of a straight brittle crack. *International Journal of Fracture* **143**, 135–142.

H. D. Bui, A. Ehrlacher, and Q. S. Nguyen (1980). Propagation de fissure en thermoélasticité dynamique. *Journal de Mécanique* **19**, 697–723.

R. Christensen and Y. Miyano (2007). Deterministic and probabilistic lifetimes from kinetic crack growth - generalized forms. *International Journal of Fracture* **143**, 35–39.

J. W. Dally, W. L. Fourney, and G. R. Irwin (1985). On the uniqueness of the stress intensity factor - crack velocity relationship. *International Journal of Fracture* **27**, 159–165.

W. Döll (1976). Application of an energy balance and an energy method to dynamic crack propagation. *International Journal of Fracture* **12**(4), 595–605.

W. Döll (1984). Kinetics of crack tip craze zone before and during fracture. *Polymer Engineering and Science* **24**(10), 798–808.

W. Döll, L. Könczöl, and M. G. Schinker (1981). Zur zeit- und temperaturabhängigen Vertreckung von Polymermaterial vor Rißspitzen bei langzeiter statischer Belastung. *Colloid and Polymer Science* **259**, 171–181.

L. B. Freund (1990). *Dynamic Fracture Mechanics*. Cambridge University Press, Cambridge.

K. N. G. Fuller, P. G. Fox, and J. E. Field (1975). The temperature rise at the tip of fast-moving cracks in glassy polymers. *Proceedings of the Royal Society of London* **A341**, 537–557.

H. Gao (1996). A theory of local limiting speed in dynamic fracture. *Journal of the Mechanics and Physics of Solids* **44**, 1453–1474.

A. A. Griffith (1921). The phenomena of rupture and flow in solids. *Philosophical Transactions Royal Society of London* **A221**, 163–198.

A. A. Griffith (1924). The theory of rupture. *Proceedings of the First International Congress for Applied Mechanics*, Delft, 55–63.

M. E. Gurtin (1979). Thermodynamics and the Griffith criterion for brittle fracture. *International Journal for Solids and Structures* **15**, 553–560.

M. E. Gurtin (2000). *Configurational Forces as Basic Concepts of Continuum Physics*, Springer, New York, NY.

H. W. Haslach, Jr. (2002). A non-equilibrium thermodynamic geometric structure for thermoviscoplasticity with maximum dissipation. *International Journal of Plasticity* **18**, 127–153.

H. W. Haslach, Jr. (2010). A non-equilibrium thermodynamic model for the crack propagation rate. *Mechanics of Time-Dependent Materials* **14**(1), 91–110. DOI: 10.1007/s11043-009-9094-9

K. Hellan (1984). *Introduction to Fracture Mechanics*, McGraw-Hill Book Company, New York, NY.

J. Lemaitre and J.-L. Chaboche (1990). *Mechanics of Solid Materials*, Cambridge University Press, Cambridge.

G. A. Maugin (1992). *The Thermomechanics of Plasticity and Fracture*, Cambridge University Press, Cambridge.

W. Muschik, W. and A. Berezovski (2004). Thermodynamic interaction between two discrete systems in non-equilibrium. *Journal of Non-equilibrium Thermodynamics* **29**, 237–255.

K. Ravi-Chandar (2004). *Dynamic Fracture*, Elsevier, Amsterdam

K. Ravi-Chandar and W. G. Knauss (1984). An experimental investigation into dynamic fracture - III. Steady state crack propagation and crack branching. *International Journal of Fracture* **26**, 141–154.

J. R. Rice (1978). Thermodynamics of the quasi-static growth of Griffith cracks. *Journal of the Mechanics and Physics of Solids* **26**, 61–78.

L. I. Slepyan (1993). Principle of maximum energy dissipation rate in crack dynamics. *Journal of the Mechanics and Physics of Solids* **41**, 1019–1033.

R. Weichert and K. Schönert (1974). On the temperature rise at the tip of a fast running crack. *Journal of the Mechanics and Physics of Solids* **22**, 127–133.

R. Weichert and K. Schönert (1978). Heat generation at the tip of a moving crack. *Journal of the Mechanics and Physics of Solids* **26**, 151–161.

J. G. Williams (1965). The thermal properties of a plastic zone. *Applied Materials Research* **4**, 104–106.

Chapter 12
Conclusion

1 Some Features of the Maximum Dissipation Construction

The maximum dissipation thermodynamic description of time-dependent non-equilibrium behavior, in general, depends on both a generalized energy and a generalized entropy production function that are defined on non-equilibrium states. The idea of conjugate variables underlying the construction of these functions is not exploited in continuum thermodynamics. The difficulties that arise in continuum thermodynamics from allowing energies to depend on fluxes are resolved in the construction proposed here by restricting flux dependence only to the entropy production function. The maximum dissipation construction then provides a thermodynamic foundation for realistic finite velocity of propagation flux evolution equations, in contrast to other empirical models. Further, the maximum dissipation postulate selects a particular non-equilibrium process from all the processes consistent with the second law and so supplements the incomplete second law.

Dissipation in this construction, $\sum(\partial\varphi^*/\partial x_i)\dot{x}_i$, directly involves the generalized thermodynamic function, in contrast to constructions postulating an additional dissipation potential. The generalized thermodynamic functions should not be mistaken for dissipation potentials, even though both the thermodynamic forces (affinities) and the dissipation through the Gibbs one-form are derived from them. The construction is predictive because it does not employ constructions, such as the dissipation potential, which are tantamount to assuming an evolution system.

The evolution by a nonlinear dynamic gradient relaxation maximum dissipation process is defined to be objective by using the Lie time derivative when the thermodynamic variables are defined in the current configuration. In the reference configuration, the Lie time derivative is the material derivative.

The maximum dissipation construction extends non-equilibrium thermodynamics to solids. The validity of the construction is supported by the fact that it generates well-known constitutive models as maximum dissipation models. These models include the Newton law of cooling, the Kelvin-Voigt model, the Freed-Chaboche-Walker viscoplastic model, the Maxwell-Cattaneo finite velocity thermal flux model, the finite velocity generalization of the Darcy law, and the Maxwell

H.W. Haslach Jr., *Maximum Dissipation Non-Equilibrium Thermodynamics and its Geometric Structure*, DOI 10.1007/978-1-4419-7765-6_12,

model. The construction therefore serves to thermodynamically justify many of these models that were originally proposed for empirical reasons. The development of well-known examples in this setting allows insight into how the maximum dissipation construction relates to traditional techniques for constructing time-dependent models.

The maximum dissipation model does not reproduce plasticity theory since plasticity theory is merely a rate-independent gross approximation to the physical behavior. For example, it is doubtful that a viscoplastic model requires the idea of a yield surface and all of its attendant mathematical constructions.

The homogeneous non-equilibrium space, the graph of the generalized energy density function, is locally a section of a contact bundle, on which section a homogenous process is geometrically represented as a path. In a non-homogeneous process, the evolution of the non-zero thermodynamic fluxes is represented on the graph of the generalized entropy function. As a non-equilibrium process approaches equilibrium, the path approaches zero on the steady state manifold while the path on the graph of the homogeneous generalized energy function tends to the thermostatic manifold. The relaxation speed is defined by the thermodynamic relaxation modulus. For a particular material, the thermodynamic relaxation modulus depends on the microstructure of the material and so is a tool to relate multiple scales in the non-equilibrium model. While the thermodynamic relaxation modulus is assumed constant in the description of the non-equilibrium response of viscoelastic materials (Chapter 4), it is assumed to depend on material structural parameters in the description of viscoplastic behavior (Chapter 5), of the dynamic response of hydrated elastin (Chapter 6), and of dynamic crack propagation (Chapter 11).

Experimental verification of a new application of the construction requires first experimentally determining, or assuming, φ and then comparing the predicted evolution of a non-equilibrium maximum dissipation process to the experimentally observed evolution.

2 Arrow of Time

The maximum dissipation assumption sidesteps the debate among cosmologists and philosophers of physics about how time asymmetry results from time irreversible physics. The arrow of time is the direction of time that is perceived by humans. Prigogine poses the question in the following manner, "How can the arrow of time emerge from what physics describes as a time-symmetrical world?" (1997, p. 2). Truesdell (1984, pp. 373–376) was aware of the difficulty in his critique of Onsager's assumption of microscopic reversibility. The thermodynamic arrow of time is that resulting from the increase in entropy. Thermodynamic theories are unable to explain why the arrow of time seems to move in a preferred direction, i.e. why entropy increases. Many thermodynamicists treat the idea that entropy must increase as merely a statement of what is physically impossible (e.g. Ericksen, 1991).

The thermodynamic arrow of time is thought to take the universe from a more ordered to a less ordered state. The interpretation that entropy measures disorder seems to be based on the statistical mechanics growing out of the work of Boltzmann. In the Boltzmann statistical mechanics idea of equilibrium (e.g. Nash, 1970), phase space is divided into macrostates composed of microstates all having the same energy and number of particles. Each microstate is assumed to have the same probability of occurrence. The macrostate with the largest volume is equilibrium, and so the equilibrium state contains the maximum number of microstates compared to other macrostates. The entropy of the non-equilibrium or equilibrium macrostate is generalized to be the logarithm of the volume of the macrostate. Since processes are assumed to spontaneously move to equilibrium, entropy increases. The entropy, S, at equilibrium is related to the number, W, of microstates in the equilibrium macrostate by $S = k \ln W$, where k is the Boltzmann constant from the ideal gas law, because the entropy is a maximum at equilibrium and because the equilibrium macrostate contains the maximum number of microstates. This computation assumes that S is only a function of W.

In the maximum dissipation construction for homogeneous systems, entropy is treated merely as the thermodynamic conjugate to temperature. However, in a non-homogeneous system, entropy production becomes a tool on the macroscopic level to predict the behavior of the body. No physical interpretation is provided.

One solution to the asymmetry of time in thermodynamics, without an appeal to entropy, is to require that a process always must approach equilibrium (Brown and Uffink, 2001). So in this view, the existence of equilibrium states causes the time asymmetry. Brown and Uffink would make this requirement an additional principle of thermodynamics to give time a specific direction. Unfortunately, Brown and Uffink do not distinguish between relaxation and forced processes; the latter do not approach equilibrium.

The maximum dissipation criterion accomplishes the same result because a relaxation non-equilibrium process must approach equilibrium. Therefore the arrow of time problem does not arise. The maximum dissipation construction does not assume that the non-equilibrium state is close to equilibrium. The construction assumes that an equilibrium state exists for an isolated system in the requirement that there is a long-term energy density function. Only an unforced system approaches equilibrium. A forced system can exhibit nonlinear dynamical behavior such as bifurcations and chaos.

The anthropic principle, supported by cosmologists such as S. Hawking, suggests that our universe, with its increase in entropy, is one of the few in which human life can exist. Its physics may be very unusual compared to other possible universes. So, the arrow of time is a result of human perception of the gradient of entropy (Price, 1996, p. 97). The real question is not why entropy must increase, according to the philosopher Price, but why was entropy so low to begin with. It is unlikely that thermodynamics can answer such a question.

Prigogine (1997) says that the arrow of time should not be viewed as a phenomenological construct imposed by humans, but can be found in the physics of non-equilibrium dynamics. An observer should not need to be postulated to impose irreversibility as in quantum mechanics. Progogine argues that dynamical mechanisms such as bifurcations and chaos are important in describing non-equilibrium behavior. "Near-equilibrium laws of nature are universal, but far from equilibrium, they become mechanism dependent" (Prigogine, 1997, p. 65). Therefore he argues that the distance from equilibrium is an essential parameter. Further, he argues that since free energy, $F = U - \theta S$, is a minimum at equilibrium and since the entropy production dS/dt has a minimum at steady states and is zero at equilibrium, the fluctuations of statistical mechanics are damped near equilibrium but are more active far from equilibrium leading to complex behavior such as is described by nonlinear dynamics. So, Prigogine's solution involves probabilities. Prigogine might object to making a new law of thermodynamics that requires approach to equilibrium since the principle would perhaps not be consistent with chaotic behavior, for example.

One suggestion of those in the tradition of Prigogine is to use the level of entropy production as a measure of irreversibility (Demirel and Sandler, 2004). They hint that the level of irreversibility is somehow related to the distance from equilibrium; "...the level of irreversibility of any step in a process is ... related to its distance from global equilibrium; this distance may be treated as a parameter of the process." However, they do not offer a means to compute the distance.

Many descriptions of non-equilibrium thermodynamical behavior depend on the state being close to equilibrium, (e.g. Garcìa-Colìn and Uribe, 1991). A metric in a mathematical model of non-equilibrium thermodynamics is required to make precise what "close" means. Several authors, including Gibbs (1873) and Tisza (1966), have pointed out that there is no metric on the traditional equilibrium Gibbs energy surface, the graph of the internal energy function, $\mathcal{U}(\mathcal{S}, \mathcal{V})$, in $\Re^3$. However, Weinhold (1975–1976) constructed an abstract vector space, based on the thermodynamic variables, in which a metric exists. But this vector space construction does not give a metric on the graph of the generalized thermodynamic function that can define closeness of a non-equilibrium process to the thermostatic manifold. The definition of a metric on the submanifold of thermodynamic non-equilibrium states that would give the distance to equilibrium for a given non-equilibrium process is an open question.

A possible alternative for a metric on thermodynamic space is to define a Finsler structure on the contact manifolds of thermodynamic variables. The Finsler geometry on an n-dimensional manifold, M, is induced by a metric defined from a real map on the tangent bundle, $F : T_*M \to \Re$, such that $F(p_1, \ldots, p_n, q_1, \ldots, q_n) \geq 0$ is positively homogeneous of degree one in the q_i, in the sense that

$$F(p_1, \ldots, p_n, \lambda q_1, \ldots, \lambda q_n) = \lambda F(p_1, \ldots, p_n, q_1, \ldots, q_n),$$

for $\lambda > 0$. A Finsler function, $F(p_1, \ldots, p_n, q_1, \ldots, q_n)$, arises in variational problems expressed in the form: optimize an integral $\int_a^b F(p_1, \ldots, p_n, q_1, \ldots, q_n)dt$. The homogeneity condition guarantees that the value of the integral is independent of the parametrization and also allows the function to be defined on an associated projective space. The map F produces the tensor, $g_{ij} = \partial^2(\frac{1}{2}F^2)/\partial p_i \partial p_j$. The scalar product defining the metric is $\sum_{i,j} g_{ij} dp_i \otimes dp_j$. If F^2 is quadratic in the p_i, this construction produces a Riemannian metric (Chern, 1992, 1996a, 1996b). The Finsler function produces a metric on the tangent space to the thermodynamic contact manifolds, but may not produce a metric on the submanifolds defined by the graphs of the generalized thermodynamic functions.

To measure when a non-equilibrium process in the maximum dissipation construction is close to equilibrium, one tempting, but speculative, strategy is to define a Finsler geometry in terms of a measure of the magnitude of dissipation. The current dissipation should decrease to zero as the state approaches equilibrium. To apply this strategy, let ω_G be the Gibbs one-form on thermodynamic space. The maximum dissipation construction implies that for tangents $\boldsymbol{t}_p$ to the path representing a process, the function $\omega_G(\boldsymbol{t}_p)$ measures the dissipation. Recall that $\omega_G = 0$ on equilibrium states. The variational principle that applies to the maximum dissipation construction is that a non-equilibrium process follows that path which maximizes dissipation. In other words, the problem is to optimize the integral $\int_a^b -\omega_G(\boldsymbol{t}_p)dt$. The positive real valued function, $-\omega_G$, is Finsler since $\omega_G(\lambda \boldsymbol{t}_p) = \lambda\omega_G(\boldsymbol{t}_p)$. Only admissible non-equilibrium paths are used in defining the metric. The questions to be investigated regarding such a construction have mathematical as well as thermodynamic interest.

References

H. R. Brown and J. Uffink (2001). The origins of time-asymmetry in thermodynamics: The minus first law. *Studies in History and Philosophy of Modern Physics* **32**(4), 525–538.

S. S. Chern (1992). On Finsler geometry, *Les Comptes Rendus de l'Académie des Sciences Paris* **314**, 757–761.

S. S. Chern (1996a). Finsler geometry is just Riemannian geometry, *Notices of the AMS* **43**(9), 959–963.

S. S. Chern (1996b), Riemannian geometry as a special case of Finsler geometry. In *Finsler Geometry*, eds. D. Bao, S. S. Chern, and Z. Shen, Contempory Math., Vol. 196, American Mathematical Society, Providence, RI.

Y. Demirel and S. I. Sandler (2004). Nonequilibrium thermodynamics in engineering and science. *Journal of Physical Chemistry B* **108**, 31–43.

J. L. Ericksen (1991). *Introduction to the Thermodynamics of Solids*, Chapman & Hall, London.

L. S. Garcìa-Colìn and F. J. Uribe (1991). Extended irreversible thermodynamics, beyond the linear regime: a critical overview. *Journal of Non-Equilibrium Thermodynamics* **16**, 89–128.

J. W. Gibbs (1873). A method of geometrical representation of the thermodynamic properties of substances by means of surfaces. *Transactions of the Connecticut Academy* II, 382–404. Also in *The Collected Works*, Vol. 1, Yale University Press, New Haven, CT 1948, 33–54.

L. K. Nash (1970). *Elements of Classical and Statistical Thermodynamics*, Addison-Wesley, Reading, MA.

H. Price (1996). *Time's Arrow and Archimedes' Point*, Oxford Universty Press, New York, NY.

I. Prigogine (1997). *The End of Certainty*, The Free Press, New York, NY.

L. Tisza (1966). *Generalized Thermodynamics*, MIT Press, Cambridge MA.

C. Truesdell (1984). *Rational Thermodynamics*, 2nd ed., Springer, New York, NY.

F. Weinhold (1975–1976), Metric geometry of equilibrium thermodynamics, *Journal of Chemical Physics* **63**, 2479–2501 and **65**, 559–564.

Index

H.W. Haslach Jr., *Maximum Dissipation Non-Equilibrium Thermodynamics and its Geometric Structure,* DOI 10.1007/978-1-4419-7765-6,